Failure Diagnostics of Electrical Power Transformers

Hemchandra M. Shertukde

DDI Publishing Group

Diagnostic Devices Inc. dba DDI

PO Box 362,

Simsbury, CT, 06070, USA

Printed on acid-free paper

International Standard Book Number :- 979-8-9920188-6-8 (Hardback)

979-8-9920188-9-9 (Paperback)

Edited and Published by Callaghan Publications –

Hemchandra M. Shertukde

CALLAGHAN
PUBLICATIONS

Visit the DDI Publishing Web site at: www.diagnostic-devices.com

Acknowledgments

My gratitude to Northeast Utilities (NU) for supporting my sabbatical leave in 1995 along with University of Hartford with the untiring help of Late Dr. John P Cagnetta (erstwhile Dean of College of Engineering (COE) at the University of Hartford; 1994-1997) and Connecticut Innovations (CI) for granting me the grant G096) in association with Diagnostic Devices Inc. (DDI) for the PD diagnostic work on Power Transformers in Connecticut, USA. Over the past 30+ years, I would like to thank all the clients who agreed to use our PD diagnostic system FDD-EPT/FAN-EPT for at more than 100 sites in the USA and worldwide.

I graciously acknowledge the support and guidance of my dear wife, Rekha H. Shertukde, MPA, (96), Chief Administrative Officer (CAO) of DDI, over the past 30 years in being on my side from the time we founded our small business HMS & RHS Inc. in 1993 in the State of CT.

In addition, I would like to acknowledge the constant support of my children Amola, Karan, and Rohan, who all supported my business for the past several years. All of them have tested one or more large power transformers with FDD-EPT at different sites in the USA and worldwide.

Finally, I acknowledge and welcome the addition of two more new members to our family:

- My/our grandson Ryaan Jignesh Bhakta, and;
- My/our Daughter In Law: Supriya Rohan Shertukde

Dearly,
Hemchandra Madhusudan Shertukde

Contents

Foreword

My association with Power Transformer Diagnostics started when I worked at Crompton Greaves Ltd, Mumbai, India, as the Superintendent of the core-coil assembly department from 1977-1983. Later, in 1995, I was approached by the electrical utility: NorthEast Utilities; of the New England area in the CT, USA, to help them diagnose failures in power transformers using my track before detecting methodology. This resulted in studying the Gas in Oil (GIO) Technique followed by the Digital Gas Analysis (DGA) as postulated by IEEE, IEC, etc.

In 1996, I received a $210,000 grant from Connecticut Innovations and Northeast Utilities as a result of my first sabbatical at Northeast Utilities in the Fall of 1995. Rest is history when I patented my technology using supersonic sensors and obtained a patent from USPTO # 6,178,386; in February of 2001. In the meantime, I got involved in the writing of User's Guide for transformer diagnostics using acoustic methods: IEEE - (C.57.127) for liquid-filled transformers.

Presently, this work is extended to C.57.124 for dry-type transformers. So far, my company, Diagnostic Devices Inc. (DDI), has tested more than 400 large power transformers. This book includes many of these cases in the Case Study Chapter 8 of this book.

In Chapter 1, I will discuss the Introduction and Scope of different diagnostic techniques for Power Transformers engulfing all facets of Electrical Power Engineering. Over the years since the early 1950s, some of the electrical power transformers started to face end-of-life conditions. Due to limited financial resources faced by utilities, ways had to be found to sustain such capital goods for a longer life. This led to the innovative diagnostic techniques created by several researchers worldwide, which were employed to retain and save these capital assets. These techniques include GIO, DGA, Basic Impulse testing, switching surge testing, frequency response testing, and Partial Discharge (PD) testing.

In Chapter 2, we will discuss the history of the diagnostic techniques used for power transformers in the power industry for the past six decades or more. These include both online and offline techniques.

This is followed by discussing specific methods in Chapter 3, like Gas in Oil (GIO), Digital Gas Analysis (DGA), Frequency Response Analysis (FRA), and Partial Discharge (PD) Tests. These include both electrical and acoustic PD tests. The electrical PD tests are associated with induced voltage tests, switching tests, and chopped wave tests.

In Chapter 4, we will discuss the acoustic tests in detail, followed by the thermal techniques in Chapter 5. In Chapter 6, we discuss invasive techniques using fiber optic sensors (FOS). To effectively summarize these techniques, we will further discuss them using online and offline techniques in Chapter 7. In Chapter 8, we will discuss case studies of the efficacy of each of these methods. These case studies include varying kinds of transformers at different sites worldwide. These include about a hundred locations in more than thirty countries worldwide.

Finally, in Chapter 9, we conclude by drawing conclusions and discussing further studies conducted by several researchers in different countries.

I hope you enjoy reading this book, which is primarily a guide to using different techniques for the diagnosis of power transformer failures. There is no guarantee in accurately assessing the detection and location of PDs in electrical power transformers. The methods enumerated in this book are primarily to whet your hands-on experience in the fascinating science of finding these faults using my patented technique, which uses the Track Before Detect (TBD) approach. This TBD technique was extensively used in the early nineties during the Gulf War to detect and locate enemy aircraft, submarines, and other harmful targets. It is the basis of my Ph.D. thesis obtained from the University of Connecticut, Storrs, under the guidance of my Ph.D. advisor, Dr. Yaakav Bar-Shalom, whom I respect a lot and consider my 'God Father' in this field of Probabilistic Data Association (PDA) techniques, which have gone a long way to save capital assets and human life.

Further, extending it to the field of Electrical Power Transformers and other equipment was my effort to bring the age-old techniques used by the then soldiers and fighters by holding their ears to the ground to track galloping horses. This concept of dual application has always inspired me to make life easier and simpler to navigate in times of distress and agony.

Chapter 1:
Introduction and Scope of Failure Diagnostic Techniques in Power Transformers

In Chapter 1, I will discuss the Introduction and Scope of different diagnostic techniques for Power Transformers engulfing all facets of Electrical Power Engineering. Over the years since the early 1950s, some of the electrical power transformers started to face end-of-life conditions. Due to limited financial resources faced by utilities, ways had to be found to sustain such capital goods, especially transformers, for a longer life. This effort led to different innovative diagnostic techniques formulated by several researchers, users, manufacturers, and utilities worldwide, which were employed to retain and save such capital assets. These techniques include:

1) Gas In Oil (GIO)

GIO is a simple traditional test employed by transformer manufacturers and utilities for preventive maintenance and failure maintenance to assess the insulation integrity of the inside structure of a transformer. This is a basic chemical test to assess the insulation integrity of both oil and cellulose. Although the primary assessment is done for oil, as a secondary derivative, cellulose insulation integrity can also be evaluated and assessed. A sample of oil is extracted from the bottom of the transformer tank and sent to a chemical laboratory for different gas content analysis. The different gases to be analyzed for parts per million (ppm) in oil are:

i) Carbon dioxide (CO_2) – formed due to the burning of the cellulose insulation after a flame has started in the insulation structure.

ii) Carbon monoxide (CO) – formed due to the initial striking of fire in the insulation when a flame has started in the insulation structure.

iii) Methane (CH_4) – created by oil decomposition due to incipient faults in transformers like partial discharge

iv) Ethane (C_2H_6) - created by oil decomposition due to incipient faults in transformers like partial discharge

v) Ethylene (C_2H_4) - created by oil decomposition due to incipient faults in transformers like partial discharge.

vi) Acetylene (C_2H_2) - this is also a gas created by oil decomposition due to incipient faults in transformers like partial discharge.

vii) Hydrogen (H_2) - created by oil and cellulose decomposition due to incipient faults in transformers like partial discharge.

viii) Oxygen (O_2) – a natural gas that exists in any condition and situation in a transformer that can aid combustion.

ix) Nitrogen (N_2) – a natural and inert gas that exists in any condition and situation in a transformer that can be used for protection. The ppm for different gases is decided by an IEEE, IEC test. A threshold is set by a respective IEEE standard for each above gas. If the threshold exceeds, certain inferences are drawn on the existence of an insulation failure. There are many other methods provided by different researchers like the Roger's Ratio Test (Acetylene - C_2H_2/Ethylene-C_2H_4, Methane-CH_4/Hydrogen-H_2, Ethylene - C_2H_4/Ethane-C_2H_6) Triangle (three-point) test, Pentagon (five-point) test, etc.

Further, when cellulose or paper burns due to any insulation failure that causes a flame, it further creates gases like CO and CO_2.

2) Digital Gas Analysis (DGA)

DGA is done with the help of computer technology. A sample of oil from the referenced transformer tank is sent to the Chemical lab for analysis. The lab also has a similar computer capacity. The different ratios as above are calculated, and depending on the results obtained inferences are drawn about the integrity of the insulation structure.

3) Basic Impulse Level (BIL) Testing

This is an onsite or in-plant electrical test conducted by the transformer manufacturer as a part of the standard test or type test, depending on the provisions of the purchase order. Such a test helps in finding the integrity of the insulation structure related to the core-coil disposition within the confines of the transformer tank. This test can also be conducted onsite by the user if suitable equipment for such BIL testing is available.

4) Switching Surge Testing

This is also an onsite or in-plant electrical test conducted by the transformer manufacturer as a part of the standard test or type test, depending on the provisions of the purchase order. Such a test helps in finding the integrity of the insulation structure related to the core-coil disposition within the confines of the transformer tank. This test can also be conducted onsite by the user if suitable equipment for such BIL testing is available.

5) Applied Voltage Test

This is also an onsite or in plant electrical test conducted by the transformer manufacturer as a part of the standard test or type test depending on the provisions of the purchase order. Such a test helps in finding the integrity of the insulation structure related to the core-coil disposition within the confines of the transformer tank. This test can also be conducted at site by the user if suitable equipment for such a BIL testing is available.

6) Chopped Wave Test

This is also an onsite or in plant electrical test conducted by the transformer manufacturer as a part of the standard test or type test depending on the provisions of the purchase order. Such a test helps in finding the integrity of the insulation structure related to the core-coil disposition within the confines of the transformer tank. This test can also be conducted at site by the user if suitable equipment for such a BIL testing is available.

7) Frequency Response Analysis (FRA) Testing

Also called the Sweep Frequency Response Analysis (SFRA). Sweep frequency response analysis (SFRA) is a method to evaluate the mechanical integrity of core, windings, and clamping structures within power transformers by measuring their electrical transfer functions over a wide frequency range. Any change in the mechanical structure can effectively change the core-coil disposition, leading to a different frequency transfer function, whose response depicts changes in the frequency response over a wide range of frequencies.

8) Partial Discharge (PD) Testing

This is also an onsite or in-plant electrical or acoustic test conducted by the transformer manufacturer as a part of the standard test or type test, depending on the provisions of the purchase order. Such a test helps in finding the integrity of the insulation structure related to the core-coil disposition within the confines of the transformer tank. This test can also be conducted onsite by the user if suitable equipment for such a BIL testing is available. The acoustic PD test is conducted while the induced voltage test or applied voltage test is being conducted in a plant. However, this test can also be conducted onsite using suitable equipment. There are many such devices offered by different manufacturers worldwide.

One such manufacturer of such a PD test manufactures and sells patented (USPTO 6,178,386) devices using acoustic techniques is Diagnostic Devices Inc. in Simsbury, CT, USA (www.diagnostic-devices.com). Such a device is called *Fault Diagnostic Device for Electrical Power Transformers* (FDD-EPT).

All the above techniques will be discussed with an intention to narrate several failure diagnostic techniques for Power Transformers, discussing the methodology, need, efficiency, and cost of each test. Does the test provide suitable, proper, and simple test diagnosing the conditions of a power transformer before the incipient condition burgeons into a catastrophic situation that may lead to the loss of the capital asset? An important consideration.

In conclusion, this effort is the result of my personal experience ranging over a period of over 50+ years in the electrical industry, firstly dedicated to transformers, followed by a sustained stint of 45+ years in controls, signal processing, and systems engineering, where every effort was made to amalgamate expertise in both the areas to generate ways to perform genuine and innovative techniques to alleviate the effort to diagnose transformer failures.

It might seem how knowledge in controls and signal processing could be harnessed to understand the problem of signals generated due to failures in transformers and using techniques I used to track targets for military applications with the track before detect (TBD) philosophy, which I formulated in 1983-89 using passive methodology and applied it to electrical power transformers. It is very fulfilling. This technique also was instrumental in awarding me my Ph.D.

in 1989 from the University of Connecticut, Storrs, USA. This TBD technique was also used in the first Gulf War, and the resulting publications were instrumental in enabling the University of Hartford to award me the Bent Award for creativity in the Spring of 1997.

All the above failure diagnostic techniques will be illustrated with case studies of real-life transformer case studies with technique used, analysis, actual testing results, and results as predicted and, if found, as indicated by the diagnostic test. Each test conducted will be illustrated with relevant permission obtained from the manufacturer or utility as the ultimate user of the transformer.

In other cases, IEEE-SA may also be contacted for appropriate permission and allocate the corresponding credits to concerned parties. In some cases, it may be necessary to use the 'Fair Use' method due to the inability to reach the copyright owners due to the present COVID-19 pandemic raging the entire world. In such cases, continued effort will be made to obtain the relevant permission. I am grateful to all who agreed to grant permission to present their transformers and the related testing sites to be cited and presented in this effort to bring to the world all real-life situations and failures of power transformers due to failure in their insulation structure within the confines of the tank.

In conclusion, there is no guarantee that the techniques described will play out to the users' satisfaction unless the user is fully trained in the use of the specific techniques. Many years of practice and dedication have gone into the formulation of all the described failure diagnostic techniques for power transformers. The efficacy of such tests has been proven in practice, and considerable confidence can be rested on the prescribed methods.

Chapter 2:
History Of Diagnostics Techniques For Power Transformers

In Chapter 2, we will discuss the history of the diagnostic techniques used for power transformers in the electrical power industry for the past six decades or more. These include both online and offline techniques. Offline techniques are considered to be nearly passive, while the online techniques are hybrid in the sense they could be passive as well as active.

Offline Diagnostic Techniques For Electrical Power Transformers

The history of offline techniques for electrical power transformers started in the early 20th century when researchers found ways of deteriorating the insulation structure of electrical power transformers. These included deterioration in oil, cellulose, and insulation covering like silicate or phosphate on the Cold Rolled Grain Oriented (CRGO) core lamination material.

The first and easiest way to detect any insulation failures was conducted by performing a gas in oil (GIO) of a sample of oil collected from an operating transformer. IEEE then laid down the standards IEEE C57.104-2019 for evaluating the parts per million (ppm) of the different gases dissolved in such a sample obtained for dissolved gas analysis from oil-immersed electrical power transformers. Besides, ppm of different gases, the temperature measured [IEEE C.57.100-1986] of such gases also helps in assessing the life of the insulation structure. There are many other aspects and techniques that have been developed such as the Roger's test, etc.

Online Diagnostic Techniques For Electrical Power Transformers

These techniques are generally active, and some of them are

i. Basic Insulation Level (BIL) testing – here the full insulation capacity of the core-coil structural disposition of the core-coil assembly is measured and verified per IEEE C.57-113 standards under lightning or switching surge conditions.

ii. Applied Voltage level testing – here the full insulation capacity of the core-coil structural disposition of the core-coil assembly is measured and verified per IEEE C.57-113 standards under normal applied voltage conditions.

iii. Induced Voltage level testing – here the full insulation capacity of the core-coil structural disposition of the core-coil assembly is measured and verified per IEEE C.57-113 standards under induced voltage conditions.

iv. Partial Discharge (PD) level testing using electrical methods – here the full insulation capacity of the core-coil structural disposition of the core-coil assembly is measured and verified per IEEE C.57-113 standards under electrical partial discharge conditions. These are incipient insulation failure conditions that can burgeon into catastrophic failures of the entire transformer.

v. Partial Discharge (PD) level testing using Acoustic methods – offline as well as online methods. Here the full insulation capacity of the core-coil structural disposition of the core-coil assembly is measured and verified per IEEE C.57-127 standards under electrical partial discharge conditions using acoustic signals that are measured using supersonic sensors, which are comprised of piezo crystals. These failures are incipient insulation failure conditions that can burgeon into catastrophic failures of the entire transformer.

vi. Electrical PD level testing using Ultra High Frequency (UHF) detection methods — as the acronym suggests, high-frequency signals are actively passed through the insulation structure via windings, and characteristics are assessed using the reflected signals.

vii. Frequency Response Analysis (FRA) — here the full insulation capacity of the core-coil structural disposition of the core-coil assembly is swept with a frequency source, and a transfer function of the insulation structure is evaluated, measured, and verified per IEEE C.57-xxx standards under normal electrical source frequency conditions using a frequency signal generator and the changes in the transfer function characteristics indicate the deterioration of the structural integrity of the core-coil insulation disposition. The FRA technique was introduced in 1978 by Dick and Erven. It is used in wide-band frequency ranges to find the fingerprint of frequency response of a healthy set of core-coil assembly of a known transformer vs the frequency response of the core-coil assembly of a failed known transformer.

In summary, all techniques used in evaluation of diagnosis of power transformer failures were discussed, but the main focus was on the acoustic techniques. Further, there are no guarantees in detection and location of PD in electrical power transformers. The use of acoustic techniques and other methods is primarily the responsibility of the user, and there are many guidelines provided by IEEE, IEC, and CIGRE, most of which are discussed in the chapters ahead.

Chapter 3:
Gas in Oil (GIO)/ Digital Gas Analysis (DGA)/Frequency Response Analysis (FRA)/Partial Discharge (PD) Tests

a. GIO – Gas in Oil test
b. DGA – Digital Gas Analysis Test
c. FRA – Frequency Response Analysis Test
d. PD Tests – Partial Discharge Tests
    i. Electrical induced/applied voltage/chopped wave tests
    ii. Acoustic Tests

This is followed by discussing specific methods in Chapter 3 like Gas in Oil (GIO), Digital Gas Analysis (DGA), Frequency Response Analysis (FRA), Partial Discharge (PD) Tests. These include both electrical and acoustic PD tests. The electrical PD tests are associated with induced voltage tests, switching tests and chopped wave tests.

These tests have been discussed in detail by several bodies viz: ANSI, IEEE, IEC, CIGRE, NEMA etc. Each body has published several documents as Standards, and User's Guides to ensure that the electrical power equipment like transformers operate efficiently with a long-life span. Generally, the life span was estimated to be 40 years but lately it has been documented that the life span for distribution transformers is only 15-20 years. In this chapter, we will try to concentrate on the failures caused by PD, that can occur as incipient failures in transformers. These failures are mostly diagnosed using the above techniques.

I will describe the other methods in the following sections:

Gas in Oil (GIO)

GIO is a simple traditional test employed by transformer manufacturers and utilities for preventive maintenance and failure maintenance to assess the insulation integrity of the inside structure of a transformer. This is a basic chemical test to assess the insulation integrity of both oil and cellulose, although, the primary assessment is done for oil, as a secondary derivative cellulose insulation integrity can also be evaluated and assessed.

A sample of oil is extracted from the bottom of the transformer tank and sent to a chemical laboratory for different gas content analysis. The different gases to be analyzed for parts per million (ppm) in oil are:

i) Carbon dioxide (CO_2) – this is a gas formed due to the burning of the cellulose insulation after a flame has started in the insulation structure.

ii) Carbon monoxide (CO) – this is a gas formed due to the initial striking of fire in the insulation when a flame has started in the insulation structure.

iii) Methane (CH_4) – this is one of a gas created by oil decomposition due to incipient faults in transformers like partial discharge.

iv) Ethane (C_2H_6) - this is also a gas created by oil decomposition due to incipient faults in transformers like partial discharge.

v) Ethylene (C_2H_4) - this is also a gas created by oil decomposition due to incipient faults in transformers like partial discharge.

vi) Acetylene (C_2H_2) - this is also a gas created by oil decomposition due to incipient faults in transformers like partial discharge.

vii) Hydrogen (H_2) - this is also a gas created by oil and cellulose decomposition due to incipient faults in transformers like partial discharge.

viii) Oxygen (O_2) – this a natural gas that exists in any condition and situation in a transformer that can aid combustion.

ix) Nitrogen (N_2) – this is a natural and an inert gas that exists in any condition and situation in a transformer that can be used for protection.

The ppm for different gases is decided by an IEEE, IEC test. A threshold is set by an respective IEEE standard for each above gas. If the threshold is exceeded certain inferences are drawn on existence of an insulation failure. There are many other methods provided by different researchers like the Roger's Ratio Test (Acetylene - C_2H_2/Ethylene-C_2H_4, Methane-

CH_4/Hydrogen-H_2, Ehtylene - C_2H_4/Ethane-C_2H_6) Triangle (three point) test, Pentagon (five-point) test etc.

Further, when cellulose or paper burns, due to any insulation failure that causes a flame that further creates gases like CO and CO_2.

As we all know the famous Chemist adage: a small increase in temperature approximately 5-15 deg C can result a manifold increase in gassing rate. This can be as high as two to three times the base value of the gassing rate. This is governed by the famous Arthenius equation, which holds for most chemical reactions.

Under low energy, discharge events like PD different gases are produced. Hydrogen produced by a PD in oil could be as much as 75% of the total gases. The remaining portion is comprised of smaller portions of hydrocarbons in ascending order of $C_2H_6 < C_2H_4 < CH_4 < C_2H_2$. Inclusion of cellulose materials like paper and pressboard, low quantities of CO are produced. Discharges in cellulose can also produce large quantities of CO and H_2.

Many of these gases start to form at different increasing temperatures; Hydrogen and Methane start forming at 150 deg C. Methane, Ethane and ethylene production peaks at certain temperatures and declines as temperature increases beyond the peak. At about 250 deg C production of ethane starts. When temperature increases to 350 deg C, ethylene emerges, suggesting low temperature thermal faults will not crate any ethylene., however plenty of ethane and methane can take place. Acetylene starts above 700 deg C., which suggests that a thermal fault that surges the temperature beyond 700 deg C can produce trace amounts of acetylene. Greater amounts of Acetylene can surge at 900 deg C and above suggesting occurrence of arcing in the system.

The ethylene to ethane ratio is a good indicator of the decomposition temperature in oil. Thus T in deg C = 100 x (ethylene/ethane) + 150 deg C — *Eq. (3.1).*

Digital Gas Analysis (DGA)

DGA is done with the help of computer technology. A sample of oil from the referenced transformer tank is sent to the Chemical lab for analysis. The lab also has a similar computer capacity. The different ratios as above are calculated and depending on the results obtained inferences are drawn about the integrity of the insulation structure. Further considering the ppm of individual gases, relative ratios are more indicative of the possible ratios as follows using Roger's Ratio (RR) method. The se relative ratios of five individual gases viz:

Hydrogen (H_2), CH_4, C_2H_6, C_2H_4, and C_2H_2 . There are three ratios that RR method uses:

a) $R_1 = C_2H_2 / C_2H_4$

b) $R_2 = CH_4 / H_2$

c) $R_3 = C_2H_4 / C_2H_6$

RR method is mainly used in diagnosing and analyzing the nature of the fault and its characteristics. RR is based on total amount of individual gases, ratios and gas generation rates as well.

These ratios are grouped in four ranges as follows:

i) < 0.1

ii) 0.1-1

iii) 1-3

iv) > 3

A combination of these ratios indicates a possible fault that may exist as follows:

- (R_1 , R_2 R_3) : (0, 0, 0); implies no fault and there is normal ageing of insulation structure, either oil or cellulose
- (R_1 , R_2 R_3): (1, 1, 0) implies low energy PD
- (R_1 , R_2 R_3): (1, 1, 0); implies high energy PD but accompanied by cellulose insulation deterioration
- (R_1, R_2 R_3): (1-2, 0, 1-2); implies low energy discharges, sparking, arcing
- (R_1, R2 R3): (1, 0, 2); implies high energy discharges, sparking, arcing
- (R1, R_2 R_3): (0, 0, 1); implies Thermal fault less than 150 deg centigrade
- (R_1, R_2 R_3): (0, 2, 0); implies Thermal fault greater than 150 deg centigrade
- (R_1, R_2 R_3): (0, 2, 1); implies Thermal fault temperature range 300-700 degrees

centigrade

- (R_1, R_2 R_3): (0, 2, 2); implies Thermal fault greater than 700 degrees centigrade. Also, these problems will involve generation of CO and CO_2. This author has not used the RR method except when the client has requested for an update on trend different chapters in this book and as done in the prequel.

Frequency Response Analysis (FRA)

FRA is a frequency response of the transfer function of the electrical circuit of a given transformer on test either in the field or in a manufacturers plant. Any transformer has a comprehensive electrical circuit in the R-L-C combination comprising of resistances, inductances, series capacitances and capacitances to ground. These do not include ant test equipment capacitances like the $C_1$and C_2 that are frequently encountered when such testing equipment are connected to the transformer circuit. Basically, this is frequency sweep of the transfer function that can yield important information about any shifting of windings in the radial and vertical direction and in addition also lateral shift.

FRA is an advanced technique to assess the physical placement of active part within the confines of a transformer tank, especially the core-coil insulation and its physical displacement with each other and the tank, which is generally earthed along with the core. Generally, these aspects are compared with existing healthy units of the same rating and configuration. This technique is being developed as I write but some base conditions have been established about the methodology by both IEEE (IEEE C.57.149-2012) and IEC (IEC 60076-18). The methodology is mainly used with new transformers manufactured and are ready to be shipped to the user. Such details are then compared with data that can be generated after shipment to site, that can cause displacement in core-coil placements. This technique can also be used after a fault occurs and or after refurbishment of an old transformer. Therefore, it is utmost necessary to obtain the baseline characteristics of the transfer function of the transformer circuit that consists of R-L-C passive components. The frequency response is conducted at frequency of 50-60 c/s or at higher frequencies to recreate situations that can be experienced by the transformer during PD 50-350 kHz and even as high as 1 MHz. The transfer function is created between the bushing and ground don one side of the transformer by injecting a known signal in the above frequency ranges and the output collected at another bushing end between the live and ground points of the transformer at another voltage level. Such conditions are created under an electrical outage from the service line. A basic circuitry of a simple two winding transformer is shown in Figure 3.1 below. Generally, it will contain the R-L-C components of the core-coil structure on say primary side of the transformer and signal is collected on the secondary side of the transformer at different voltage levels with another R-L-C composition to represent the proper circuitry for assessment of the circuit parameters like C_1 and C_2 as well as insulation capacitance and tan delta parameter.

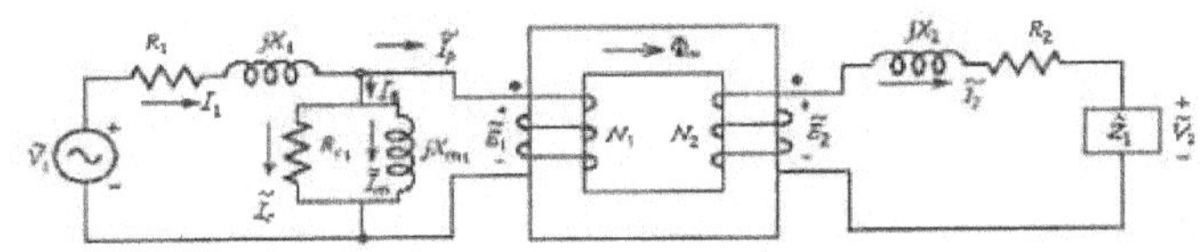

Figure 3.1 Equivalent Circuit of a simple transformer with parameters on the primary and secondary side.

An easy way is to conduct a frequency analysis of such a circuit. Simulations are conducted and base response for a new transformer is saved in the data base. Later on before shipping the transformer to the client a new frequency response of the same is obtained using suitable equipment in the market manufactured by Megger, Tettex or Newtons 4th etc.. When a failure takes place and the transformer is refurbished another FRA test is conducted and compared with the baseline trend at hand.

Recently on August 16, 2022, the author received another patent USPTO # 11,415,619; Shertukde, et. al; entitled: 'Digital Modulation/demodulation Signal Generator (DMDMSG) with active monitoring for Measurement of Power Factor and Capacitance in High Voltage Bushings, Transformers, Reactors and other Electrical Equipment with High Voltage Insulation' on Aug. 16, 2022. More details can be found on *www.USPTO.gov.*

The equipment for such an active measurement of power factor and capacitance is being developed by DDI. This equipment will enable in finding change in these two values that can provide diagnosis of failures caused by different operating conditions in the electrical power transformers when the transformer is in operation. A pole-mounted transformer was tested using this device which is called as Digital Modulation Demodulation Signal Generator (DMDMSG).

As a rule, the author has used most of these methods as a trendsetter for the subsequent application of the acoustic method in most of the test cases discussed in the sequel in Chapter 8.

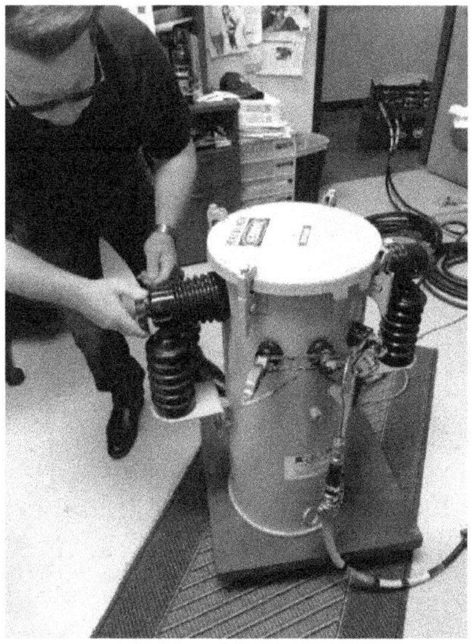

Figure 3.2 Pole mounted transformer tested with DMDMSG unit by DDI

The corresponding digital insulation diagnostic system is as shown in Figure 3.3 adapted from our USPTO patent 11,415,619, published August 16, 2022 - Hemchandra Srtukde, et.al

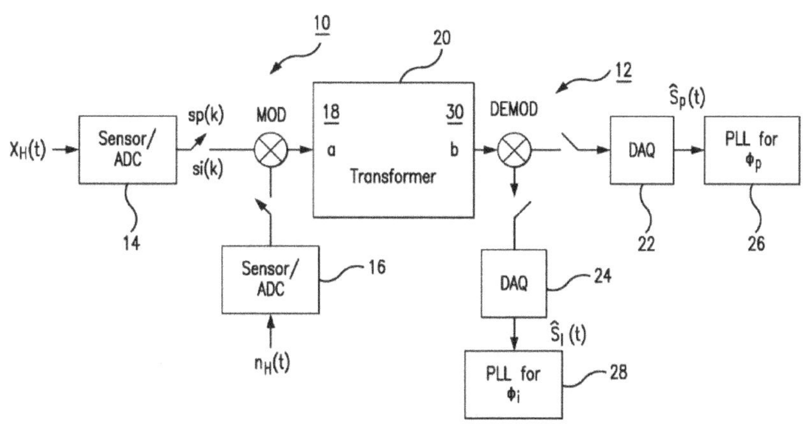

Figure 3.3Pole mounted transformer tested with DMDMSG unit by DDI

Chapter 4:
Acoustic Method

a. Offline
b. Online
c. Liquid filled transformers
d. Dry Type transformers

Process of PD formation and resulting signals generated of such an incipient fault.

When a PD happens in the confines of the transformer tank of an Electrical Power Transformer, it affects the insulation coordination of the system that is stressed electrically and mechanically as well. The signals generated are in the frequency range of 50 kHz to 350 kHz. Sensors which are resonant piezo-electrical transducers are placed on the outside surface of the transformer tank. These transducers detect the mechanical stress waves that have propagated from the PD source passing it through the different media from the source to the transformer tank which can include, core, steel, cellulose material, oil and finally through the transformer steel till it is detected by the sensor mounted on the tank surface. This signal is a broadband signal which looks like an arrowhead as shown in Figure 4.1

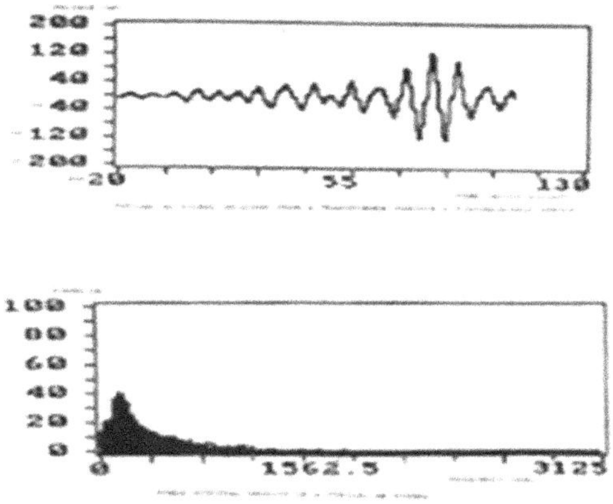

Figure 4.1 Arrowhead acoustic signal of PD.

Present methods estimate this PD location coordinates using triangulation methods. These are old and prone to errors. Considerable research has been done on the development of an effective diagnostic method to locate faults in electrical power transformers initiated by PD. When a PD forms, it is an arc and resembles a propane torch like flame that creates heat and acoustic signals as above. The heat creates disintegration of oil surrounding the arc and causes creation of gases. The electrical signal is a PD that travels through different media as a mechanical stress wave which is solved and identified using Wavelet Transform Methodology embedded in digital signal processing. Several methods estimate the travelling time difference of arrival (TDOA) between the different sensors using appropriate algorithms that use correlation of TDOA signals. Considerable research has been done over the past four decades in the development of an effective diagnostic method using Track Before Detect approach professed by the author. In addition, the general method suggested previously can be effectively used in conjunction with methods described in to obtain location of the PD faults within the confines of the transformer tank of an electrical power transformer. Conventional electrical PD tests are generally invasive in nature are available and conducted as per ANSI/IEEE standards. However, those may cause inclusion and insertion of incipient failure conditions, which over time may burgeon into catastrophic failures and prove detrimental and fatal to the life of an electrical power transformer.

In this novel method the signal received by different sensors depicts a bursty nature like the arrowhead it displays which is not statistically stationary over a given time window; which seems unfavorable to apply FFT. Also, the STFT cannot yield efficient results in locating the failures with an accuracy of 6-9" from the centroid of the failure. This will introduce errors in estimating the location of the PD failures. Instead, a Wavelet Transform (WT) is applied to the non-stationary signals as shown in Figure 4.2 below:

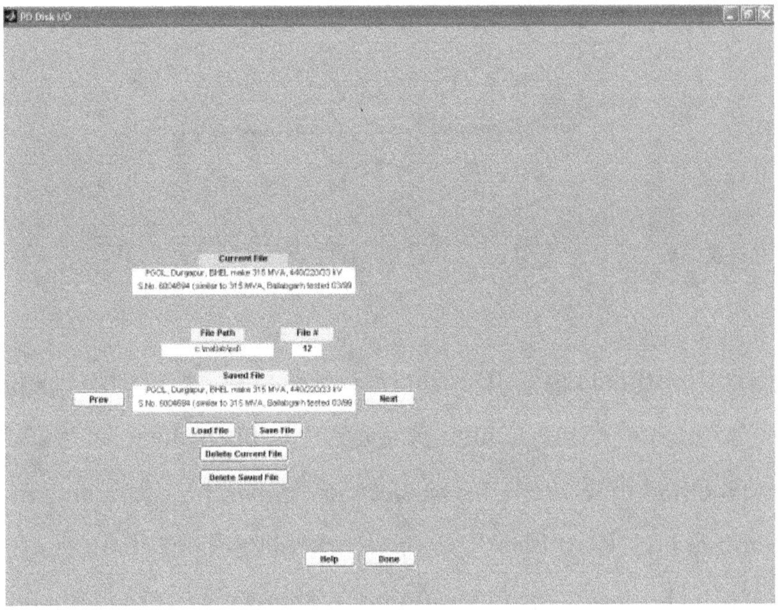

Figure 4.2. Wide Band Correlation Tracker using WT courtesy of DDI.

The signals received by several plurality of sensors are wideband signals which when operated with a Wide Band Correlation Tracker (WBCT) yields more accurate results in the determination of location of PD failures within 4-6" off the centroid of the failure location. Initially the location accuracy was 6-9" till when the data base kept increasing with more large transformers tested with FDD-EPT, as a result with the inclusion of AI the location accuracy increased to 4-6" off the centroid of the failure location within the confines of the electrical power transformer tank. Some examples that illustrate this methodology is illustrated with examples in Chapter 8. Of significance are transformers of 20 MVA, 115 kV/23.8 kV rating with low gas contents from GIO of H_2 and Acetylene. Initial results indicated some PD activity near the O/C switch side of the transformer. Results were confirmed by client to be within 6-9" of the location of estimated centroid of the PD failure.

Another power transformer tested during the same period was a 47 MVA, 345kV/13.8 kV, GSU with GIS termination, in a country outside USA. This transformer has several GIO analyses that indicated severe PD conditions. Upon testing the transformer with FDD-EPT by DDI unit, location was estimated to be in the terminal connection of the high voltage a phase with GIS termination. Test results showed levels of acoustic signals as high as 300 mili volts. Incidentally there is no direct correlation between the picocoulombs as specified for the electrical test but the author has done considerable research to achieve some semblance of connection between the two methods. Similar tests were conducted on many transformers to follow like 5.3 MVA, 15 kV rectifier transformer, 340 MVA, 500 kV GSU, 700 MVA345 kV/230 kV three-phase transformer, for a local utility in a nearby state of CT. Results further re-enforced the methodology efficacy used by FDD-EPT by DDI. The author was granted a large research funding and such methodology was published as a patent which was USPTO 6,178,386 in January of 2001.

a) Offline; onsite Method Using FDD-EPT

This method was developed using a shelf-top version of FDD-EPT by DDI using WT methodology built on the USPTO utility patent 6,178,386. This is also called as the lunch-box unit with an industrial computer that can be easily transported from DDI offices to local, domestic and international sites for service at the client's physical sub-station or manufacturing plant location. A depiction of the system is shown below courtesy DDI.

Figure 4.3 Offline; onsite Lunch-Box system courtesy of DDI

In such a system which can range from 4-24 channels in different footprints basically has two to four channels per data acquisition card and supersonic acoustic sensors that can be mounted on the surface of the transformer tank with magnetic hold-downs that can be movable or permanently placed at strategic locations. Some of these locations could be:

i) On HV side of transformer tank right opposite the hv lead points or on a corresponding turret as shown below in Figure 4.4
ii) On LV side of the transformer tank opposite the lv lead pints on the vertical face or on top of the tank as shown in Figure 4.5
iii) On the OLTC changer tank and/or O/C switch points.
iv) Opposite a line of locations about in line with the bottom core location.
v) Many other suspected locations which may yield a better detection and location as long as adjacent sensors are not within adjacent locations of 3-4' of each other
vi) Acoustic sensors are located on the transformer tank in conjunction and consultation with lineman at site who are employees of the client. Service provider should just be an advisor in the intended locations. Many clients have restrictions of this issue and should be properly checked before testing service is conducted.
vii) A site insurance policy is generally preferred by clients to protect their capital asset.
viii) Before moving sensors, proper discharging mechanisms should be adopted to make sure sensors don't carry spurious ESD charges.
ix) The onsite lunchbox should be at least 30' away from the transformer tank to keep operators out of harm's way. A failure concussion could be fatal if individuals are within this limit.
x) Some clients prefer operators use fire retardant clothing to protect from inadvertent accidents by explosions.
xi) A standard operating procedure (SOP) is specified by the service provider. Once the test is conducted a 3-D print out is created for the client as shown in Figure 4.4

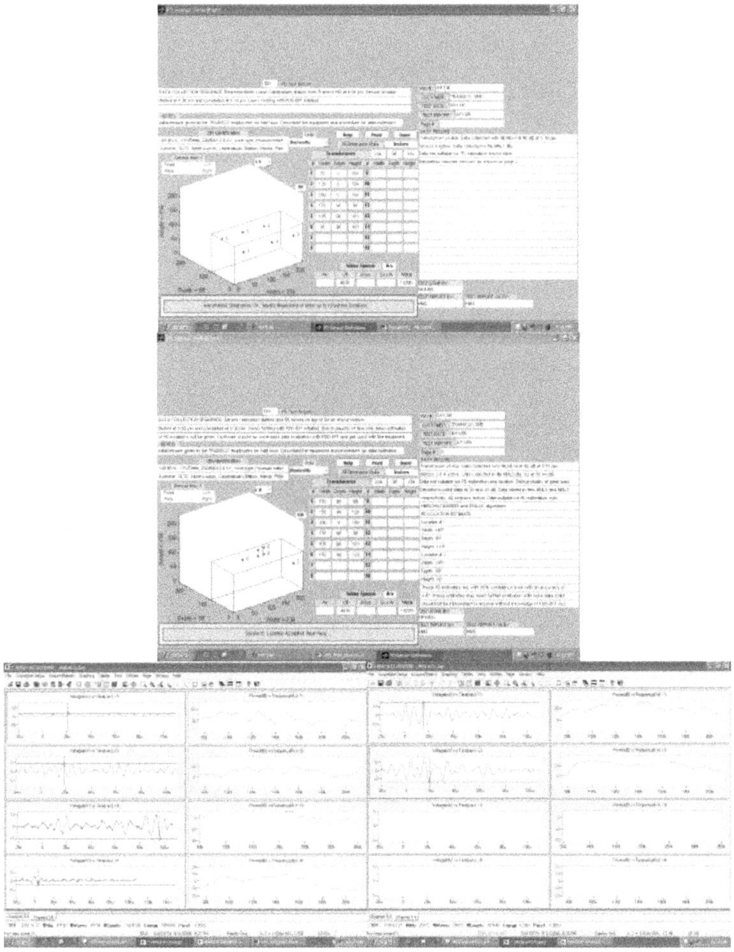

Figure 4.4 3-D diagram of sensor locations, transformer tank and location (s) of the pd failures courtesy of DDI

b) Online; offsite Method Using FAN-EPT

This method was developed using a laptop version of FAN-EPT by DDI using wide band correlation tracker - WT methodology built on the USPTO utility patent 6,178,386. This is also called as the industrial online laptop unit with an industrial computer that houses DAQ cards and the unit can be can be easily installed by DDI at local DDI office, domestic and international sites for service at client's physical sub-station and lastly at manufacturing plant location. A depiction of the system is shown below courtesy DDI.

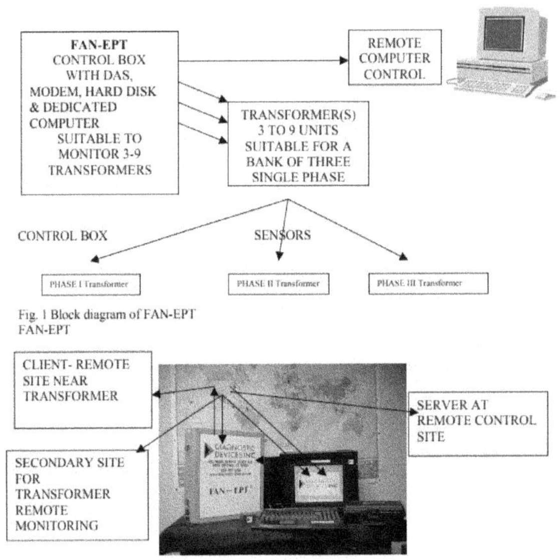

Figure 4.5 Online; offsite Laptop industrial computer system courtesy of DDI

In such a system, which can range from 4-8 channels in different footprints basically has two to four channels per data acquisition card and supersonic acoustic sensors that can be mounted on the surface of the transformer tank with magnetic hold-downs that can be movable or permanently placed at strategic locations. The operation of this system is via a master slave arrangement as shown above in Figure 4.5 achieved via a T-1 line, or wireless 5G/10G connection or via cloud.

Some of these sensor locations could be:

1. On HV side of transformer tank right opposite the HV lead points or on a corresponding turret as shown below in Figure 4.6

Figure 4.6 Sensors located on HV side of transformer tank courtesy DDI

2. On LV side of the transformer tank opposite the lv lead points on the vertical face or on top of the tank as shown in Figure 4.7

Figure 4.7 Sensors located on transformer tank courtesy DDI

3. On the OLTC changer tank and/or O/C switch points.
4. Opposite a line of locations about in line with the bottom core location.
5. Many other suspected locations which may yield a better detection and location as long as adjacent sensors are not within adjacent locations of 3-4’ of each other
6. Acoustic sensors are located on the transformer tank in conjunction and consultation with lineman at site who are employees of the client. Service provider should just be an advisor in the intended locations. Many clients have restrictions of this issue and should be properly checked before testing service is conducted.
7. A site insurance policy is generally preferred by clients to protect their capital assets.
8. Before moving sensors, proper discharging mechanisms should be adopted to make sure sensors don’t carry spurious ESD charges.
9. The onsite lunchbox should be at least 30’ away from the transformer tank to keep operators out of harm’s way. A failure concussion could be fatal if individuals are within this limit.
10. Some clients prefer operators use fire retardant clothing to protect from inadvertent accidents by explosions.
11. A standard operating procedure (SOP) is specified by the service provider. Once the

test is conducted a 3-D print out is created for the client as shown in Figure 4.8

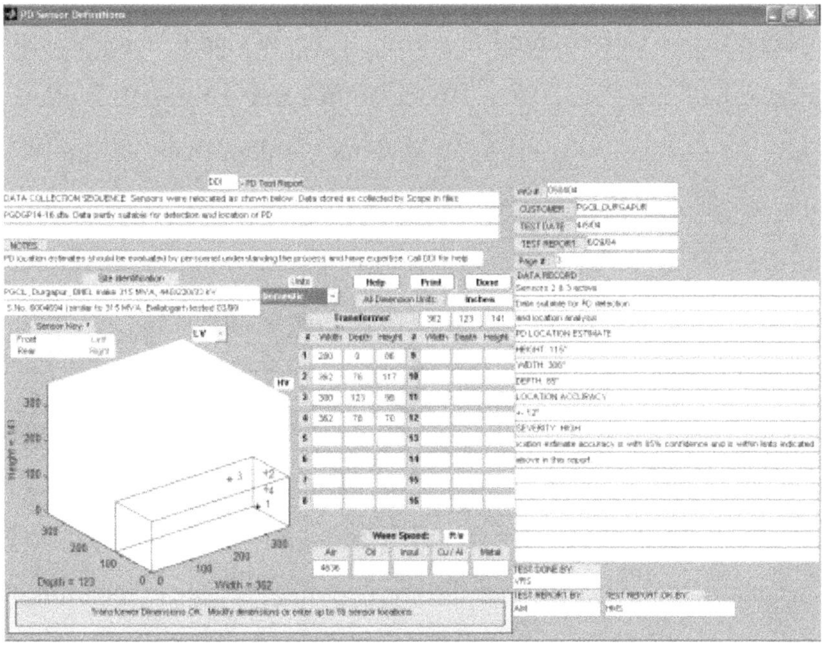

Figure 4.8 3-D diagram of sensor locations, transformer tank and location (s) of the pd failures. Courtesy DDI and respective client

There is a 24/7 connection between the master and slave locations. Any failure is instantly communicated via telephone or email that is sent to the concerned operators for further analysis and maintenance action. A very prominent case study is described in Chapter 8, which involved 3 three-phase 525 kV, three phase transformers at a generating station which had one faulty transformer and two other healthy transformers with the same rating installed at the site on the west coast of USA. The said transformers were monitored online with a FAN-EPT device which was installed at the transformer site, DDI and at the manufacturer's site which was outside of USA. Results were monitored 24/7 and continued for six months at which point stable operation of all units was reported by the client on the west coast.

Problem Formulation

A passive system sown in Figure 4.9, there are only received signals. There are no active signals transmitted by the sensors as in pinging experience in radar signals etc. Plurality of sensors is a must in this case at least three as three co-ordinates are to be estimated for the failure condition within the confines of the transformer tank. As this is digitally developed system the minimum number of sensors is generally four. With the PCI footprint of data acquisition card, a maximum of sixteen channels is an optimum and efficient FDD-EPT that is developed by DDI. The software utilizes the Wideband Cross-Correlation Processing using WT and AI for acoustic signals of a PD source in the bandwidth of 50 kHz-350 kHz. If one of the received signals is chosen as the reference signal that exceeds the threshold for all other sensors the data from all the remaining sensors is then collected and stored with proper time stamps and details of sensor locations with respect to a chosen origin in the 3-D space to represent the transformer tank. This reference signal is then scaled and translated, plus compared with the remaining received signals and the correlation of received signals with scaled and delayed versions of reference signal forms a wideband cross ambiguity function. These functions are plotted in time and frequency domain with peaks indicating the relative time-scale and time-delay of the multipaths along which the source signals travel in the line of sight of a particular sensor. Estimation of these Relative Normalized Time Difference of Arrivals (RNTDOAs) are used to project the geometric location of the failure point due to PD; with respect to the chosen origin. The mathematical formulation is illustrated below:

Let x(t) be the reference PD signal at the source. The attenuated, scaled and delayed versions of this PD signal corrupted by noise will be received by the plurality of other acoustic sensors, which are located on the transformer tank. This is depicted in Figure 4.9.

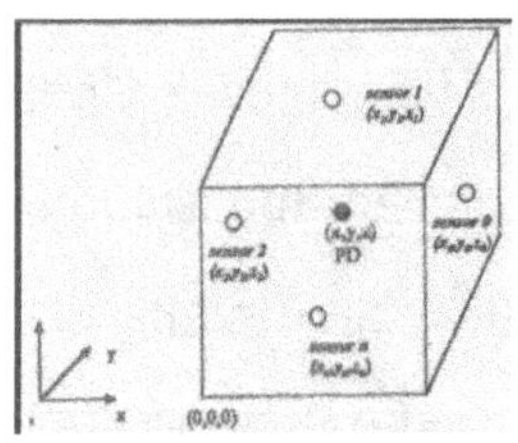

Figure 4.9 Wideband Matched Filter and sensor locations to obtain failure locations courtesy of DDI

We can now develop the following model for these received signals at several sensors as:

$$r_i(t) = \frac{\alpha_i}{\sqrt{s_i}} x\left(\frac{t-\tau_i}{s_i}\right) + n_i(t); where\ i = 0,1,\ldots.,,n-1$$

$$= x_i(t) + n_i(t) \qquad Eq\ (4.1)$$

The noise $n_i(t)$ is modeled as independent, identically distributed Gaussian random variable with mean zero and variance ${\sigma_i}^2$ and is uncorrelated with the received signal $x_i(t)$.

We can choose the signal received at sensor zero be the mother wavelet and let $W_{r0},\ r_i(s,tau)$ be the Wavelet Transform of signal of the signal $r_i(t)$ with respect to $r_0(t)$. The continuous wavelet transform computes the inner product of the function $r_i(t)$ with the scaled and delayed version of the mother wavelet, $r_0(t)$, i.e.

$$W_{r_o} r_i(s.\tau) = \frac{1}{\sqrt{s}} \int r_i(t) r_0\left(\frac{t-\tau}{s}\right) dt \qquad Eq\ (4.2)$$

In order to allow reconstruction of $r_i(t)$ from its WT with $r_0(t)$, the mother WT $r_0(t)$ must satisfy the following conditions:

$$C_{r_o} = \int \frac{|R_0(w)|^2}{|w|} dw \ < \infty \qquad Eq\ (4.3)$$

Therefore, any real-time signal received in a system with finite duration and zero average value qualifies as a mother WT. Using the scale/delay property of WT yields:

$$W_{x_0} x_i(s,\tau) = \frac{\alpha_0 \alpha_i}{\sqrt{s_0 s_i}} W_x x\left(s\frac{s_0}{s_i}, \frac{\tau + s\tau_0 - \tau_i}{s_i}\right) \qquad Eq\ (4.4)$$

Since $W_x x$ will have a maximum value at (s, tau) =(1,0). This equation (4) will have a peak when:

$$s = \frac{s_i}{s_0} \text{ and } \tau = \tau_i - \frac{s_i}{s_0} r_0 \qquad Eq\ (4.5)$$

Where tau refers to RNTDOA wrt reference time domain of PD source. If we change the reference time domain to the time data is required a set of new equations that need to be developed and have been developed since then by the author and are available with DDI and can be shared only with clients with proper proprietary rights satisfied by both parties.

The data acquisition hardware comprised of:

1) Data Acquisition industrial PC expansion system

2) PCI data acquisition cards

3) Data acquisition system with an interface box

4) Supersonic piezo electric crystal acoustic sensors

5) The number of digital acquisition channels per PCI card is guided by the following:

a. The desired minimum sampling of data per channel. This is the sampling frequency and is guided by the Nyquist criterion. Thus, for a signal bandwidth of 50 KHz to 350 kHz the minimum frequency of sampling is twice the maximum signals frequency in this case that comes out to be 700 KHz. We chose a sampling speed of 1 MHz. With the advent in DSP this frequency can go as high as 100 MHz.

b. The switching speeds of the associated hardware components

c. Clock synchronization of data acquisition by the DAS board with external multiplexer. The DAS board by itself is completely unaware of the external multiplexing. As a result, the data acquired by the DAS board may not be at the same time when the data is made available by the analog data multiplexer for data acquisition. In order to ensure contemporaneous data availability and data acquisition, synchronization of the system main clock is required. Multiplexer circuit is the heart of the DAS I/F board. It basically consists of two analog multiplexers. All channels exhibit break-before-make switching action to prevent momentary shorting when switching channels and the related commutation that may take place with adjacent circuits. Upon receiving a trigger signal from the DAS board a set of events occur and then the DAS I/F board is ready for actual real time data acquisition and it then awaits for the trigger from the DAS board.

d. Localization of the channels with respect to physical location of the sensors on the transformer tank mounted with magnetic hold downs.

Estimating the Location of a PD Source

Figure 4.11 shows an example of a transformer tank, PD source inside the confines of the transformer tank and sensor locations on the transformer tank. The sensors are located with a chosen origin and the PD location to be estimated is also shown.

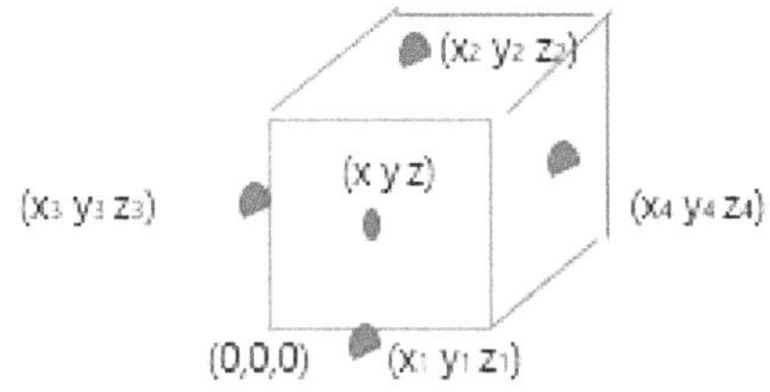

Figure 4.11 Transformer, PD source and sensor locations — only four sensors shown above

Let there be (n+1) number of sensors located on the transformer tank. Let the ith sensor location be denoted as (x_i, y_i , z_i) with respect to the origin (0,0,0) chosen in a 3-D condition. When the various travel times (RNTDOAs) have been determined it is necessary to convert those to distance. This necessitates these be multiplied by velocity of signal travel in different media. If the speed of sound in oil is C feet/sec, then:

$$(x - x_i)^2 + (y - y_i)^2 + (z - z_i)^2 = C^2 \tau_i{}^2 \qquad Eq\ (4.6)$$

Equation (6) will give the estimate of the PD source with similar equations to be written for different media path, from which ultimate average estimate of the PD location can be assessed.

Evaluation and Expected Results

The WT techniques, with the corresponding data acquisition hardware/software, are estimated to locate the partial discharge faults within 6-9" of the PD source centroid. Alternate methods can be employed to estimate the location of PD source. Over the years significant system efficiency has been improved and the time taken to estimate PD sources has significantly improved. The latest development concentrated on implementing the DAS Interface Board and the Decoupling/Amplification Boards on the PCB PCI boards and a friendly user-friendly Graphical User Interface (GUI). More details can be found on DDI website: www.diagnostic-devices.com.

A set of GUI figures is shown below courtesy DDI for the HMSGHAT2000DDI:

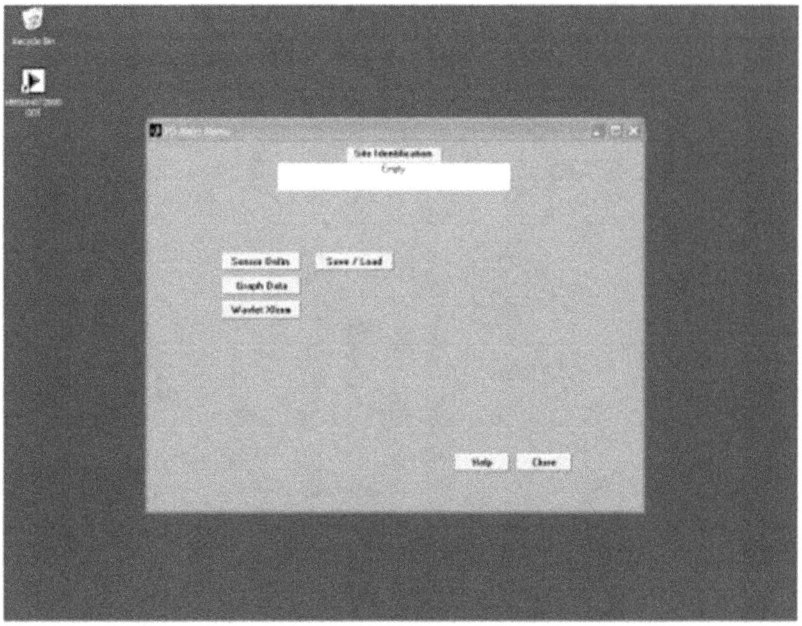

Figure 4.12 Initial data insertion window.

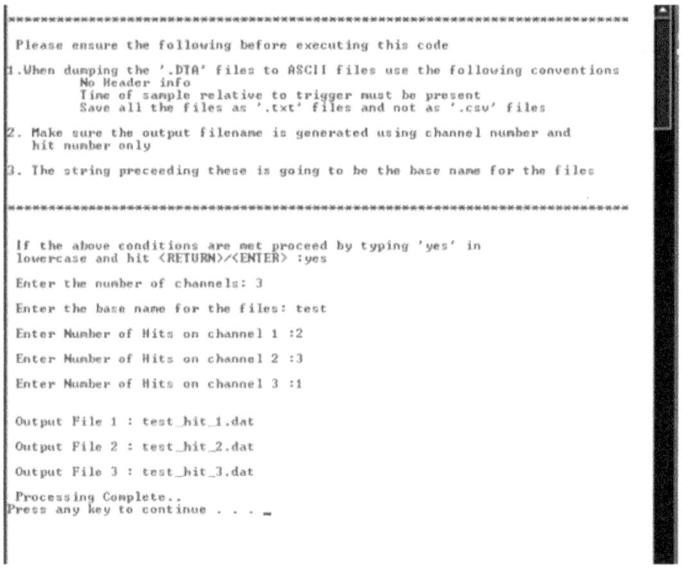

Figure 4.13 Data Processing sequence of HMSGHAT2000DDI courtesy DDI.

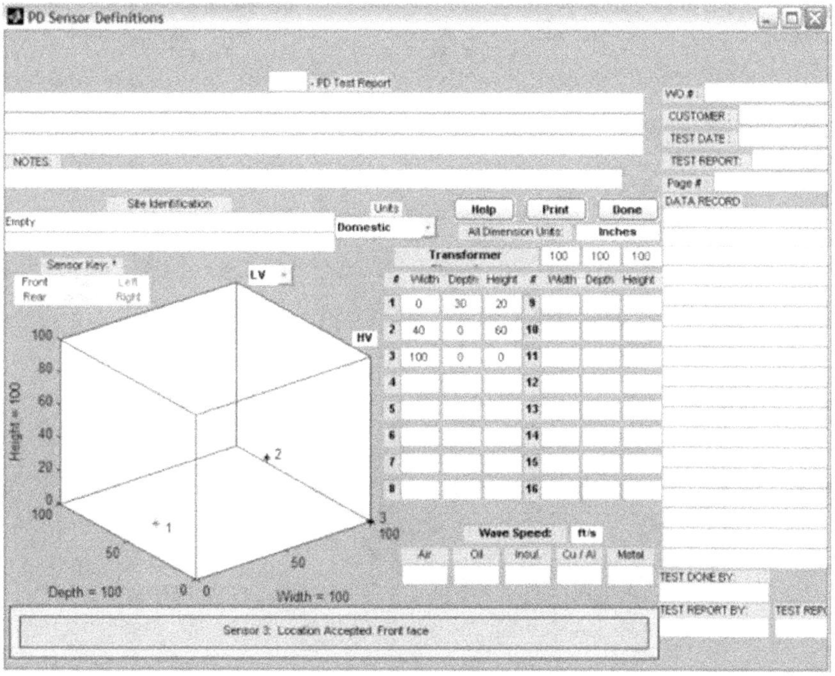

Figure 4.14 Initial 3-D screen to start data acquisition courtesy of DDI.

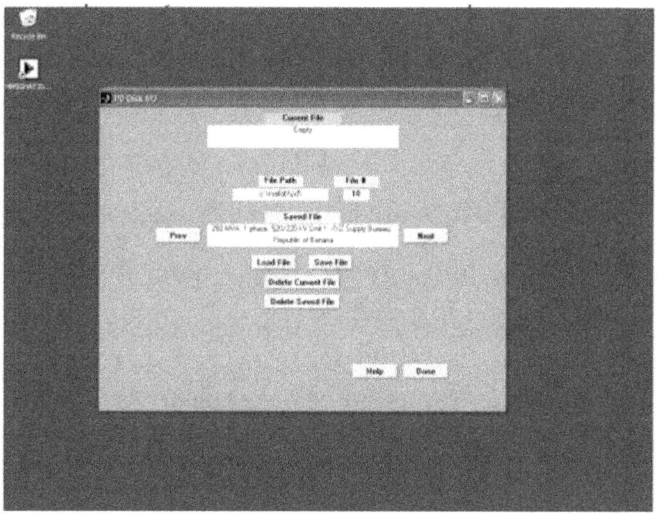

Figure 4.15 Initial data entry sheet about transformer site etc. courtesy of DDI

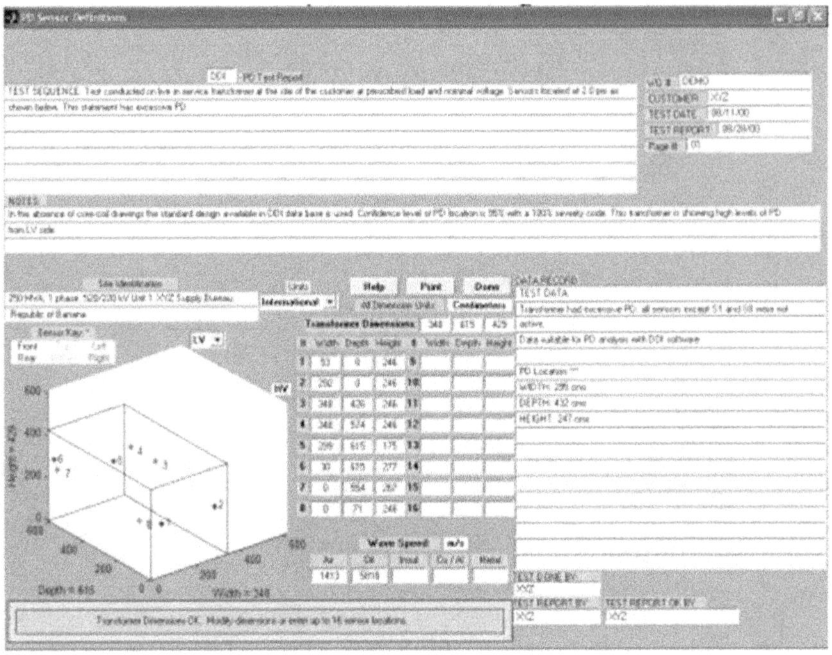

Figure 4.16 Final 3-D diagram of transformer with al details regarding sensor locations, failure location etc. courtesy of DDI

The above process using FDD-EPT or FAN-EPT; especially the former unit generally takes a few hours. Continuous online system FAN-EPT also takes similar amount of time. In the worst-case scenario it may take a little longer but not more than two working days. The last onsite testing done out of the country took a bit longer because of the nature of the beast. It had a set of 3 inverter transformers of a 25MVA rating with 25/345 kV voltage rating.

Some examples of onsite testing are shown below for several cases in Figures 4.17 and 4.18 courtesy of DDI.

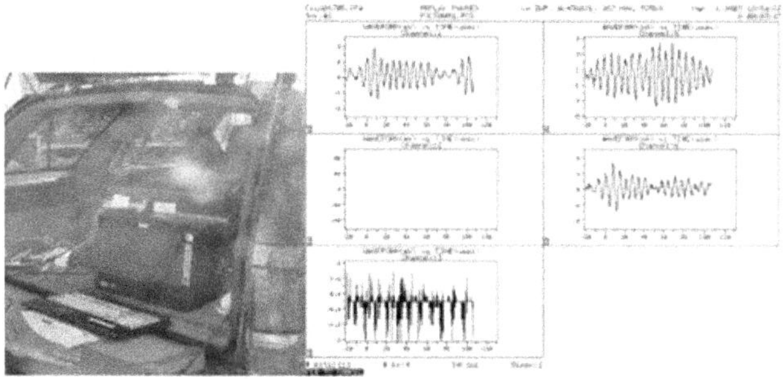

Figure 4.17 Onsite testing with a lunch-box unit, courtesy of DDI

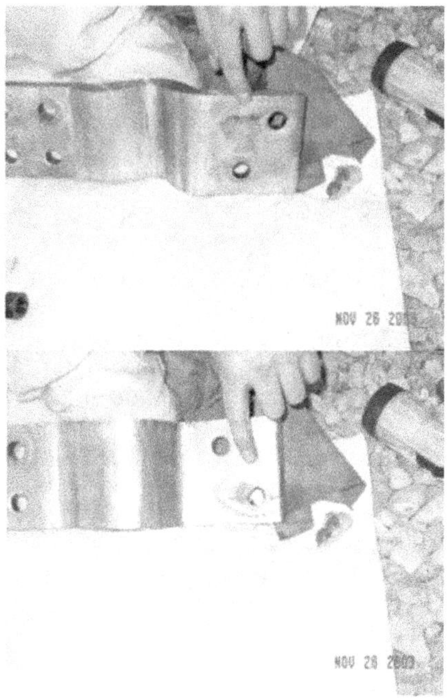

Figure 4.18 Actual failures of a LV-side lug connections of receiving station transformer courtesy of DDI.

The entire work resulted in a US patent # 6,178,386 awarded to the author as below as in Figure 4.19, in 2001:

Figure 4.19 US PTO # 6,178,386 that was issued on February 21, 2001

Many more large power transformers were tested including oil-filled, liquid-filled and dry type with wound cores for the solar and wind energy sectors. The sensors are installed on a metal tank inside which the dry-type transformer was located. This site was located on the west coast of USA. However, due to confidentiality of terms of service the author cannot elaborate on the actual tests conducted and results obtained on testing these transformers. However, the rating of these transformers were 5MVA at 13.8 kV, 3-phase, 60 cc/s with wound cores. All the results were obtained by using signal processing techniques as illustrated in the DDI website and using WT as below in Figure 4.20:

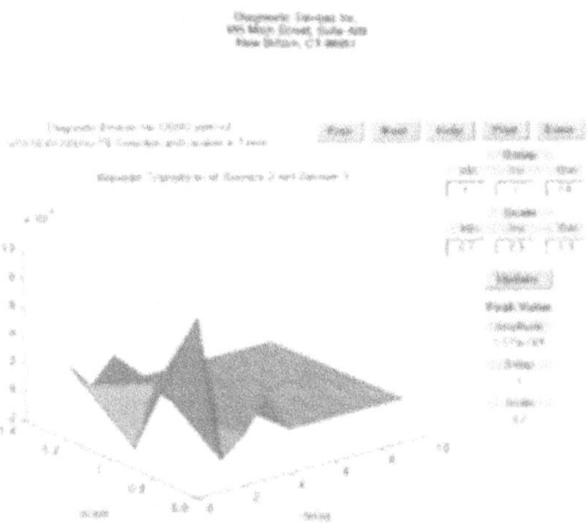

Figure 4.20 WT method of detecting and locating pd failures as in USPTO # 6,178,386 courtesy of DDI.

The technical specifications of FDD-EPT are provided below courtesy of DDI:

FD²-EPT™ HARDWARE SPECIFICATIONS

BRIEFCASE SYSTEM GENERAL

4-8 CHANNEL SYSTEM

2-4	Data acquisition cards with 8 bit A/A (>48 dB) or better 1-8 MHz sampling or better, with suitable data flow rates
4-8	DDI special Acoustic Sensors with line amplifiers optional
1	Briefcase system, lap-top computer with 32 MRAM, 4Gbyte HDr, 3.5" floppy drive, 10.5"LCD panel, keyboard
120'-240'	signal cable (RG 58) with bnc/bnc termination
2	High vacuum grease couplant tubes
3	

Accuracy

PD detection level	
	Greater than 150 pC when measured with a suitable electrical system for insulation failure
PD location accuracy	12"-18" about the centroid when transformer internal drawings not provided by customer 6"-8" about the centroid when transformer internal drawings provided by customer for insulation failure and using WTPDDDI2000

Data Acquisition Board

Channels

Number	**Two, simultaneous (real)**
Input Impedance	50 ohm (DC coupled) shunted by approx. 20 pF
Input Capacitance	20 pF, approx.
Maximum safe input Voltage at 50 ohms	5 V (rms) continuous, 250 V transients for less than 100 micro secs.

ADC

Type	8-bit or better, flash
Integral and differential linearity	+/- 0.6 LSB
Integral and differential Non-linearity	+/- 0.94 LSB with no missing codes
Aperture jitter	15 ps rms
Input slew rate	440 V/microsec
SNR	48 dB or better, typical

Accuracy

Gain	+/- 5% of FS
DC offset	+/- 5% of offset
Real-time Sampling period	Time base accuracy (+/- 1 ps) + ADC apperture jitter

FD²-EPT USER'S MANUAL - VERSION 2.0 ______________ Page 16 of 88

(+/- 17 ps) = +/- 18 ps total 9+/- 0.18% @ 64 MHz)

Pacer clock

Internal	Frequency: up to or less than 64 MHz Stability: 100 ppm over temp. and aging (1 year)
External	Type TTL-level, up to or less than 64 MHz Max input voltage: +/- 4 V, with 50 ohm termination

General

Input ranges:	+/- 100 mV up to +/- 4 V (16 ranges)
Onboard memory buffer	1M bytes
Power Consumption	2.7 A at + 5.0 V, typical 0.3 A at + 12.0 V, typical 0.2 A at -12.0 V, typical 0.2 A at 28.0 V, typical
Operating Temperature	0 to + 55° C

Acoustic Sensors

Gain with integral amplifier	40 dB +/- 1 dB
Peak sensitivity	30 dB re 1V/microBar
Noise (rms)	< 1.0 micro V
Dynamic range	> 80 dB
Output Voltage	> 15 V pp into 50 ohms
Power required	28 V at 20 mA
Temperature	-45 to + 80°C
Pressure	< 400 psi hydrostatic pressure

Computer

System Requirements

- Pentium, Pentium Pro, Pentium II, Pentium III, Pentium IV or AMD Athlon based industrial computer (briefcase, lunch-box or 19" rack mount type)
- Microsoft Windows 95, Windows 98 (original and second edition), Windows Millennium Edition (ME)
- CD-Rom drive (for installation from CD)
- 64 MB RAM minimum, 128 MB RAM recommended
- Disk Space varies depending on size of partition and installation of on-line help files.
- 8-bit graphics adapter and display (for 256 simultaneous colors)

Other recommended items may include:

- Microsoft Windows supported graphics accelerator card, printer, and sound card
- Microsoft Word 7.0 or Office 2000 is recommended to run HMSGAT2000 notebook.

Chapter 5: Thermal Techniques

5.1 Thermal Techniques

With the advent of thermal techniques, it has become necessary to assess the conditioned monitoring of transformers with the use of thermal cameras and forward-looking infrared (FLIR) sensors and cameras. Ever-changing research and development activities can generate new methodologies to develop new applications to solve these problems.

5.2 Thermal Cameras

Thermal Cameras help to find hot spots within the confines of the transformer tank. It can also find hot spots external to the transformer. The camera is used from a distance of 30" from the transformer tank and any live connections. These cameras are used very recently over the past decade to find failures due to hot spot. Nowadays, these are also used fixed on drones to obtain a live hot-spot picture of the transformer and its fittings. IEEE-PES-TC has not created any new User's Guide for this application. The use of these techniques has become more popular over the past decade. Constantly increasing the applications will make these techniques more popular as these are very small and light in size, and with the use of cloud things are more global.

5.3 Hotspot Detection

IRT identifies abnormal temperature hotspots within the transformer. These hotspots may indicate issues like insulation degradation, overheating, or partial discharge. By monitoring these hotspots, maintenance teams can take timely corrective actions. Many such failures close to the surface of the tank can be easily sorted out, but the difficulty lies when the location of the thermal fault is further inside the core-coil insulation disposition. However, recent research and development has helped to alleviate this problem. Restriction, however, lies with transformers that are of dimensions greater than 30' on all x-y-z dimensions. Such equipment is now developed in the Near Infrared (NIR) bandwidth.

Over the past few years, the request from clients to have a handy camera that uses thermal technology to get an initial idea of the possible hot spots within the confines of the transformer tank has significantly increased. Just recently, as of 3/10/24, fresh new inquiries have been received by DDI to render help in assessing failures in large power transformers using thermal camera hand tools. To alleviate such emergency issues, DDI is working feverishly to address the situation.

5.4 FLIR Cameras

Forward-Looking Infrared (FLIR) cameras are used to find failures within a transformer tank for signals in the infrared bandwidth. These cameras can also be used to find failures external to the transformer tank and any lie connections. The cameras are held at a distance of 30' from the transformer tank. These cameras have been used very recently over the past decade to find failures due to hot spots. Nowadays, these are also used fixed on drones to obtain a live hot-spot picture related to infrared signals of the transformer and its fittings. IEEE-PES-TC has not created any new User's Guide for this application.

5.5 Hotspot Detection

IRT identifies abnormal temperature hotspots within the transformer. These hotspots may indicate issues like insulation degradation, overheating, or partial discharge. By monitoring these hotspots, maintenance teams can take timely corrective actions.

5.6 IRT

Condition health monitoring of transformers is crucial in electrical power industries to prevent sudden breakdowns, power outages, reduce downtime, minimize maintenance due to downtime, and extend the life span of machines and equipment. One effective technique in vogue is Infrared Thermography (IRT). IRT detects variations in temperature precisely.

Here is how IRT works:

5.6.1 IRT

In this method, heat signatures are captured using thermal cameras. Such a methodology allows real-time acquisition of temperatures without having direct contact with the hot substance.

Here is how IRT works: Slight temperature variations within the tolerance are acceptable, but any abnormal temperature rise can indicate an impending potential fault. This is similar to the incipient PD failures in transformers, except the signatures are different, which can lead to catastrophic failures of transformers. These unusually high temperatures can damage transformer insulation, either cellulose or oil, and reduce the life span of the capital asset.

These hot spots can indicate the incipient faults burgeon into catastrophic failures in insulation or cellulose or oil and core insulation. Maintenance teams can monitor these conditions in the health of the capital asset and save costly expenses to get the equipment back in action and prevent further losses in power restoration to clients and end users. There are two ways of analyzing thermal images.

A. Matlab-based Analysis: Image processing techniques and digital signal applications can be implemented in MATLAB. These algorithms identify patterns, compare temperature distributions and highlight anomalies.
B. Internal problems due to hot spots in oil/Esther or cellulose can be verified using GIO or DGA or Roger's Method, etc. Such techniques have been discussed in the prequel and sequel as well.

Over the past two decades, the request from clients to have a handy camera that uses infrared technology to get an initial idea of the possible hot spots within the confines of the transformer tank has significantly increased. Just recently, as of 3/10/24, fresh new inquiries have been received by DDI to render help in assessing failures in large power transformers using infrared camera hand tools. To alleviate such emergency issues, DDI is working feverishly to address the situation.

Chapter 6:
Invasive Techniques – Using Photo Sensors

In Chapter 6, we discuss invasive techniques using Fiber Optic Sensors (FOS). To effectively summarize these techniques. These FOSs are placed inside the tank confines at predetermined locations so that the location of the FOS does not alter or hinder the insulation coordination per the original design. An example is the best way to illustrate such a condition and is succinctly discussed in the sequel in Chapter 8 where some case studies have been discussed in the use of FOS technology in the diagnosis of failures of electrical power transformers [Ref. 1].

FOS technology was initially utilized for failure analysis in data transmission and telecommunications. Incidentally, the sensors developed using fiber optics have undergone rapid advances, improving sensing performances and capability to function in remote and harsh environments of equipment under test and operation. Fiber-optic sensors are cost-effective, miniature in size, lightweight, and immune to electromagnetic interference.

The failure condition generates a frequency band of signals that are picked up by the FOS, which travel through the different media between the location of the sensor and the source of the failure within the confines of the transformer tank. FOSs are immune to electromagnetic signal interference. FOSs have high sensitivity and specificity characteristics and small dimensions that can interfere with the insulation coordination of the entire transformer. Thus, FOSs are highly suitable for monitoring power transformer conditions monitoring in continuous time and online. Many such systems are developed in the market, one being a suitable system developed and marketed by one of the reputed manufacturers of power transformers. Such failures can be caused by PD, dissolved gases, increased temperatures, and cellulose failure due to sparks caused by loose contacts.

FOSs use the physical properties of light as it travels along a fiber to detect changes in temperature, dissolved gases emanated in the proximity of the sensor, strain and other parameters. The latter is generally associated with the bending and twisting of winding copper conductors due to unusual forces that can exist in the transformer winding configurations. The primary dissolved gas that can be detected is H_2 which is a good indicator of problems that can be burgeoning inside a power transformer. This technique for finding power transformer failure diagnostics is fairly new and has not been set in a framework that can be universally used for failure diagnostics in power transformers. However, considerable research has been done and different methodologies have been suggested by many researchers as in reference [x]. Here, the working principle is explained and the different applications of the methodology highlighted, one of which is for power transformer failure diagnostics and is mostly and online system.

Application to dissolved gas analysis in oil and assessment of abnormal temperature increases in oil and cellulose due to incipient pd failure, which can, over time, burgeon into a catastrophic fault. PD mostly causes gases to dissolve in oil which can be tracked by the FOS and evaluated for the safe levels of ppm that are allowed for the safe operation of a power transformer.

FOS is mounted at strategic places inside the transformer tank, one at the bottom of the tank where the valve to obtain oil samples is located and the other at the top of the tank in line with hot spot and winding hot spot locations, which are mostly in line with WTI. The choice of the optic fiber is decided by the Bragg Gratings (BG) chart, which is available from the manufacturer of the FOS. BG is based on a permanent refractive index change in a special multi-component silicate glass after exposure to UV radiation, followed by a thermal development. This Bragg Glass enables the fabrication of phase volume diffractive gratings with an absolute diffraction efficiency exceeding 99% and thermal stability of 400 degrees centigrade. It also has a laser thermal damage threshold of 40 J/cm^2 for 8ns pulses and tolerance to CW laser radiation in the near IR region at least up to several tens of kilowatts per square centimeter. Such volume BGs provide narrow spectral selectivity down to 20 pm and narrow angular selectivity down to 100 micro radians. These elements have low losses and can be used for a variety of applications in optoelectronics, analytical, and electrical power and defense industries.

A general picture of an FOS is shown below in Figure 6.1

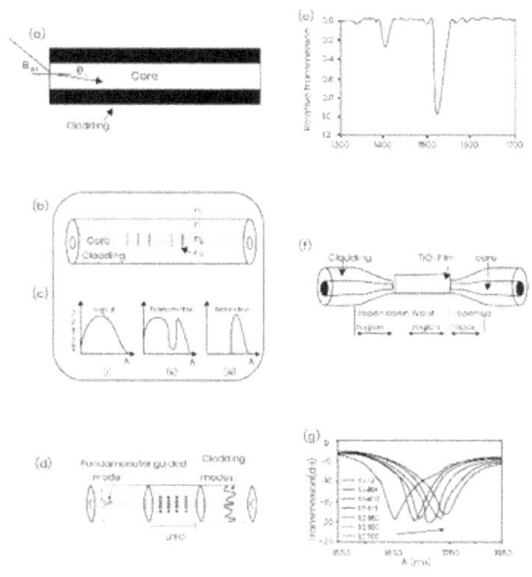

Figure 6.1 Typical FOS sensor courtesy [Ref. 1]

A fiber-optic sensor is a sensor that uses optical fiber either as the sensing element or as a means of relaying signals from a remote sensor to the electronics that process the signals. Fibers have signal characteristics, as shown in Figure 6.2.

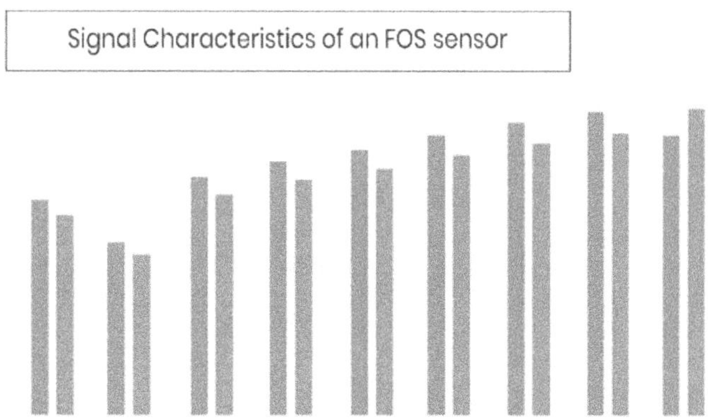

Figure 6.2 Signal Characteristics of an FOS sensor [Ref 1]

The global fiber optic sensors market is sustained by the growing demand for several high-end industrial applications together with sophisticated functions. Some of its applications are in the detection and location of failures in power transformers and the technology is very promising but is not sustained yet in the transformer market.

This technology and methodology is not used frequently in power transformer failure diagnostics as it can disturb the core-coil insulation coordination within the confines of the transformer under consideration. The method has many restrictions, and a careful understanding of all issues has to be considered by the proponents of this technology. The author is also not fond of this method and has considered it only in the last stages of transformer diagnostics.

Chapter 7: Offline Vs Online Techniques

We will further discuss these offline and online techniques in Chapter 7.

7.1 Offline Vs. Online Techniques

7.1.1 Offline Technique

Such a device is used onsite in the field or at a manufacturer's plant. For PD testing of transformers using acoustic techniques, DDI has developed and sold a Fault Diagnostic Device for Electrical Power Transformers (FDD-EPT). This is a patented device (USPTO # 6,178,386; Shertukde et al.). This patent was awarded in February of 2001 to the author and his co-inventor and assigned to the University of Hartford (UHART) in Wes Hartford, CT, USA. UHART then awarded the license to market this device to DDI.

Over the past twenty-two years, the author has worked towards perfecting a partial discharge (PD) detection and location system, which is both novel and economical. Guided by the US patent # 6,178,386 and co-invented by the first author and licensed by his employer to the small business in Simsbury, CT, the PD detection and location system is indigenously made by the employer of the second author and sold worldwide to reputed transformer manufacturers and utilities.

The system consists of an efficient hardware system efficiently supported by a state-of-the-art algorithmic approach that accounts for the largest transformer tank size without losing the physical attributes of the signals generated by the PD sources in a multi point failure conditions.

FDD-EPT uses state-of-the art methodology of digital signal processing with Wavelet Transforms (WT) and with supersonic acoustic sensors to acquire pd signals. These acoustic sensors have piezo electric crystal and are mounted on the external metallic surface of the transformer tank with magnetic hold-downs that can sustain a wind force of 25 lbs/sq. Inch.

Electrical Equipment like transformers are continuously plagued by internal incipient faults like partial discharges that can eventually burgeon into catastrophic faults, resulting in the property damage of an end user. Thus

PD = PD

PARTIAL DISCHARGE = PROPERTY DAMAGE

Many utilities, as well as manufacturers, have fallen prey to such a calamity either because of an increase in the age of the equipment and arduous operating conditions or genuine negligence on the part of management and operators. Such financial disasters are easily prevented, and the principals from Diagnostic Devices Inc. in New Britain, CT, USA have commissioned such a system in the USA for a power utility on the West Coast.

Now, I will describe the working of this online monitoring system and explain its utility in saving capital assets. Efforts will be made to interpret and analyze the real-time and online data acquired by such a system from the two transformers being monitored as we write this paper. Measurement parameters of both systems will be compared and analyzed. Real-time examples will be discussed to discern difficult data. The results will be compared with those obtained by an online, onsite system called the Fault Diagnostic device for Electrical Power Transformers (FDD-EPT). This is also called as the online, onsite system. Here, the possibility of the data-gathering personnel being in harm's way of a troublesome transformer unit is larger. In the online-onsite system, the data-gathering personnel is generally present in the proximity of an online transformer, and therefore, the risks of personnel injury are greater than usual when some other means of monitoring are used. As such, an online offsite system serves a better purpose in assessing the health of a troublesome faulty transformer.

7.1.2 Online Technique using acoustic PD system

7.1.2.1 Hardware

This online monitoring system to monitor partial discharges in electrical power transformers has the capability of monitoring remotely. It can also be controlled via a modem with an audio alert system. This online monitoring system has a control box to be mounted near the transformer, with a remote computer acting as a client. The client is connected to the server via a modem.

This connection between the client and the server can also be achieved using wireless modems. The limitations of the latter are the physical distances between the client and the server. The control box and remote computer can be configured in a master/slave operation. The sensors are mounted on the transformers either temporarily with magnetic hold-downs or permanently and the sensors are connected using co-ax cables/leads to this control box. The remote computer is a lunch-box industrial computer.

Details of the physical layout of the system are shown in Figure 7.1:

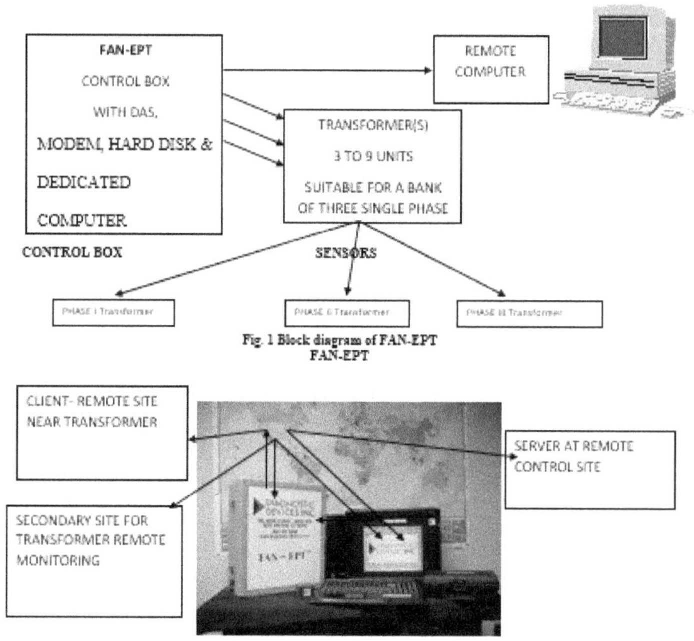

Figure 7.1: Block Diagram of FAN-EPT (world map in the background)

The hardware of the online, offsite system can be later used to conduct the location of the PD source with the additional upgrade of the location algorithms for either an eight to ten-inch PD location accuracy or four-six-inch location accuracy of the PD location estimates. The sensors can be mounted when the transformer is energized; therefore, it is not necessary to take a power outage. The power to our control box (115V AC OR 220 V AC) can be provided from a local control box on the transformer or from the local site power outlets or suitable power supplies.

7.1.2.2 Software

The online offsite system is only an online system for monitoring the detection of PD. It does not perform location analysis of the PD. The location of PD for a six-channel system can be done with suitable PD location software available in the market. This helps in conducting a 3-D location of the single PD source or multiple PD sites from the data acquired by the sensors and the system hardware. This system, which is only for detection purposes, will also alert the operator of the existence of a PD by setting off audible alarms connected to each sensor. These alarms will sound when the threshold signals are exceeded.

Such a system was used to monitor two 180 MVA, 525 kV GSUs at a generating station on the west coast of the USA. The client-server system consisted of the client unit or the control box located on the wall of the well surrounding the transformer, as shown in the site picture as in Figure 7.2.

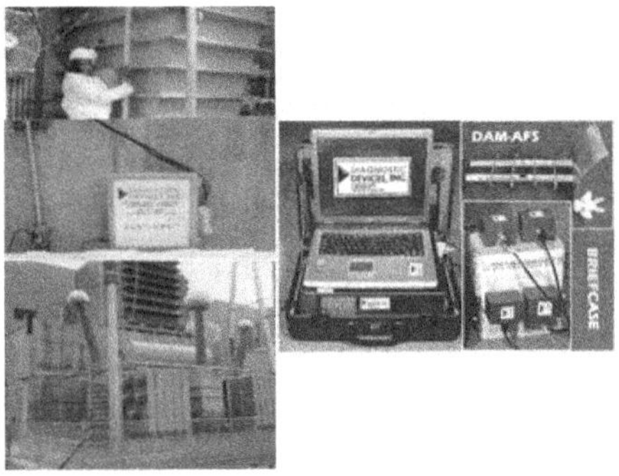

Figure 7.2. Site picture of an online, offsite client-server system

Six sensors to two such transformers (one healthy and the other unhealthy) were mounted on the tanks. Picture shows one of our operators mounting the sensors. Typical signals obtained from the six sensors are shown in Figure 3 (attached below). Channels 2 & 5 are from sensors on the healthy transformer (unit # 2), and sensors 1,3,4 & 6 are from sensors on the unhealthy transformer (unit # 1). Signals show considerable PD activity at sensors four and six on unit # 1, the PD-prone unit. It is imperative to understand the signal characteristics of PD and its time characteristics. The occurrence of the PD signal is always asynchronous.

If the signal obtained by the measuring device is synchronous, it is more often than not some noise. The shape of a PD signal typically will have an arrowhead shape, and its bandwidth for a typical ultrasound sensor with a piezo-crystal and integral amplifier is between 50kHz and 350 kHz. The amplitude measurements are shown in Figure 7.3. A corresponding power spectral density can also be plotted or illustrated using a suitable graphical user interface to understand the signal-to-noise content of the measurements.

Typically, the system hardware and software filters are efficiently tuned to match the above bandwidth requirements. A common myth about acoustic testing is that the sensors are not very sensitive or the signal cannot propagate through copper and insulation media of the core-coil assembly of the transformer. Generally, sound travels faster in metals and slower in oil and insulating solids.

However, a clear understanding of longitudinal and transverse wave propagation can help to relegate this recurring doubt about signal propagation. The data collected by the online offsite system in the above step is then fed to the locator algorithm with three-dimensional details of the transformer tank. The software called the WTPDDDI2000 then generates the best estimate of the PD location. Software also helps to isolate multiple PD sites in the confines of the transformer tank. This software uses the Wavelet Transform methodology to estimate the PD location coordinates. Further details can be obtained in [1] and [2].

A 24/7 monitoring of these transformers has been going on since April 7, 2001. A daily record of measurements is maintained. The hardware in the control box can store data for up to a period of one month. This monitoring was finally concluded in the month of December, 2001, with encouraging results for the customer. The initial apprehensions regarding the health of the transformer have now been cleared. The transformer has not exhibited any dangerous levels of PD, as shown in Fig. 3 for the unhealthy transformer during the initial monitoring period:

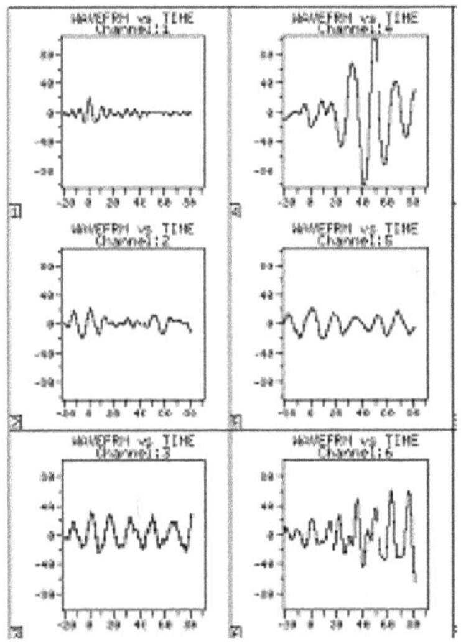

Figure 7.3. Measurements, from six *sensors*, monitoring two 180 MVA, 525 kV GSU transformers

A typical set of data acquired from a transformer under test at a site using the online, onsite system shows a good characteristic of the arrow-shaped PD signal. The travel of the signal through different media is also illustrated by the long travel time till the maximum is achieved. This is the online offsite system. The data shown in Figure 3 clearly indicates the existence of PD in GSU unit #1. The arrowhead signal attributed to the PD activity is especially seen in measurements from channels 4 & 6, respectively. It is imperative to note that this is an acoustic signal detection and location system, and there is no hard correlation between the electrical PD signals in relation to pico coulombs or microvolts.

This can be verified by empirical studies conducted at sites where the transformer being tested was being subjected to an induced voltage test simultaneously. Some of the results for different electrical testing conditions are shown in Table 1 below for comparison. The lowest value of the corresponding PD on the electrically triggered instrument was about 75 pC, which corresponds to an SNR value of – 6 dB for the acoustic system. The comparable amplitude of the acoustic signal for such an SNR was about 5 to 8 mili volts. This empirical data will further help to formulate and develop a correlation between the electrical and acoustic systems for the measurement parameters. Efforts are in progress to build such a database in collaboration with some of the research partners as listed in the acknowledgement section in the sequel.

7.2 Conclusion

Additional work reported in [3] further strengthens the myth that the present terminology used in the industry for online monitoring and diagnostics has very little difference. These authors feel that at the present both these activities are intertwined, and there is no clear difference between the two.

Table 7.1. Test data for a 525 kV Transformer. *Test* conducted per IEEE guidelines

Induced Voltage Test	Pico Coulombs	Micro Volts	Acoustic Test mili volts
1.0 pu	500pC	1000	39
1.5 pu	750pC	1800	80

The purpose of both these activities in the utility industry is mainly to assess the health of the capital asset, namely the electrical power transformer. The above information in sections 1 and 2 provides credence to the contention that online monitoring of the health condition of high-value capital assets like Electrical Power Transformers is possible in a real-time situation. The method described above illustrates the importance of a non-invasive method of detecting and locating faults in electrical Power Transformers due to incipient faults like partial discharges.

Such a system can provide continuous data on the health of not only one transformer unit but also several units located in close proximity inside a substation or a power generating station. Successful commissioning and continuous operation of such a system were illustrated and described. It is shown with evidence that such a system is safer to use than an online, onsite system due to the nature of the online/offsite monitoring achieved by the client-server system.

This new system, called *Briefcase*, was independently evaluated by one of customers in China. Hunan Power Bureau Personnel verified the testing of the FDD-EPT device and found independently that this method gives excellent results. They [4] found that the estimation of PD locations is within half a meter of the suggested centroid.

Notes:

As you have seen, there is a picture of a lunch-box industrial computer unit as the server and the client in a NEMA-4, IP 55 box. A different control box can be provided in the control room of the local transformer site, serving as another server. Control box Units and servers can also be provided with internal or external modems. A provision for an RS-485 for multi-drop is also possible. Remote control from a server can be achieved also by using cellular modems or TCP/IP for LANs or WANs if a network access is available.

Chapter 8: Case Studies

In Chapter 8 we will discuss case studies the efficacy of each of these methods. These case studies include varying kinds of transformers at different sites worldwide. These include about hundred locations at more than thirty countries worldwide.

8.1 GIO Test

A sample of oil is collected from the bottom of a transformer tank and analyzed for the gases dissolved in the transformer oil. Degradation of oil from arcing which can produce high concentrations of Hydrogen (H_2). Thermal degradation of oil-impregnated cellulose, which produces carbon monoxide (CO), and carbon dioxide (CO_2). In addition, PD in transformers produces both hydrogen (H_2) and acetylene (C_2H_2). However, the mere GIO technique does not provide the location of the pd fault. Inline spectroscopy is needed for more complex GIO. Measuring hot spots in windings directly with nonconducting temperature sensors in direct contact with the transformer winding conductor, which can be either copper or aluminum.

8.2 Acoustic Method – offline, onsite using lunchbox unit: FDD-EPT

This case study illustrates the interpretation of results obtained from testing transformers covering a volt-amp rating from 5MVA to as high as 750MVA, both as single-phase and three-phase transformers. The single-phase transformers were mostly common with the 340 MVA rating as GSUs. The voltage rating extended from 4.8 kV to 500 kV. The major categories encountered were:

A. Distribution Transformers – Northeast Utilities, CT, USA
B. Sub-Station-Transformers - Taipei Taiwan
C. Generator step-up Transformers (GSUs) – Klamath Falls, OR
D. Unit Transformers – NTPC, India; INEL. Italy
E. System Service Transformers - LADWP, USA
F. Rectifier Transformers – Badlapur, India; Switzerland; Canada

G. HVDC Transformers – Ballabgarh, India
H. Phase-shifting Transformers – Canada
I. Shell type GSUs – IEM, Mexico
J. EHV inverter transformers – Inertkirchen, Switzerland

Results for all these test cases and more locations will be discussed in Chapter 8. The results obtained were primarily of the faults caused by PD type. However, further validation of the results yielded some other kinds of incipient faults caused by very high oil temperature (Roger's Test-hotspot in oil). Very high amounts of H_2 (cellulose insulation failure); high settling (loose contacts); floating loose parts in OFAF-cooled transformers; high ppm of CO and CO_2 (hot spot in winding-burning of cellulose and further deterioration); etc. A thorough database of failure signals has been collected for all the above configurations and suitable Artificial intelligence (AI) methodology developed by DDI since 2000 and onwards till date. Above test cases have been documented in the past and as late as February of 2020 during and post-COVID. This database was also developed to generate the standards for several IEEE User's Guides and guidance used in the manufacture of such commercial devices by DDI as FDD-EPT and FAN-EPT, respectively. Many other companies have developed their own investigating devices for PD and its location but have not been that successful. All major clients finally come to DDI for a dependable evaluation in terms of detection and location of PD. A list of customers is listed on the DDI website: www.diagnosti-devices.com.

Some DDI Customers from the USA and Europe are shown below:

DYNEGY PROLEC SOUTHERN CALIFORNIA EDISON

DELTA STAR INCORPORATED ABB

The major contribution of the USPTO # 6,178,386 patent is the combined hardware-software package developed that is a part of both the FDD-EPT unit and FAN-EPT devices. The software package is coined ass HMSGHAT2000DDI. This software uses Wavelet Transform methodology combined with signal processing and Track Before Detect (TBT); with a database that has been embedded with an AI approach.

Some of the earlier test cases are discussed below in a nutshell:

8.2.1 PD Tests conducted prior to 2000

8.2.1.1 5 MVA, 23 kV/4.8 kV, Distribution Transformer

This transformer had several gas-in-oil (GIO) analyses data which indicated the existence of several pd conditions as incipient faults. There were several PD tests conducted using the acoustic method for a period of about four months to ascertain the effects of several environmental conditions like, snow, gloomy days, very low temperatures, freezing rain, etc. Also taken into consideration were loading conditions, no-load and over voltage conditions caused by different On-Load Tap Changer (OLTC) settings. Results of tests were documented in {x] PD estimate accuracy was validated within +- 8" of the centroid of the failure location and confirmed and validated by the client. Several samples of the failure locations were provided which matched with the physical locations suggested by the 3-D report provided by DDI.

A picture of the transformer under test is shown below:

Figure 8.1 450 MVA Core type transformer

8.2.1.2 25 MVA, 115 kV/23.8 kV, Substation Transformer outside of USA

This transformer was a typical example of a transformer overloading conditions. The transformer at times would be overloaded for two hours in a day at an overload condition of 110%. The results of the test are depicted on the website:www.diagnostic-devices.com.

A picture of the transformer under test is shown below:

Figure 8.2 Mobile Transformer 25 MVA

8.2.1.3 47 MVA, 345 kV/13.8 kV, GSU with a GIS termination at a client site outside of the USA

This transformer had several GIO analyses that suggested severe PD conditions. During the pd test conducted with FDD-EPT estimated the fault locations at the terminal connection of the HV red phase (a-phase). Test results showed levels of acoustic signals as high as 300 milivolts which is a considerably large value of PD signal. There is no direct correlation between the pico coulombs as specified for the electrical test.

A picture of the transformer under test is shown below:

Figure 8.3 47 MVA, 345 kV/13.8 kV, GSU with a GIS termination

8.2.1.4 340 MVA, 500 kV single phase GSUs at a client site out of USA

There were 13 such transformers that were tested with the eight channel FDD-EPT device purchased by the client from DDI. All units has failure points in the HV bushing turret. A failure in the cellulose structure of the line lead indicated the failure points that were reported in the 3-D test reports that were created by the client using the software HMSGHAT2000DDI.

A picture of the transformer under test is shown below:

8.2.1.5 25 MVA, 25 kV Mobile Transformer, at a site in the USA in a plant

This transformer was tested using a six-channel FDD-EPT shelf-type unit, and failures were detected and located. These failures were found in the delta connections in the winding leads due to improper and faulty connections due to brazing and rounding of the edges covered with appropriate layers of winding paper.

A picture of the transformer under test is shown below:

Figure 8.4 340 MVA, 500 kV single phase GSU

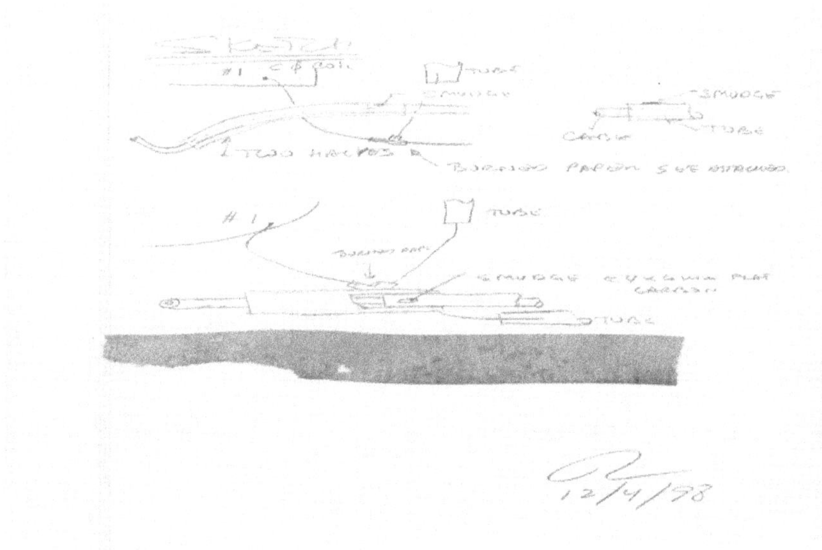

Figure 8.5 25 MVA, 25 kV Mobile Transformer

Pictures of failures found are also shown below in Figure 8.6

A picture of the failures under test is shown below:

Figure 8.6 PD failures under test of mobile transformer

The main customers for this FDD-EPT device are utilities from USA and worldwide. It was also found that some transformer manufacturers were interested in this new non-invasive technology using supersonic acoustic sensors and WT methodology in signal processing.

8.2.2 PD tests of transformers after 2000

Many large power transformers were tested for PD after 2000 in several countries worldwide like the USA, China, India, Italy, Canada, Mexico, Italy, Switzerland, Finland, Germany, Taiwan, etc. In total, 300 large power transformers at more than 100 sites were tested for PD. In all the test cases described below, the FDD-EPT device performed admirably.

a) Phase shifting Transformer 675 MVA,

b) Single-phase BHEL test transformer

c) Three-phase Germany

d) Three-Phase Finland

e) Three Phase Switzerland

f) Three-phase China

g) Three-Phase Test Transformer, China

h) Three-Phase, USA West Coast

i) Three Phase, Distribution Transformer, East Coast, USA

j) Three Phase GSU, West Coast, USA

k) 180 MVA, 345 kV, GSU, USA

Fig. 8.7 90 MVA, Shell tested for PD. Courtesy DDI

Fig 8.8 180 MVA, 345 kV, GSU

Fig 8.9 300 MVA, Sub-station Transformer

Fig. 8.10 180 MVA, 345 kV GSU tested for PD using FDD-EPT

Fig. 8.11 180 MVA, 345 kV GSU revealed failure of cleats of OLTC lead outs

Figure 8.12 LV Turret view where failure of Zenox connections was detected and located by DDI with FDD-EPT

These transformers were receiving station transformers in a power station where several transformers had failed due to some suspected failure in the LV turret, as shown above in Figure 8.12. Several transformer tanks had buckled due to the concussion caused by the failures. These suspected failures were probably expected to be due to Zenox connections at the LV leads. Site location was in the north-east side of USA. Site evaluations with briefcase unit of FDD-EPT located failures within 6-9" of the centroid and were verified by the client. Onsite repairs were conducted and transformers were restored in service within a very short time, with less than a total of about two days of downtime. The confidence level for such failure detection and location was greater than 90%. The placement of sensors by the client's lineman is shown in Figure 8.13 below. The ease with which this can be achieved is easy to mount the sensors with magnetic hold-downs. The sensors can withstand a force of 25 lbs/sq. Inch. It also illustrates that there is no need to take a service outage, and without crossing the voltage barrier, sensors can be easily mounted on the transformer tank.

Figure 8.13 Placement of sensors on transformer tank by client lineman

(a)

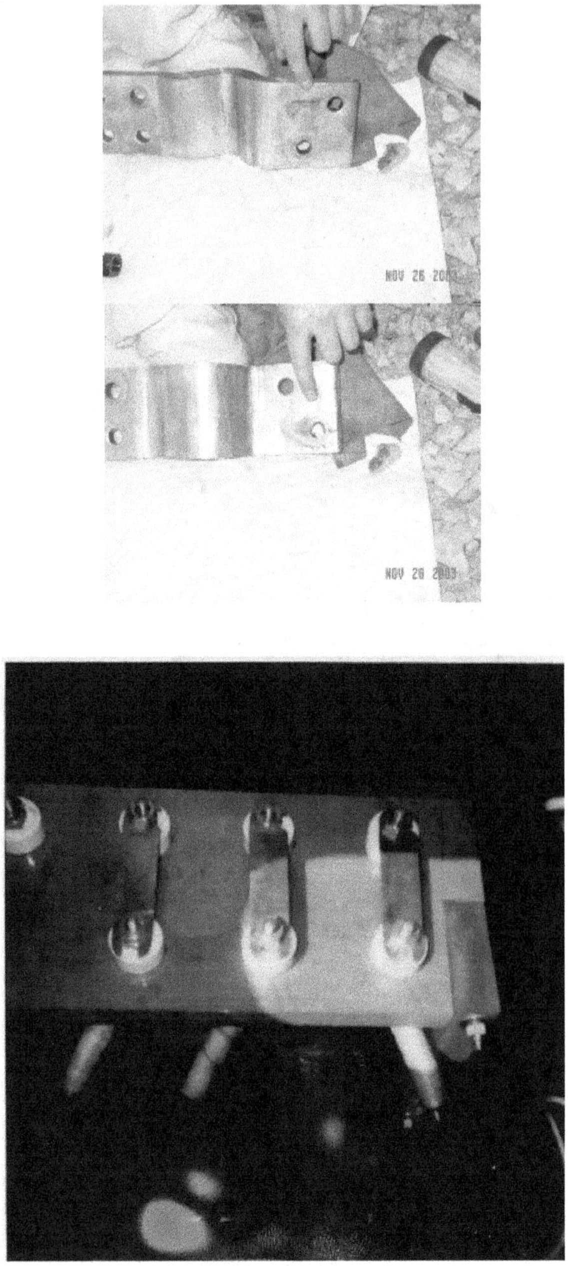

Figure 8.14 (a)Failure at LV Delta connection lugs of a 33 KV primary voltage GSU and (b) 33 kV LV Dsitribution transformer

(b)

The above Figure 8.14 (a) shows the failure at an LV Delta connection of a receiving station transformer on the west coast of the USA. The GIO tests showed a trend of increase in settling over a period of three weeks from GIO tests. Indications were that there might be some loose contacts causing sparking and, thus, some possible PD. Sensors were mainly placed on the LV side of the transformer tank. Failure locations were evaluated, and the client provided the coordinates of the failures with respect to the chosen origin on the tank bottom end. Within a few hours, the client lowered the oil in the transformer tank after taking an outage and verified the location at the LV connection lug flexibles, as shown in Figure 8.14 (b) above. Surface of the lugs was properly cleaned and smoothened and transformer put back in service in half a day's time. This yielded a considerable saving in power loss delivery time which increased reliability on the part of the utility. Use of FDD-EPT briefcase was advisable and useful to the client. Client provided excellent feedback and testimonials for the performance of FDD-EPT and its sensitivity and specificity.

Figure 8.15 Failure detected in 500 kV GSU in HV HiserCap windings

In the Figure 8.15 above, a single phase 525 kV GSU is shown with internal winding structure in an interleaved configuration with a center line lead and HV winding split into two halves. These windings are also called as High Series Capacitance insulation structures popularly called as HiserCap winding. This enables the voltage distribution along the length of the line from line to neutral uniform and with an alpha factor close to one (1). In this transformer and 13 other of similar construction at the site, failures were found at the line end inside the HV side turret and at the center line lead connection. The client purchased a 16-channel system and verified the failure locations within 6-9" of the centroid of several locations. These transformers were from the old Westinghouse designs of the 80s when the author was trained at the Hamilton site in the EHV design and manufacture of power transformers during his days with CGL, India. A configuration of the interleaved winding is discussed below for more insight into such configurations and the hazards that can be faced while coordinating the core-coil insulation structure.

Figure 8.16 650 kV, 650 MVA test transformer in a manufacturer's plant in Southeast Asia.

The large 650 kV single phase transformer shown in Figure 8.16 above was a test transformer at an outside-the-USA site in South-East Asia. The transformer had very large sparking signals emanating when testing started on a particular transformer. This would happen when the test generator was started in the SOP during the applied or induced voltage stage. The failure showed up at the LV side terminals and were confirmed to be external to the transformer confines and thus concerns eliminated which were conveyed to the client and results confirmed by the client as an external source of sparking.

Figure 8.17 Locations of Mobile Transformer failure points verified by client.

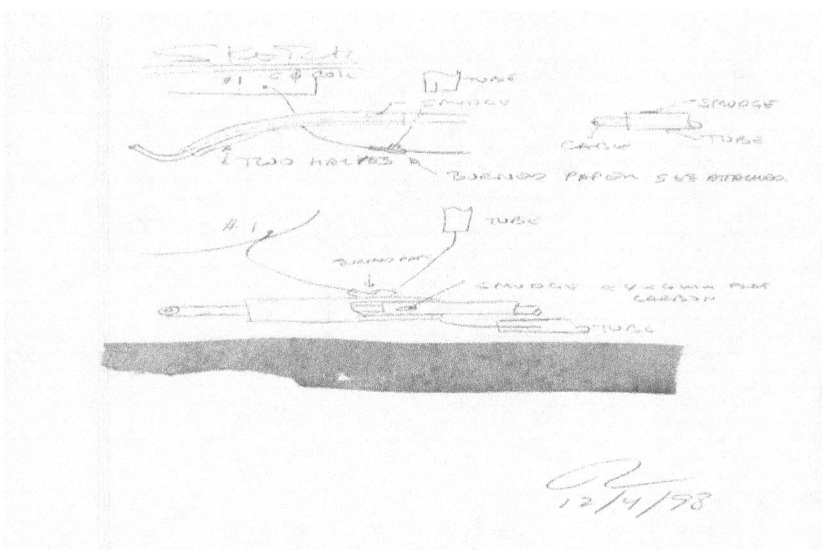

Figure 8.18 Aluminum crepe paper failure at LV joints

In the Figure 8.18 shown above is a mobile transformer with a sequence of three snapshots showing the actual placement of sensors on the outside of the transformer tank in the foreground. At the top left is a snapshot that shows the failures at the LV side lead connections of the delta winding, and at the top right, a snapshot of the sensor locations on the opposite side of the transformer tank of the same mobile transformer. Clearly, the results that were detected and located by using the shelf-top version of FDD-EPT system by DDI estimated the failures with an accuracy of 6-9" of the different centroid locations, in this case, about six connections on the LV side delta.

Actual failure point details, as shown in Figure 8.17 above show the failure burnt portion of the aluminum crepe paper in the foreground. The hand-drawn figures above show what was supposed to be in place and the conditions of the brazed joints. In all cases, the failures indicated sharp points that caused some discharges that caused cellulose failure and burning causing CO and CO_2 in the media surrounding the joints. Joint covering was removed, and sharp points in joints were smoothened and retaped. After retesting the failure points resolved by client and proper working PD-free was validated for this mobile transformer.

Figure 8.18 Failure detected in HV HiserCap winding for a 500kV GSU spare transformer in the HV turret

Figure 8.19 FDD-EPT – Briefcase – lunchbox version by DDI in testing operation in field:

FDD-EPT Briefcase lunchbox version is shown in Figure 8.19 above was used to test a rectifier transformer at a site outside the USA in the up-state north countryside. These were in very cold conditions at site and the briefcase FDD-EPT unit withstood the environmental changes at about -20 degrees C with a considerable windchill conditions. Personnel from DDI had to wear extreme cold condition gear to protect themselves and a mobile heated unit was available for further personnel protection. Failures were at the LV side connections, which had multiple cleats with a considerable overlap and crisscrossing of leads. Failures were found at these locations and were verified by the customer and validated.

Figure 8.20 161 kV Transformer tested with FDD-EPT usage at a local site.

The transformer shown in Figure 8.20 above was system transformer at a local site south of USA being tested for PD failure at the repair facility of one of the utilities. Although the title indicates it was some four weeks ago, it was actually in one of the earlier days of the servicing days of DDI with FDD-EPT lunch-box unit. The transformer had severe PD failures indicated by DGA results and were obtained again on induced voltage tests conducted during the FAT sequence witnessed by the client and the end user. The occurrence of pd faults using FDD-EPT system clearly showed results at 1.2 pu in terms of acoustic dBa but not high enough for the FDD-EPT system by DDI for actual pd detection and location sequence as illustrated by the User's Manual in Appendix I. The test had to be conducted at higher pu levels for a very few seconds, as agreed by the client and end user. This was done at 1.3 pμ for a very few seconds, which yielded useful data to be analyzed by the HMSGHAT2000DDI software package, as also illustrated in Appendix II. The 3-D diagram of the failures obtained cannot be shared with the August audience but can be shared by DDI with suitable permission obtained from the end user.

One of the clients who purchased one FDD-EPT system from outside of the USA had their testing team come over to USA for a week's training in the operation and use of the FDD-EPT system. A complete User's Manual is available in Appendix xxx as provided by DDI. The team was brought to one of the local substations at the author's University to demonstrate the use of the FDD-EPT system. The transformer was a 5 MVA, 13.8 kV/0.440 kV, 3-phase, 60 c/s distribution transformer. The Figure 8.21 shows one of the DDI test team personnel illustrating the placing of the supersonic sensors on the surface of the tank without taking an outage. The transformer was used to train the personnel from the international client for hands-on training of the actual FDD-EPT system. The training lasted for a week as planned, and all aspects of the User's Manual for FDD-EPT, as illustrated in Appendix I were demonstrated to the client test team. Since this was a healthy transformer, none of the tests conducted showed any failures due to PD or any other incipient faults. The exercise of training is provided by DDI to its clients who purchase the FDD-EPT system or even if the clients purchase the FAN-EPT system.

Figure 8.21 5MVA distribution transformer tested at local site while training to client personnel for one week offered by DDI with employee from DDI.

A recent case of a 110 MVA 3-phase transformer 132/115 kV, DYn11 with DETC on the HV winding and OLTC on the neutral end of the lv winding with +- 17 steps, Esther (FR3) filled insulating liquid was tested for pd over the past two weeks in May of 2024. The pd levels were found by other methods and showed an activity of up to and beyond 500 pC in all phases. The estimated location was not obtained and customer approached DDI to detect and locate PD. The PD test was conducted over two days with different conditions as requested by the customer as follows:

1) Induced Voltage test

2) Induced Voltage test on each phase with other phases grounded

3) Applied voltage test

Some initial data on DGA and other pd tests conducted were provided by the customer as follows:

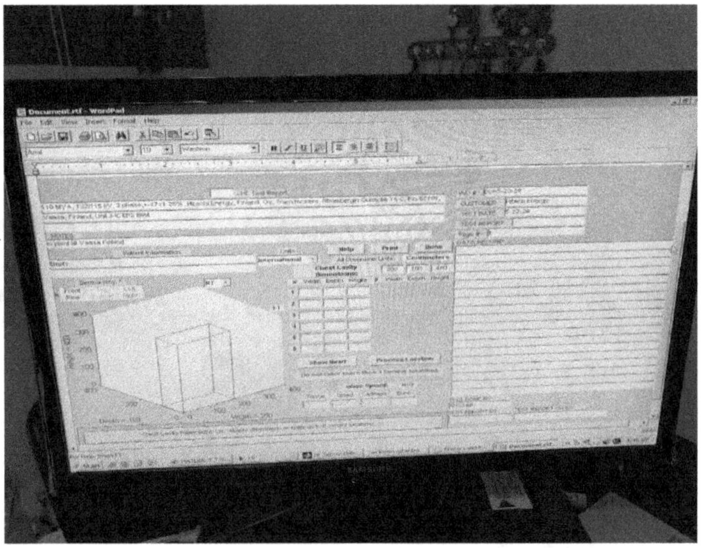

Our 3-d diagram to start data collection and analysis using FDD-EPT is shown above:

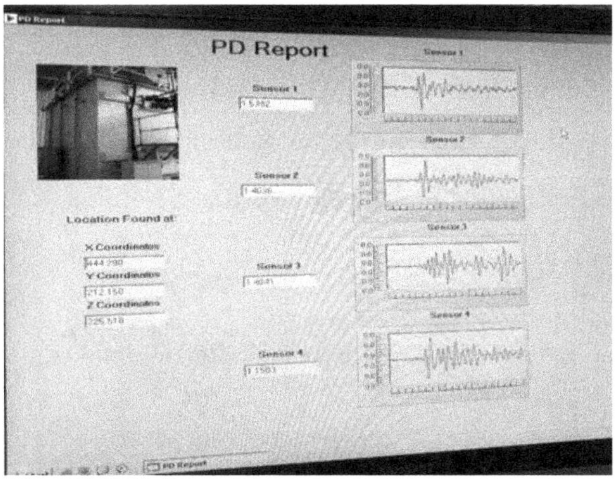

Figure 8.23 3-D pd report with failure location and signals at different active sensors for pd location for 110 MVA Transformer using FD-EPT and HMSGHAT2000DDI

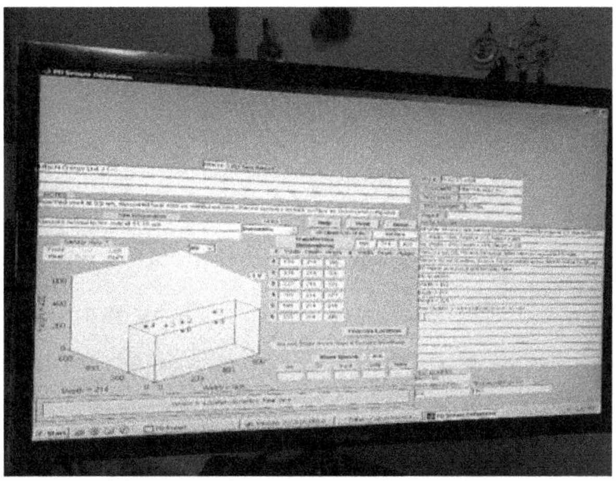

Figure 8.24 3-D pd report with failure location and signals at different active sensors for pd location for 110 MVA Transformer using FD-EPT and HMSGHAT2000DDI

Sensor signals as shown above clearly indicate the presence of some PD activity. It appears from the DGA results from the recent past show some cellulose burning with CO and CO_2 levels and ratio coming close to 3.5 per IEEE standards. Accordingly, on further analysis using HMSGHAT2000PDDDI SW, the final test report was submitted to the customer as below:

The PD failures indicate at points where some superconducting material is placed around the core to suppress sharp points created by laminations that form the core shape. This transformer was also filled with Esther and some other locations where the HV line leads come close to each other at the delta line ends located at the bottom part of the core-coil disposition. Customer has confirmed these locations and has taken corrective action to remove these failures. A second unit of the same rating will be redesigned to take care of these failures. Some of the above customer details had to be expunged to maintain non-disclosure aspects of the PD test agreement.

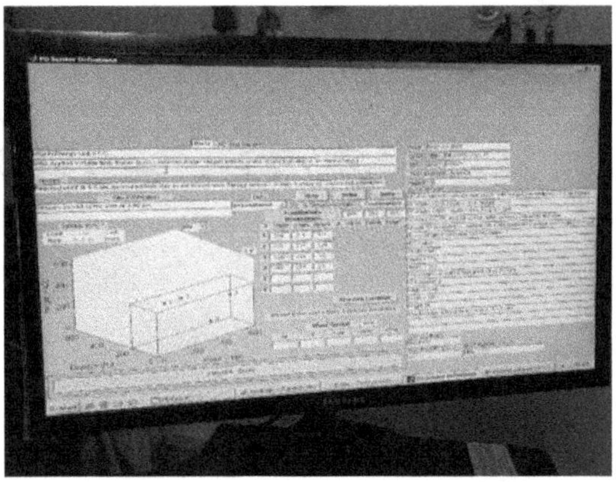

Figure 8.25 Final pd location for 110 MVA Transformer using FD-EPT and HMSGHAT2000DDI

The above test cases, overall, provide a good indication of the efficacy of the FDD-EPT and FAN-EPT systems by DDI. On average, the specificity numbers are 92%, and sensitivity numbers are 88%, yielding a PD location accuracy of +- 6"-9" of the centroid of the PD failure. Over the past 26 years, the diagnostic system developed by DDI has yielded a consistent performance with minimal change in their overall firmware but a considerable amount of inclusion of data analytics and Artificial Intelligence (AI) techniques to be in line with the Industry 4.0 standards for a futuristic paradigm shift in the detection and location methodology of PD.

DDI and the author are ever grateful to the opportunity provided by several customers worldwide, the University of Hartford, and IET for requesting a manuscript for this study on PD detection and location of such incipient failures that can burgeon into catastrophic failures as shown below in Fig. 8.26.

Chapter 9: Summary and Conclusions

This book was written by the author at the invitation of Christoph Von of IET, UK, in 2022. The author has extensive experience in the acoustic method of evaluating power transformer failure using the acoustic method to detect and locate PD. The work started in 1981 when the author worked for Crompton Greaves Limited in Mumbai, India. During my stay at Westinghouse, Hamilton, Canada, I had the opportunity to study extra high voltage insulation coordination in large power transformers and newly developed Gapped Core Shunt Reactors.

While learning all the finer nuances of insulation co-ordination for the core-coil structure immersed in oil in the transformer tank, every effort was made to understand the cause of failure that can start as an incipient fault and burgeon into a catastrophic fault that would totally destroy a capital asset to the tune of millions of dollars. It came to a full circle when the technology track-before-detect that was developed for tracking aircraft and submarines using acoustic sensors was finally applied to power transformers in a static domain. A sabbatical leave sponsored by erstwhile Northeast Utilities (now Eversource) in CT led to further research and development with a grant of $225,000 from Connecticut Innovations + Northeast Utilities and College of Engineering, University of Hartford; resulting in a filing of a utility patent that was published in February of 2001 by USPTO: 6,178,386. This was followed by the creation of Diagnostic Devices Inc., which markets the commercialized product and services of Fault Diagnostic Devices for Electrical Power Transformers (FDD-EPT). DDI, since then, has tested and served more than 300 large power transformers at more than 100 sites worldwide. The clients include transformer manufacturing companies, utilities, and research laboratories. A set-up of FDD-EPT lunch-box unit is shown below:

Figure 9.1 FDD-EPT lunchbox unit, courtesy of DDI

Figure 9.2 Online FAN-EPT unit courtesy of DDI

All the offsite and online systems developed by DDI on the passive acoustic methodology have proved to have excellent specificity for PD location and sensitivity for PD detection. Failure diagnostics of power transformers has, over the past quarter century, saved several large power transformers from catastrophic failures, thus saving multi-million capital assets at over 100 sites worldwide like in Switzerland, Finland, Canada, USA, China, India, Italy, Mexico, UK, Taiwan, Philippines, and several other countries. Some of the cases in real life power transformer failure diagnostic have been discussed earlier in Chapter 8. Various sizes of transformers have been illustrated for detection and location of power failures due to PD. It has been found that the PD diagnostic methodology patented by the author has very distinct advantages as follows:

i) Easy and non-invasive

ii) Can be conducted without taking an electrical outage of the transformer

iii) Can be easily conducted during the induced voltage test in a factory environment.

iv) Can be easily conducted in field with the supersonic sensors easily placed on the surface of the exterior of the transformer tank with magnetic hold-downs that can withstand a force of 25 lbs/sq inch.

v) Several test cases discussed earlier also illustrate the use of FAN-EPT for online detection and location of pd with efficient use in conditioned monitoring that has become so popular in the prevention of incipient faults and prevent catastrophic failures of power transformers and saving large amounts of capital asset monies. The most glaring test case was of 525kV GSU at a large utility on the West coast of USA. The unit was monitored online for four months with a master-slave FAN-EPT device by DDI. One site was right at the location of the transformer, the second one was at the offices of the transformer manufacturer in a country outside of the USA, and the third location was in the offices of DDI located on the east coast of the USA. The results provided accurate conditioned monitoring of large power transformers with the capability to send failure alerts with an outgoing email as well as a telephone call to the respective Control Engineers at all the above sites.

The PD test report is a 3-D printout of the actual testing procedure with details of transformer, site location, voltage levels, locations of the sensors, and the location coordinates of the failure points detected with location dimensions coordinates and accuracy details. Generally, the sensitivity is about 88% and specificity is about 92%. This has been validated by customers via their own testimonials as below:

"75 sites all over the world and 12 countries. World-renowned Transformer manufacturers and large utilities all over the world."

— FD2-EPT

"Three customers in the North American Continent including world-renowned transformer manufacturers and electrical contractors."

— FAN-EPT

"...Your test pinpointed the problem very accurately in this case. We might have saved $20,000 and 5-6 weeks in shipping time..."

— Mr. George Reitter, Corporate Technical Director of Delta Star, Inc. in CA, USA

"...The ability to locate partial discharge breakdown within eight to ten inches will prove to be cost-effective in many ways. The low cost and flexibility of the test system will make it a valuable tool for the utility industry..."

— Mr. Joseph Barresi, Director, T&D maintenance of Northeast Utilities, Systems in CT, USA

"We at North American Transformers are very impressed with the efficacy and the sensitivity of your FD2-EPT .....I would not hesitate to recommend your system as a detecting tool for factory or field testing..."

— Mr. Bertrand Poulin, Manager, Research, Development and Testing, North American Transformer, Rockwell Automation. (Now with ABB, Quebec, Canada)

... Your system is user-friendly and easy to operate and use in the factory setting......We are glad we availed of your services and we have no reservations in recommending your product and services to other transformer manufacturers and users like the utilities in USA and the World...

— Mr. Emilio Morales, Design Manager, Ohio Transformer, Palmetto, FL, USA

FDD-EPT by DDI is an extremely efficient system that detects and locates failure in power transformers. It offers the electrical power industry and offers a complete and comprehensive system that has the following salient advantages and features:

1. Real-time diagnostics
2. Most accurate with fault diagnostic accuracy of within 6-9" of the centroid of the failure location.
3. Lest costly by several folds, almost ten times less than the electrical system used to detect such failure in electrical power transformers.

4. Knowledgeable technical support; collectively, about 140 years with emphasis to electrical power transformers, design, manufacture, testing and testing features.
5. Expert training provided by DDI's engineering staff trained in transformer design, manufacture, and production of HV and EHV Power Transformers
6. Flexible data acquisition channel sizing from 4-24 channels in several footprints and data acquisition rate to 100 MHz
7. Proven performance at more than 100 sites worldwide with efficacy numbers for the sensitivity and specificity numbers for detection and location of PD failures.
8. Reduction in maintenance cost by a factor of three in the labor as well as material. This has been documented in several test cases described earlier in several chapters here in the prequel.
9. The efficacy of the detection and location system used and manufactured by DDI has increased considerably with the introduction of AI and data analytics built on the database, which has developed over more than two decades.
10. The efficiency of such passive systems is now in terms of sensitivity up to 92% and specificity up to 88%. The confidence level has also increased to 95%. This has physically translated into a 6-9" failure location estimation of the centroid of the actual location. Such verification has also been obtained, achieved, and verified by all kinds of end users, manufacturers, etc.
11. The techniques developed by the use of passive supersonic sensors and WT methodology have now been accepted worldwide and have found their place in the published documents of several industrial associations worldwide.
12. The author has developed this scheme and system for diagnosing failures in power transformers mainly due to PD over the past 28 years and has definitely made a main contribution to saving capital assets of utilities and manufacturers worldwide.
13. Many utilities have acclaimed this technique to be more successful than other methods employed in the electrical power industry and as such many utilities and end user's will seek the services of DDI for the detection and location of PD failures by purchasing either physical systems like FDD-EPT/FAN-EPT or testing services in field or manufacturers facility.

14. Since the method used by the author has been in place for the past several decades the electrical power industry has benefitted considerably yielding in a saving of more than $20 million in capital assets. This number may be manifold as it pertains to the actual testing and service provided by DDI. Other non-documented cases may not be included in this estimate of savings.
15. The system and process laid down by DDI for the use of onsite, offline system: FDD-EPT; and use of offsite, online system: FAN-EPT is presently renowned worldwide and accepted for use by the electrical power industry for the detection and location of failures in electrical power transformers.
16. In the last few years, the author has worked with other research colleagues to obtain an analysis of transformer failures by employing active methods to assess the insulation characteristics like power factor (pf). In this respect, a patent was awarded by USPTO #11,415,619 on August 16, 2022.
17. This patent is now being developed into a Digital Modulation Demodulation Signal Generator (DMDMSG) to assess the quality of insulation due to failures and, thus assess the life of the expected life of the transformer after some failure has happened.

In summary, it appears to the author that the different techniques used in the diagnosis of failures in electrical power transformers have their own merits. Some are in place for as long as 100 years and are well-established, like those followed by the author, are in their infant stage and have existed for the past 35-odd years but have been relatively more successful in establishing a base condition for conditioned health monitoring that can help utilities and manufacturers in assessing the life of the capital asset transformers both by conducting preventive maintenance and breakdown maintenance.

References

[1] Optical Fiber Sensors: Working Principle, Applications – Elsherif -2022-Advanced Photonics Research – Wiley Online Library.

[1] R. H. Shertukde and H. M. Shertukde, "Manufacture of Fault Diagnostic Device for Electrical Power Transformers Using DSP Technologies", Proceedings, ICSPAT96, October 7-10, 1996, Boston, MA.

[2] H. Shertukde, "A Generic Fault Diagnostic Device for Several Industrial Applications", Proceedings ICSPAT99, Orlando, FL, November 1999.

[3] H. M. Shertukde and R. H. Shertukde, 'Fault Area Network for Electrical Power Transformers – A Novel Tool for Online Monitoring of Large Power Transformers', Proceedings, ISEI02, April 2002, Boston, USA

[4] Fang Qing, Liu Weidong, Deng Wei, Zhang Huan, Xiong Yun Lee, 'Partial Discharge of Ultrasound Technology Research", Hunan Electric Power Company EHV Authority, Changsha, Hunan, PRC, June 2011.

[1] H. M. Shertukde, Y. Bar-Shalom, "Target Parameter Estimation in the near-field with two sensors", IEEE Transactions in Acoustic Speech and Signal Processing, August 1988.

[2} Randy K. Young, "Wavelet Theory and Its Applications", Kluwer Academic Publishers, MA, 1994.

[3] Partial Discharge Detection Using Acoustic Emission – Report published by Physical Acoustics Corporation, P.O. Box 3135, Princeton, NJ, 08543

[4] John Reason, "Online Transformer Monitoring", Special Report, Electrical World, pp 19-26, October 1995

[5] Dennis Berent, "Acoustic Monitoring and Gas-In-Oil Analysis for Transformers", Report # 62PAIC95, pp 8-3.1-8.3-6, Doble Engineering Company

[6] Hemchandra M. Shertukde, Hisham Alnajjar and Uday Prabhune, "Fault Detection

Device for Electrical Power Transformers Using Novel DSP Scheme", Proceedings ICSPAT'96, pp 1219-1223, October 7-10, 1996, Boston, MA

[7] R. H. Shertukde and H. M. Shertukde, 'Manufacture of Fault Diagnostic Device for Electrical Power Transformers (FDD-EPT)", Proceedings, ICSPAT'96, pp 1229-1233, October 7-10, 1996, Boston, MA

[8] Hemchandra M. Shertukde, "Detection of Partial Discharges Using a Novel Fault Diagnostic Device for Electrical Power Transformers (FDD-EPT), Proceedings, TechCon 98, February 12-13, 1998, New Orleans, LA

[9} Hemchandra M. Shertukde, Detection of Partial Discharges Using a Fault Diagnostic Device for Electrical Power Transformers:, Proceedings Doble Client Conference, March 30-April 3, 1998, Boston, MA

[10] Peter M. Eleftherion, "Partial Discharge XXI: Acoustic Emission-Based PD Software Location in Transformers", Proceedings IEE Insulation Magazine, 0883-7554/95, 1995

Bibliography

[1] R. H. Shertukde and H. M. Shertukde, "Manufacture of Fault Diagnostic Device for Electrical Power Transformers Using DSP Technologies", Proceedings, ICSPAT96, October 7-10, 1996, Boston, MA.

[2] H. Shertukde, "A Generic Fault Diagnostic Device for Several Industrial Applications", Proceedings ICSPAT99, Orlando, FL, November 1999.

[3] H. M. Shertukde and R. H. Shertukde, 'Fault Area Network for Electrical Power Transformers – A Novel Tool for Online Monitoring of Large Power Transformers', Proceedings, ISEI02, April 2002, Boston, USA

[4] H. M. Shertukde and R. H. Shertukde, 'Monitoring vs. Diagnostics - FAN-EPT: – A Novel Tool for Online Monitoring and Diagnostics of Large Power, Proceedings, IEEE T&D Conference, October 9, 2002, Yokohama, Japan.

[5] Hemchandra M. Shertukde and Rekha H. Shertukde, "Condition Based Health Maintenance of Large Power Transformers Using online and Remote Diagnostics', Proceedings, ISH2003, Delft University, Amsterdam, Holland, August 28, 2003.

[1] Seminar on Target Tracking with Passive Sensors, Chung Shan Institute of Technology, Taipei, Taiwan, May 1995.

[2] Seminar on "Manufacture of Fault Diagnostic Device (FD2-EPT) Using state-of-the-art DSP Technology", UTRC research center January 1997 (also given to UCONN E&SE Dept. March 1997, CT Professional Engineers Group, National Engineers Week 1998).

[3] Workshop on Diagnostic Techniques for High Voltage Electrical Equipment, in conjunction with TjH2b & Diagnostic Devices Inc., December 1997.

[4] IEEE T&D Conference, Industry demonstration of online, offsite PD monitoring system, FAN-EPT, Nov 1, 2001, Atlanta, GA.

[5] ECNE Conference, March 14-15, 2002, demonstrated the Online, onsite PD monitoring

system FDD-EPT for Diagnostic Devices Inc. Nashua, NH.

[6] Seminar on Partial Discharge Detection and Location in Oil-filled large power Transformers, IEEE-TC, March 2003, Raleigh, NC.

[7] Partial Discharge Detection and Location in oil-filled transformers, CWIE Conference in Chicago, IL, October, 2009.

[8] Real-time fault diagnostics for Power Transformers Using Supersonic Sensors and Wavelet Transforms with NU for $100,000 from CII (G096) and awarded (matched by NU for $110,000 and Univ. of Hartford for $12,000), 1996

[9] Transformers: Theory, Design and Practice with Practical Applications: New and Improved look at Transformers with emphasis on partial discharge, core designs ... finite element analysis and 18-slot designs - Paperback (Aug. 8, 2010) by Hemchandra Shertukde, VDM-Mueller. ISBN # 978-3-639-25244-6

[10] H. Shertukde et al, "Guide for detection of Acoustic Emissions from Partial Discharges in oil-immersed Power Transformers, IEEE guide # c.57.127. Since 1995 and being balloted for final approval in 2000. Approved 2007 as User's Guide C.57.127

[11] H. Shertukde et al, "Guide for location of Acoustic Emissions from Partial Discharges in oil-immersed Power Transformers, IEEE guide # c.57.127.xxx (draft # 10 -drafts being compiled since April 1995 to present).

[12] H. Shertukde, R. Shertukde, "Design and Manufacture of a Fault Diagnostic Device for Electrical Power Transformers (FD2-EPT) and success stories for online monitoring with FAN-EPT", in preparation to be submitted to the IEEE Transactions on Power Apparatus.

[13] H. M. Shertukde, "Efficient Wide-Band Cross-Correlation Tracking Using Wavelet Transforms", Proceedings ICSPAT '94, Dallas, Texas, October 1994.

[14] Uday Prabhune, Hisham Alnajjar and H. M. Shertukde, "Real Time Test Equipment for Partial Discharge Analysis", Proceedings Connecticut Symposium on Microelectronics and Optoelectronics, Storrs, CT, March 21, 1996.

[15] H. M. Shertukde, Hisham Alnajjar and Uday Prabhune, "Fault Detection Device for

Electrical Power Transformers Using Novel DSP Scheme", Proceedings ICSPAT'96, Boston, MA, October 1996.

[16] R. H. Shertukde and H. M. Shertukde, "Manufacture of Fault Diagnostic Device for Electrical Power Transformers (FD2-EPT)", Proceedings ICSPAT'96, Boston, MA, October, 1996.

[17] H.M. Shertukde, H. Alnajjar, U. Prabhune and S. Sambanther, "Real-time Fault Diagnostics Using Supersonic Sensors", Proceedings, ICSPAT97, Sep 14-17, 1997, San Diego, CA.

[18] Hemchandra M. Shertukde, "Detection of Partial Discharge Using a Novel Fault Diagnostic Device for Electrical Power Transformers", Proceedings TechCon 98, New Orleans, LA, February 1998.

[19] Hemchandra M. Shertukde, "Detection of Partial Discharges Using a Fault Diagnostic Device for Electrical Power Transformers", Proceedings Doble Client Conference, March 30-April 3, 1998, Boston, MA.

[20] Hemchandra M. Shertukde and John G. Lackey Interpretation of results obtained from testing of oil filled Electrical Power Transformers in Field for several Utilities in the world", Proceedings IEEEIS98, Washington DC, June 1998.

[21] Hemchandra M. Shertukde, "A Generic Fault Diagnostic Device for Several Industrial' Applications", Proceedings ICSPAT99, Orlando, Florida, November 1999.

[22] Hemchandra M. Shertukde, et.al, "Future of Tracking and Fusion" Workshop on Tracking and Fusion, May 17, 2001, CA.

[23] Hemchandra M. Shertukde, et. al. al. Al; "User's Guide for Detection of PD in oil-filled transformers", January 2001, IEEE/PES publication.

[24] User's Guide for Location of PD in oil-filled transformers:, October 2000-2002 (St. Catharines, Canada, Nimegan, Holland, Vancouver, Canada – published for IEEE/PES in 2007).

[25] H. M. Shertukde and R. H. Shertukde, "Fault Area Network for Electrical Power Transformers – A Novel Tool for Online Monitoring of Large Power Transformers", Proceedings,

2002 IEEE International Symposium on Electrical Insulation, Boston MA, USA, April 7-10, 2002.

[26] Hemchandra M. Shertukde and Rekha H. Shertukde, "Monitoring vs. Diagnostics - FAN-EPT: – A Novel Tool for On-Line Monitoring and Diagnostics of Large Power Transformers", Proceedings, IEEE T&D Conference, Yokohama, Japan, October, 2002.

[27] Hemchandra M. Shertukde and Rekha H. Shertukde, "Condition Based Health Maintenance of Large Power Transformers Using On-Line and Remote Diagnostics", Proceedings, XIII th ISH Symposium on High Voltage Engineering, Delft, The Netherlands, August 25-29, 2003.

[28] Hemchandra M. Shertukde and Rekha H. Shertukde, "Case Study of Condition Based Health Maintenance of Large Power Transformer at Rihand Sub-Station Using On-Line FDD-EPT", Proceedings, ISEI 2004, Indianapolis, USA, September, 2004.

[29] Hemchandra M. Shertukde and Rekha H. Shertukde, "Case studies – Practical Results of online Transformer Conditioned Monitoring Using Passive Supersonic Acoustic Sensing Techniques and Wavelet Transform Methodology for more than sixty-four sites all over the world" (1995-2004), Proceedings, CPRI Workshop, Feb 10-11, Bangalore, India

[30] Hemchandra M. Shertukde and Rekha H. Shertukde, "Lessons learned in the tallying of theoretical predictions to real-life testing of Power Transformers for Partial Discharges with an Acoustic System", Proceedings, CPRI Workshop, Feb 10-11, Bangalore, India

[31] Hemchandra M. Shertukde and Rekha H. Shertukde, "Novel tool for Online monitoring of Partial Discharges in Electrical Power Transformers using acoustic supersonic sensors: Proceedings TechCon'05, Darling Harbor, Sydney, Australia, May 2005.

APPENDIX - I

Appendix I and II appear almost identical. Are you sure these were the appendices I sent to you?

UTILITIES MENU HMSGHAT2000DDI™ SOFTWARE

DIAGNOSTIC DEVICES INC.
94 Wolcott Road, Simsbury, CT 06070

Advancing Signal Processing Beyond The Year 2005

Copyright DDI 2005, FDD-EPT

www.Diagnostic-Devices.com

HMSGHAT2000DDI™ SOFTWARE

Part No. 6301-7001
Version E5.20

Installation, Operation and User's Reference Manual

INTRODUCTION

HMSGHAT2000DDI is a 32 bit WINDOWS, Data Acquisition and Replay program capable of running 's DiSP, SAMOS, PCI-2, , SENSOR HIGHWAY or SPARTAN AT/2000 based products. HMSGHAT2000DDI uses full WINDOWS resources including: Setting of any WINDOWS available screen resolutions, Printing, Networking, Multi-Tasking, Multi-Threading, etc. It is capable of operating in WINDOWS 2000/XP/Vista operating systems. HMSGHAT2000DDI is fully compatible with 's standard (DTA) data files. This allows you to replay and analyze all your previously collected AE files. HMSGHAT2000DDI is easy to learn, operate and use. The software has all the acquisition, graphing and analysis capabilities that you have come to expect in your AE system, plus many more new and enhanced features to ease your data analysis and visualization tasks. Multiple copies of HMSGHAT2000DDI can be run at once, one for acquisition with full display and analysis capabilities and one or more as Replay for additional analysis purposes.

Some of the general AE display and analysis capabilities found in HMSGHAT2000DDI includes the following: HMSGHAT2000DDI provides a framework for easily adding graphs and graph screens to the main system WINDOW. HMSGHAT2000DDI has a very flexible line listing (dump) capability, which can be called at any time, in any screen, which can be scrolled up or down. HMSGHAT2000DDI has various user selectable tool bars including; Setup icon toolbar, acquisition control, line dump, status toolbar and statistics tool bar. HMSGHAT2000DDI has many built in, enhanced features including; Hit-Event Linking, Graph zooming and panning, more flexible graph setup, more graphs, and built-in filtering functions (graphical filters and Post Filtering). Printing capabilities include: Print a graph, a screen or multiple screens to a WINDOWS or network printer, to the clipboard or to a JPEG file.

Some of HMSGHAT2000DDI graphing capabilities include:

- Exceptional 2D and 3D graphing capabilities. Multiple graphs can be displayed on a screen, limited only by the resolution of the screen itself.
- Graphs are individually sizable on a screen, making for a very flexible arrangement. The user can set up one (or more) large graphs for visualization with multiple supporting small graphs alongside or around the main ones.
- Multiple Graphs are arranged on a screen, however the user desires. Multiple screens can be setup, each accessible by selecting a user labeled tab. Therefore, a user can set themes for screen layout (e.g. Location & Cluster analysis, waveform analysis, AE activity analysis, AE Feature correlation analysis, Alarm Analysis, themes to name a few)
- Many different types of graphs can be set up including; histograms, point plots, 3-D graphs, waveforms, FFT's, multiple plots on a single graph with coloring options, etc.
- All graphs can easily be enlarged to full screen (maximized) by the touch of a button.
- All graphs have full Cursor Readout capability, either continuous or point by point movement.
- All graphs (2-D and 3-D included) can be infinitely zoomed and panned for close-up analysis.

In terms of Location determination, the standard version of HMSGHAT2000DDI has zonal and linear location modes, while the optional "Full Location" software adds 2D Planar Location capability to the list. In addition, there is also a 3D Location option available and a Spherical 3D location option. Key features regarding location include the following:

- Multiple (8) location groups standard, (32 optional), each capable of using all the AE channels in the AE system. Mouse oriented sensor setup includes; Set sensors manually (click and drop), automated sensor setup for triangular setup, and easy sensor editing drag and drop, or table entry based sensor position.
- Multiple 1 dimensional, 2 dimensional and 3 dimensional location modes.
- Flexible Clustering, cluster reporting and Cluster statistics available on any point plot graph screen (not just limited to location).
- User selectable location structure visualization to ease setup. Options include; Plate, vertical vessel, horizontal vessel, sphere, and Free, each with specific setup menu's geared for easy setup of that structure, including automated and manual placement of sensors, weld lines, nozzles, etc.
- Visualization overlay planes to ease setup and viewing of location setup and results. Overlay planes include; Sensors, grid, attenuation map, welds and nozzles.

- Advanced Location setup capabilities provides improved event detection classification, and accurate source location techniques.

The HMSGHAT2000DDI program runs under WINDOWS 2000/XP/Vista operating systems.

System Requirements

HMSGHAT2000DDI software requires a DiSP, SPARTAN AT-2000, PCI-2, , SAMOS or SENSOR HIGHWAY System with Pentium III (or equivalent). Although lesser system computer configurations will operate HMSGHAT2000DDI, it is recommended that at least a 2 GHz processor, 1Gigbyte RAM, 100 gigabyte hard disk and 17" display monitor with 1280 x 1024 resolution be utilized. These days this type of computer power is easily available at low cost. Please consult the factory if your computer is below these standards.

If HMSGHAT2000DDI is being installed explicitly for replay and analysis, it is also recommended to use a similar computer base as that described above.

Software Installation and Licensing

The HMSGHAT2000DDI program is generally pre-installed on all AE system computers from the factory. In special instances, such as software upgrades, however, you will need to perform the installation yourself from the supplied CD. Likewise, if your AE system was built at our factory it will come pre-licensed. You will generally need to perform licensing yourself only when installing HMSGHAT2000DDI on a new computer.

If you are installing HMSGHAT2000DDI yourself, you will need to follow the simple installation instructions below. Software installation and licensing is a two step process which will be discussed in the next two sections for an example of "HMSGHAT2000DDI for DiSP". Other products such as PCI-2 and SAMOS will be very similar.

Software Installation

To fully install HMSGHAT2000DDI, you must perform the following steps.

a) Exit all running applications.
b) If you have a License Key already plugged into your system please remove it now. This is a green or gray USB dongle that is used in licensing. It should not be plugged in during HMSGHAT2000DDI installation.
c) Run SETUP.EXE from the root folder of the HMSGHAT2000DDI installation CD. You can find this through Windows Explorer.
d) Upon executing SETUP.EXE you will follow a typical Windows installation process. Select all the default folders (highly recommended but not mandatory), if desired.
e) Install the PCI-DSP4 driver if you are doing acquisition. See the section below titled "Installing the Data Acquisition Board driver if you are doing acquisition".
f) Install the Audio driver if using an audio board. See the section below titled "Installing the Audio board driver if using the Audio Board".
g) If you have a License Key then plug it into a USB slot now. Windows may need to install the driver for the Key at this point. Simply follow the prompts and accept the default values.

Software Licensing

Licensing Basics

systems come pre-configured with licenses (either internal-file or USB Key based). If you received such a system then you should not have to do anything beyond plugging in the License Key (if you were issued one). If you install HMSGHAT2000DDI yourself on a new computer then you will need to license it. The section "Licensing Instructions" below gives instructions on how to do so.

HMSGHAT2000DDI can be licensed using either of two methods.

- <u>License Keys</u> – These are green or gray USB dongles that can be purchased either with your system or at a future time. They can be used to run the licensed program on ANY computer they are plugged into. This makes them very portable. Note that these are not the same as the black or purple "Hardware Security Keys" used by older versions of HMSGHAT2000DDI.
- <u>License Files</u> – This is HMSGHAT2000DDIs default licensing system. If you have not purchased a Key you can acquire a License File from Customer Service by following the steps in the section titled "Licensing Instructions" below. This will license a single copy of HMSGHAT2000DDI to the computer of your choice.

Licensing Instructions

<u>If you have a Licensing Key (Green or Gray USB Dongle).</u>

a) Plug the Key into a USB port. If this is the first time you have done this on this particular USB port then Windows may want to install the driver. Let it do so now. Follow the prompts and accept the defaults. It should only take a few moments.
b) Run HMSGHAT2000DDI.

<u>If you don't have a Licensing Key.</u>

a) Run HMSGHAT2000DDI. You will most likely see a dialog box like this one.

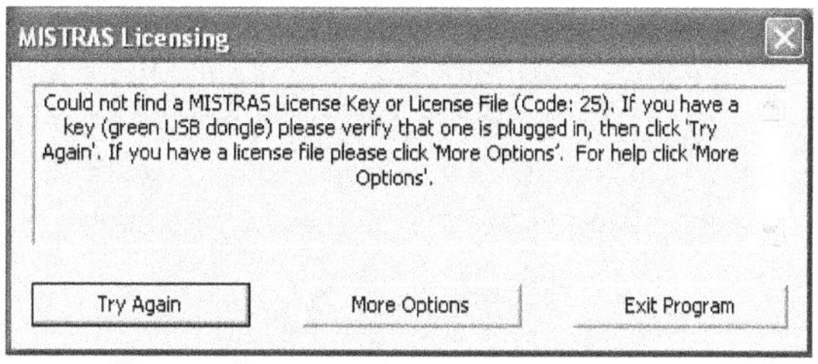

Figure 1. Initial Licensing Screen

b) Click "More Options". This will bring up the " Licensing" dialog box shown here.

Figure 2. MISTRAS Licensing Screen

c) The center section labeled " License File" guides you through the process of acquiring and installing a license file. It is a two step process as follows.
   a. Contact Customer Service – You will need to contact Customer Service to acquire a license file. The easiest ways are via email or by phone. In the sunken area of the menu is a block of information that Customer Service will need in order to create your license. You can copy/paste this into an email if you are using this method (be sure to enter your name, company and sales order#), or simply have the dialog box open while you call so the information will be at hand. Once Customer Service has this information they will create a license and email it to you. You can exit HMSGHAT2000DDI at this point if you wish, while your license is being made for you. If so, you can return to this dialog box later by following the same instructions above.
   b. Install your License File – Once you have received your license file from Customer Service click the "Install License File" button and browse to the file. You should receive a message indicating success, at which point you can click the "Start HMSGHAT2000DDI" button at the bottom of the screen (which previously was "Exit HMSGHAT2000DDI").

d) HMSGHAT2000DDI is now licensed.

Upgrading an HMSGHAT2000DDI License

If at any time you decide you would like to purchase additional options for HMSGHAT2000DDI such as "3D Location", "HMSGHAT2000DDI Post" or "LeakTEC/Online Crack Alarms" (there are many more options to choose from), you can purchase them by contacting Customer Service. You will not need to re-install HMSGHAT2000DDI or return your License Key (if you use one) to the factory. Customer Service can upgrade your license via email.

- If you use a License Key then Customer Service will send you a special file called a "Patch" file.
- If you use a License File then Customer Service will send you a new one to replace it.

In either case once you have received the file you will need to open the 'Upgrade License' dialog box. This can be accessed from one of two places.

- If HMSGHAT2000DDI is displaying a licensing error message on startup such as the one in 'figure 1' above then click "More Options" to open the " Licensing" dialog box. From here click the "Upgrade License" button found at the bottom left portion of the screen.
- If you can access HMSGHAT2000DDI normally then open the "Options" dialog box from the "Help" menu. Click the "Upgrade License" button.

This will open the "Upgrade License" dialog box shown here.

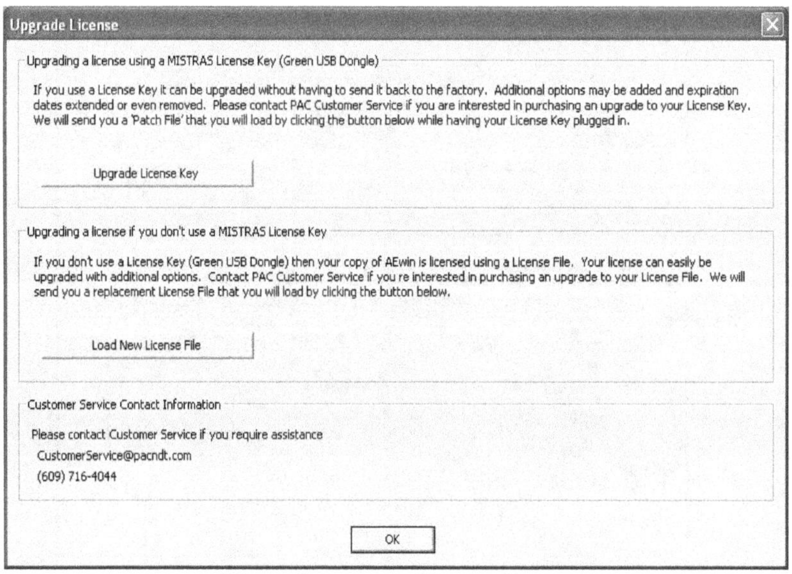

Figure 3. Upgrade License Screen

- If you are upgrading a License Key then make sure it is inserted into a USB slot. If you have more than one License Key make sure that only one is inserted at any one time. Click "Upgrade License Key" and select the "Patch File" emailed to you by Customer Service.
- If you are upgrading/replacing your License File then click "Load New License File" and select the License File emailed to you by Customer Service.

That's it. HMSGHAT2000DDI may need to close. If it does not, or when you re-start it, HMSGHAT2000DDI will use your newly upgraded license.

Licensing FAQ and Troubleshooting

Q: My license has stopped working when I upgraded HMSGHAT2000DDI. What do I do?
A: Starting in mid 2009 HMSGHAT2000DDI was upgraded to a new licensing system that is more flexible and user-friendly. The old "Hardware Security Keys" (black or purple USB dongles) have been replaced by " License Keys" (green USB dongles). Likewise the old "lservrc" license files have been replaced as well. If you have an old "Hardware Security Key" or "lservrc" license file then this will need to be replaced with a new version. Please contact Customer Service for assistance.

Q: HMSGHAT2000DDI won't install and gives the following error message: "There is a problem with this Windows Installer kage. A program required for this install to complete could not be run. Contact your support personnel or kage vendor.".
A: Make sure you run "Setup.exe" to install HMSGHAT2000DDI, not "HMSGHAT2000DDISetup.msi".

Q: After plugging in a License Key Windows gives an error message like one of these:

- "You must be a member of the Administrators group on this computer to install this hardware: USB Device".
- "You do not have sufficient security privileges to install hardware on this computer. Please contact your site administrator, or log out and log in again as an administrator and try again.".

A: The first time you plug a Key into a particular USB port Windows will attempt to install the driver for it. If you do not have sufficient privileges to do so you will get this warning. In this case you will need to log in as an administrator or contact your network or IT administrator for assistance. To avoid this problem in the future, be sure to plug the Key into all USB ports one by one (when you have admin privileges) to install the driver for all ports.

Q: Can I use my License Key in multiple computers?
A: Yes. License Keys are portable. Simply exit HMSGHAT2000DDI on your first computer, remove the Key, plug it into your other computer and start HMSGHAT2000DDI there. You will have the same options as before.

Q: Can I install my License File on multiple computers?
A: No. License Files are created for a single computer. You can purchase additional license Files or Keys for replay purposes from Customer Service.

Installing the Data Acquisition Board driver if you are doing acquisition

The various Data Acquisition Boards (PCI-DSP4, AEDSP, etc) and Audio Monitor Cards (ISA and PCI) require a driver in order to function properly in Windows.

Normally, when Windows is first booted after installing the card it will prompt for a driver for it. Point Windows to the appropriate driver folder on the CD and follow the instructions.
The driver folders are located on the CD at the following locations:

- – Drivers\AEDSP.
- **All other boards** – Drivers\Pci.

If for some reason you need to install a driver manually then read the document (drivers.doc or readme.txt) found in the appropriate driver folder. Use the procedures described there only if the above automatic driver installation does not work. It is not normally required to install the driver this way. But if problems arise, that is how to do it.

Installing the Audio board driver if using the Audio Board

The Audio Monitor Board (ISA and PCI) require a driver in order to function properly in Windows.

Normally, when Windows is first booted after installing the card it will prompt for a driver for it. Point Windows to the appropriate driver folder on the CD and follow the instructions.
The driver folders are located on the CD at the following locations:

- **ISA Audio Board** – Drivers\IsaAudio.
- **PCI Audio Board** – Drivers\Pci.

If for some reason you need to install a driver manually then read the document (drivers.doc or readmr.txt) found in the appropriate driver folder. Use the procedures described there only if the above automatic driver installation does not work. It is not normally required to install the driver this way. But if problems arise, that is how to do it.

Warning Notes

Please note the following warnings:

1. If the software is installed on a computer without AE hardware boards (e.g. PCI-DSP4 boards) installed, the program will run in replay only mode.
2. If the software is installed on a computer with AE hardware (e.g. PCI-DSP4) board(s) installed, it will not run in acquisition mode unless the Windows drivers for those boards are installed.
3. If a audio board is installed in the system, the software will not be able to control it unless its drivers are also installed.
4. If the entire system came from , the drivers and software are pre-installed at the factory.

Installing a Shortcut Icon for HMSGHAT2000DDI on the Desktop

Once the HMSGHAT2000DDI software installed as described in the previous sections, and assuming that all the default selections have been selected, the software can be started by the following mouse operated (activated by pressing the left mouse button) sequence:

Start → Programs → Physical Acoustics → HMSGHAT2000DDI for DiSP

Other forms of HMSGHAT2000DDI, such as SAMOS and PCI2 will have similar shortcuts.

An HMSGHAT2000DDI desktop Icon can easily be created by performing the same sequence above except pressing the right mouse button when pointing to HMSGHAT2000DDI. Upon doing this a menu will appear. You can either select "create shortcut" (which will create a second copy of the HMSGHAT2000DDI program title) and drag it to the desktop or you can select "Send to", then select the "desktop (create shortcut)" menu button which will place this extra copy directly on the desktop. Once created, the Icon title can be modified by left-clicking on the title. Now the program can be run directly by double clicking the HMSGHAT2000DDI Icon on the desktop.

HMSGHAT2000DDI LAYOUT (.LAY) FILES AND DATA (.DTA) FILES

This section discusses the various file types associated with HMSGHAT2000DDI. The most important files you will be working with everyday with HMSGHAT2000DDI are the Setup Files (called Layout files in HMSGHAT2000DDI) and the Data files. Your working copy of HMSGHAT2000DDI will have examples of both these layout files and test data files. You will be creating, using and modifying these files in the course of your work.

The initialization or setup files, referred to as "Layout" files, distinguished by the suffix .LAY, hold all AE test setup information for the HMSGHAT2000DDI software. A .LAY file stores the hardware setup, graph setup, acquisition setup, filters, alarms and location setup. Other program settings, such as graph item colors and various preferences are stored in the Windows registry.

A .LAY file can be loaded from disk by means of the "Open Layout" option in the "File" Menu of the HMSGHAT2000DDI program. Once loaded, all the initialization information from the layout file immediately becomes effective for system operation. A .LAY file is written to disk by means of the "Save Layout" or "Save Layout As" option in the File Setup menu. All the initialization information currently established in the system is written to the file. A .LAY file can have any name followed by the .LAY extension.

There is one special .LAY file known as the Autoload file. It is distinguished by the special name Layout.LAY. The Autoload file, when present, is automatically read from disk when the HMSGHAT2000DDI program begins to run. If the Autoload file is not present, the system is set up with default values held in the HMSGHAT2000DDI program itself. The Autoload file is a convenience feature to speed your work.

AE Test Data Files are identified by the filename extension .DTA. You may specify any name for a data file using standard Windows file naming conventions. This is done in the Start acquisition dialog box from the Acquire Menu. If a new name is not entered, the system will use the default data file name e.g. TEST0000.DTA for the first data written after bootup. The next file will be TEST0001.DTA, then TEST0002.DTA, and so on. It is convenient to use numbers at the end of the filenames (e.g. ABCD000.DTA) as the system will then increment the numbers to help keep the files in order and organized.

Layout and data files may be written to any disk drive and any specified subdirectory. A good practice is to save both .LAY files and .DTA files to the same drive and subdirectory, keeping related test files together. It is also a good practice to name the files very similar (e.g. test.LAY and test0000.DTA) as this will help keep the layout files organized with the data files.

There has been an intentional name and file-extension change from the "Initialization" (.INI) file of the "Legacy" (DOS) based programs to the "Layout" (.LAY) file for setup files utilized by HMSGHAT2000DDI. The file extension for a data (.DTA) file remains consistent. There is a good reason for this. Due to the extensive changes in migrating to Windows, and the many, many extra capabilities that have been added to the setting up of the system, it is impossible to keep consistency in the setup files between the previous legacy software and the HMSGHAT2000DDI software. Therefore, to avoid confusion, the extension file name has been changed. Now whenever a user sees an INI file, they will know that this can only be used with the previous legacy software, while the .LAY layout file can only be used with the HMSGHAT2000DDI software.

However, with the DTA file, the compatibility remains between the data file structure and AE .DTA data files can be read and displayed by either the DOS or the HMSGHAT2000DDI software.

OPERATING HMSGHAT2000DDI

The purpose of this section is to provide a quick software operational run-through, to provide a familiarization with operating and using HMSGHAT2000DDI. This is the most important part of this HMSGHAT2000DDI software user's manual as it will provide a view of the important screens and commands necessary to get HMSGHAT2000DDI setup, and running in both Replay and Acquisition modes.

Starting HMSGHAT2000DDI

The first step in the process of using HMSGHAT2000DDI is booting up the program on your AE system. As the HMSGHAT2000DDI program is a standard Windows program, it is started in the same way as any other WINDOWS program. This is accomplished either by using the "Start" command or by selecting the HMSGHAT2000DDI Icon from the WINDOWS desktop.

It is assumed that the AE computer is booted up and in its opening screen which is the desktop layout showing all the available program icons to select. If the HMSGHAT2000DDI icon (as shown on the right) is on the desktop, all you need to do is select this by double clicking the left mouse button, once the mouse is positioned over the icon. If the icon is not on the desktop you can create the shortcut by referring to section 1.2.6.

Alternatively you can start HMSGHAT2000DDI by selecting the following sequence starting with the Start menu at the bottom left side of the screen:

Start → Programs → Physical Acoustics → HMSGHAT2000DDI

The boot-up sequence for HMSGHAT2000DDI starts by loading the executable program HMSGHAT2000DDI.exe. Once HMSGHAT2000DDI is loaded, the program searches the startup "AE Data" subdirectory for the default Layout (setup file) called "Layout.LAY". Upon finding this file, HMSGHAT2000DDI loads it and becomes fully configured and ready to run in accordance with the setup information in that default Layout.LAY file. If no Layout.LAY is found, HMSGHAT2000DDI chooses a common default starting setup that is ready for use. In either case an HMSGHAT2000DDI startup screen similar to figure 4 will appear.

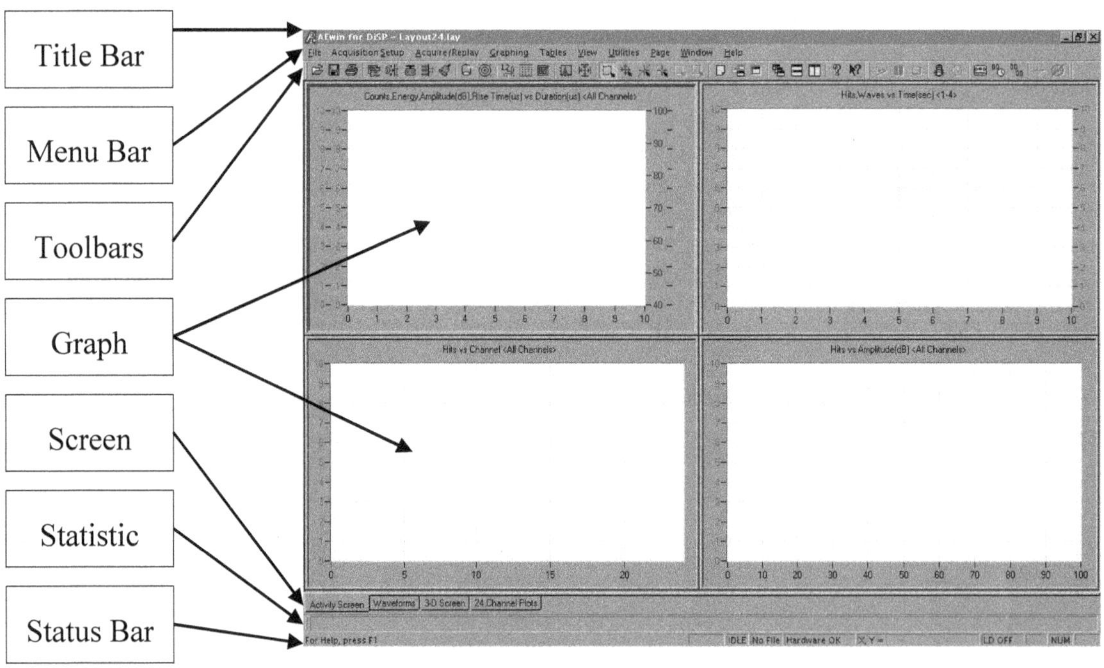

Figure 4. HMSGHAT2000DDI Opening Screen

HMSGHAT2000DDI Screen Layout Familiarization

Figure 4 shows a typical 4-screen layout for HMSGHAT2000DDI. Above and below the graph area are various informational, status and setup toolbars. Each bar is labeled with a name. Depending on the preferences selected for your program, you may have other toolbars or be missing some. The "View" menu in the main menu bar area controls which bars are on or off. The following is a short description of each area identified on the HMSGHAT2000DDI screen. This section will get you familiar with the layout and the overall function. Further chapters will provide detailed insight into the use of each of these functions.

Title Bar: This is a typical WINDOWS Title Bar with all the standard capabilities. At the leftmost position is the Physical Acoustics Logo (Logo) Icon from which you can manage the HMSGHAT2000DDI program with functions such as minimize, restore, maximize, move, size or Close the program altogether. At the far right are the application buttons for Minimize, Restore and Close. Back on the right hand side next to the Logo Icon is the program name (in this case it is HMSGHAT2000DDI for DiSP) and next to that is the name of the current layout file being used by the program. If an asterisk (*) appears next to the file name, it indicates that one or more settings have been changed since the last time the layout file was loaded or saved.

Main Menu Bar: The main menu bar provides a series of text commands running across the line, including: File, Acquisition Setup, Acquire/Replay, Graphing, Tables, View, Utilities, Page, Window, and Help. Most of these items contain pull-down sub-menus of their own, many of which in turn contain graphical dialog boxes and prompts of their own. You narrow down the range of choices until you arrive at the specific choice you want.

To display a pull-down sub-menu from a main menu entry, simply move the mouse pointer to the name of the menu item on the menu bar and click the left mouse button. This process is known as "clicking" (or "left-clicking") on an item, and will herein be referred to as such. To use the keyboard, simultaneously press the ALT key and the underlined letter of the menu item name at the same time. Once you have opened a pull-down menu item from the main menu, you can use the left and right arrow keys to move through each entry, allowing you to see the pull-down menu for each main menu entry. You can also use the up and down arrow keys to move through each selection of the pull-down menu. To select an item from a pull-down sub-menu, click on the desired action, or use the up and down arrow keys to select an action, and then press ENTER. Pressing the underlined letter of a menu item will also select that item. If an item in the pull-down menu appears "grayed out," this action is not available in the current software or at the current time. Use the <ESC> key to remove the pull-down menu display.

Most of the selections in the pull-down sub-menus will display a dialog box requesting information necessary to carry out the command you requested. A dialog box contains areas where you enter information, select options or commands, or activate controls to execute or cancel a command. There are several types of boxes and buttons available within a dialog box. These boxes and buttons are described briefly below and will be described in further detail in subsequent sections of this manual. It is important that you be familiar with dialog boxes and be comfortable using them, for they are the means by which most functions are carried out in the HMSGHAT2000DDI software.

Icon Toolbars: Many of the menu items accessible from the Main pull-down menus can also be accessed by use of the Icon Toolbar and Acquisition Control Toolbar just below it. Figure 5 shows the Icon toolbars.

Figure 5. Icon Toolbars

To determine the function of any of the Icon's simply move the mouse cursor over top of the desired icon. Upon resting there for approximately 1 second the Icon's function will appear. To select an icon as a shortcut command simply click on it with the left mouse button.

Graph Area: The graph area is the most important area of the screen. This is where the AE information in the form of graphs and line listings appear regarding the AE examination status. The graph area can be set up in many ways with a minimum of 1 graph to a maximum only limited by the graph visibility. More about the graph setup, functionality and flexibility will be discussed later in the Graph setup menu.

Screen Page Tabs: HMSGHAT2000DDI can show many different graph screens, these are called screen or graph "Pages". Accessibility to viewing any screen is provided via the Page tabs located just below the graph area. The Page tabs can be named to be synonymous with the graph theme as shown in figure 4 with the 4 Page tab names, Activity Screen, Waveform Screen, 3D Screen and 24 channel plots. Using the right mouse button and clicking on a page will display a 'context' menu from which pages can be added, inserted, deleted or renamed. The context menu is also a quick way to add new graphs and close existing graphs.

Statistics Bar: This toolbar area is reserved for display of test statistics. Test statistics allow the user to view key Acoustic emission test indicators including: Cumulative AE counts, Total AE Hits, Total AE Events, Total # Waveforms, Cumulative Counts, Cumulative Energy, Time of Test, Disk Free Se, Parametrics and others. These are very useful activity indicators during a test and very useful summary statistics after a test. The user can select any or all of these to customize his statistics display bar.

Status Bar: The Status bar provides useful information regarding HMSGHAT2000DDI and test status. On the left is a Text bar providing status and help information regarding the functions being carried out. Near the center is a status bar. In figure 4 the word "Idle" appears in this bar. Other status information such as "Test Paused", "Replay", "Test Stopped" "Test Active", "Abort Test", etc. are displayed in this text field. Next to the Test status field is a filename field informing the user the name of the file being replayed or the name under which an AE test is being saved. Next to that is a system diagnostics text field. In figure 4, this status is "Hardware OK". Next to that is a cursor position, showing the current location of the Cursor on the graph. When selected, the X, Y position will be read out in actual screen units.

Navigating the HMSGHAT2000DDI Menus

This section provides a detailed reference of HMSGHAT2000DDI's major menu items and dialog boxes. The first time through, we suggest you just look at the menus (to get familiar with them) and use the setups provided with your HMSGHAT2000DDI installation, that have been predefined by ; later, you can make your own choices within the menus to control system operation.

From the Opening Screen (Figure 4), pay attention to the Main Menu Bar near the top of the screen. This specific area is shown again below. Each main menu item will be described. This brief introduction to navigating the menus is provided in order to get ready for guiding the user in the sequence of starting and replaying a test.

Figure 6. Main menus and toolbars

File Menu

New Layout

The 'New Layout' menu item is used to remove all graphs and pages and restore the graph layout to a default setting of one page and one simple graph. This command does not alter any other settings.

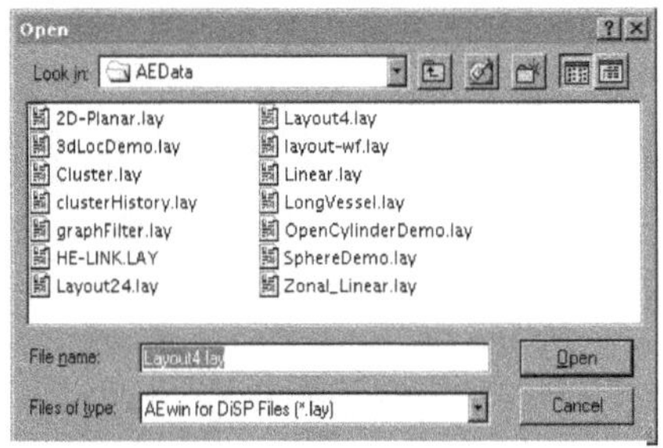

Figure 7. Open Layout...

Open Layout...

The 'Open Layout...' menu item is used to load a specified layout file's setup information into HMSGHAT2000DDI.

Save Layout/Save Layout As...

The 'Save Layout...' and 'Save Layout As...' menu items are used to save the setup information in HMSGHAT2000DDI to a specified layout file. When the 'Save Layout As...' item is clicked the 'Save As' dialog is displayed. Here you can enter a new layout filename to create or an existing one to overwrite. Once this is done you will see the layout filename you entered on the upper left corner of the screen. You can use 'Save Layout...' from then on to update the file with any changes you have made since you have last saved.

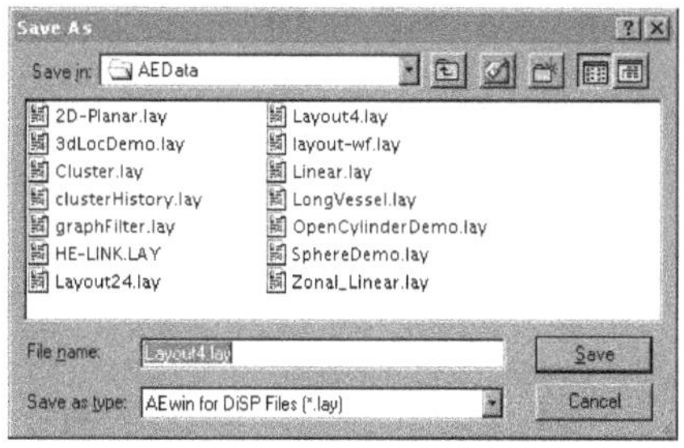

Figure 8. Save Layout...

Specify Data Folder...

The 'Specify Data Folder...' menu item opens a dialog box of the same name. It is used to select the default folders for storage of Data and Layout files. These are the folders that are displayed by default in the various Open/Save Layout and Acquire/Replay dialog boxes. The 'Layout File Folder' is also where HMSGHAT2000DDI looks for the autoload 'Layout.lay' file on startup. By default these are both set to the AEData folder found in the installation folder of HMSGHAT2000DDI.

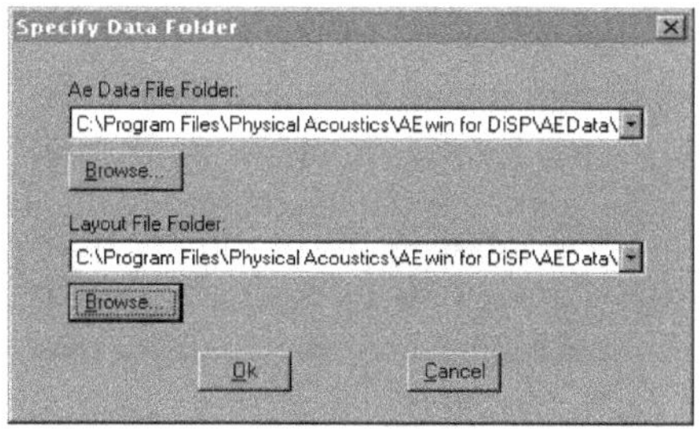

Figure 9. Specify Data Folder

Print Page(s)/Print Graph...

The 'Print Page(s)...' and 'Print Graph...' menu items are used to print a graph or multiple pages of graphs. 'Print Graph...' is used to send the currently selected graph to the printer, zoomed as much as possible while still retaining aspect ratio. 'Print Page(s)...' is used to send a selection of pages to the printer one page at a time, each page zoomed as much as possible while still retaining aspect ratio. You can select 'All pages' or a range of pages.

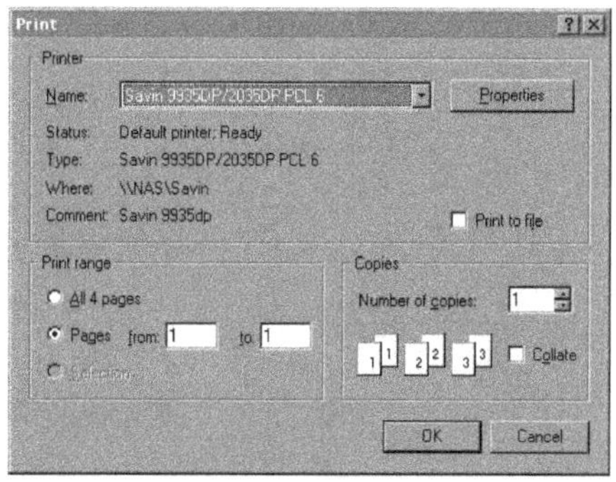

Figure 10. Print...

Export to JPG...

'Export to JPG...' is used to save pages as JPEG files in the same way that 'Print Page(s)...' sends pages to the printer. Here you enter a base filename and a destination folder. Each page will be saved as a separate JPEG file with its name formed from the combination of the base name and the page's title. e.g. If you had 2 pages named 'waveforms' and 'activity' and you set the base name to be 'TEST' then the two files will be named 'TESTwaveform.JPG' and 'TESTactivity.JPG'.

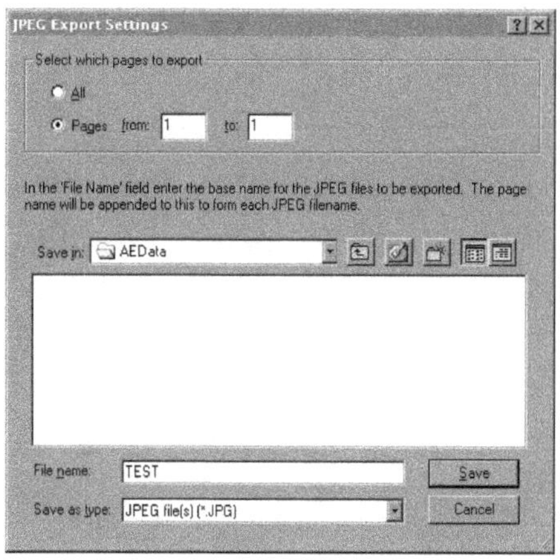

Figure 11. Export to JPG

Copy Screen to Clipboard

'Copy Screen to Clipboard' is used to save an image of the entire screen to the clipboard. This image can then be pasted in other applications such as word processors and image editors.

Exit

'Exit' is used to terminate the HMSGHAT2000DDI program. If you have made any unsaved changes to your current layout HMSGHAT2000DDI will prompt you to save your layout first, exit without saving or cancel.

Acquisition Setup Menu

Hardware...

This is the dialog box where all hardware settings are selected before the test. There are a number of property pages that each deal with a different part of hardware setup. Each is described below. Note that there are some differences in this dialog box between products. For instance 'HMSGHAT2000DDI for Spartan' has a Gain setting but no Waveform Setup whereas 'HMSGHAT2000DDI for DiSP' has no Gain setting but can have Waveform Setup if the Waveform option has been installed. Features specific to one or more types of hardware are noted by the names of the applicable hardware types. These types are briefly described below.

Name	Description	Waveforms
ICC	2 Channel card for **Spartan** systems. *	Not Available
DSP4	4 Channel card for **Spartan** systems. *	Not Available
AEDSP	2 Channel card for systems.	Standard
PCI-DSP4	4 Channel card for **DiSP** systems.	Optional
SAMOS	8 Channel card for **SAMOS** systems.	Standard
PCI-2	2 Channel card for **PCI-2** systems.	Standard

* ICC and DSP4 cards can be used simultaneously in the same system.

Standard Channel Setup

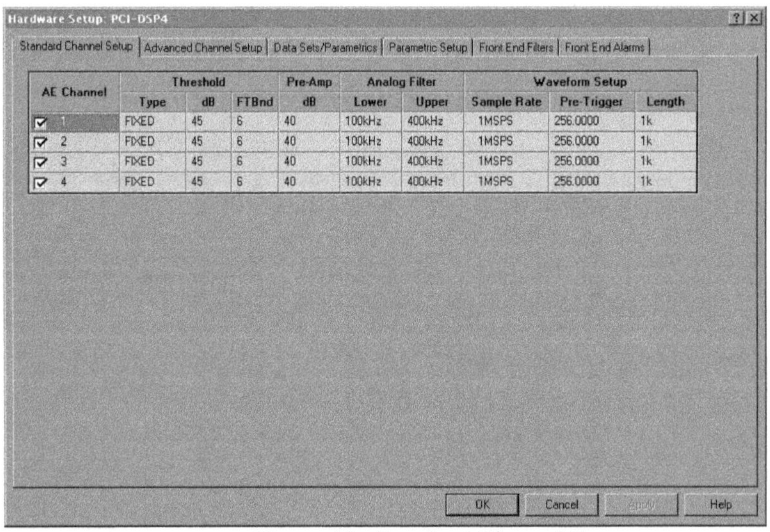

Figure 12. Standard Channel Setup

In the 'Standard Channel Setup' page you can enable or disable each channel and edit their common settings. If your system has the waveform option installed you can also edit each channel's waveform settings.

<u>AE Channel</u> – You can enable a channel by clicking its checkbox.

Threshold – 'Fixed' threshold is used when you want the threshold to remain unchanged at the 'dB' level you set (in the dB column). 'Floating' threshold is used under conditions of high, varying background noise. When floating threshold is selected, the value shown in the 'dB' column represents the threshold level at the start of the acquisition, the threshold will vary to 6dB (or FTBnd value on systems that have this column) above the current ASL reading for that channel. The threshold will not go below the initial value specified in the 'dB' column. Threshold is the prime variable that controls channel sensitivity.
Gain (dB) [ICC/DSP4, SAMOS, PCI-2] – Enter the hardware gain here.
Pre-Amp (dB) – Enter the pre-amp gain of your sensor's pre-amplifier here.
Analog Filter [not available for ICC] – The Lower and Upper drop-down list boxes allow you to select from the available analog filter values for low pass and high pass filters on each active channel. This is done by simply selecting the desired value from the drop down list boxes. Note that if waveform analysis is desired, these values should be kept below ½ of the Sample Rate. Make sure to raise your Sample Rate if this becomes an issue.
Sample Rate [only with waveform option] – This is the rate at which the data acquisition board samples waveforms on a per second basis. A sample rate of 1 MSPS (Mega Samples Per Second) means that one waveform sample is taken every μsec. A sample rate of 2 MSPS means that one waveform sample is taken every ½ μsec., etc. The sample rate can be selected by clicking on the down arrow on the drop down list box, then clicking the mouse on the desired sample rate.
Pre-Trigger [only with waveform option] – This value tells the software how long to record (in μsec.) before the trigger point (the point at which the threshold is exceeded). The user may enter a value in μsec in the pre-trigger edit box. The minimum allowable pre-trigger value is zero. The maximum allowable pre-trigger value is calculated by dividing the hit length by the sample rate in MHz. If, for example, the hit length was 1 k (1 k = 1024) and the sample rate was 4 MHz, then the maximum allowable pre-trigger value would be 1024/4 = 256 μsec.
Hit Length [only with waveform option] – This determines the size of a waveform message. Clicking the down arrow on the drop-down list box for Hit Length will display a list of allowable hit lengths from 1 k to 15 k. At a 4 MSPS sampling rate, a hit length of 1 k will allow up to 256 μsec. of data, a hit length of 2 k will allow 512 μsec of data (2 * 256), and so on.

Timing Parameters

In the 'Timing Parameters' page you can enable or disable each channel and edit Peak Definition Time (PDT), Hit Definition Time (HDT), and Hit Lockout Time (HLT).

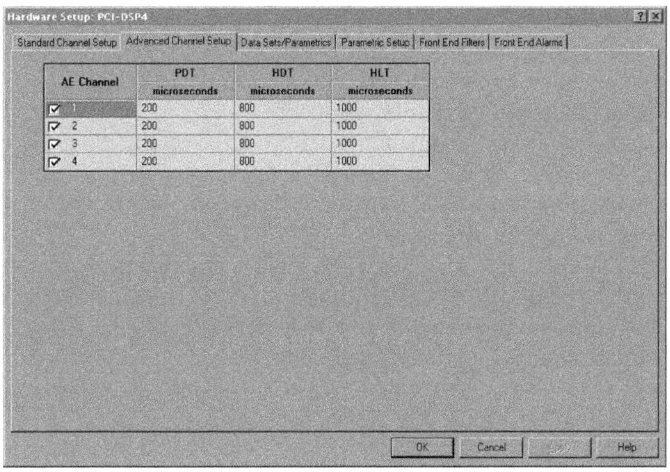

Figure 13. Timing Parameters

PDT, HDT and HLT are timing parameters of the signal measurement process. In brief, a proper setting of the PDT ensures correct identification of the signal peak for risetime and peak amplitude measurements. Proper setting of the HDT ensures that each AE signal from the structure is reported as one and only one hit. With proper setting of the HLT, spurious measurements during the signal decay are avoided and data acquisition speed can be increased. Recommended values for general-purpose testing are:

	PDT	HDT	HLT
Composites, Non-Metals	20-50	100-200	300
Small Metal Specimens	300	600	1000
Metal Structures (high damping)	300	600	1000
Metal Structures (low damping)	1000	2000	20000

Data Sets/Parametrics

HMSGHAT2000DDI features user-selectable data sets. This means that if you do not plan to use all the measurable AE parameters, you can leave some out and enjoy the compensating advantages of higher data throughput speed and more economical use of storage se. Selecting Data Sets/Parametrics from the Hardware Setup dialog box displays the Data Sets/Parametrics page that follows.
Note that the above figure is specific to the DiSP product. Your product may have more/fewer options available.

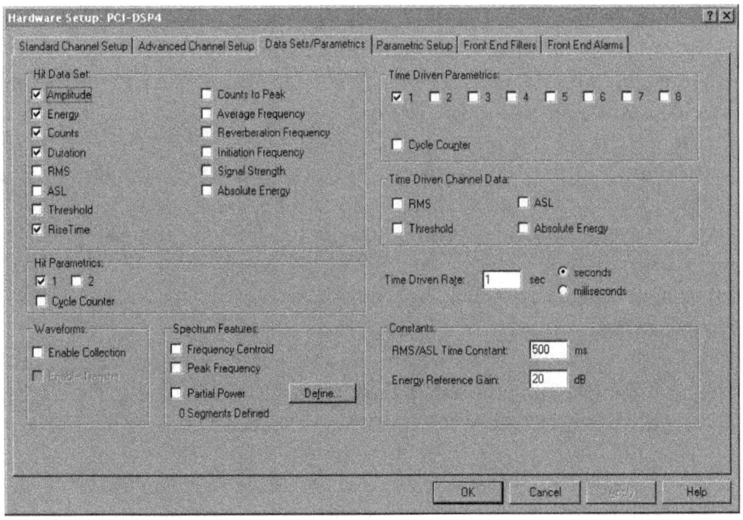

Figure 14. Data Sets/Parametrics

Hit Data Set

This box allows you to select which measured signal features are to be included in the reporting of each AE hit. The features to be included are so marked with a check in the checkbox next to the feature. Features can be included/excluded from the data set by selecting or deselecting the checkbox. "Time of test" is not shown because it is always included in the data set. A short description each hit feature in the Hit Data Set is as follows:

1) Amplitude - highest voltage in the AE waveform, expressed on the dB_{AE} amplitude scale.
2) Energy - time integral of the absolute signal voltage. The reported magnitude, depends on the value selected for Energy Reference Gain. Proportional to Signal Strength.
3) Counts - number of times the signal crosses the detection threshold.
4) Duration - time from first to last threshold crossing (µs).
5) RMS - root mean square voltage during the timeframe of the hit, referred to the input to the signal processing board.
6) ASL - RMS, converted to the dB_{AE} scale ($0dB_{AE} = 1\mu V$ at the sensor, before any amplification).
7) Threshold - detection threshold, on the dB_{AE} scale.
8) Rise Time - time from first threshold crossing to highest voltage point on the waveform (µs).
9) Counts to Peak - number of threshold crossings from first to highest voltage point on the waveform.
10) Average Frequency - Counts divided by Duration, divided by 1000 (thus, kHz). Note that this is not a spectral domain calculation, but a calculation from time domain features.
11) Reverberation Frequency - (Counts - Counts to Peak) divided by (Duration - Risetime). .
12) Initiation Frequency - Counts to Peak divided by Risetime.
13) Signal Strength - time integral of the absolute signal voltage, expressed in pVs (picovolt-seconds) referenced to the sensor, before any amplification. Proportional to Energy.
14) Absolute Energy - time integral of the square of the signal voltage at the sensor before any amplification, divided by a 10kΩ impedance and expressed in aJ (attojoules).

Hit Parametrics

This box allows you to select which parametric inputs are to be included in the description of each hit. A maximum of two parametric inputs (e.g., external load signals), or one parametric input plus the cycle counter, can be included in the Hit data set. These items, however, must also be included in the Time Driven Data Set, otherwise an error message will appear.

Waveforms

This box is available if you have the waveform option installed.

Enable Collection – This checkbox is used to signal the data acquisition board(s) to acquire waveforms. If this checkbox is unchecked then the board(s) will not be able to collect waveforms for the duration of the test.

Enable Transfer – This checkbox is used to signal the data acquisition board(s) to transfer collected waveforms to the software for processing/display and (if auto-dump is enabled) saving to a data file. If this checkbox is unchecked then the waveforms will not be visible to the user and won't be saved in a data file until the checkbox is turned on. Unlike Enable Collection this checkbox can be toggled on/off at any time during a test. When Enable Transfer is OFF the system can experience a boost in performance especially at high data rates if you are saving to a data file. Turning Enable Collection OFF can grant a greater performance boost at very high data rates as the system concentrates on hit-based data.

Spectrum Features

This box is available if you have the waveform option installed. Spectrum Features are derived from the power spectrum of the waveform associated with a hit:

1) Frequency centroid - the center of mass of the power spectrum graph.
2) Peak frequency - the point where the power spectrum is greatest.
3) Partial Powers - see 'Partial Power Features' below.

You must check Enable Collection in the Waveforms box for these features to be usable. Transfer of waveforms can still be set off so that the system can record spectrum features without recording waveforms.

Partial Power – Partial Power is a feature which is derived from the power spectrum of the waveform associated with a hit. It is a percentage and is calculated by summing the power spectrum in a specified range of frequencies (up to four of these ranges can be specified), dividing it by the total power in another range of frequencies, and multiplying the result by 100.

To use partial power you must

1) Check the partial power checkbox.
2) Define at least one partial power segment. To do this press 'Define'. This brings up the Spectrum Segment Setup dialog box described below.

Spectrum Segment Setup is used to define the partial power features. For each of the 4 definable segments the dialog box has a checkbox to enable the segment, an edit box for defining the start frequency and an edit box for defining the ending frequency.

The group box titled 'Total Power Range/Equal Sing Setup' contains two edit boxes. These edit boxes serve a dual purpose. The main reason for them is to define a range of frequencies over which the power spectrum will be summed.

The result of the summation is used in the calculation of total power for the partial powers calculation. Their other purpose is to allow the user to spread the frequency range evenly among the enabled segments. The sing will occur when the user presses the 'Apply Sing' button which is located to the right of the edit boxes.

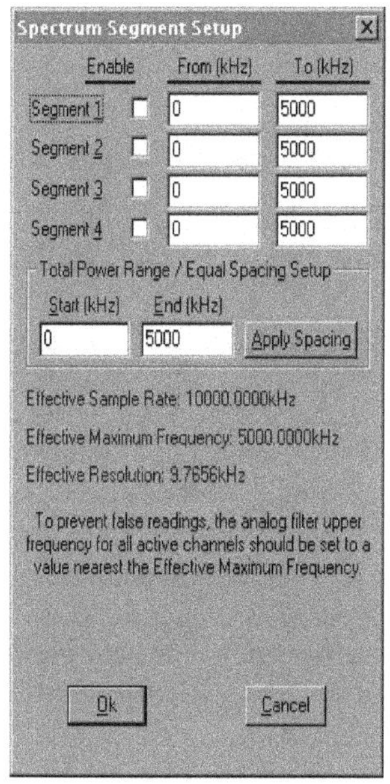

Figure 15. Spectrum Segment Setup

CAUTION: If the edit boxes are used to set up equal sing, always make sure the edit boxes contain the Total Power range values when the dialog box 'OK' button is pressed. Otherwise the partial powers will be based on the equal sing range.

The dialog box also displays information based on the sample rate and hit length of the first active AE channel:

1) Effective Sample Rate – The sample rate of the decimated waveform (the FFT calculation is always for 1k of samples). For a 1k hit length the effective sample rate is the same as the hardware sample rate. For a 2k hit length it would be half the hardware sample rate. Etc.
2) Effective Maximum Frequency – This is one half that of the Effective Sample Rate and is the maximum frequency which can be defined for any segment.
3) Effective Resolution - This is the Effective Maximum Frequency / 512 and is equivalent to the frequency range of a single FFT bin.

The system does not automatically set the input filters based on the sample rate to reduce aliasing, so the dialog box will remind the user to set the input filter to a value closest to the Effective Maximum Frequency.

The Spectrum Segments are global for all active channels, if the sample rate and/or hit length is not the same for all active channels a warning will be displayed to alert the user.

Time Driven Parametrics and **Time Driven Channel Data**

Additional to the hit descriptions, HMSGHAT2000DDI allows the recording of RMS, ASL, Threshold, Absolute Energy, Cycle Counter and Parametric inputs, at regular time intervals even in the absence of AE activity. These are called Time Driven messages. This box allows you to select a set of features for them.

Time Driven Rate – This controls the frequency of the readings and the data to be included in the records. Time driven data is useful for leak monitoring and for recording changes in load and background noise levels. Also, time driven data can be used for updating hit data graphs, involving parametric values or time, even when no hits are coming in. The interval you set must lie in the range 10 msec to 1,800 sec and must be a multiple of 10 msec. Typical values are in the range 1000-60,000 msec (1–60 seconds). For frequent parametric value updates, you may like to choose a short interval, but this leads to lengthy disk files and data listings, so it is not necessarily the best choice. When the time driven data set is used to update graphs, it is a good idea to set the rate less than or equal to the 'Seconds Between Updates' field selected in the Display Mode dialog box.

Constants

RMS/ASL Time Constant [not in DSP4/ICC] – The time constant can be set from 10ms to 1000ms in steps of 10ms. Previous versions of the program had a fixed time constant of 500ms.

Energy Reference Gain [PCI-DSP4 and AEDSP only] – In some instruments, such as SPARTAN, Energy is a gain dependent quantity. However, some systems (like DiSP and) don't have a gain setting so the Energy was designed to give a reading as if 20dB of gain were applied. The Energy Reference Gain allows you to match the value of Energy reported by system to that of a SPARTAN-2000 or other instrument which has programmable gain. The value of the Energy Reference Gain has a valid range of 0 to 60 dB.

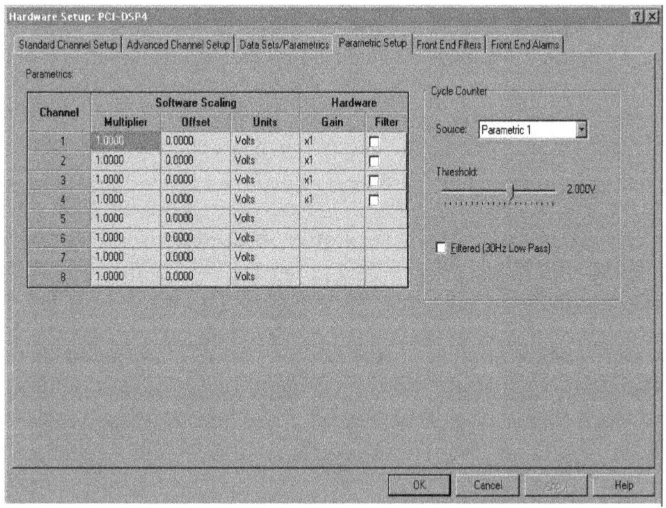

Figure 16. Parametric Setup

Parametric Setup

The Parametric Setup page allows you to scale the voltages measured at the parametric input, such that the system displays the corresponding load or pressure values. This page also allows you to tell the system what units (kN, psi) you want displayed on your graph axis and printouts, and it allows you to select from four available filters on gain.

The Parametric Scaling process used by HMSGHAT2000DDI is: Displayed Value = (Measured Voltage x Multiplier) + Offset

The default values set up in the delivered Autoload file are 1.00, 0.00 and "Volts" so that the system will display the raw voltage delivered to the Parametric Input socket on the rear panel.

To scale the output of a linear load-measuring transducer, the easiest way to find the correct Multiplier and Offset is to go into a simulated test and turn on the Line Listing Display (F7). From the display read the parametric voltages for two widely separated loads. Suppose these readings are V1 at load L1, and V2 at load L2. The correct values for Multiplier and Offset are then:

Multiplier = (L1 - L2) / (V1 -V2)
Offset = (V1*L2 - V2*L1) / (V1 - V2) = L1 - Multiplier * V1

Once you click OK to exit this dialog the scaled values of the parametric input will now appear on all graph displays and data listings. Minimum and maximum values for graph axes should be specified in the scaled values, and the entered unit will appear on the graph (e.g., 0–1000 psi). The raw data on disk will contain the original voltage readings and the scaling will be done again each time the file is replayed.

The Hardware section allows you to select from one of four available values for a hardware gain filter. Most cards (all except ICC) contain a programmable amplifier as well as a 30 Hz Low Pass filter for better resolution and less noise in the data collection process.

The **Cycle Counter** section allows you to control the source and threshold of the cycle counter [PCI DSP-4, AEDSP only].

- The Source can be selected as parametric 1-4 (on most data acquisition boards).
- The Threshold slide control allows you to set the threshold for the cycle counter. The cycle counter counts every time a signal crosses the threshold. The threshold can be set within a -10.00 volts to +10.00 volts range. This voltage should be set based on the load signal coming into the system. It should be set low enough to assure constant, accurate triggers.
- The Filtered checkbox allows you to enable an independent 30 Hz low pass filter on cycle counter.

Front End Filters

Selecting this action allows you to set up the "front end filters." These filters apply accept/reject criteria to the hit descriptions as the data is processed in the data acquisition board, so that rejected hits or waves do not even reach the main processor board.

There are two kinds of front-end filters: **AE Hit Filters** reject hits and any waveforms associated with them based on the criteria you set. **Waveform filters** work the same way but reject only the waveforms, allowing the hits to remain. The setup is identical for both.

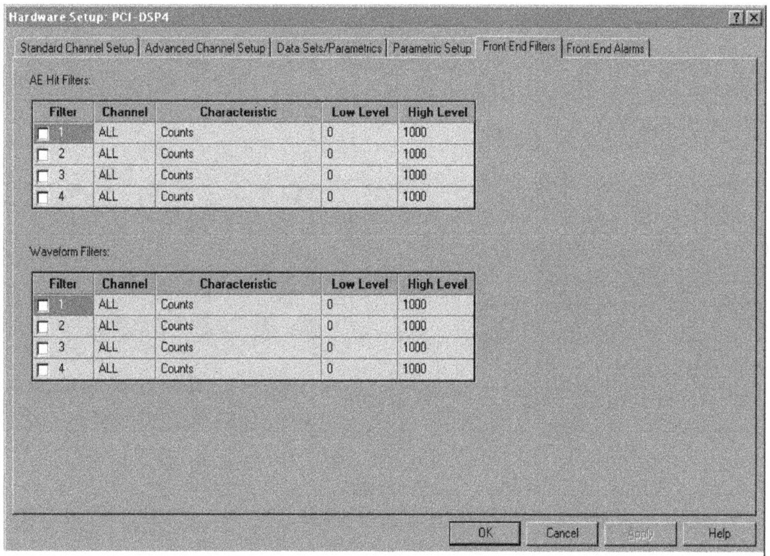

Figure 17 Front End Filters

There are four filters available at the present time. All four filters will always be displayed. To activate a particular filter simply check the 'Filter' checkbox associated with the filter you want to use.

Use the Channel drop-down list box to select which channel (or all channels) to filter on.

Use the Characteristic drop-down list box to select an AE Characteristic to filter on. Each hit will be individually screened to see whether the measured value of the specified characteristic falls in the acceptance range from LOW LEVEL to HIGH LEVEL, inclusive. For the message(s) (hit and/or wave) to be accepted, the hit must fall in the acceptance range for all active filters affecting that channel. Change the filter LOW and HIGH levels by entering new values in the text boxes. If a hit is not within the acceptance range, the message(s) will be eliminated from the front end signal processing.

Front End Alarms

There are four alarms defined in the alarms page. This page is almost identical to the front-end filter page with the exception that the user may select the type of alarm to be Single Channel Single Hit or Single Channel Cumulative. An alarm is generated if one or more of the alarm criteria is exceeded. Upon alarm, the PC computer speaker will sound, and an alarm output on the I/O connector will indicate an alarm. In addition, an alarm popup box will appear on the screen indicating if the alarm is a warning or a trigger, which alarm was exceeded and the value that triggered the alarm. To reset the alarm click the 'Clear All Alarms' button on the alarm popup box. See section 3.3.3.6 entitled 'Display Last Alarm' for more information.

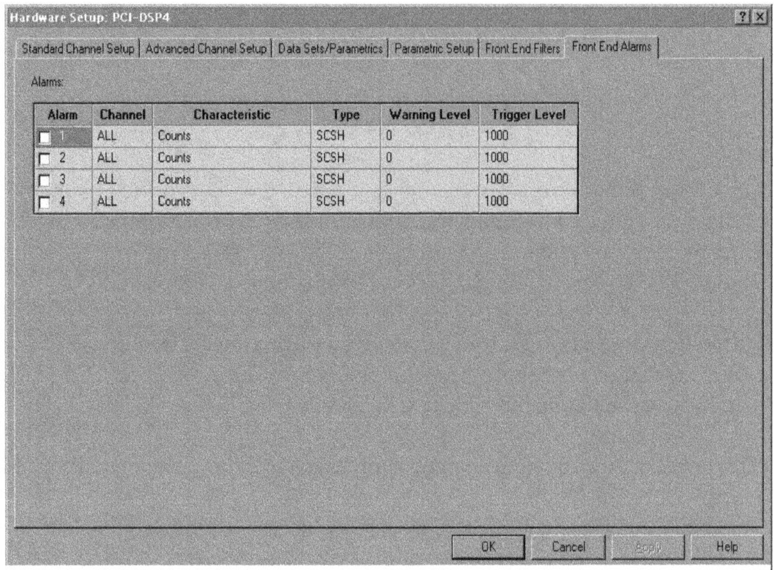

Figure 18. *Front End Alarms*

DeltaT Front End Filters (only)

Delta-T Front End Filters are used to eliminate "on board" AE hits which are part of the same event but outside the acceptance range selected in the DeltaT Filters Setup page.

A DeltaT Filter is located within each (AEDSP) board, so that this filter can only eliminate "on board" AE hits. This type of filter is very useful for cases where linear location is used. It will successfully filter hits that come from the outer areas of the sensors and beyond. Delta-T Front end Filters are very useful in 2 channel systems.

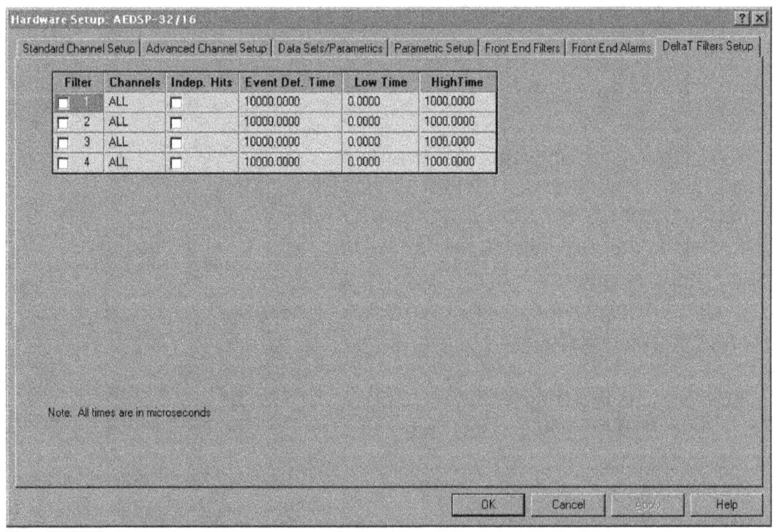

Figure 19. *DeltaT Front End Filters*

In cases where more than two channels are needed (such as planar location) a front end filter cannot be used since the hit information is not available from other boards. For this situation a "high end" filter is needed and Graph Filters can be used for this purpose. See the section on Graph Filters for more information.

The DeltaT Filters page is very similar to the Front-End Filters page discussed previously. Four different filters are available to allow programming of each board differently. Alternatively, the first filter can be chosen as an "ALL," which will have all boards filter using the same set-up parameters.

Each menu item and its function is discussed below.

Filter: This toggles between On and Off just like the Front End Filter page does.

Channel: This selects the channel numbers for the DeltaT filter. Because the DeltaT Filters operate on a per board basis, *pairs* of channels must be selected (e.g., 1-2, 3-4...) or ALL channels may be selected.

Indep. Hits: Allows independent hits (those hits not associated with an Event) to be accepted or rejected for transmission to the high-end, at the discretion of the user.

Event Def. Time: This is the maximum event time (in μsec) required in order for the AEDSP-32/16 to identify paired hits on the board as events. This should be set at the maximum time expected for an AE signal to propagate from the first to the second sensor plus 10% over range.

Low Time: This is the lower limit of acceptable Delta-T values in μsec. Below this is not an acceptable difference and the hits will be discarded. Negative numbers can be used. For delta-T filters, a hit occurring closer to the even channel is assigned a negative delta-T.

High Time: This is the higher limit of acceptable Delta-T values in μsec. Above this is not an acceptable difference and the hits will be discarded. As above, negative numbers indicate that the hit occurred closer to the even channel

The information set up in the DeltaT Filter page is downloaded into each specified AEDSP board with the rest of the set-up information. Upon acquisition, each AEDSP board has software event detection and filter routines implemented as follows:

As hits are received from both channels, the board has an event detection queue whereby it checks hit pairs (hits on opposite channels) against the Event Definition Time to determine if it is an event. To provide for asymmetric acceptance ranges between two sensors, the delta-T is always calculated from the lower numbered sensor to the higher numbered sensor. Hits whose delta-Ts are longer than the Event Definition time are considered independent hits. In this case, the first hit will pass to the high end as an independent hit.

In those cases where an event is detected (hits that have a delta-T within the Event Definition Time) its Delta-T is checked against the high and low levels for that board. If the Delta-T is outside of the range set in the "Low Level" and "High Level" fields (less than the Low Level and higher than the High Level), then both hits will be rejected and neither hit nor waveform data will be passed to the high end. If the Delta-T is within the "Low Level" and "High Level" values then the hits (and waveforms) are allowed to pass to the high end.

Delta-T filters are activated independently of any location processing. This allows use of Delta-T filters in tests with extremely high data rates, whereby the data is stored and analyzed later with the slower location algorithm.

PCI-2 Waveform Streaming (PCI-2 Only)

Background

The PCI-2 AE System on a Card is capable of streaming waveform data continuously into physical memory on the PC. It uses an onboard DMA controller to read samples from each channel's 4K FIFO and transfers them into a circular buffer in the computer's memory. When a stream to disk event is detected (this being a threshold crossing, an external digital trigger, a manual trigger, or a set periodic interval), the waveform is pulled out of this buffer and written to the output data file. This data file (with a WFS extension) is created separately from the normal acquisition data file (with a DTA extension) used to store collected AE data. Once the output file is created it can be viewed and analyzed with the WaveViewer program supplied on the installation disk.

Waveform Streaming Setup

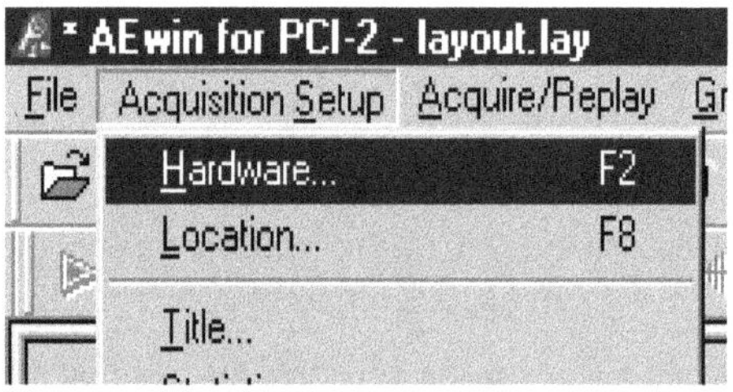

Figure 20. – Hardware Setup

Waveform streaming settings are located under the Hardware menu item under the Acquisition Setup menu (Hardware Setup Figure 20) or by pressing the F2 accelerator key on the keyboard. This will display the Hardware Settings dialog box. The waveform streaming setup is accessed by clicking on the Waveform Streaming tab near the top of the Hardware Setup dialog box.

Per Channel Waveform Streaming Settings

AE Channel for Streaming:
Selects a channel for streaming. When streaming is triggered, data from all selected channels will be stored to the file. Click the box to toggle this setting. A checked box indicates that a channel is enabled for streaming. A channel must be enabled for AE data collection to be able to stream. The program will enforce this by automatically enabling any channel that is enabled for streaming. If it is undesirable to have AE data from a channel that is streaming, go to the Standard Channel Setup and set the channel threshold type to 'Fixed' and raise the threshold level to its highest value (typically 99 dB).

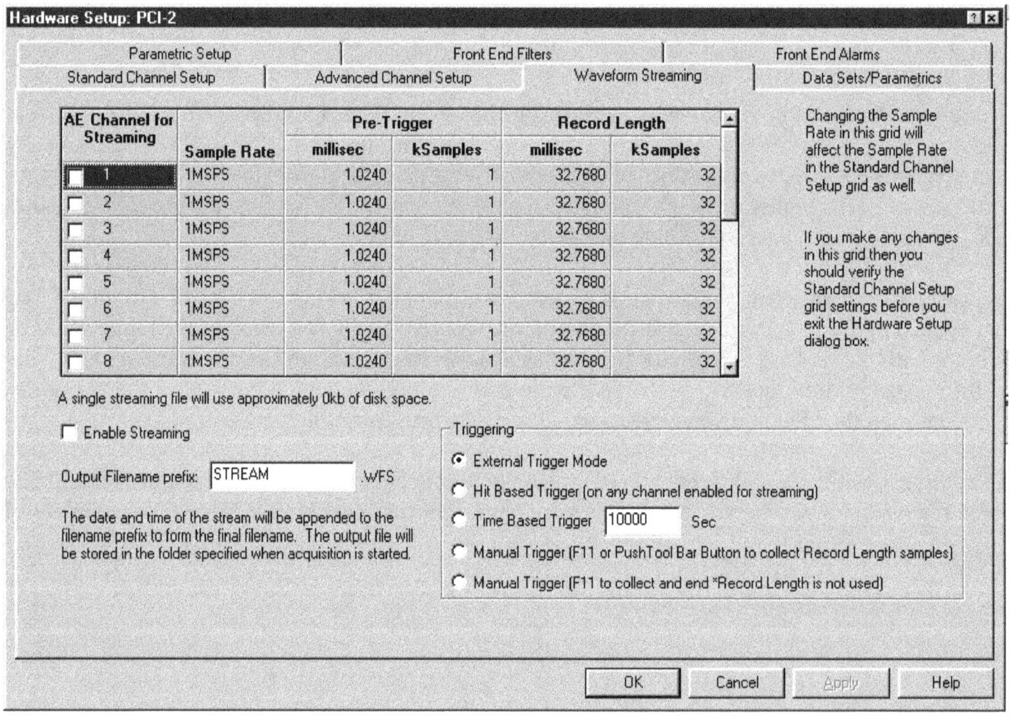

Figure 21. – Hardware Setup Dialog, the Waveform Streaming Setup Section

Sample Rate:
Selects the sample rate. This setting is the same as the one in the Standard Channel Setup. If it is changed here, it will automatically be changed in the Standard Channel Setup (and vice versa) because AE and streamed waveforms Are always collected at the same sample rate. Channels on the same board must have the same sampling rate (channel 1 = channel 2, 3=4…) but different rates are allowed between boards.

Also, the system cannot transfer more than 10MSPS total. This means 1 channel can be enabled at 10MSPS or 2 channels at 5MSPS... Always try to keep the total MSPS for all enabled channels as low as is possible. Changing the sample rate also changes the time value for pre-trigger and record length.

Pre-Trigger:
This is the amount of data that occurred prior to the actual trigger of the streaming that should be saved to the file. It cannot exceed either 32k samples or 4 seconds whichever is smaller. The pre-trigger amount cannot exceed the record length, because the record length includes the amount of pre-trigger. The value can be entered as either kilo-samples or milliseconds. When one value is entered into its field, the program will automatically calculate the other value. The program always stores the value as the number of kilo-samples, therefore it may not be possible to obtain an exact value in milliseconds.

Record Length:
This is the total amount of data that will be streamed once the stream is triggered. The value can be entered as either kilo-samples or milliseconds. When one value is entered into its field, the program will automatically calculate the other value. The program always stores the value as the number of kilo-samples, therefore it may not be possible to obtain an exact value in milliseconds.

Waveform Streaming Trigger Modes

Each time a valid trigger is detected, a new file is opened for streaming. Any trigger condition that is met while a stream is in process is considered to be invalid and is ignored.

External Trigger:
In this mode the streaming is triggered when an external pulse is applied to Pin 12 of J6 (digital input #2) of the PCI-2 board that contains channels 1 and 2. The pulse should be TTL level, normally low. It should go to a high state for at least 1 microsecond before returning to low. Theoretically it is possible to trigger this input with a pulse that is only 50 nanoseconds wide. Precautions should be taken to make sure that the input signal is clean or false triggers could occur. The signal must return to a low state for at least 10 microseconds before returning high to trigger another stream. If a stream is in process when the trigger input goes from low to high, the transition will be ignored and will not affect the current operation. The input must go from a low state to a high state while streaming is idle in order to trigger any streaming.

Hit Based Trigger:
In this mode the streaming is triggered whenever an AE hit is detected on any of the channels that are enabled for streaming. Additional hits that occur while streaming is in process do not affect the triggering.

Time Based Triggering:
In this mode streaming can be set to occur on a regular periodic basis. Make sure that the record length is less than the specified interval. Otherwise, for example, if the stream is to be 1.5 seconds long and the interval is set to every 1 second, the interval from 1.5 to 2.0 seconds will not be stored to a file. The first interval begins when acquisition is started.

Manual Trigger:
There are two types of manual triggers. Either will start when either the F11 key or the 'Trigger a Waveform Stream' acquisition toolbar button is pressed.
In the first case the stream will terminate when the number of samples specified in the record length field is reached. In the second case either the F11 key or the 'Trigger a Waveform Stream' acquisition toolbar button needs to be pressed to terminate the stream.

General Waveform Streaming Settings

Enable Streaming:
This must be checked in order for streaming to be available during acquisition. This is a quick way to enable and disable the feature while still preserving all of the settings.

Output Filename Prefix:
This specifies the first part of the name of the output data file. It can be changed to any valid Windows filename characters. It is recommended that this be used to differentiate the data from different tests or experiments. The second part consists of the date and time that the stream was triggered.

Filename example: **`STREAM20040804-090744.WFS`**
`PREFIXYYYYMMDD-HHMMSS.WFS`

Here, STREAM is the default prefix. Following the prefix is the time and date of the stream. Four digits of year (Y), two digits of month (M), two digits of day (D), a hyphen, two digits hour (H, 0-23), two digits of minutes (M) and two digits of seconds (S). In the case of the example, the stream was started on Aug 4, 2004 at 9:07:44am.

Suggested Procedure for Configuring Waveform Streaming

- Select the channels for streaming from the first column by clicking on the check box.
- Select the sampling rate for each channel. Channels located on the same AE acquisition board must have the same sample rate. The program will enforce this.
- Select the pre-trigger. This can be a specified in milliseconds or K samples (1K samples = 1024 samples). It is easier to specify milliseconds, but if the sampling rate is subsequently changed, the milliseconds value will be recalculated and displayed.
- Select the recording length in milliseconds or K samples (1K samples = 1024 samples). There is no set limit on recording length other then the hard disk available. But the amount of physical memory on the computer does affect the ability to view the data using the WaveViewer program. For file lengths greater then 1.5Gbytes at least 256 Mbytes should be installed. As with pre-trigger, subsequent changes to the sampling rate will force the millisecond value to be recalculated and displayed.
- Select the Triggering Mode
- Specify the Output Filename Prefix
- Click the Enable Streaming Check Box. This box must be checked for streaming to be enabled.
- Press the 'Apply' button to verify the settings, fix any errors that show up.
- Go back to the Standards Channel Setup and verify that the waveform settings are acceptable.

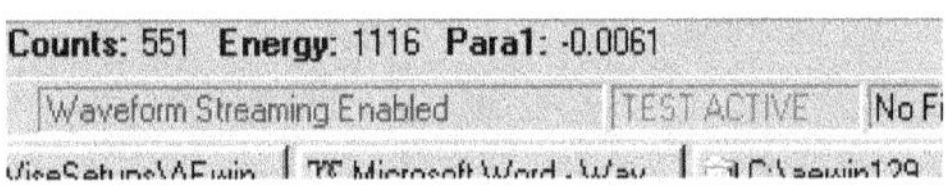

Figure 22. – Status bar indicating that streaming is enabled

After configuring all options the system can be put into acquire mode to begin data collection. If waveform streaming is enabled, the status bar at the bottom of the HMSGHAT2000DDI window will display "Waveform Streaming Enabled".
When a waveform save is triggered the text in the status bar will indicate it.

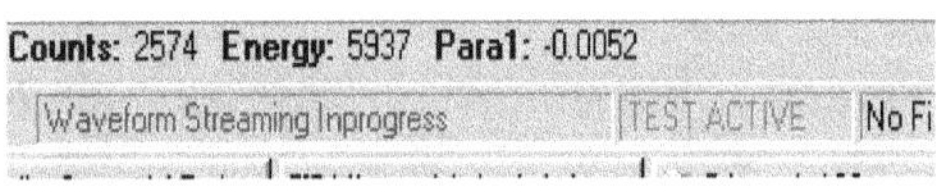

Figure 23. – Status bar indicating that streaming is in progress

When the waveform save is complete the text in the status bar indicate it.

This message will persist for 4 seconds before reverting to "Waveform Streaming Enabled" text.

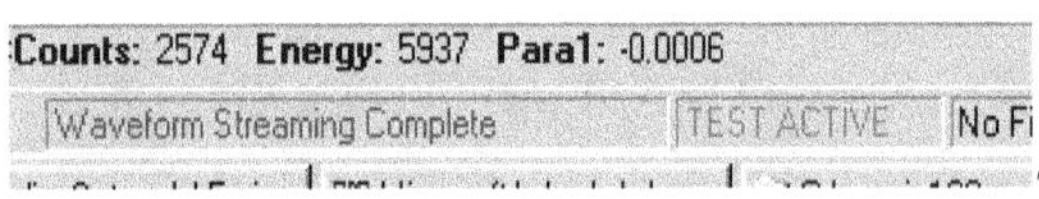

Figure 24. – Status bar indicating that streaming is complete

The file is placed in the last folder that was used to save a DTA file. If an acquisition is done and Autodump is enabled (producing a DTA file), the streaming data file will be created in the same folder. If Autodump is off, the streaming data will be placed in the last folder that was used for AE data. The streaming file is always created and there is no way to disable it. The destination folder for the streamed data cannot be specified without also saving AE data.

Viewing a Streamed File

Once a WFS file is created it can be viewed using the WaveViewer software kage. This is included on the PCI-2 installation CD. Please see the WaveViewer User's Guide for complete instructions on using the software and viewing waveforms.

Waveform Streaming Performance and Trouble Shooting

The rate at which waveforms can be saved to a data file is a function of many different variables. The card is capable of streaming up to 20 Mbytes per second. This would correspond to 10M samples per second to memory. Unfortunately the path from memory to the data file is dependent on hard disk performance and CPU utilization, so it is not possible for us to say what the highest achievable rate will be. The DMA controller requires physical non-paged memory in order to operate. If the sampling rate of the PCI-2 card is set to 100Ksamples per second and the duration of the waveform is greater than 4 seconds then the amount of memory required is (100Ksamples)*(2 bytes)*(4 seconds)*(2 channels) = 1600K. If the sampling rate of the card is set to 5M samples per second, then the physical non-paged memory requirement increases to (5Msamples)*(2 bytes)*(4 seconds)*(2 channels) = 80,000,000 bytes. So there must be enough physical memory to map the buffer. The normal AE hit and waveform processing of the card also required physical memory as well. If HMSGHAT2000DDI locks too much physical memory, then overall OS performance will decrease. Because of these difficulties HMSGHAT2000DDI will not allocate more than 20% of physical memory for its use. If there is insufficient physical memory available on your system to allocate the waveform DMA buffers the following error message will be displayed during waveform streaming setup:

Insufficient Physical Memory, Reduce Recording Length

If this happens, you can reduce the sampling rate or the waveform length until there is enough physical memory for the buffers. Another option is to increase the physical memory installed in the system.

Figure 25. – *Buffer overrun error*

For long duration waveform saves the possibility exists that the CPU will not be able to write the waveform data to the output file at an equal or greater rate than the waveform is being read from the PCI-2 card. If this happens then the buffer overrun error message will appear.

There are several solutions to this problem. First is to reduce the CPU utilization by turning off non-essential items. Disable the line listing display. Reduce the number of plots. Disable AE hit waveforms. Don't save AE data to disk. Other options include reducing the sampling rate of the channels or the number of channels being written to the data file. Upgrade you hard disk or add more physical memory.

Channel Group Setup… (for Non-HMSGHAT2000DDI Products)

The 'channel group setup…' dialog box is used to define and edit channel groups. A 'channel group' is simply a named list of channels. You can select this group in graph setup instead of selecting channels individually. The names of any channel groups selected for a graph are displayed on the graph's title for easy identification.

To create a group, click the Add Group button. This brings up a dialog box shown at right. From here you can name the new group and select its member channels.

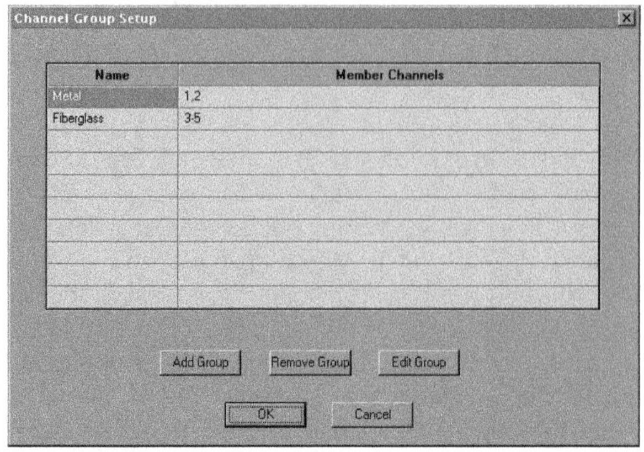

Figure 26. Channel Group Setup

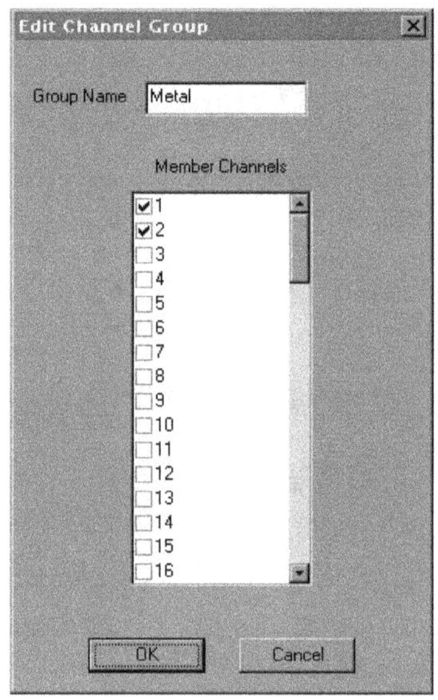

To modify an existing group, highlight it and click the Edit Group button. This brings up the same dialog box. To delete an existing group, highlight it and click the Remove Group button. Note that when you remove a channel group it is automatically removed from any graphs it was selected in.

Figure 27. Edit Group

Location…

The Location Setup dialog box is used to define and enable event groups (location groups). Use of Location involves more than just this dialog, however. For a full description of the Location dialog as well as other Location details see section 4.

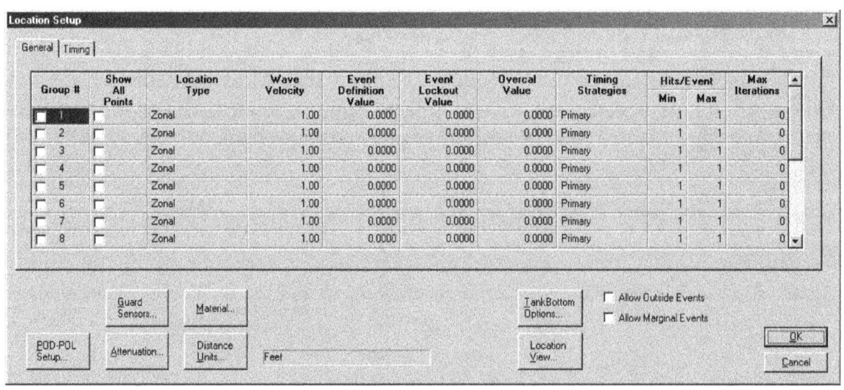

Figure 28. Location Setup

Title…

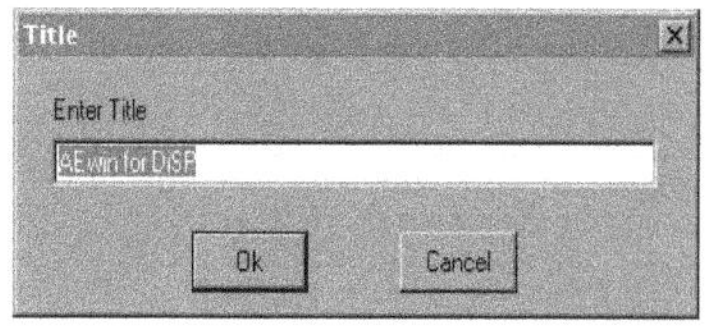

Figure 29. Title

'Title…' is used to set the Test Title. It can be displayed on the statistics bar using the Statistics menu described below.

Statistics…

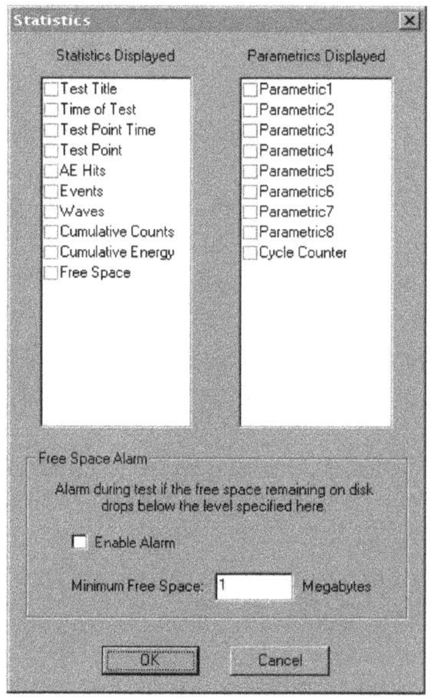

Figure 30. Statistics

'Statistics…' is used to toggle the various test statistics on or off, as well as to customize the Free Se Alarm. Any number of statistics can be on at any one time.

- Test Title: Displays the test title set in the 'Title' dialog box.
- Time of Test: Displays the elapsed time since the start of the test or replay.
- Test Point Time: Displays the elapsed time since the last test point (user-requested time mark).
- AE Hits: Displays the total number of AE hits recorded so far during the current test or replay.
- Events: Displays the total number of located events recorded so far during the current test or replay.
- Waves: Displays the total number of waveforms recorded so far during the current test or replay.
- Test Point: Displays the total number of test points (user-requested time marks) recorded so far during the current test or replay.
- Cumulative Counts: Displays the total number of counts recorded so far during the current test or replay.
- Cumulative Energy: Displays the total number of energy counts recorded so far during the current test or replay.
- Free Se: Displays the current amount of free se left on the disk on which the data file you are replaying or acquiring.
- Parametrics: Displays the value of the specified parametric from the most recent hit or time data message recorded or replayed during the current test or replay.
- Cycle Counter: Displays the value of the cycle counter from the most recent hit or time data message recorded or replayed during the current test or replay.

Free Se Alarm

The 'Free Se Alarm' is designed to warn the user in the event that the amount of free disk se remaining has run low during a test. To enable this alarm, check the 'Enable Alarm' checkbox and set a minimum threshold (in Megabytes), below which the alarm will sound.

Logistics Information…

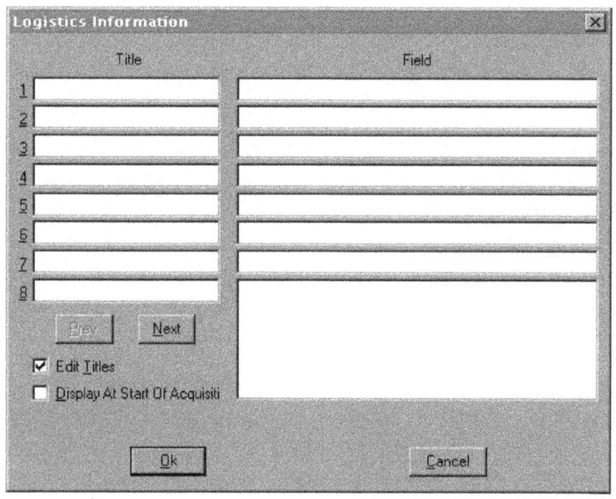

Figure 31. Logistics Information

The 'Logistics Information' dialog box is used to store information about a test. The Title column contains eight editable fields to be used as descriptions of the information that is to be entered by the tester into the field to the right of the title field (Field column). Examples of useful titles would be date of test, location of test, tester name(s), item being tested, identification or serial number, weather condition, etc.

The Field column also contains eight editable fields that may be filled in now, or at test time, with information about the test to be performed. This information can then be preserved in the Layout and Data files.
In addition to these two columns of fields, there are several check boxes and buttons located at the bottom of the screen:

Edit Titles - When this item is checked, the information in the Title column fields can be accessed and edited. When this item is not checked, the information in the Title column fields will be for display purposes only, therefore, all text in the Title column will be grayed out and inaccessible.

Display at Start of Test - Checking this box will cause the logistics information dialog box to automatically be opened after initiating a new test. When this convenient option is checked, you will not have to remember to update the logistics information each time a test is begun, because the program will automatically prompt you for logistics information each time a new test is begun.

Audio Monitor...

The audio monitor board contains the (audible) alarm and audio controls. The circuitry is designed to produce audible tones from high frequency acoustic emission signals. This circuit can be selected to operate in either a homodyne or heterodyne mode. Operation in heterodyne mode allows the user to select the frequency of the AE signal to be monitored. Homodyne operation does not require frequency tuning and allows audio monitoring of AE signals at any frequency.

To activate the dialog box for setup of the Audio Monitor board select 'Acquisition Setup' then 'Audio Monitor...'.

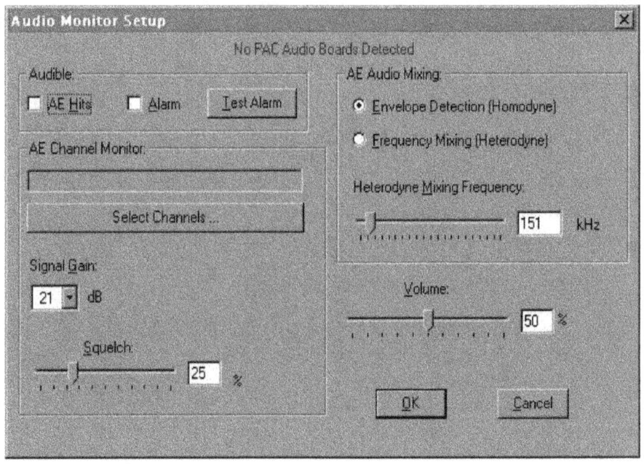

Figure 32. Audio Monitor Setup

In the group titled 'Audible' there are two checkboxes. The first one (AE) enables the board to emit a sound when AE hits are detected. The second one (Alarm) enables the board to produce a tone whenever a front end alarm warning or trip level is exceeded (these alarms are defined in the 'Acquisition Setup', 'Alarms' dialog box). By clicking on the 'Test Alarm' button you can hear how the alarm will sound. Click it a second time turn it back off.

The group titled 'AE Channel Monitor' contains controls for selecting monitored channels as well as setting gain and squelch. Click 'Select Channels' to choose which channels are to be used by the audio monitor. The 'Signal Gain' combo box contains a list of valid gain levels for the AE channels. This allows small amplitude AE signals to be monitored. 'Squelch' acts like a threshold circuit to eliminate the sound of background noise. Valid entries are from 0 to 100%. This value should be set to a level so that under conditions of no AE signals there is no audio output. Then as soon as a signal received is above the threshold as set by this control the audio output for that signal will be heard. The scroll bar located under the squelch control can be used as a mouse interface. Clicking on either arrow changes the squelch by 1%, clicking in the area between the thumb and an arrow changes it by 10%.

The group titled 'AE Audio Mixing' contains mutually exclusive checkboxes for selecting the method by which the AE tones are generated. When 'Envelope Detection (Homodyne)' mode is selected each hit produces a click sound. When 'Frequency Mixing (Heterodyne)' mode is selected each hit is mixed with a signal whose frequency is entered in the edit box below the mode selection controls. This limits the audio to hits whose frequency is 10kHz above the selected frequency. For example if the mixing frequency was set to 150kHz then AE signals greater than 150kHz and less than 160kHz will be heard.

The last control is for volume. It controls the volume of both the AE tones and the alarm tone. It ranges from 0 (off) to 100% (full volume).

The changes made to the dialog box will go into effect immediately. Pressing Cancel will restore the audio settings to those in effect before the dialog box was displayed.

Note that the AE detection by the audio board is independent of the threshold levels of each channel so it is possible to set the board up to hear AE which is below the threshold of hits detected during a test.

The top line of the dialog box will display Audio Board Version X.X if a audio board is detected. If one is not, it will show No Audio Boards Detected.

Auto File-Close…

In most tests all data from test-start to test-end is saved to a single data file. In long tests it is sometimes desirable to instead split the data into multiple files . When the Auto File-Close feature is active during acquisition the program can automatically stop the test and start a new test under user-defined conditions. The default result is a numbered set of data files. Optionally it can close the file and begin saving data to another file without stopping the test. The result in this case is a numbered set of 'continued' data files which can be replayed together as if they were one file.

The program will name new files by incrementing the name of the first file. (i.e.: if the name entered for the first file is test0001.dta, then subsequent files will be named test0002.dta, test0003.dta, etc.)

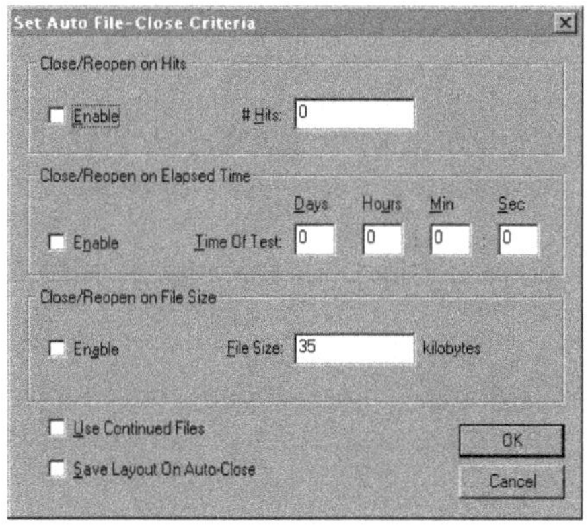

Figure 33. Auto File-Close

If checked the program will also re-save the current layout every time a file is automatically closed. This saves the name of the new file in the layout as the default data file name for future tests.

Enabling and Customizing Auto File-Close

You can enable the Auto File-Close feature and set the conditions under which it will work in the 'Set Auto File-Close Criteria' dialog box seen in the figure above. There are three conditions under which an Auto File-Close operation will take place:

a) <u>#Hits:</u> Whenever the total number of hits in the test equals the #Hits value or a multiple of it, an Auto File-Close operation is performed. (i.e.: If #Hits is 1000 then an Auto File-Close will happen at 1000 hits, 2000 hits, etc.)
b) <u>Time of Test:</u> Whenever the elapsed time equals the 'Time of Test' or a multiple of it, an Auto File-Close operation is performed. (i.e.: If 'Time of Test' is 1hr 10 min then an Auto File-Close will happen every 70 minutes.)
c) <u>File Size:</u> Whenever the current file size equals 'File Size' then an Auto File-Close operation is performed. (i.e.: If 'file size' is 64kb then you will end up with a string of files, each one approximately 64,000 bytes in size.). Note that file sizes will vary especially at high data rates.

You can enable any or all of these conditions.

Using Continued Data Files

If the 'Use Continued Files' box was checked the program will close the file in 'continued' form and open a new file, without stopping the test. The resulting set of files are referred to as a continued file set. They differ from normal data files as follows:

- Every file in the file set (except the last one) has an internal link to the next one. This means that when you replay one of these files it will look for the next file(s) and replay them also. (It can only find them if they are in the same folder.)

Replaying Continued Data Files

a) To replay the entire file set in order: Replay the 'first' file (the one with the name you gave it when you started the test) in the file set. The entire file set (if present in the same folder) will replay as if it were one file.
b) To replay the file set from a certain file onward: Replay the file from which you want to begin replay. The file set from that point on will replay as if it were one file.
c) To replay a single file only: Since the files are linked the program will always try to find and replay all files that come after the selected one in order. To replay only one file you must first copy it to another folder (or move/delete the others) then replay it as you would any other. Eventually it will reach the end of the file and

will warn you that it can't find the next file. If you want to eliminate this message from future replays of this file you must remove its link using the Time Order/Repair utility (see the next section for details).

Note that if the program encounters a 'gap' in the file set (i.e.: a missing file) it will display the "File Not Found" error described above. It can't replay beyond this point because the link is, in effect, broken. You can pick up where it left off by starting a new replay of the next existing file in the set.

Removing the Link From a Continued Data File

To remove the link from a continued data file (other than the last file which has no link) run the Time Order/Repair utility on it. This will allow you to replay this file without the "File Not Found" warning being displayed. Doing this will prevent any files 'after' this one from being replayed as part of the complete file set, so be sure to make a copy first. Time Order/Repair will ask you if you want to replace the continue message with an abort message. Click 'Yes' in response.

Auto Alarm Reset

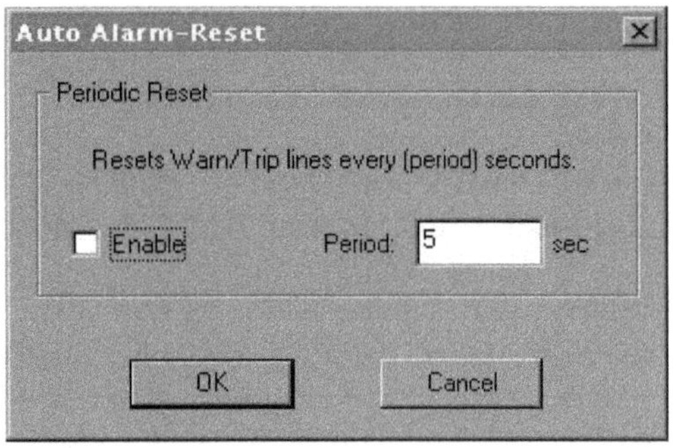

Figure 34. Auto Alarm-Reset

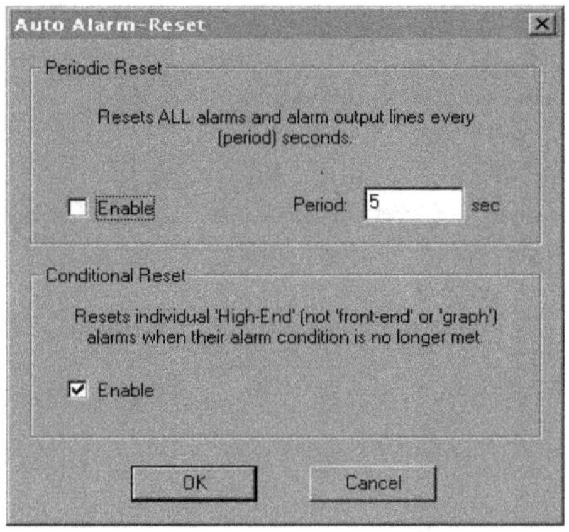

Figure 35. Auto Alarm-Reset

The Auto Alarm Reset dialog box is used to automatically reset alarms based on a user set criteria. This dialog box can be accessed through either the Acquisition Setup menu and the Alarm Status/Test dialog box. There are two types: Periodic and Conditional.

<u>Periodic:</u> If the Periodic Reset feature is enabled it will (during a test) reset ALL active alarm Warn/Trip lines every time a period of time elapses equal to that set in this dialog.

<u>Conditional:</u> The Conditional Reset feature is used in conjunction with 'High End' alarms. High-End alarms include the 'Free Se' alarm (set in the Statistics Setup dialog box), LeakTEC and On-Line Crack alarms (if this option is enabled). If this feature is enabled, a high-end alarm is automatically reset when the condition that caused its triggering no longer exists. i.e. : If the 'Free Se' alarm is triggered because the amount of free se on the disk dropped below 5 Megabytes and the user then deleted enough files to raise the free se remaining above 5 Megabytes the alarm would automatically reset.

Alarm Status/Test

The 'High End Alarm Status/Test' dialog box is used to view and alter the status of various alarms in HMSGHAT2000DDI. A table of alarm types is displayed either on the left or the top of the dialog box. Each cell represents one type of alarm (such as Free Se Low), or condition (such as 'Not in Acquisition'). If a cell is checked it means that that alarm is triggered or that condition is true. You can reset an alarm simply by unchecking it. Note that doing so will not prevent that alarm from triggering again in the future (or automatically clearing itself if Alarm-Reset is enabled).

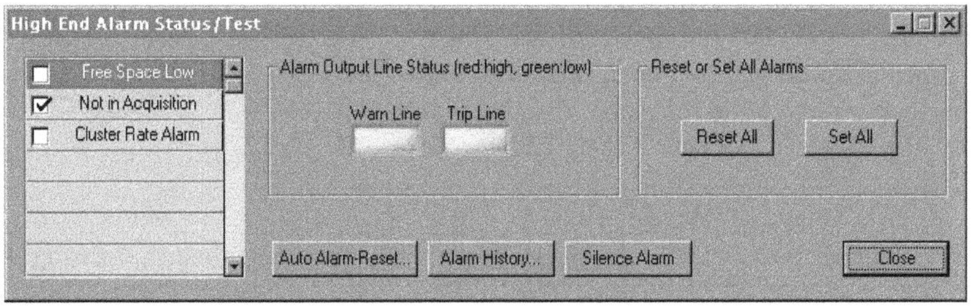

Figure 36. High End Alarm Status/Test

You can reset or set ALL possible alarms by clicking the appropriate button on the right side of the dialog box. In the center of the dialog box are colored indicators that display the status of the Alarm Output Lines (warn and trip). They are colored green if they are reset (low) or red if they are set (high). You can silence the audible 'beeping' sound without resetting the alarm(s) by clicking 'Silence Alarm'.

ASL Alarm...

ASL Alarms are used in conjunction with Internet Monitoring. Unlike Front-End or Graphical Alarms, ASL Alarms do not sound an audible alert or display a dialog box when triggered. Instead they save their alarm data in the STA files.

You can set alarm criteria on a channel by channel basis. Any channels not checked are not evaluated for alarms. The Low Level and High Level you set define the acceptable (inclusive) non-alarm range.

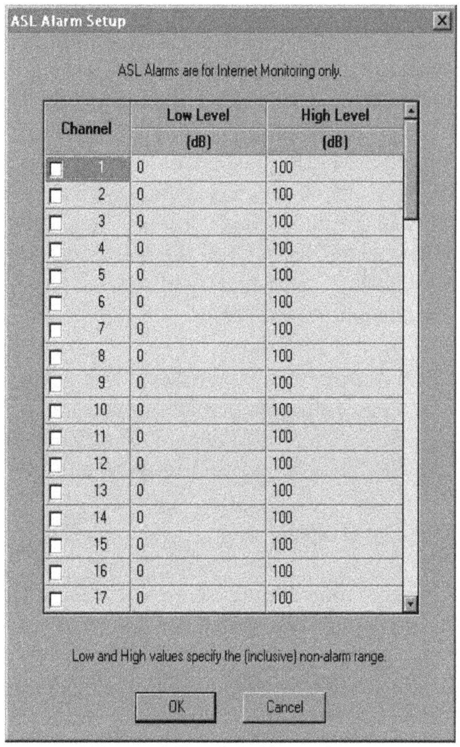

Figure 37. ASL Alarm Setup

Voltage Time Gate Setup...

Voltage Time gate is used for parametric based AE data filtering. It is controlled by parametric 1 and allows the user to eliminate the collection of all AE data that occurs when the parametric falls outside a pre-selected range of values and other criteria.

AE data collection is either Enabled or Disabled. It specifies what to do when the voltage-time gate is generated.

Each threshold is specified with a level in Volts and >= or <=, which indicates whether the signal has to be above or below the level in order to satisfy the threshold. Activation of each threshold is independent of the other; the user can activate either, both or no threshold at all.

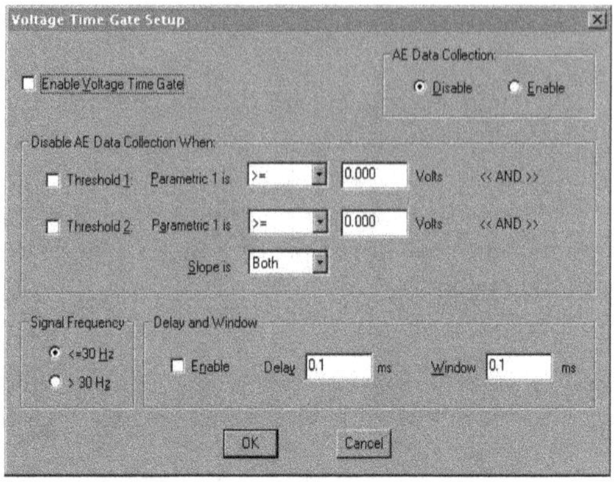

Figure 38. Voltage Time Gate Setup

The slope control can be POSITIVE, NEGATIVE or BOTH. It is satisfied if the parametric is rising, falling or at any time.

The slope and all active thresholds must be satisfied in order to generate a voltage-time gate. This allows the user to restrict the gate to crossings of thresholds with the parametric specifically changing a certain way (rising or falling).

Delay and Window specify an optional time window (the 'time' part of voltage-time gate). When enabled they cause the gate to be a time window of controlled duration, starting with some delay after the moment that the threshold and slope criteria are met.

Signal Frequency specifies a low-pass or high-pass filter that is applied before the parametric is fed into the voltage-time gate processing circuit. This provides the ability to respond to signals that change slowly (most cases) or quickly (fatigue cycles, etc). There are two settings, one for parametric signals below 30 Hz and one for above 30 Hz.

Internet Monitoring Setup...

This dialog box allows you to setup the Internet Monitoring feature of HMSGHAT2000DDI. This is used in conjunction with other 'Remote Monitoring' applications purchased separately. Please refer to their documentation for details.

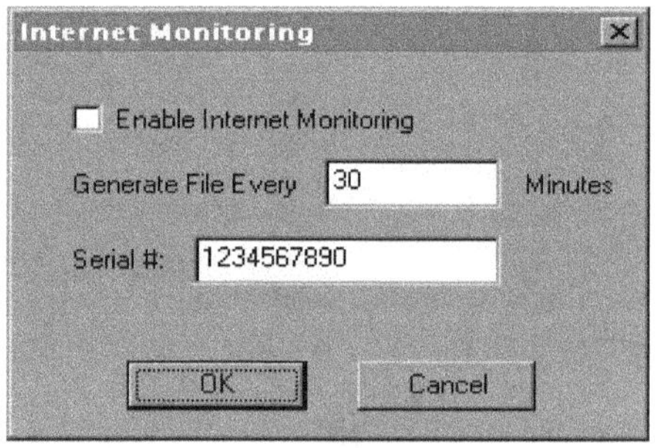

Figure 39. Internet Monitoring

From this dialog box you can:
- Enable/disable internet monitoring.
- Set the time interval between file generations.
- Set the serial number for your unit.

Display Mode...

Display mode allows you to select the form of display for the AE data, as well as the frequency of graph updates, via the Acquisition/Replay Modes dialog box.

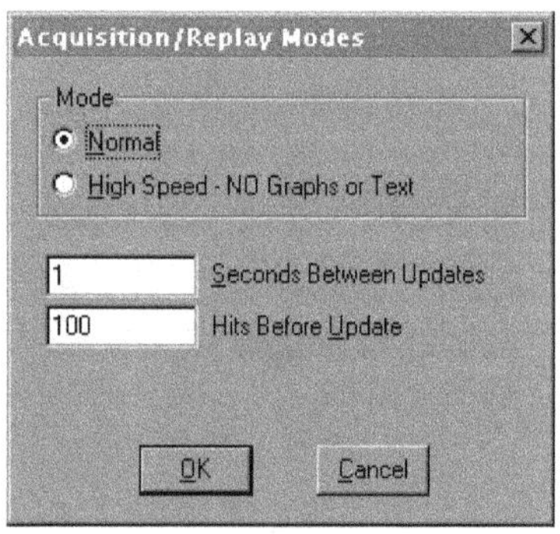

Figure 40. Display Mode

There are two Acquisition/Replay modes available in HMSGHAT2000DDI:

- Normal: This is the standard mode for acquisition and replay. Statistics, graphs, tables and the line display are all active in this mode.
- High Speed: This mode is used to achieve maximum data acquisition speed in tests where a very high emission rate is expected. Graphs, tables

and the line display are inactive in this mode. Only selected statistics will be displayed and updated; the priority task is transferring data to disk for analysis later.

Seconds Between Updates, Hits Before Updates

During data acquisition and replay, graphs and tables are redrawn (updated) at intervals determined by these settings. They are updated whenever the specified time has elapsed or the specified number of hits has occurred since the last update, whichever comes first. Depending on the anticipated data rate and test duration, you will want to select an update rate that is fast enough to keep you current without having the screen constantly redrawing. Type in the desired values in the text boxes provided. Recommended values for initial use are 3 seconds, 100 hits. Since screen updates affect overall data acquisition rate, in high data rate situations, the user may want to use larger intervals.

Line Display…

The Line Display shows a data listing in numeric format. This listing shows all the AE parameters, the parametric and other test messages as numbers in a row. There is one row for each message, plus a title row every time it is needed. The title row shows the user which parameters are being displayed and where they are being displayed. This depends on which active parameters were selected for display in the Line Display Setup dialog box.

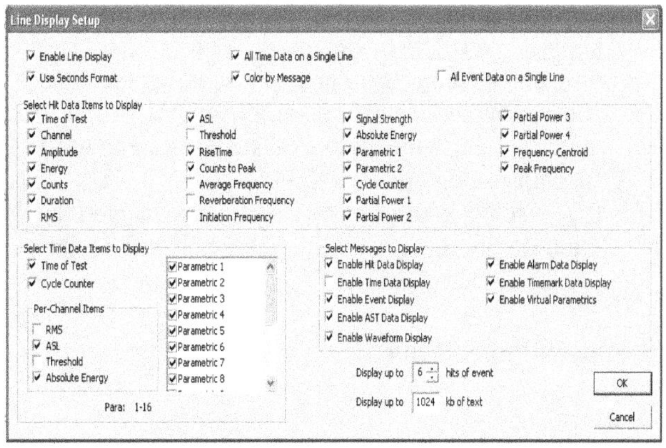

Here is a typical illustration:

*Figure 41. Line **Display** Setup*

ID DDD HH:MM:SS.mmmuuun PARA1 PARA2 CH RISE COUN ENER DURATION AMP

- The ID is the message ID. This will be 1 for a data message, 2 for a time driven message, 172 for a waveform message and 128, 129 or 130 for a test control message. A test control message is a RESUME, PAUSE, or STOP message.
- The TIME column is the time of hit read from the time of test counter. This time is displayed in days, hours, minutes and seconds from the start of data acquisition, with resolution to 250 nanoseconds.
- The PARA1 column is for parametric number one. The listed reading is scaled according to the values entered for Multiplier and Offset in the Parametric Scaling dialog box.
- The PARA2 column is for parametric number 2. Again the reading is scaled according to the values entered in the Parametric Scaling dialog box.
- The CH column is for the channel number if the message was of type 1 or an AE data set.
- The RISE column is for RISE TIME. The range of this variable is from 0 to 65535 μseconds.
- The COUN column is for threshold crossing COUNTS. The range of this variable is from 0 to 65535 counts.
- The ENER column is for ENERGY. The range of this variable is from 0 to 65535 energy counts.
- The DURA column is for signal DURATION. The range of this variable is from 0 to 16,777,762 μseconds.
- The AMP column is for AE AMPLITUDE. The range of this variable is from 0 to 100 decibels.

In the Line Display Setup dialog box you can:

- <u>Enable Line Display:</u> Note that the display can be visible (see section on View menu) but not enabled. Turning the display off can boost performance on very high data rate tests.
- <u>Use Seconds Format:</u> With this option you can display TIME in seconds (total) rather than HH:MM:SS.mmmuuun format.
- <u>All Time Data on a Single Line:</u> With this option Time Data messages (ID 2) are displayed using only a single line (each) rather than several shorter lines.

- All Event Data on a Single Line: With this option Event Data Messages (ID 128) are displayed using only a single line (each) rather than several shorter lines.
- Color By Message: With this option you can have the various messages colored by message ID. This can make the display easier to read, though it is not recommended for very high data rate tests.

The Line Listing Display can be made visible by pressing F7 or through the View menu.

Preferences...

The Preferences dialog box is used to modify settings that are not specific to any layout. These settings are remembered from session to session so you generally only need to set them once. They are:

- Only Show Installed Channels: With this item enabled, Hardware Setup will only display entries for channels that represent hardware installed on the unit. If it is disabled then Hardware Setup will display entries for up to 80 channels, including those that do not exist. This can be useful if you want to create layout files for use on other machines.
- Pre-enable Autodump: With this item enabled the 'Auto-Dump' checkbox in the Data Storage (Acquire) dialog box will be checked by default, otherwise it is unchecked by default.

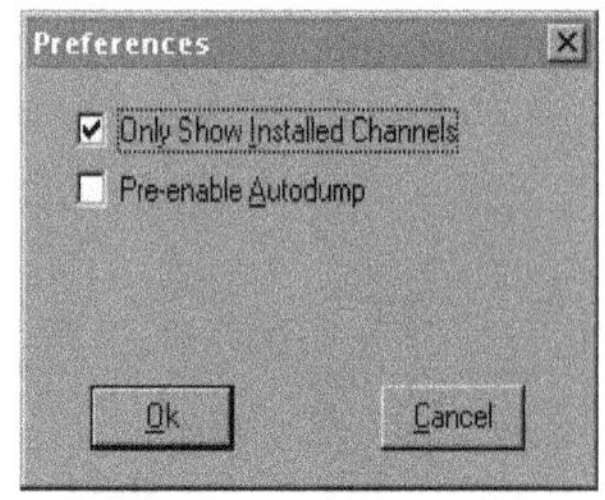

Figure 42. Preferences

Acquire/Replay Menu

Acquire...

An AE test is begun via the Acquire... menu item or by pressing F9. This will display a dialog box prompting you for information before the test begins.

This dialog box is very similar to the Save Layout dialog box discussed earlier, in which you are prompted for a filename, in this case to store AE data. The exception here are the checkboxes in the upper left corner:

- Auto-Dump: If this box is checked, it means that Autodump is ON and the test data will be recorded to the filename you specify in this dialog box. If you want a permanent record of your data, make sure that Auto-Dump is ON. If it is OFF, no data will be written to disk and your record will be limited to the real-time graphic displays.
- Enable Event Linking and Cluster/Area History: If this box is checked then support for these features will be enabled in acquisition. If it is not checked then event linking and cluster/area history will not function for the duration of the test, but the system will exhibit better performance at high data rates. If you expect high data rates then be sure to uncheck this and save the data to a file (Auto-Dump on). You can then use Event Linking and Cluster/Area History during a replay if needed.

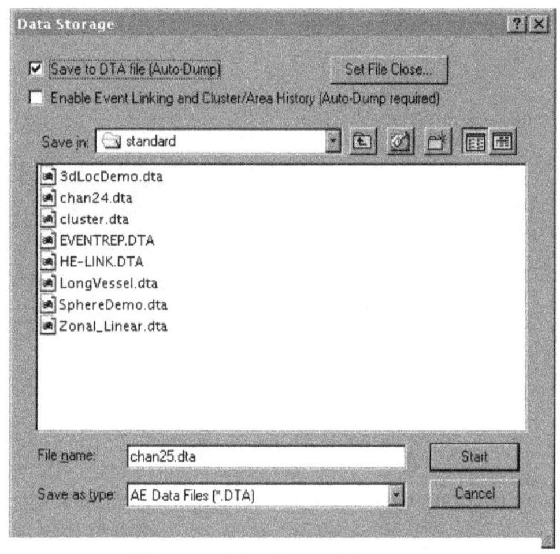

Figure 43. Data Storage

If Auto-Dump is ON, the program will save data to the filename specified. The program will then automatically increment the last letter or digit of the filename specified in the Test Storage dialog box each time you enter acquisition with AUTODUMP ON without specifying a different data filename.

The data file is automatically closed and reopened every second. This assures that the worst case data loss is one seconds worth if power fails. A manual update to disk can be accomplished by pressing pausing/resuming the test.

Replay…

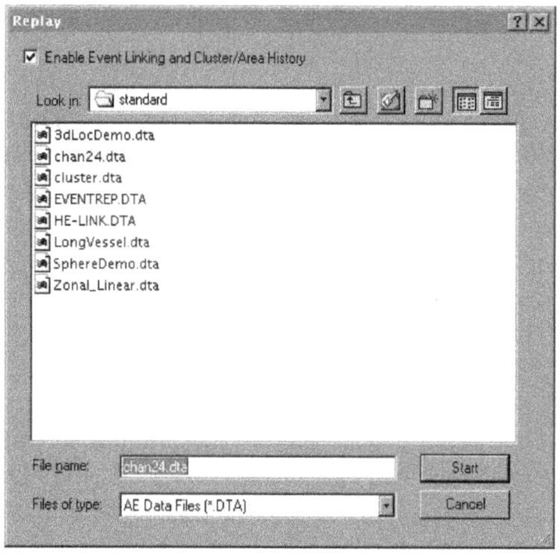

Figure 44. Replay

This box is essentially the same as the Acquire dialog box discussed in the last section. Because Autodump is not needed here, it is not an option in this dialog box.

Once the filename has been selected, the controls are essentially the same as for Data Acquisition. During the replay, the system reads data from the disk, updating the graphs/tables at the rate specified in the Display Mode dialog box. The replay automatically ends when it finishes replaying the requested data file. If the file specified data file is a 'Continued file' (described in sections 3.3.2.8.2 and 3.3.2.8.3) then it replays the next file in its continued file set.

Resume/Pause/Stop/Abort

These four commands are used to control the test or replay.

- <u>Pause:</u> Click Pause to temporarily halt a test or replay.
- <u>Resume:</u> Click Resume to continue a paused test or replay.
- <u>Stop:</u> Click Stop to terminate a paused test or replay. This is the proper way to end a test. Any buffered data is saved (if auto-dump is on).
- <u>Abort:</u> Click Abort to terminate a non-paused test. This is a quick way to exit a test and can be used even if the hardware is not responding. It is recommended for use only in emergencies or when data isn't being saved to disk as any buffered data is lost.

Time Mark

Click this command to add a Time Mark (a.k.a. 'Test Point') to the data during test. This message represents a point in time of significance to the tester (i.e.: beginning of loading cycle, important observation, etc.). It is saved with the data for future replays and is displayed on all time-based graphs (graphs that have 'time' as their x-axis parameter) as vertical bars. It is shown on the Line Listing Display as an ID 211 along with its time of test.

The statistic 'Test Point' displays how many Time Marks (Test Points) have been added to the data.

Turn Waveforms ON/OFF

Click this during data acquisition to toggle waveform transmission on or off. These commands are only available if you have the Waveform option installed.

Display Last Alarm…

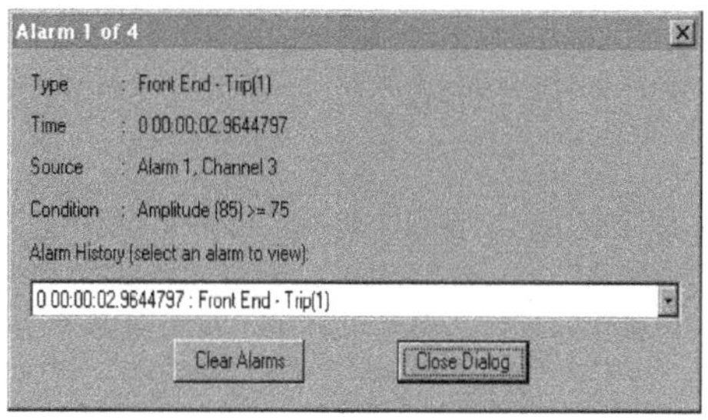

Figure 45. Display Last Alarm

Click this to view the Alarm dialog box. This displays information on all alarms detected during the current test (or previous test if no longer in test). Use the drop down list box in the lower left corner to select which alarm to view. You can Clear (all) the alarms by clicking the 'Clear Alarms' button. This resets any alarm output and turns off the audible alarm (if on).

This command is available only if there has been alarm activity during the current test (or the last one if ended).

AST

The purpose of the Auto Sensor Test is to verify that nothing has gone awry with the sensors before or after a test. The Auto Sensor test will run a short acquisition test, while pulsing through to each sensor sequentially, using the current settings in the Test and Hardware Setup menus. An AST test should be run before and after an AE test and results compared to make sure there were no sensor sensitivity changes. Also, in order for AST to work, 1220A-AST preamps or Integral Sensors with the AST option (e.g., R15I-AST) are required. The AST Dialog Box is used to setup and begin the AST.

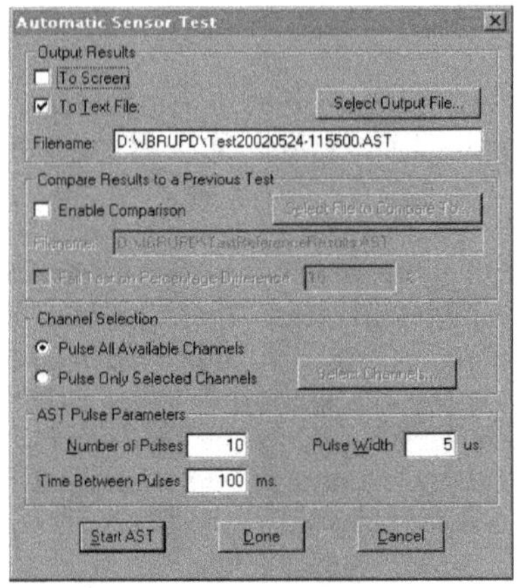

Figure 46. AST

The results of the AST will be displayed on the screen, and/or saved to a file. This is done by selecting the "output Results to Screen" and/or the "Output Results to text file" check boxes. If the "Output Results to text file" box is chosen, you may accept the default filename for the text output file or select a new output filename, either by typing the filename in the Filename text box, or by selecting to overwrite an already existing filename through the Select Output File button.

The results of the AST will be compared to a previous baseline test run if the "Enable Comparison" box is checked. A filename must be supplied either by typing a filename in the Filename text box provided, or by selecting a filename through the Select File button.

The AST pulse is generated automatically by the system following the pulse characteristics shown in the bottom of the dialog box. These characteristics may be changed by typing the appropriate values in the text boxes provided. The characteristics and their legal ranges are:

Number of Pulses: This is the number of times a single sensor will be pulsed. Each active sensor set in the Channel Settings dialog box will be pulsed this number of times. This number must be in the range of 10 to 250.

Pulse Width: This time, specified in μsec, determines the length of the pulse. It must be in the range of 1 to 20.

Time Between Pulses: This time, specified in msec, determines how far apart the pulses are.

Selecting Done will save AST settings, and exit this dialog box without starting the test.

Selecting Start AST will start the Auto Sensor Test. If "Enable Comparison" was chosen, the files supplied in the filename box will be read and used to calculate a percent change for delta-T, amplitude, and energy for all active sensors defined in the Channels Settings of the Hardware Setup Menu.

The results will be displayed, on the screen, written to a file, or both, depending on the options selected in the AST Dialog Box. On screen will be displayed the test setup parameters (Pulses Output, Pulse Width, Pulse Interval) and the actual results one line per receiver channel, one page per pulser channel.

The report file will contain this information plus the date and time of the AST and the baseline report file used for comparison, if one was specified.

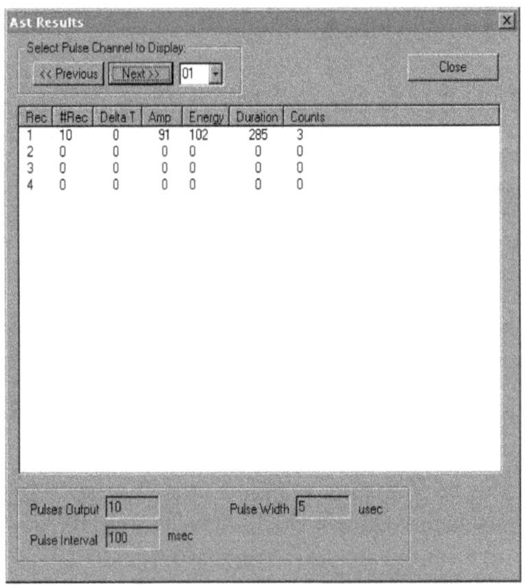

Figure 47. AST Results

Pause Replay On Timemark/Pause

These menu items are used to automatically pause replay whenever a timemark or pause message is replayed. You can toggle these items on or off individually. They allow you to quickly jump to various pre-set points of interest in your data (time marks and pauses set during acquisition). You can resume replay in the normal fashion by pressing F9 or clicking Resume.

Graphing Menu

Graph Setup...

The 'Graph Setup...' dialog box is used to customize an existing graph (created using the 'New Graph' or 'Copy/Paste Graph' commands). You can enter this dialog box by selecting (right clicking) a graph and selecting 'Graph Setup...' from the Graphing menu or by right-clicking the graph and selecting 'Graph Setup...' from the menu that pops up. The appearance of this dialog box will vary depending on the type of graph. The example at right is representative of a simple activity graph.

Section 3.7 details the procedure for creating and customizing graphs. Below is a brief description of the various graph types available and the major pages of the graph setup dialog box.

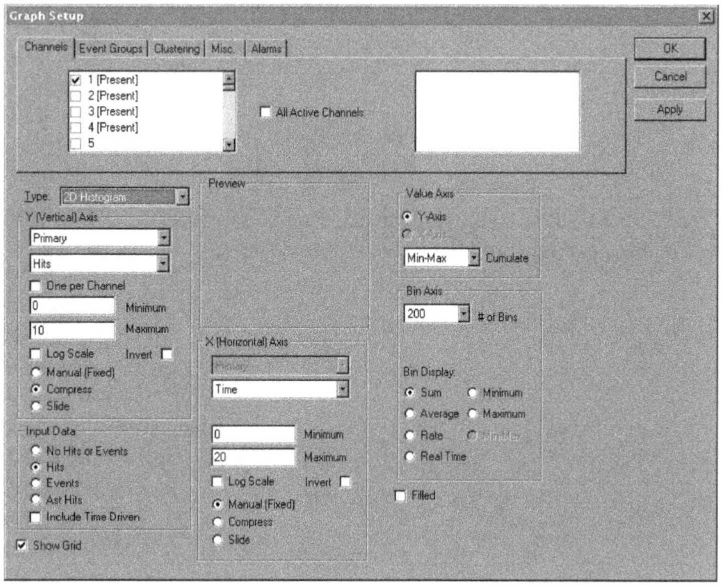

Figure 48. Graph Setup

Graph Types:

- 2D Histogram – This is the standard Bar (if 'Filled') or Line (if not 'Filled') graph. You can define multiple plots or use the 'One per Channel' checkbox to automatically create one plot for every enabled channel (max 6). The X-Axis is known as the 'Bin' Axis and is divided into a user-defined number of 'bins'.
- 2D Scatter (Point Plot) – This is the basic point plot graph. Multiple plots can be defined.
- 2D Binned Scatter – This variant of '2D Scatter' uses a different strategy for data display. It divides the graph area into a grid measuring 200 bins high by 200 bins wide. Thus, every dot represents not a discrete point but actually a small range of values. The advantage of this over '2D Scatter' graphs is that it can display large amounts of data quickly. 2D Scatter graphs can slow down when burdened by more than 100,000 points or so, and will eventually run out of memory around 650,000 points (depending on the setup). This can be a problem during extremely long tests and replays. 2D Binned Scatter graphs do not suffer from this 'data overload'. The disadvantage is a slight loss of resolution. Due to the binned structure there are only 200 possible values across an axis for a total of 40,000 possible distinct points. To calculate the resolution of a graph axis divide its full scale by 200.
- 2D Colored Scatter – This variant of '2D Scatter' is nearly identical in operation to '2D Binned Scatter' but uses a single plot which is colored by value. On the right-hand side panel you choose the 'Color Feature' (counts, duration, etc), define the various color level maximums (or choose 'compress') and set the display mode. The color levels define what color a point on the graph is based on its value. For instance assume you set the color feature to be energy and set level 1 (yellow by default) to be 1000 and level 2 (purple) to 2000. If you get a single hit with energy 600 the point displayed will be yellow in color. If you got a single hit with energy 1001 then this point would be purple. Sum, Average, Minimum and Maximum modes display (the color of) the sum, average, minimum or maximum values respectively of all points that occupy the same grid location. 'Count' mode ignores the color feature and simply displays a count of the number of points that occupy the same grid location.

- 3D Histogram – This is the standard 3D bar graph. It operates in much the same way that 2D Histograms do. Both the X and Z axes are 'Bin' axes.
- 3D Scatter (Point Plot) – This is the standard 3D point plot graph. It operates in much the same way that 2D Scatter graphs do except with a depth (Z) axis.
- Waveform – Standard single plot Waveform graph.
- Power Spectrum – Standard single plot Power Spectrum graph.
- Spherical Location – This graph type is used exclusively for the display of spherical events.
- Cylinder Location – This graph type is used exclusively for the display of cylindrical events.

The 'Channels' page (at right) allows you to select which channels and/or channel groups will be active on your graph. Select channels from the left panel and channel groups (if any are defined) from the right panel.

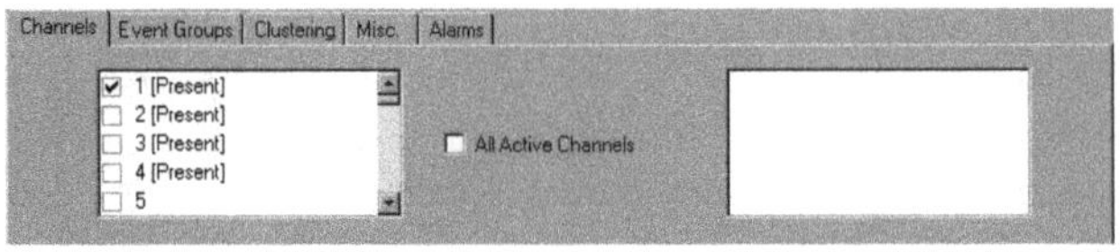

Figure 49. Channels

The 'Event Groups' page (at right) allows you to select which event groups will be active on your graph.

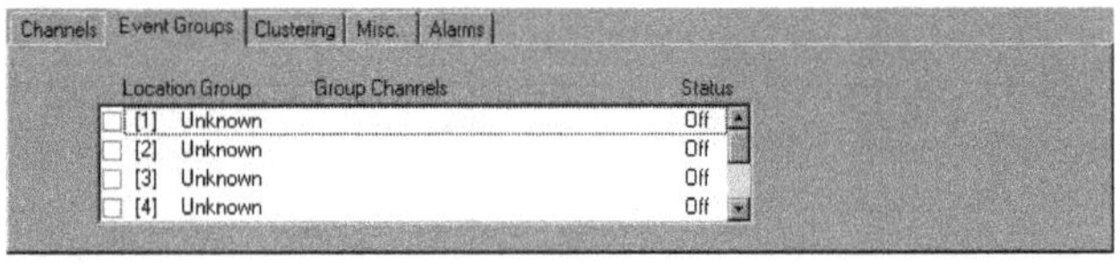

Figure 50. Event Groups

The 'Clustering' page (at right) allows you to customize your graph to display cluster information. See section 3.8 for details.

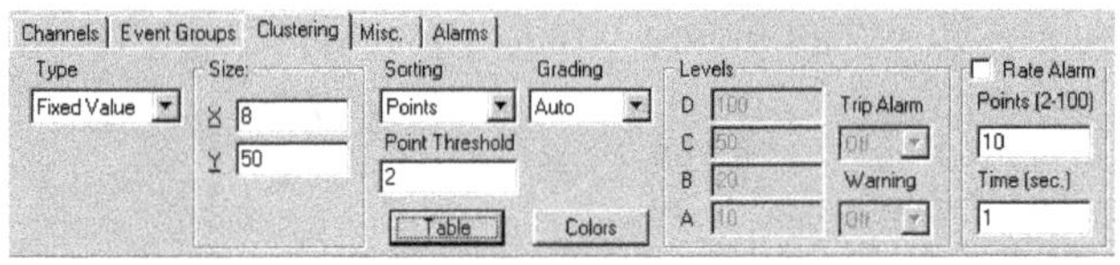

Figure 51. Clustering

The 'Misc' page (at right) allows you to set various graphing options that don't fall into any other category. From here you can turn on the 'color legend'. This displays a small chart on the right of the graph describing what each color means. You can also designate your graph to be a 'history' graph. See section 3.10 for details on how to use the Cluster/Area History feature.

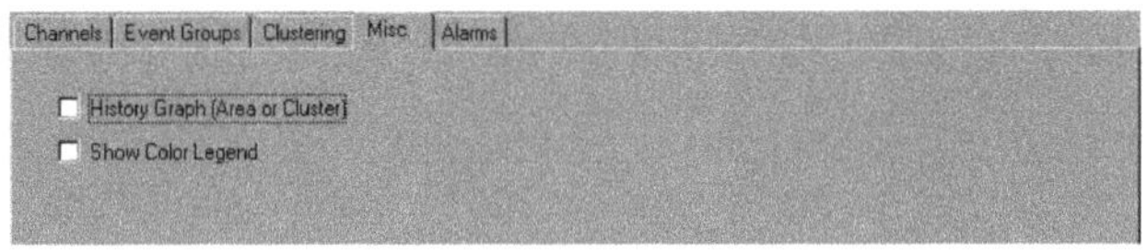

Figure 52. Misc.

The 'Alarms' page (at right) allows you to enable and customize Graph Alarms on your graph. See section 3.11 for details on Graph Alarms.

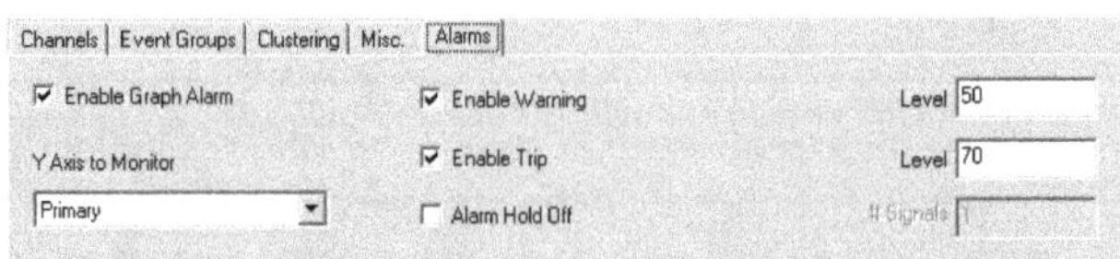

Figure 53. Alarms

Graph Filters…

Graph Filters are used to restrict the data that gets plotted on a particular graph by specifying a graph parameter (such as amplitude, energy, duration, parametric) and a range of values to either plot (accept) or not plot (reject). The parameters can be same as or different from the graph axis parameters of the graph to which they are applied. They only affect the data being plotted, they do not affect the data going to the disk. The Graph Filter dialog box is launched by either right-clicking on a graph and selecting 'Graph Filters' or by selecting Graph Filters from the Graphing pulldown of the main menu.

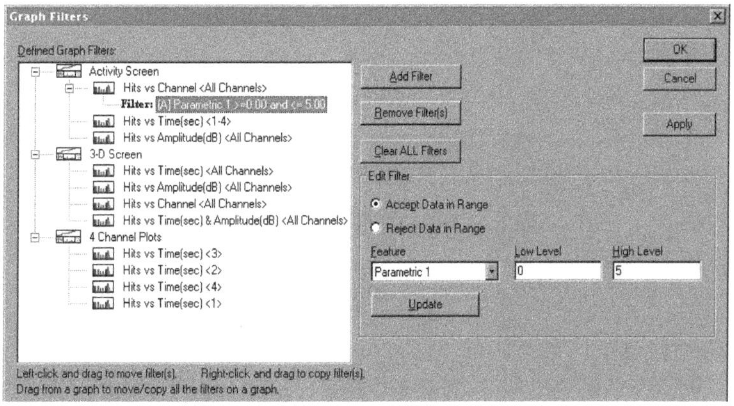

Figure 54. Graph Filters

The left side of the dialog box contains a tree control showing the 'Defined Graph Filters'. This control displays information in three levels. At the highest level are the currently defined pages of graphs. The next level shows all the graphs that belong to a page. The final level shows all the filters that are defined for the graph. If a given level has a '+' sign to its left then it contains items which are hidden. Left-click with the mouse on the '+' to expand the level. Expanded levels have a '-' to the left of them. Left-click on the '-' to compress the level. The text for the screen is the same text that is used on the 'Page Bar' at the bottom of the page. The text for the graph is the same as the caption above the graph. When the dialog box is first displayed it highlights the graph that is currently selected. So if the dialog box is displayed by right-clicking a graph and then selecting 'Graph Filters' the graph that was clicked on will be the graph that is highlighted. Filters are displayed with the format [A/R] Feature >=Low Level <=High Level. [A/R] will be [A] for accept filters and [R] for reject filters. The buttons to the right of the tree control are enabled or disabled depending on which item is highlighted in the control.

Add Filter: Will be enabled when a graph or filter is highlighted. Selecting it creates an empty filter under the highlighted item. The user must then use the controls in the 'Edit Filter' group to define the filters. The 'Update' must be pressed to change the filter.

Remove Filter(s): This will be enabled as long as a filter is defined in any Screen or Graph that is highlighted or if a filter is highlighted. Pressing it will remove either the single filter that is highlighted or all the filters that belong to a highlighted graph or all the filters in all the graphs of a highlighted screen. When the button is pressed another dialog box will appear that will prompt for verification. The filter removal operation can be canceled at this point.

Clear ALL Filters: This will be enabled as long as a filter is defined. Selecting it will remove all filters. When the button is pressed another dialog box will appear the will prompt for verification. The filter removal operation can be canceled at this point. The filter is defined by using the controls in the Edit Filter group. These controls will only be active if a filter is highlighted in the 'Defined Graph Filters' control.

Accept/Reject: Determines if the data will be accepted (plotted) or rejected if it passes the filter test (i.e. falls in the range of the low and high levels).

Feature: Selects the feature used as a filter. The choices are the same as the graph parameters. Note that all graph parameters are listed here. The user must make sure that the selected parameter makes sense. Only hit data can be filtered (all time driven data will plot). If the graph is an event graph, then only the features from the first hit sensor will be used in the test and the event will plot based on the result of the test on the first hit data.

Low/High Level: Determines the range of the filter. The range is inclusive of the limits (as denoted by the <= and >= in the graphical representation of the filter in the 'Defined Graph Filters' control). Any floating point value can be entered and the range is not validated.

Update: This button must be pressed to change the values in the 'Defined Graph Filters' control. The Apply button takes the filters as they are defined in the 'Defined Graph Filters' control and applies them to the system. Pressing the Cancel button after pressing Apply will not restore the filters to the state they were in before the dialog box was displayed.

Drag and Drop:
The text below the 'Defined Graph Filters' control explains how to move or copy a filter or group of filters.
If a filter is dragged with the left mouse button it can be moved to another graph by dropping (release the mouse button) on a target graph.
If a filter is dragged with the right mouse button it can be copied to another graph by dropping (release the mouse button) on a target graph.
If a graph is dragged then all the filters in the graph will be moved/copied to the target graph.
If the target graph is off the display, hold the dragged item very close to the top or bottom of the control to initiate auto-scrolling.
If the target graph is in a collapsed page or graph ('-' to the left of the target), hold the dragged item over the collapsed page or graph and it will automatically expand.

Tips:
- Reject is used over accept. If one filter on a graph is defined as [R] Amplitude >=0 <=50 and another filter on the same graph is [A] Amplitude >=45 <=50 then no hit will be plotted. The [A] filter winds up filtering out all hits with Amplitudes > 50 and <45. The [R] filter removes all the hits in the [A] filter range because it includes the entire range of the [A] filter. Think carefully about what is to be accomplished and test the graph first whenever possible.
- No more that 20 filters per graph can be defined.
- The graphs indicate if they have any filters defined by placing 'Flt' on the caption at the top of the graph. To quickly make sure that there are no filters, bring up dialog and check the Clear Button status. It will be disabled (grayed out) if there are no filters on any graphs.
- Don't forget to press the Update button whenever a filter is added or modified.

Colors

The 'Colors' submenu is used to set the various colors used in HMSGHAT2000DDI.
- Cluster Colors: Set the colors of the various cluster grades used on cluster graphs.
- Colored Scatter Graph: Set the colors of the various levels used by Colored Scatter Graphs.
- Plot Colors: Set the colors used by the plots of all other graph types. 'Plot 1' is used by all graph types other than Colored Scatter Graphs. 'Plot 2' through 'Plot 6' are used only by graphs that can use multiple plots (2D Histogram, 2D Scatter, 2D Binned Scatter).
- Sensor Color: Set the color of sensor icons used on event-based graphs.
- Structure Color: Set the color of the structure used on event-based graphs.
- Legs Color: Set the color of the legs used on spherical location graphs.
- Graph Background Color: Set the background color of the graph frames.

New/Move/Close Graph or Table

- New Graph: Use this command to create a new graph. The new graph will be a copy of the currently selected graph. If no graph exists then the new graph will be a simple 2D Histogram of hits vs. time. Use graph setup to customize it.
- Move Graph or Table: Use this command to move a graph or table using the mouse.
- Close Graph or Table: Use this command to close the selected graph or table.

Cut/Copy/Paste Graph or Table

- Cut Graph or Table: Use this command to cut a graph or table, presumably for later pasting on another page.
- Copy Graph or Table: Use this command to copy a graph or table, presumably for later pasting on the same or other page.

- Paste Graph or Table: Use this command to paste a cut or copied graph or table onto the current page.

Export Data to File...

This option on the right-click menu allows you to export the numerical values used in plotting a 2D Histogram onto a text file. These values are outputted to the corresponding x-y values found on the graph and are properly tabbed. This allows the user to import the values onto an Excel graph for instance for proper analysis. When the option has been selected on the right-click menu, the following dialog box will appear:

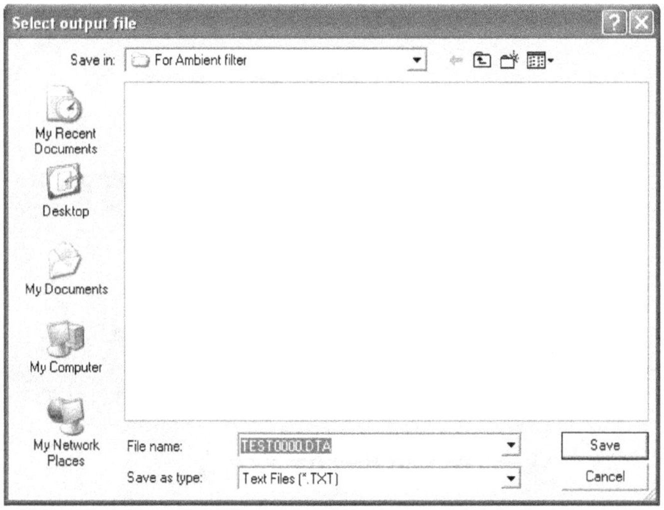

*Figure 55. Select **Output** File*

Here, the user can save the text file containing the numerical data onto any location of their choosing.

Note – This option is only enabled for 2D Histogram graphs. If the user attempts to right-click on any other type of graph, this option will be disabled.

Turn Waveform Threshold Lines On/Off

Waveform graphs can be displayed with horizontal lines denoting the threshold and a vertical line denoting the time of first threshold crossing. Use these commands to turn the display of these threshold lines on or off on all waveform graphs. You can also turn threshold lines on or off on individual graphs by right clicking the graph and checking (or unchecking) 'waveform thresholds'.

Tables Menu

With this menu you can create specialized tables. A table is a window that displays mainly text-based information instead of graphical information. Like graphs, tables can be moved, resized and closed from commands on the 'Graph' menu.

View Menu

With this menu you can select which toolbars and information bars you would like to display on the screen. There are fixed positions for each of these toolbars defined on the HMSGHAT2000DDI screen, but some (Toolbar, Acq. Controls, Line Listing Display) can be moved according to the user's preference by grabbing (left clicking the mouse) the handle (vertical bar) at the left side of the toolbar and moving it to the desired location. The settings on this menu are saved internally (i.e.: not in the layout file).

Utilities Menu

This menu contains a number of useful 'utility' programs for post-test analysis of datafiles. See the Appendix at the end of this chapter for details.

Page Menu

With this menu you can create insert, delete and rename screen pages. This is an HMSGHAT2000DDI screen page management menu. You can also perform these functions with the icon keys and a right mouse click on the screen page tab.

Window Menu

This is a typical "Window" menu that you would find on any common WINDOWS program and therefore the menu contents should be familiar. It allows you to arrange your graphs in cascade or tile patterns as well as edit their window settings. Select 'Maximize Graph' to enlarge the active graph , and 'Restore Graph' to return it to its normal size.

Help Menu

With this menu you can get some on-line help to the operation of certain features of the software. In addition, the "Options" selection informs the user which options have been installed in his/her HMSGHAT2000DDI software and the "About HMSGHAT2000DDI" informs the user to the release version and date of HMSGHAT2000DDI.

Acquiring AE Data with HMSGHAT2000DDI for the First Time

This section is provided to help you understand the sequence of getting into and running an AE Test for the first time using the provided Layout (.LAY) files and also gives you insight into a more detailed sequence of carrying out an AE data acquisition test. Before starting in acquisition, you should have your AE system already set up as per the "Getting Started" section of your AE system manual. This usually entails, connecting sensors, preamplifiers and cables to the AE system, connecting up your AE system, computer keyboard, mouse and monitor and booting up your system into HMSGHAT2000DDI as previously instructed. If you are not ready to connect up the AE system to conduct some type of AE test or at least any AE experimental test on a laboratory test, then skip this section and proceed to the next major section on Replaying an AE test with HMSGHAT2000DDI.

The key to carrying out an AE test is in setting up the hardware set-ups, graphics, location setups and display modes to your particular requirements prior to entering the test. Although there are many set-up items and they might seem confusing for first time users, supplies good default set-ups for most typical tests and they are an excellent starting point. As you get more familiar with the system you can modify these set-ups a little bit at a time and save your new layout (.LAY) setup files for your next AE test. This small step-by-step "Trial and Error" approach builds up your knowledge base until you are totally familiar with all the set-up options and what they mean. Until you get this knowledge please depend on the prepared .LAY files.

The following are the steps that should be taken when carrying out your first AE test using the supplied layout files. First we will outline the steps and next we will carry them out.

1. Load the appropriate system Layout (.LAY) file into the system in order to set-up the system to your desired operation and display conditions. We will be using the standard **Layout4.LAY** (or **Layout2.LAY** on Spartan and PCI-2 systems) layout file in this example which sets up 4 (or 2) channels for acquisition.
2. Enter the Acquire menu.
3. Select the datafile name which you want to save to disk.
4. Start Acquisition.
5. Look at various screens, graphs and data listings during acquisition.
6. Exit data acquisition.

The following is the step by step detail on starting and running your first AE test. We assume that you have sensors set up and will be exciting them in order to generate signals for AE generation or have them attached to a structure, bar or plate with a pulser or some type of simulated AE source.

Loading a Layout File into HMSGHAT2000DDI

From the Menu bar, use your mouse to select **Files** then select **Open Layout...** A menu window called **Open** comes up. With your mouse, select the **Layout4.LAY** (or **Layout2.LAY**) file from the list of layout files so that this file name appears highlighted in the file box next to **File Name:**. This selects it as the layout file that will set up the AE hardware for a 4 (or 2) channel operation with a good, complete default set-up. To finish the sequence, move the mouse to the **Open** box area on the right side of the **Open** window box and click on it (left mouse key) or press <ENTER>. Upon executing this menu, the selected Layout file is read into the program. If you had made any changes to the current layout file, a message box will pop up asking you if you want to save the those changes. If yes, select yes, if no, select no. HMSGHAT2000DDI will then load and configure itself to the selected layout's setup. This completes the reading of the LAY (HMSGHAT2000DDI setup) file.

Entering Acquisition

Entering acquisition involves steps 2, 3 and 4 of the above sequence. or the Mouse operated icon. To begin, left-click the mouse on the **Acquire/Replay** main menu item then click on the **Acquire** selection. Alternately, you can enter the Acquire function by pressing the **F9** function key (on the keyboard), or you can select the Acquire icon in the Acquire/Replay toolbar. The Acquire icon is shaped like a traffic light with the green light showing indicating that HMSGHAT2000DDI is ready to go into acquisition. This toolbar is usually to the right of the main icon toolbar or just below it. If you do not see this small toolbar (like that shown below), it may not be turned on from the **View** menu. To activate it, simply select the Tools menu and make sure the "Acq. Controls" entry is checked. If not, click on it and it will be activated and will show up with the other toolbar icons.

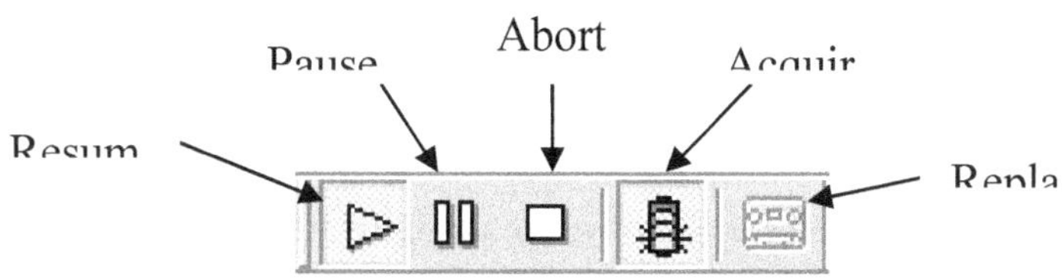

Figure 56. Acquire/Replay Icon Toolbar

At this point the Acquire function, a **Data Storage** files menu box will appear to signal you to enter a data file name and whether you want to save the file to disk. (An example of the Data Storage file menu box is shown in figure 60.)

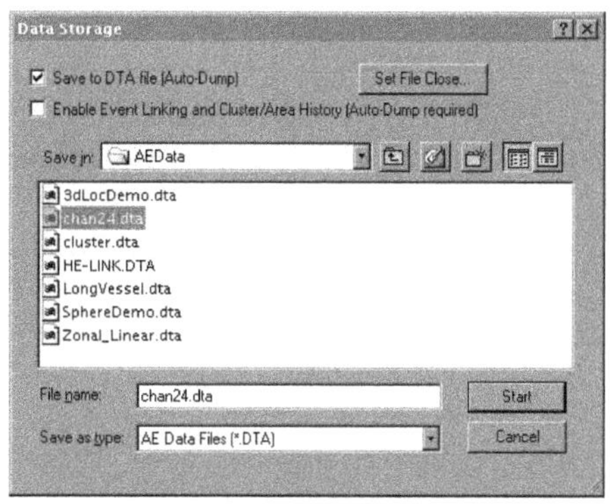

Figure 57. Data Storage

First, make sure the "Save to DTA file **(Auto-Dump)**" box is checked if you want to save the data you will be collecting in this exercise. In HMSGHAT2000DDI, this is always automatically checked and you usually have to deselect (uncheck) it to prevent the generation of an AE data file.

Next, check the name shown in the **File Name:** box. If that name is acceptable then you are ready to start the test by clicking on the **OK** button. If you want a different name then enter the new name by typing it in when the **File Name** box is highlighted. Once you enter the name (you do not have to enter the .DTA extension) you can hit <ENTER> on the keyboard or click the **Start** button (located to the right of the Filename) with the left mouse button to begin acquisition. At this point HMSGHAT2000DDI enters the acquisition screen and waits for you to press the <ENTER> key to start the test. Press <ENTER> to start Acquisition.

The system is now in data acquisition and is collecting and displaying AE information and displaying it on the graphs. Tap the sensor against the table your instrument is set up on or scratch it with your fingernail.

At first you will see no change in your Data Acquisition Graph but once the amount of time specified in **Test Setup → Display Mode** menu parameter "Seconds between Update" time has elapsed (it is usually set for 1 – 2 seconds), you will begin to see a visual representation of your data. Continue tapping the sensors to see the changes to the graphs to verify that you are getting waveforms and graph updates. An example plot showing the first AE graph screen page (selecting the tab, Activity Screen), is shown in figure 61. You should be seeing data visualization like that shown, where graph 1 shows a group of 4 point plots including Counts, Energy, Amplitude and Rise Time (on the Y-axis) versus Duration (on the X-axis), graph #2 (to the right) is showing #AE hits and #waveforms collected versus time,

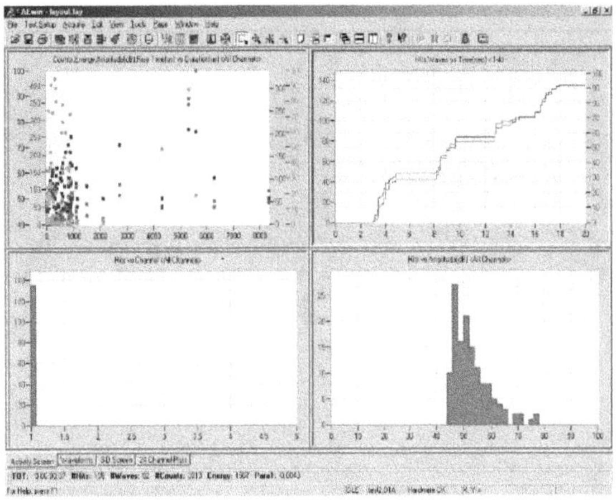

Figure 58. *Example Activity Screen*

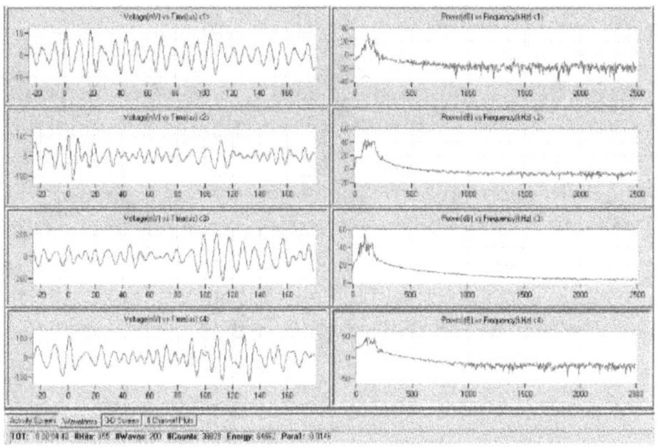

Figure 59. *Example Waveforms Screen*

Graph #3 (below graph 1) is showing Hits versus channel and Graph #4 (lower right) is showing the Hits versus Amplitude (or Amplitude Distribution).

Actions within Data Acquisition

There are several actions you can take in acquisition to view the graphs and other modes in the instrument. Try these items to see what happens.

Selecting and Viewing the Different Screen Pages

You can scroll between screens to view the activity between the different AE graph screen pages. This can be done simply by left clicking the mouse on any of the AE graph screen page Tabs located just below the graphs on the left. In the example Layout file, we have defined four different graph screen pages. They are the Activity Screen (shown in figure 58), Waveforms (figure 59), 3-D Screen (figure 60) and the 4 channel plots (figure 61).

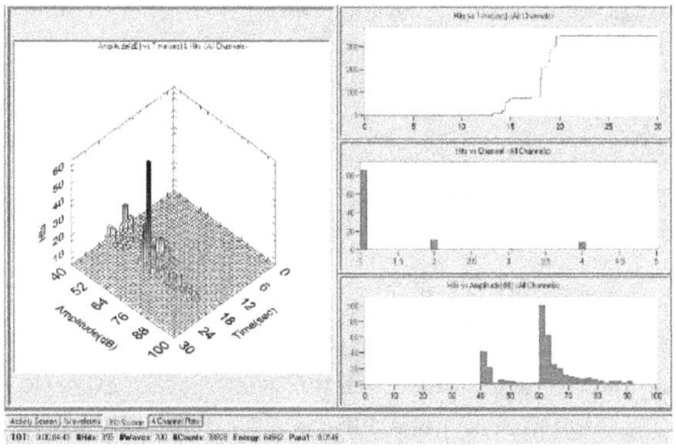

Figure 60. Example 3-D Screen

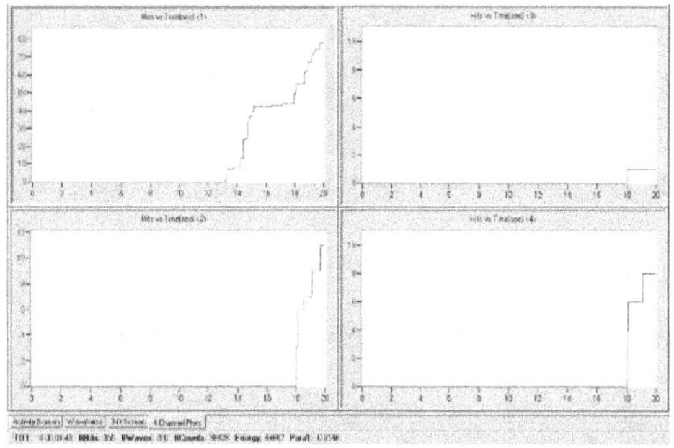

Figure 61. Example 4 channel screen

These graph screen pages give a good idea to the capabilities and the graph types that you can create with HMSGHAT2000DDI. On the Activity Screen page, you can see multiple point plots on one graph, multiple line graphs (AE Hits and Waves versus time) as well as bar charts. On the Waveforms Screen page (figure 59), you can see waveforms on each of 4 channels with the associated FFT located alongside the waveform for each of these channels. On the 3-D Graphs screen page (figure 60), you can see a nice 3-D plot with 3 associated graphs alongside. This graph shows that you can setup 3-D graphs (which can be rotated) with HMSGHAT2000DDI. It also shows that you can size your graphs differently. On the 4 channel plots (figure 61) you can set up graphs with multiple plots on them, one for each channel so you can link AE with other test parameters. This can all be done in acquisition. As you observe and change graph screen pages, the test continues.

There are many other things that can be done within acquisition. This includes maximizing a graph to full screen, zooming in on certain areas of a graph, viewing cursors and cursor readouts, printing graphs, saving graphs to a data file or the clipboard, just to name a few.

Zooming and Panning on a Graph Region

To zoom a graph, please note the graph associated toolbar icons for zoom, pan, and cursor select. This set of toolbars is near the center of the icon toolbar. This toolbar section is shown and defined in figure 62.

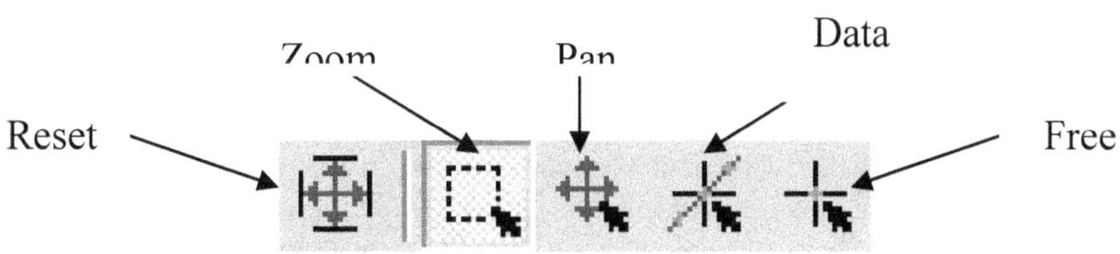

Figure 62. Zoom, Pan, Cursor Control Toolbar Icons

To use, select a graph by clicking on it. A good graph to demonstrate these functions on is the Activity Screen Point plots (see upper left graph of figure 58). Select the graph by clicking on it. Next, click on the zoom icon. This is probably already selected. Next, with the mouse cursor, define a zoom rectangle by pointing to the starting point on the screen where you want one corner of the zoom rectangle to be. Next, press and hold the left mouse button and drag the mouse to the point where you want the zoom box to end. You will see the dotted box as you drag the mouse. As soon as you let go, the graph will zoom and only the area that you selected in your zoom rectangle will be displayed in that graph. You will see that the axis scales have changed as well as the number of points in the point plot. You can now pan the plot (move around the zoomed area) by selecting the Pan icon button, moving your mouse pointer to the center of the graph screen that you have zoomed, click the left mouse button and move the mouse around. You will see the graph pan as you move the mouse. You can continue panning in one direction by letting go of the left mouse button, repositioning the mouse and pressing down on the left mouse button again, and moving the mouse. You can even pan un-zoomed graphs (but there is not a lot of sense in doing that). Zooming and panning can be done on any 2 dimensional graph.

Once done and you want to return that graph back to the unzoomed state, simply press the "Reset Axes" icon. This is the leftmost icon shown in figure 62. This will bring the graph and its axes back to the normal setting.

Zooming, Panning and Rotation of 3-D Graphs

Zooming, Panning and Rotation of 3-D graphs can be accomplished but not in the same way as that for 2-D graphs. For 3-D graphs, this requires a combination of the mouse and keyboard to accomplish. Before starting this exercise, click the Graph Screen Page tab "3-D Screen" and follow along to get familiar with the controls. You should see a graph like that shown in figure 60.

Rotating the graph is the easiest function. All you need to do is place your mouse pointer somewhere near the middle of the 3-D graph, press the left mouse button and while holding the left mouse button down, move the mouse slightly around in a circle. You will see the rotation of the graph change. You can position it anywhere you want.

To zoom, simply place the mouse pointer somewhere near the center of the 3-D graph and when ready, press the **ALT** button on the keyboard, and while holding that down, press and hold the left mouse button down. Then slightly move the mouse down (while holding the ALT button and the left mouse button depressed). As you move the mouse down the 3-D graph will zoom. As you move the mouse up, the graph will shrink.

To pan the 3-D graph, place the mouse pointer somewhere near the center of the 3-D graph, depress and hold the **SHIFT** button on the keyboard down and at the same time depress and hold down the mouse left button. Then slightly move the mouse around to see the graph pan.

Once done and you want to return that graph back to the un-zoomed state, simply press the "Reset Axes" icon. This is the leftmost icon shown in figure 62. This will bring the graph and its axes back to the normal setting just like in the case of the 2-D graphs.

Using the Position Readout Cursor

Two cursor control icons are shown in figure 62. The cursor readout controls are active in both acquisition and replay just as zoom and pan controls are. Each cursor provides a slightly different function. The "Data Cursor" identified above in figure 62, provides a discrete cursor movement only to legal data points on the screen. For example, for a waveform or a point plot, the Data cursor would only be able to identify actual existing data points. This therefore follows actual points on the screen. The other cursor, called the "Free Cursor", identified as the rightmost cursor in figure 62, will be able to be moved to any location on the screen and provide a continuous position that is graph based and not data based.

The Cursor readout position is located on the status bar at the lower right side of the HMSGHAT2000DDI screen. The position readout is identified by the **X,Y =** readout box. The position readout will be in actual graph value readings such as Hits, Amplitude, etc. If the graph is a multiple plot graph, the data point read out will be for the exact graph point, so it will be correct regardless of how many plots are on a graph. If the graph is a Binned Scatter or Colored Scatter graph then the readout will display the range of values represented by the data point.

The cursor readouts can be used on any graph screen. Simply select a graph screen by clicking on it. For this example, let's use the Point plot again on the Activity screen. First click on it. Next, select one of the cursors. Once a cursor is selected, it usually is drawn with the center at the origin of the graph (the lower left corner). Note the small hollow circle which identifies the center. Next you need to position the mouse pointer over the center of the cursor and grab the cursor by pressing and holding the left mouse button down. Once done you can move the cursor around the graph. If you have selected the Data cursor, you might have a little trouble moving it around the first few times but you will get used to moving from point to point. The "Free Cursor" is a little easier to operate. Anyway, as you are moving the cursor around (with the left mouse button depressed), you can see the graph coordinates changing as you move. This is a very handy feature to have.

Maximizing a Graph to Full Screen

Sometimes you have many graphs on a screen and it is a little difficult to see the axis values clearly. You can expand any graph to full screen by right-clicking the mouse in the desired graph area, and selecting the menu item "Maximize Graph" from the graph menu. You can select the same function from the Main Menu bar, **Window → Maximize Graph** selection. To restore the graph to its normal size and place on the screen, simply, right-click the mouse in the graph area and select "Restore" or **Window → Restore Graph**.

Printing any or all Graphs

You can print out any graph, any graph screen or all graph screens, using the various Print commands located in the File menu. To print a given screen of multiple graphs or to print out all screens, simply select the "Print Page(s)" from the files menu. To print out a single graph first click on the graph, then select the "Print Graph" menu item from the files menu. You can also export a graph or screen to a JPEG file by selecting the "Export to JPG" menu item on the files menu or you can "Copy Screen to Clipboard".

Pause, Resume, Abort The Test

The Pause, Resume and Abort functions are provided on the Acquisition/Replay Icon toolbar as shown in figure 56. The icons are fashioned after similar visual controls used in the Audio tape recorder business. To pause your test, simply press the Acquisition/Replay icon labeled **Pause**. In pause you can see the test status box change to "Test Paused". You can also see the Time of Test clock stop. At this point, the system will stop collecting AE data, will flush all buffers to disk, but the test will not be exited and the AE data file will remain open. Someone might want to Pause during a test to ignore some event that might add noise to the test (such as pressure or loading ramp-up's, etc.). To resume, just press the **Resume** icon button located next to the Pause button. This restarts acquisition and the Time of Test clock starts back up again showing the new and current time, taking into account the pause time. To Abort the test, simply click the button labeled **Abort** (see figure 56). Selecting this function stops the test and saves only the data to the data file that already exists on disk. The only difference between the Abort and the standard exiting of a an AE test is that the buffer contents are ignored in the case of an abort while ALL data is saved and written to the AE data file in a normal test exit. This function was mainly implemented for SPARTAN products as it could take the ICC's many seconds (when AE data rates were high) to empty their buffers. The abort function was convenient in those cases where the user wanted to quickly stop a trial test without waiting for data in the buffers to clear. The abort function is not needed in newer products as the transfer rates are very fast.

Exiting Data Acquisition

To exit data acquisition, first Pause the test by pressing the **Pause** icon button (see figure 56). Then stop the test by pressing the **Abort/Stop** key (the Abort/Stop key becomes a STOP function key, whenever the system is in Pause). Upon receiving the Pause, all buffers are flushed to disk and upon receiving the Stop, the AE data file is closed and HMSGHAT2000DDI exits acquisition. The Status box (lower center of screen) indicates "IDLE" meaning that HMSGHAT2000DDI is ready to perform another test or a replay.

If you have made any setup changes before, during or after acquisition and you want to update the .LAY (layout) file (which you should), you may do so by going to the file menu and choosing **Save Layout...** or **Save Layout as...** menu item. "**Save Layout**" will overwrite your existing layout file with the new information, while "**Save Layout as...**" will allow you to rename the layout file as a new file. Then you can follow the guidelines to save the new .LAY file. Enter a name such as TEST1.LAY. It is not recommended to overwrite the existing Layout.LAY file since you may always want to keep this default layout file in case of problems in the future. As computer experience dictates, you should always save your work before carrying out a new task such as acquisition in case for any reason the power or computer fails or that you simply forget after the test to save the LAY file. This is always a good practice.

Replaying AE Data

This section helps you understand the sequence of preparing for, recalling and replaying your previously recorded AE test using the provided system Layout (.LAY) files and AE data (.DTA) files. It also gives you insight into a more detailed sequence of data analysis and manipulation of data during replay.

Replaying an AE test is almost identical to that of acquiring data. As a matter of fact many of the same steps that have been carried out to perform data acquisition are repeated. The following are the sequential steps that should be carried out when replaying an AE test using the supplied Layout and AE data files. First we will outline the steps and next we will carry them out. For more details on the actual operation of software and menu items please refer to the proceeding sections of this manual.

1. Read the appropriate system Layout (.LAY) file into the system in order to load the system to your desired set-up and display conditions. We will be using the **Layout4.LAY** (or **Layout2.LAY** on Spartan and PCI-2 systems) layout file in this example.
2. Enter the Replay function.
3. Select the data file (with .DTA extension) name which you want to replay. We will be calling the datafile "Chan24.DTA" for our replay.
4. Start Replay.
5. Move between the various screens, graphs and data listings during and after replay.
6. Carry out some data manipulation functions during replay.
7. Exit Replay.

The following is the step by step detail of replaying an AE data file using the provided **Chan24.DTA** data file. For a replay test, you do not need to have any sensors connected to the instrument since you will be playing previously recorded AE tests.

Loading a Layout File into HMSGHAT2000DDI

This is identical to section 3.4.1 but is repeated here. From the Menu bar, use your mouse to select **Files** then select **Open Layout...** A menu window called **Open** comes up. With your mouse, select the **Layout4.LAY** (or **Layout2.LAY**) file from the list of layout files so that this file name appears highlighted in the file box next to **File Name:**. This selects it as the layout file that will set up the AE hardware for a 4 (or 2) channel operation with a good, complete default set-up. To finish the sequence, move the mouse to the **Open** box area on the right side of the **Open** window box and click on it (left mouse key) or press <ENTER>. Upon executing this menu, the Layout.LAY layout file is read into the program. If you had made any changes to the current layout file, a message box will pop up asking you if you want to save those changes. If yes, select yes, if no, select no. HMSGHAT2000DDI will then load and configure itself to the Layout.LAY setup. This completes the reading of the LAY file.

Entering Replay

Entering replay involves steps 2, 3 and 4 of the above sequence. To begin, click the mouse on the **Test Setup** main menu item. On the submenu, click on **Display Mode**. Make sure the **Normal** mode is selected. Also verify that the **Seconds Between Updates** is set for 1 or 2 seconds (this selects the graph update rate) and that the **Hits Before Update is 1000**. If not, set these values by clicking the mouse onto the particular box whose value you want to change and edit those numbers by typing new ones and deleting old one's with the delete or backse keys on the keyboard. If you want, set in your desired values. When done, click the **OK** window on the bottom of the menu to execute the changes.

Now you are ready to enter the begin replay. This can be started by clicking that icon (that looks like a cassette tape, see figure 56), or pressing the <F10> Function key on the keyboard or by selecting **Replay** from the **Acquisition/Replay** menu. Once selected, a **Replay** window comes up to signal you to select a data file name for replay. Check the name highlighted in the **File Name:** box. If that name is acceptable then you are ready to replay the test by clicking on the "Start" button. If you want a different file name to replay, then use the mouse to select it in the listed data file names. In this case you want to select the name **Chan24.DTA**. Click the mouse on that entry. Then, to begin replay you can hit <ENTER> on the keyboard or click the "Start" button on the menu. At this point HMSGHAT2000DDI begins replay and is displaying AE information as it appeared during data collection on the test.

Actions within Replay

All the actions carried out during acquisition can be carried out during replay. Please refer to section 3.4.3 for details relating to replay actions. There are several actions you can take in replay to view the graphs and other modes in the instrument. Try these items to see what happens.

Exiting Replay

There is no need to exit replay, when the file is finished replaying, HMSGHAT2000DDI returns to the Idle status waiting for the user to continue manipulating the graphs in replay, enter replay to analyze another file, make changes to the setup before continuing or enter acquisition.

Using Line Listing Mode

Line (Dump) listing mode is a very familiar feature and function of 's AE software. has provided a "line by line" listing capability with all its AE products for many years. This has also been referred to as "Line dump" in the past but now we refer to it as "Line listing mode". The line listing mode allows the user to monitor any and all types of messages being processed by the AE instrumentation. The information is in a "line by line" text format listing offering an additional monitoring capability over the graphing of AE data. With Line listing mode, the user can watch messages related to AE Hits, Events, Time driven data (parametrics), comments, Auto Sensor Test (AST), waveforms and administrative messages. Watching the line listing scroll by, shows how the data coming in is being processed, giving the user an idea as to the AE activity and severity. In HMSGHAT2000DDI, the Line listing has been improved with the following features over the capabilities found in our other legacy software programs:

- the ability to scroll the line listing up or down (used to only scroll down),
- the ability to select which data can be shown on the line listing
- flexible display of the line listing with the AE graphs, the line listing can be adjusted to take up a small part of the screen, operate within a floating textbox or take up the entire screen.

These features do add some extra utility to the line dump listing that makes it even more important and useful in AE testing. The rest of this section explains how to use these capabilities in HMSGHAT2000DDI.

The line listing mode can be used in acquisition or replay, and offer the same flexibility in both, however, it is recommended to use this mode sparingly in acquisition as its use might slow down the data acquisition speed by a significant amount. It is therefore recommended to use this when data acquisition rates are low, before a test as a prescreening method or after a test in replay.

The line listing mode color-codes the messages in order to make them easier to spot as they are scrolling by. Please note the following message types and colors as identified below.

Line Listing Message Type	Line listing Color
AE hit data	Blue
AE Event data	Violet
AE waveforms	Blue
Time Driven Data (e.g. Parametrics, ASL, RMS, etc.)	Green
Administrative messages (Pause, Resume, Alarm, etc.)	Red

Enabling Line Dump Listing for Viewing

In normal use, HMSGHAT2000DDI is kept in graph mode and the line listing is not displayed (as shown in figure 56). In order to view the line listing, two steps need to be performed. First, one has to "Enable" the Line listing mode, and then the line listing mode needs to be selected for viewing. This is done as follows (whether in acquisition or replay).

To Enable the line listing mode, simply click the mouse as follows;

Test Setup → Line Display → Select "Enable Line Display" Checkbox

To Display the "Line Display", simply press the **F7** function key or click the mouse as follows;

View → Line Listing

As mentioned above, the function key **F7** provides a shortcut to toggling the line listing on and off, once the Line Listing mode has been "Enabled" in the Test setup menu.

Note that whenever the line listing is displayed, it displaces the existing graphs based on its size and location. To restore the graphs simply press one of the "Tile" functions (Tile horizontally or vertically).

The line listing shown above shows various messages. The top line shows a message ID 2 or Time Driven message, while the message ID 1 is an AE hit message and ID 173 is a waveform message. As can be seen, all messages start with a time of test stamp and all data is in time order. In addition, the Time driven message shows some time driven based parametric readings while the AE hit message shows various AE hit based features.

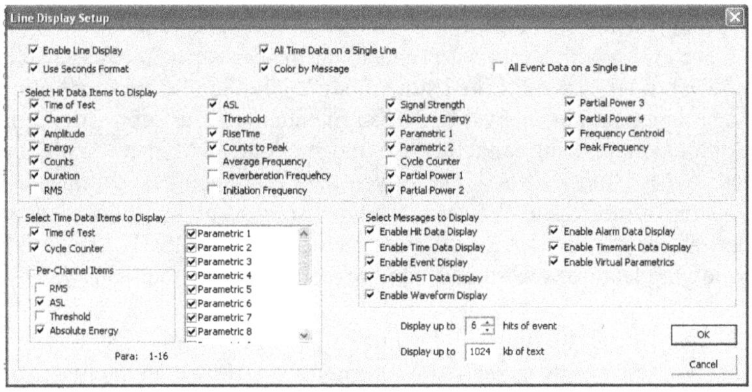

Figure 63. Line Display Setup

Modifying The Line Display

Many times the user wants to view a limited set of messages in the line listing screen. For example, maybe it is only important to look at AE hit messages or even just a few features in an AE hit message, even though it is desired to collect much more comprehensive data then that to be viewed. The Line Display Setup menu (click the mouse on Test Setup menu, then select the Line Display menu) allows the user to select exactly which data is to be viewed in the line listing display. The figure at right shows the Line Display Setup menu. There are 4 sections to this menu. The Enable Line display at the top of the menu was discussed in the previous section, make sure the checkbox is clicked if you want to enable the line display mode. The next section is called the "Hit Data Items". These are all the AE features that could be displayed on the line listing, if all the features were enabled in the **Test Setup → Hardware → Data Sets/Parametrics** menu. Simply check those AE hit features you want to view in the line display, even though all the features selected in the Data Sets/Parametric menu will be processed to graphs and the data file.

The "Time Data Items" (shown on the bottom left of the Line Display Setup menu), allows the user to select which Time driven data will be displayed with the Time driven data message. The final "Messages" area (to the bottom right of the Line Display Setup menu), allows the user to select which messages will be displayed. In all cases, just click those checkboxes to enable their display in the Line Listing display.

This menu is accessible in acquisition or replay and can be changed at any time.

Sizing and Moving the Line listing Display

The line listing display, when selected to be viewed, pops up as a full width, anchored text box and anchor's itself just below the HMSGHAT2000DDI toolbar. However, the Line listing display can be moved, floated or re-sized, simply by using standard WINDOWS functions. To float, just "double click" on the border of the window. Once floated, the Line Listing Display window can be sized using typical WINDOWS sizing capabilities of moving the mouse to the edge of the window border until the double sided arrow appears. Then the user can left click the mouse and drag the window side to be larger or smaller as desired. The user can also move the floating Line Listing Display window anywhere on the screen simply by left clicking the mouse when the mouse cursor is in the title bard (the blue title bar at the top of the Window) and can also anchor the Line Listing Display anywhere on the screen top, bottom, side, by moving one of the Line Listing Display Window edges to intersect with the side where it is to be anchored.

Once the Line Listing Display has been moved, this position will be remembered.

Clustering Setup and Operation

The purpose of clustering is to automatically identify and associate groups of hits or events which meet user controlled "sing" criteria. These clusters are drawn as rectangles around the data on a point plot screen to visually identify active region(s). Cluster data can also be shown in a tabular form, allowing a quantitative assessment of the activity. Clustering offers the user computer-aided processing to identify regions of AE activity.

Cluster analysis can be activated on any 2D point plot (2D Scatter, Binned Scatter or Colored Scatter), regardless of whether it is a Hit or Event based graph. Clustering is also available on the Spherical, Conical & Cylinder Location graphs types & 2D Histogram graphs allow 1-dimensional clustering along the X Axis. Clustering on a location graph identifies active areas that require follow-up for defect growth, while clustering on a Hit based graph usually identifies hits & events with similar AE feature pattern's which can lead to hit classification of that data.

Clustering Settings

Clustering is controlled from the Clustering tab at the top of the Graph Setup dialog box shown below. There are a number of parameters related to clustering that can be adjusted including the type, cluster size, point threshold, sorting & grading options. In addition there are options for generating alarms based on clusters based on either the cluster grading (level alarms) or cluster growth (rate alarm). Each of these settings are described below.

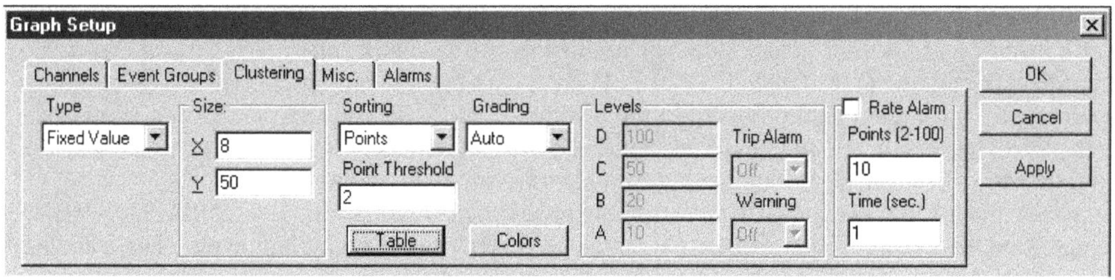

Figure 64. Clustering setup menu (at top of Graph Setup Menu)

<u>Type & Size:</u> Type controls how we want the Size parameter interpreted. Setting Type to None turns clustering off. Using Fixed Value allows you to enter the cluster dimensions you want directly into the Size X & Y field in graph axis units. In this case, this is the distance that 2 points must be within to form a cluster. Once the cluster is formed, the distance from the center of cluster to any new points must be less than half this value. Axis % allows you to enter the cluster size you want as a percentage of the graph axis range. For example, a Size X entry of 10 means the cluster can be as large as 10% of the X axis. For example, if Size X were 10 & the X Axis Maximum & Minimum are set to 10 & 50, then to form a cluster 2 points must be within 10% of 40 from one another or within a distance of 4. For graphs using autoscaling & Axis %, the internal size used does not change as the graph scale changes. Instead it is fixed at the value corresponding to the percentage of the initial axis limits.

<u>Point Threshold:</u> Point threshold is the minimum number of data points that must be in the cluster before it is reported. Internally the software maintains a list of up to 100 clusters on a graph, but it only report clusters that are above the threshold up to a limit of 26 reported clusters. If your Input Data is set to Hits, the number of points in the cluster is the number of hits. If the Input Data is set to Events, then it corresponds to Events. The minimum number of points that can be in a cluster is 2, but clusters containing just 2 points are not usually of much interest. You can increase the Point Threshold beyond 2 (up to 100) to decrease the number of clusters reported in the Cluster Table.

Sorting: Sorting controls how clusters are reported in the cluster table. The available choices are Points, Energy, Counts, Counts to Peak. The cluster table displays clusters sorted from highest to lowest by one of these quantities. In order for Counts To Peak to be useful, you must have actually collected Counts To Peak as part of your data set. Otherwise the software simply treats it the same as sorting on Counts. When a cluster table is displayed in the current graph window, you can use Ctrl-F6 to control the sort. Pressing Ctrl-F6 toggles the sorting through the list of available options.

Table: The Table button brings up the Table Setup dialog box shown at the right. The Display checkbox controls whether or not a cluster table is shown in the graph window. The F6 key is a keyboard shortcut for turning the cluster table on & off. The Placement setting controls where the table appears relative to the graph. The Shift-F6 key shifts the table through each of the 4 positions. The Window % setting controls how much of the graph window allocated to the cluster table. If the text in the table grows too large for its window, scrollbars will appear to allow you to see all the text.

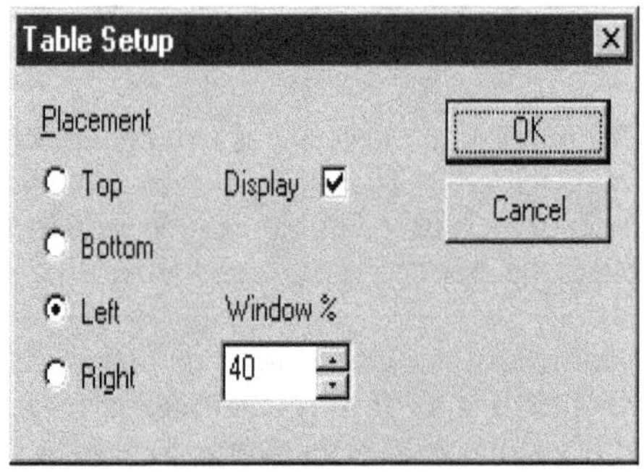

Figure 65. Table Setup

Colors: The Colors button brings up the Cluster Colors dialog box shown at the right. You may also access this dialog directly using the Graphing pulldown Colors submenu. Look for the Cluster Colors menu option. This dialog allows you to configure what colors are used in the cluster table & when drawing the cluster rectangles on the graph. These colors are used by all graphs doing clustering. Double clicking one of the entries will cause a standard Windows color dialog to appear allowing you to select a new color for the selected grade.

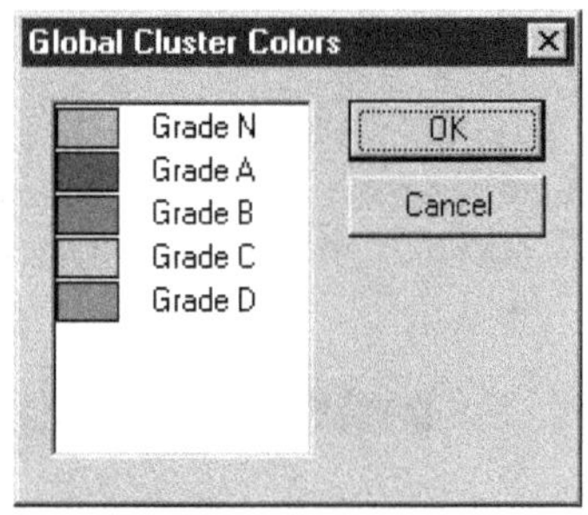

Figure 66. Global Cluster Colors

Grading: The Grading selection controls how clusters are assigned a color. After the clusters are sorted, a grade level is assigned to each cluster based on the sorted quantity. Choosing None causes all clusters to be classified as N (not significant) & they are drawn & reported in the Grade N color. Choosing Auto grades clusters by their percentage of the maximum observed in the quantity used for sorting the clusters. It just divides the maximum observed value into 5 equal ranges (<20%, <40%, <60%, <80%,<=100%). For example, if the clusters are sorted on Points & the largest clusters had 250 points in it, then the grading levels used would be 50, 100, 150 & 200. If the Grading is set to Manual, then the Levels section will enable & you can enter your own grading levels. That can be convenient when you want to compare the cluster results from 2 different graphs as it allows you to enforce the same grading levels on both graphs.

Levels: The Levels section is used when the Grading parameter has been set to Manual. It disables when Grading is set to Auto or None. Clusters are drawn in up to 5 colors corresponding to 5 different grades. This section controls the boundary levels between the manual grades. If you are sorting on Points, have Grading set to Manual & entered levels of 10, 20, 50 & 100, clusters with less than 10 points are graded N & by default colored green (see Colors to change). Clusters with 10-19 points would be graded A, 20-49 points are B, 50-99 are grade C and any with 100 or more points are considered grade D.

Warning & Trip Alarm: The Warning & Trip Alarm settings allow the user to generate alarm messages in the line display & set the Alarm Warning and/or Alarm Trip output lines high when a cluster reaches a user determined grading level. The Warning & Trip Alarm settings require that Grading be set to Manual before they will enable in the dialog box & they only have an effect during acquisition. Both Warning & Trip Alarm can be set to Level A, B, C or D or to Off. The Warning setting must always be lower than the Trip Alarm setting if the Trip Alarm setting is not set to Off. During acquisition when a cluster level alarm is generated, an alarm message will be saved to the DTA file. If the line display is enabled, a text message containing the details will be displayed & the alarm dialog box appears with the details. The Alarm Warning & Alarm Trip lines will be set high as appropriate. The output voltage of the Alarm Warning & Alarm Trip lines are never altered in replay mode,. When a DTA file is replayed, the same cluster level alarm messages will reappear that were observed when the file was recorded. These messages are not controlled by the current Warning & Trip Alarm settings during a replay.. The appearance of these alarms is only controlled by the settings that were in use when the DTA file was recorded.

Rate Alarm: The Rate Alarm can be used to monitor a graph for concentrations of activity that occur at high rates. This type of cluster alarm works identically during acquisition & data replay & does not have to be enabled during data collection. The checkbox allows for simple on/off control. The Rate Alarm has 2 parameters, Points & Time. The Points setting specifies how many points must be observed within the Time period to set off the cluster rate alarm. The number in the Points field can range from 2-100, while the Time setting is specified in seconds & can be any floating point number. For proper functioning, a Time value greater than 0 should be entered or else the alarm will trigger anytime data is added to a cluster. Likewise a large entry like 3.4E+38, the floating point maximum, causes problems as well as it would be so long that the alarm would never trigger. If the software receives enough points in one cluster within the alarm time, a cluster rate alarm will be generated. For the settings above, that would be 10 points within 1 second of one another.

The cluster rate alarm is implemented to not generate extraneous messages. If the user says to check for 10 events in 1 second, the cluster has to collect at least 10 points before the rate is ever checked. Once the 10th point comes in, the software adds 1 sec. to the time of the first point & checks to see if the 10th occurred before or after that time. If it occurred before that time, then the rate is higher than the user setting & an alarm message is generated. When the cluster gets an 11th point, its time will be compared with that of the 2nd data point. If the new data maintains the rate above the user cluster rate & the cluster has already generated an alarm message, the cluster stays in the alarm state and does NOT generate a new alarm message. If a cluster is in an alarm state & a new data point comes in after the alarm time, then the cluster alarm state will be reset to no alarm. If the cluster later on gets more data that is again above the user specified cluster rate, it will generate another cluster alarm message. Basically, if the system is flooded with hits, one message will be generated & nothing else until the hit rate drops below the user threshold to reset the cluster & then the hit rate comes back up. When a cluster rate alarm is generated, the software will generate a status message in the line display reporting the cluster ID & current cluster information.

Clustering Example

With so many settings, it may be useful to see an example of how to set up clustering. The software includes example layout and data files, **Cluster.LAY** and **Cluster.DTA**. The steps in setting up and running a clustering application are described below.

Step 1. If you wish, start by using New Layout under the File pulldown menu. Reconfigure the default Hts vs. Time 2D Histogram to meet your requirements. For example, you could set up a Y-Position versus an X-Position location graph set up as a point plot, or set up an AE feature versus AE feature (e.g. Amplitude versus Duration, etc) plot. When we plot an AE feature versus a feature, we usually call it a "Correlation" graph, referring to the way we are trying to "correlate" one AE feature against another to see how and if they are related. Our example layout **Cluster.LAY** and data file **Cluster.DTA,** shows an example of a Duration versus Amplitude point plot. If you were starting with a new layout file & graph, you would set the graph type to 2D Scatter & set the X & Y Parameters to Amplitude & Duration. The example layout uses Manual scaling on both axes, with the Y Axis limits set to 0 & 600, while the X Axis limits are set to 0 & 120. The Input Data is set to Hits. Alternatively you can just recall the example layout file & use the graph settings in it.

Step 2. Next we must configure the Clustering function. We do this in the Graph Setup menu by selecting the tab labeled "Clustering" at the top of the Graph Setup menu (see figure 64). Start by picking Fixed Value from the Type dropdown listbox.

Step 3. Next set the Size X setting to 8 & the Size Y setting to 50.

Step 4. Leave the Sorting set to Points, but change the Point Threshold to 3 to eliminate the smallest clusters.

Step 5. Select Auto from the Grading dropdown list. Leave the Rate Alarm disabled & ignore the Table & Colors buttons for the moment. Clicking on the Graph Setup OK button will save the settings. If this is a new layout for you, you may wish to use Save Layout under the File pulldown menu before starting a replay.

Step 6. Now we are ready to run a test and view clustered data. Use the F10 key or select Replay under the Acquire/Replay pulldown menu to open a DTA file for replaying. Select the Cluster.dta data file and click on OK or press Enter to replay the data. If you use the setup as provided in the Cluster.lay layout file with the Cluster.dta data file, you will see results similar to those shown below.

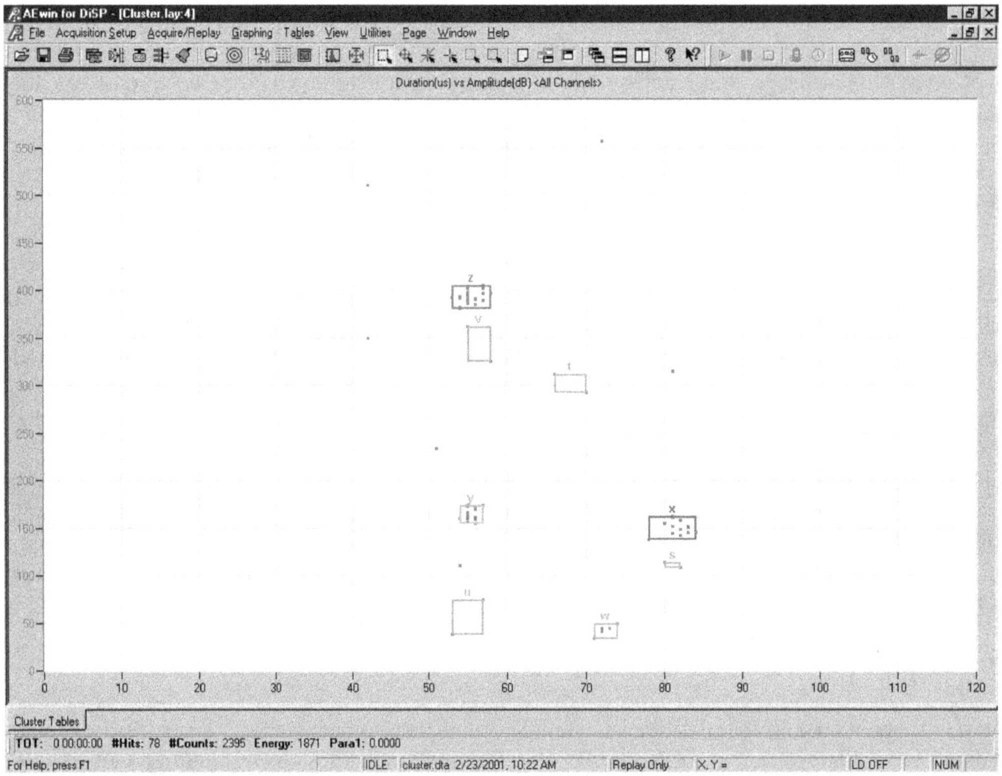

Figure 67. Cluster Graph

From the graph, we can see that we have identified 8 clusters based on the cluster configuration that we have set up. We can also see that there are cluster letters associated with each cluster and that each cluster is colored. Clusters are labeled with a letter starting with "z" as being the most intense and "a" being the least intense. Up to 26 clusters can be reported with the clustering capability built into the software.

Step 7. You can adjust the colors of the clusters by using the Colors submenu of the Graphing pulldown menu. Look for the Cluster Colors menu option. When you select it, the Global Cluster Colors dialog box will appear as shown in Figure 66. You can also get to this dialog box using the Colors button on the Clustering tab of the Graph Setup dialog box. The Global Cluster Colors dialog box controls the color of each grade level. If you double click one of the entries a standard Windows Color dialog box will appear to allow you to select a new color for that particular grade level.

Step 8. This graphical representation of the clustered data is nice, but we may also want to see a more quantitative tabular listing. Press the F6 function key from the keyboard. By pressing the F6 key, you toggle the cluster table on/off. This cluster table shares the window with the cluster graph. When we use F6 to turn the cluster table display, we get the table as shown in figure 68.

The Cluster table has the following columns. The leftmost shows the cluster letter, starting with z as the highest. Next is the cluster ID, this is the identification number assigned when the cluster was formed. As new points are added to the graph, the a-z cluster letter can change, but the cluster ID always stays the same. The next column is the Points. This indicates how many points (hits or events) are in a given cluster. The Total Energy is the accumulated energy in the cluster from all the hits or events. The Total Counts is also the accumulated number of AE Counts for the cluster. The Xrange & Yrange min & max values identify the boundary points of the cluster while the Avg. X and Y values identify the center of the cluster. Each line of the cluster table is colored by the grade level of the cluster so it matches the color of the cluster box on the Cluster graph.

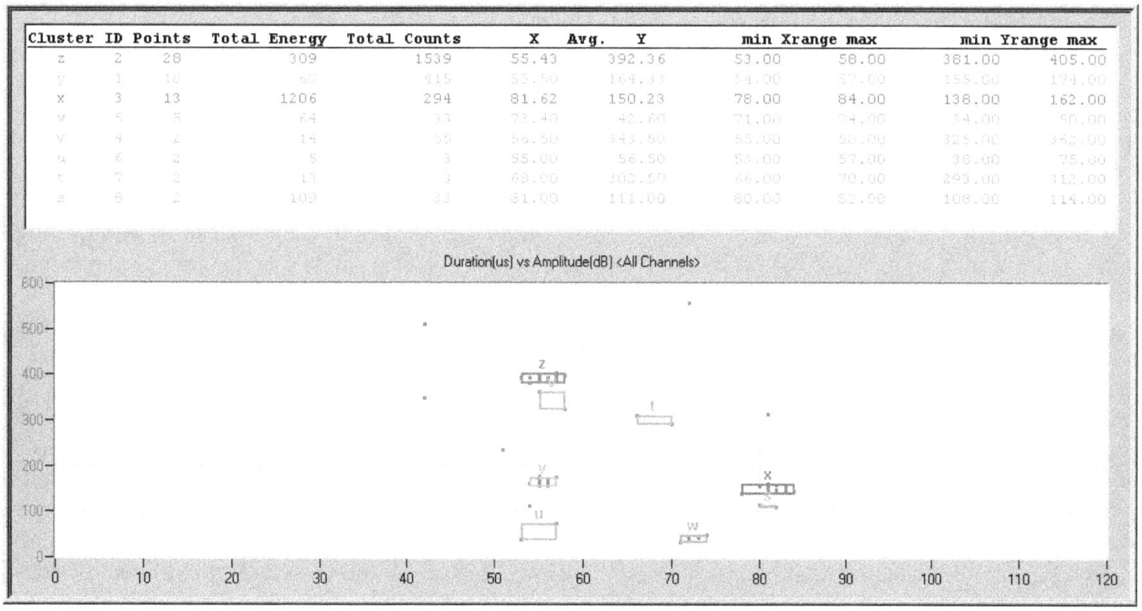

Figure 68. Cluster Table

Step 8. Try pressing shift-F6 while the table is displayed & the cluster graph is the active graph. Shift-F6 causes the cluster table to shift its position around the graph from top to bottom & then from left to right.

Step 9. Try pressing Ctrl-F6 to control the sort order in the cluster table. Each time you press Ctrl-F6, the table will switch between sorting based on Points, Energy & Counts & immediately update.

Step 10. The final step in the process of setting up and running the clustering function is adjustment of the setting for optimal results. This is accomplished by changing some of the clustering configuration settings & replaying the test until you feel that you have the most logical results. If you have too many small clusters, make your Size X & Y settings bigger. Alternatively, you can adjust your Point Threshold higher to only display the largest clusters & ignore the smallest clusters. You might wish to try setting Grading to Manual & modifying the Level A thru D values to adjust the grading results.

In our setup, we set the cluster threshold at 2 which is a bit low. We can also see by figure 68 that there are 4 of the 8 clusters with only 2 hits within the cluster. Let's get rid of these by going back to the graph setup menu and changing the Point Threshold to be 3 instead of 2. After we click the OK button, we'll find the cluster table has been updated & now only hold 4 of the original 8 clusters. Note also that those points that are not part of the cluster seem to be randomly sed, which tells us that it does not seem that these points or hits should have been identified as a cluster. Experiment further to get familiar with the clustering function.

Graphing in HMSGHAT2000DDI

The graphing capability in HMSGHAT2000DDI is very impressive, as it should be since data visualization is very important in AE testing. HMSGHAT2000DDI is able to provide multiple graphs, multiple plots within a graph, multiple graphs on a screen (or "Page" using Windows terminology), and multiple screens (Pages) full of graphs. The graph types can multi-dimensional (2D and 3D) graphs, multi-colored graphs and multiple graph types including: Point plot, Histograms, Waveforms and Spectrum graphs. Graph axes can be set up to be fixed scaling (always stays at a fixed minimum and maximum value), automatic scaling (compression or sliding scale) logarithmic or linear scaling. The graphs can be set up to plot anything versus anything, limited only by the operators imagination.

Before getting into setting up Graphs in HMSGHAT2000DDI, it is important to define the terminology used in graphing on HMSGHAT2000DDI. This will be used throughout the manual.

Plot: A plot is defined as a single graph within a graph (or graph window). Since a graph can be set up with multiple graphs inside, we need to distinguish the difference between the word graph as a graph window and the fact that there can be multiple graphs in a single graph window. We therefore call this a plot. So now we can say there are multiple plots within a graph or graph window.

Graph: A graph is a single Window on a Screen or Page. A graph is controlled by the Graph setup menu. The graph can have multiple plots of the same type within it, but these are all set up by a single specific graph setup menu.

Screen or Page: A screen or Page is the graph(s) that show up in the display area of the HMSGHAT2000DDI screen. This the area between the top and bottom tool bars (see figure 4). A screen or page has a page tab located at the bottom of the screen to select between the screen pages.

Now that we have definitions we can provide an example. Figures 58 –61 show examples of different screens. Four are shown, Figure 58 shows what we have labeled as an activity screen. The page tab at the bottom of the screen has been labeled "Activity Screen" to let the user know what types of graphs are set up on this screen page. Figure 59 shows a waveform and FFT screen page. Its tab also identifies it as "Waveforms". Figure 60 shows an example of a 3D screen page with the appropriate Screen Page tab labeling. Figure 61 shows an example of some channel by channel line plots with an appropriate screen page tab.

In regards to graphs, figures 58 – 61 show multiple graphs on each screen. Figures 61, 63 and 64 shows a 4 graph screen while figure 59 shows an 8 graph screen. There is no limit to the number of graphs that can be placed on a screen, the only limitation is a physical one, pertaining to the readability of the graphs. As far as multiple plots on a graph, the top two graphs of figure 58 shows graphs with multiple plots. On the top left plot there are 4 different point plots on one graph, while in the top right graph there are two line plots.

The figures 58 – 61 which are provided by the example **Layout4.LAY** and the data file **Chan24.DTA** give a good idea as to the functionality of HMSGHAT2000DDI as far as graphing is concerned. Since all the major examples for graphs and graph setups can be found in this **Layout4.LAY** file, it provides an excellent reference for how to set up the different types of graphs. Please use this example as such. The following subsections will provide information on how to generally set up a graph and screen page.

Graph Setup Procedure

The following is the step-by-step process in setting up graphs in HMSGHAT2000DDI. This assumes that you are starting from a very simple single screen, single graph arrangement, into a multiple graph, multiple screen setup.

1. Before starting, give a little thought to the number of screen pages your application will need, what the theme will be for each screen page (for tab labeling considerations) and how many and what type of graphs you would like on each screen. The better you plan this out, the easier and faster the graphing setup process will be. Use as an example the theme provided (and theme tabs) in the **Layout4.LAY** (or **Layout2.LAY** on Spartan or PCI-2 systems) layout file whose graphical results are shown in figures 58 – 61.
2. Find an existing Layout file that gets you partially to your goal. You can then edit that layout file to meet your exact needs. It is always easier to start with an existing layout. You can easily copy and paste graphs that have already been setup, saving a lot of time. If this is not possible, select **File → New Layout** which will start with 1 screen page (labeled Page 1) and one graph (a Hits versus Time graph).
3. Create the number of screen pages that you need by adding and deleting screen pages, then edit the Page tab titles to the desired name or description of what that page will contain.
4. Create the number of graphs you need in each screen page by adding or deleting graphs.
5. Size your graphs as desired and tile them as desired so that they are laid out in the order that you wish.

6. Individually setup each graph in each graph screen to get the desired result. Enter the graph setup menu for each graph by right clicking the mouse when it is inside the graph, then select the first pop-up menu called "Graph setup" to enter the graph mode.
7. Once you are done setting up your graph(s), test them out in acquisition or replay to make sure they act as expected. There will always be the need for some adjustment, you will rarely get things perfect the first time.

Screen Page and Tab Management

As with all Windows operations, there are various ways to add and edit Screen pages. The most convenient is by Right-clicking your mouse on an existing page Tab, adjacent to where you want to add the page. A Page/Graph setup menu will appear. To add a New Page after the selected page, simply select **the New Page** menu item. To insert a new page before the selected page, select the **Insert Page** menu item. In either case, a new page and page tab will appear. The page tab will be labeled as "Page X".

There are other ways to add a page including selecting the **New Page Icon** from the icon toolbar (looks like a blank page), which will always place a new page at the end of the screen page tab list. You can also go to the main menu bar and select the **Page** Menu to make the **New Page** and **Insert Page** selections.

Once you have created a new page, it will be empty (no graphs). You will add the desired graphs and will label the page tab. You may add as many graphs as you wish to the screen page.

To label the page tab simply right-click on that tab and select the Rename Page menu item. This will allow you to enter a page name. There are no major restrictions to naming a page tab. However do not make the page tab title too long as it might take up too much se.

There is one other screen page editing function, which allows you to remove or delete a screen page. Simply right click on the screen page tab and select the **Delete page** function. The screen page and tab will immediately disappear. If you had graphs in the screen page that you deleted, they will be deleted as well.

Adding, Placing and Sizing Graphs on a Screen Page

Once you have a Screen Page set up, you need to begin adding and placing graphs on it. If you know how many graphs you want and what types they will be you can quickly add and edit the graphs. If all the graphs are to be the same style (e.g. waveform graphs, one for each channel), you should start by adding a graph, defining it and then copying it as many times as you need. Then you can size the graphs and create the layout (position of the graphs) as desired and then you can edit the graph setup to finalize the graph function. If each graph is to be different, you can either copy similar graphs and paste them into the new screen page or you can simply add as many graphs as you want on the screen and edit the graph setup for each graph added.

To add a graph to the graph screen, simply right-click you mouse anywhere in the screen page and select the **New Graph** menu item. Alternatively you can left-click the new graph icon from the icon toolbar menu. This icon shows 3 staggered graphs. Additionally you can add a new graph by selecting the **Graphing → New Graph** main menu selection. Whichever way you select a new graph, a single graph will appear. To add multiple graphs on the screen, simply perform this function multiple times.

Once you have selected the number of graphs you want on the screen, you need to tile them. Tiling allows you to place them next to each other so that each part of every graph can be seen. There are two tiling options. You can use **Tile Horizontal** or **Tile Vertical**. The most convenient way to perform this function is to select one of these functions from the icon tool bar. The Tile Horizontal icon is shown as two graphs stacked on top of one another while the **Tile Vertical** icon shows two graphs side by side. Alternatively you can select these from the **Window** main menu. Tile Horizontal will favor a length-wise positioning of the graphs and will usually make the graphs longer than taller. Tile Vertical will give preference to a vertical direction. Choose one of these to place the graphs in a nice ordered manner.

Once the graphs are placed on the screen, we can alter their sizes. We do this using the standard Windows conventions of moving the mouse to the edge of a graph until we see a double arrow and then left-click the mouse and move the edge in or out to make the graph larger or smaller. We can also move graphs by right-clicking on the graph and selecting the **Move Graph** menu item. An example of sizing graphs can be seen in figure 63 where we have sized the 3D graph to be larger than the associated graphs.

The final function that we need to perform is that of editing or setting up the graphs to perform the desired display function.

Graph Setup Function

This section discusses the graph setup menu and how to set up a graph. First, we select the graph that we want to modify. We do this by right clicking the mouse when it is inside the graph region of the graph that we want to work on. Once we do that a menu appears. This menu gives us many choices but we will select the top one called "Graph Setup". Upon selecting this a graph setup menu appears like that shown to the right.

This shows us how we have setup the Graph setup menu for a Hits versus Time graph. It gives us a lot of insight into how the graph setup menu works and how to set it up. Lets note first the layout of the menu. We see at the top of the menu, a selection area for Channels, Event Groups, Clustering, Misc and Alarms. These settings allow us to configure those items. When in the Channels Tab, we can select which channels and/or channel groups will be displayed on the graph that we are setting up. In this case we are selecting all active channels for viewing in the graph. You can see that we have no groups set up in Event Groups if you select that tab. Clustering allows you to set up parameters that link closely sed hits/events together for point plot graphs. Misc allows you to set various options like turning on/off the legend control. Alarms allow you to create and customize graph alarms on this graph.

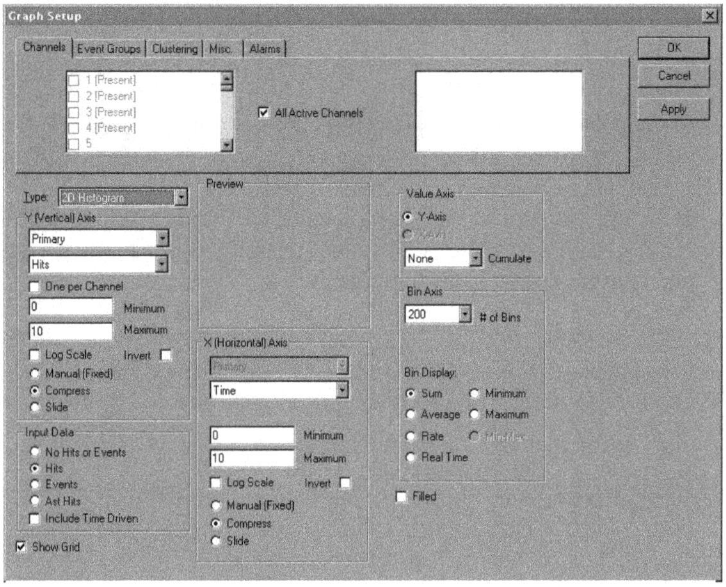

Figure 69. Graph Setup Menu for Hits versus Time Graph

The next menu area is the Graph Type selection box. This is located just below the Channel, Event Group and Clustering selection area. Various graph types can be selected included 2D Histograms, 2D Scatter (Point Plots), 3D Histograms, waveforms, Power Spectrum plots, etc. With the selection of the graph type the graph setup menu below this may change to offer the special configurations needed by that graph type. For the hits versus time application we will use the 2D Histogram.

Below the Graph Type selection, is the Y-Axis setup menu area. Going down one item at a time in this menu is the Plot # selection within the graph. This selection allows you to select more than one plot to be graphed. The one being selected for setup is the one shown. We have only selected one plot in this graph called the "Primary".

Next is the Y- Axis parameter to be plotted on the graph. By clicking this, you can see many selections. However, we have selected Hits. This will display the number of hits (on all channels) detected, on the Y-Axis.

The checkbox "One per Channel" allows us to select a unique per channel plotting color that will give us a multiple color by channel plot. In our case for hits and all active channels, if we selected this checkbox, we will get a multiple plot graph, each plot a different color based on the hits from that one channel, up to 6 plots on one graph. If not checked, we will only get one color plot.

The next two edit boxes allow us to set the starting Y-axis settings (minimum and maximum).

The next checkbox "Log Scale" allows us to set a log or linear graph scale.

The next 3 radio buttons are related, allowing you to select one type of graph scaling. "Manual (Fixed)" will force the graph limits as set above to remain fixed. "Compress" will compress the graph so that the minimum graph value remains but the maximum keeps changing as the graph limit is exceeded. And the "sliding" selection will allow both the minimum and maximum limits to change or slide as data is received that is beyond the current limits.

The next box allows us to select what type of Input data we are plotting. In our case since we are plotting hit data, we want to choose the Hit data for plotting.

Below this is a checkbox for displaying the a grid on the graph.

The next column to the right, shows the graph preview at the top and then the X-Axis settings.

The X-axis settings have similar setup boxes as were described for the Y-Axis and will not be described here. Just note how they have been set up to display the Time parameter and that we have set up the minimum and maximum graph values and a compression scaling so that we can see all data from the start of the test.

The last column on the right changes according to the graph type being set up. If a 3-D graph were being set up, this would be the column for the Z-axis setting. If a Point plot were being set up, there would be no extra column as the points do not need to be conditioned for display.

Since we have selected a graph type of 2-D Histogram, a "Value and Bin Axis" setting menu appears. The Value axis sets which axis the histogram or bars will grow. For our example we are selecting the Y-Axis as the Hits will grow in the vertical direction. We set a cumulate mode if we want the data in the histogram to build upon the previous bin or bars.

The bin axis settings allow us to set the resolution and how the data that makes up a bin get counted in the histogram bar (or what we call a bin). In general, the larger the number of bins, the finer the bars. If your bin axis is time you usually want to have a very high number. If your bin axis is channels, you would like to match the number of bins and channels. In terms of the bin display, you can choose whether you want to sum up all the data in the bin or if you want to average the data in the bin or if you want a rate to be displayed (e.g. Hits/second). To understand, this assume that we have a test for 100 seconds and that we have 100 bins. Each bin will contain one second's worth of data. We can decide to "add it up" and display it as a bin sum, average the amount of data received in that time period or display a rate. In this case since the bin size is 1 second, all the numbers would be the same for this setting, but for any other case, they would be different.

Learn more about these settings by making changes and seeing the effect.

At the bottom of the third column is a histogram setting that allows you to fill the histogram bars or keep them hollow.

Please note and try to make sense of all the selected menu settings. Although it might look complex at the start, if you have gone through these settings, one at a time you will see that all is quite simple. Exit the graph menu by selecting the OK button at the top right of the Graph setup menu.

By looking at the top of the graph we see its graph title of "Hits vs. Time (sec) <all channels>". This summarizes what is being displayed. First is the graph type "Hits vs. Time (sec)". Next we are displaying data from all channels.

Now that we are finished, we see that we have set up a Hits versus time graph and are ready to test it. Test it and make any corrections, until you get the desired result. If we have other graphs, we set them up in a similar way, one at a time until all the graphs are set up.

Now we are ready to proceed with a test or a replay. As a reminder, always save your Layout file before trying it out. It is not fun to have all your work disappear because power went down or something happened to lose the layout.

Hit/Event Linking and the Data Lookup Display Dialog Box

Hit/Event Linking is a way to retrieve hit/waveform or event information on specified points on all types of scatter graphs ("2D Scatter", "2D Binned Scatter", "2D Color Scatter", "3D Scatter", "Spherical Location", etc.). Once a point or area is selected a dialog box is displayed showing a mini-linedisplay containing the hit or event messages linked to that point (or all points in that area). The associated waveform(s) or power spectrum(s) (if any) are displayed above it. Furthermore, a simple location plot is available with most location modes that displays the location of all selected event points. The user can hide/display and customize this plot as desired.

During hit/event linking small cursors (crosshairs) are placed on all other valid 2D scatter/binned/color graphs highlighting the selected hit or the first hit of the selected event.

Hit/Event linking can be performed at any time during or after a test that has autodump ON or a replay. A data file MUST be present for this feature to be available. Changing location settings will disable hit/event linking until a new test (with autodump on) or a replay is performed.

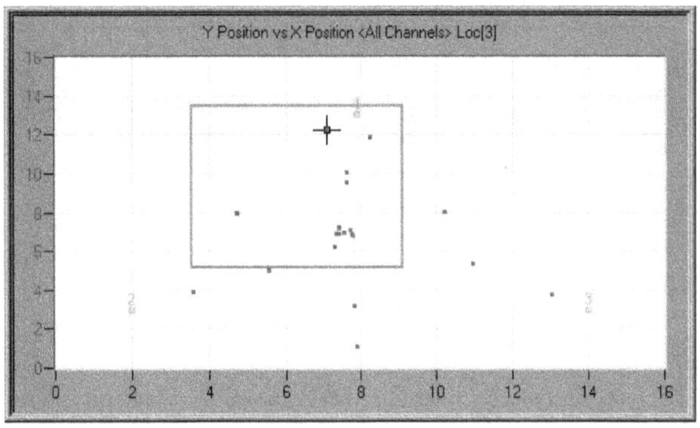

Figure 70: Area-Based Hit Event Linking

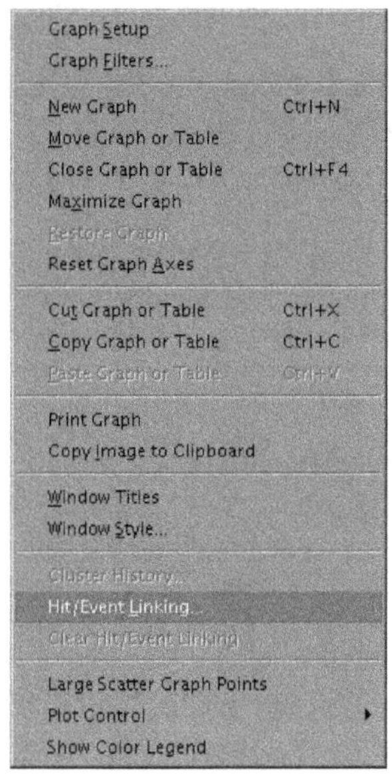

Figure 71: Point-Based Hit Event Linking

Procedure

1) Start a test with autodump ON or begin a replay.
2) Select a graph to work with by left-clicking on it. Its border will darken.
3) Select the point(s) to work with. There are two ways to do this:
   a) Point-based: You can select an individual point by using the data cursor. Activate the 'Data Cursor' from the toolbar. Position the cursor over the desired point. Then right-click the graph and select 'Hit/Event Linking' from the menu that pops up.
   b) Area-based: You can designate a rectangular area on a 2D graph. Activate 'Area Hit/Event Linking Mode' from the toolbar. If you are working with a 2D Scatter or 2D Binned Scatter graph that has multiple Y axes defined it will prompt you to select which Y Axis to work with. Then use the mouse to select an area on the graph (as if you were zooming).

   This begins the Hit or Event linking process. Depending on the graph and the size of the datafile this may take some time. Once finished, the Data Lookup Display dialog box will be displayed and the cursors will be positioned on the appropriate graphs.
4) The cursors can be cleared by right clicking any graph and selecting 'Clear Lookup Cursors', or by exiting the dialog using the 'Exit and Clear Cursors' toolbar button. Exiting the dialog using the 'Exit' toolbar button will leave the crosshairs on the graphs.
5) You do not have to exit the dialog to perform another hit/event link operation. Repeat steps 2 and 3 at any time to refresh the dialog with new data.

Controls

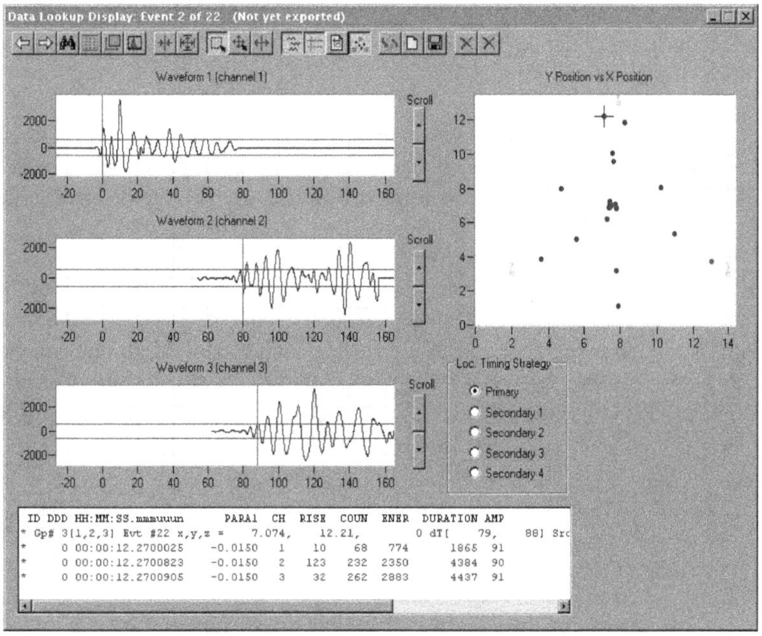

Figure 72: Data ***Lookup*** *Display Dialog Box*

The Data Lookup Display dialog box has a number of controls, most of which are located on a toolbar at the top of the dialog. These are described below. As you move the mouse cursor over each toolbar button a 'tooltip' will popup to display the button's name. Some of these controls are only available in event linking. Other controls are only available if the Waveform Option is present.

Prev/Next Buttons: If there are multiple hits/events associated with the selected point(s) then you will be able to view all of them. Use the prev/next buttons to step through them sequentially. The cursors (crosshairs) on any 2D point plot graphs will move appropriately.

Select Button: Instead of using the Prev/Next buttons to step through a large number of hits/events, the 'Select' button can be used to jump to the desired hit/event by simply entering its hit/event number.

View All Hits/Events Button: You can view all selected hits or events together by clicking the 'View All' button. This will popup a dialog box with a large line display showing all hits or events associated with the selected point or area. If you are working with a Hit-based graph (as opposed to an Event-based one) then you can export these hits to a new data file by clicking "Export to DTA". This cannot be done with events here. You can export events one at a time from the main dialog box. See section 3.9.3 below.

Set Power Spectrum Width Button: Clicking this button will open a dialog box allowing you to set the power spectrum width. You can set a specified width or accept the default.

Location Graph Setup Button: Clicking this button will open the 'Location Graph Setup' dialog box. From here you can customize the location graph, by selecting axis parameters, limits and autoscaling.

Reset Unexported User Changes Button: Clicking this button will reset any unexported changes you have made by use of the time-of-arrival cursor. Once you 'Export' the current hit/wave/event the change becomes permanent in the exported file and cannot be reset.

Reset Axes Button: Clicking this button will allow you to reset the axes of one or more graphs. A dialog box will be displayed for you to specify which graphs. You can also reset a waveform graph by right clicking it and selecting 'Reset this graph's axes'.

Select Zoom Mode Button: Clicking this button will select the 'Zoom' mode for all graphs and clear the previously selected mode. In Zoom mode you can use the mouse to select an area of a graph to view in greater detail. Use the 'Reset Axes' button to view the original graph again.

Select Pan Mode Button: Clicking this button will select the 'Pan' mode for all graphs and clear the previously selected mode. In Zoom mode you can use the mouse to drag the viewable window of graph to view data that is currently beyond the graph axes limits. Use the 'Reset Axes' button to view the original graph again.

Select Cursor Mode Button: Clicking this button will select the 'Select Cursor' mode for all graphs and clear the previously selected mode. In 'Select Cursor' mode you can use the mouse to drag the red 'Time of Arrival' cursors on waveform graphs forward or backward to effectively change its time of arrival. While in this mode the mini line-display is replaced with a table showing the time of arrival for each graph. If you do this and then click the 'Force Recalculation of Event' button the event location will change and the new (modified) location will be displayed on the location graph as a black X. Use the 'Reset Unexported User Changes' button to undo these changes (but only BEFORE you export them).

Synchronize Waveform Display Button: When this button is checked the waveforms will be displayed in the same time reference. i.e. If the second waveform in an event has an arrival time 20 us after the first waveform then the entire waveform will be positioned 20 us to the right so its time of arrival (red line) will be positioned at 20 us. If the button is not checked, then all waveforms are left justified on the graphs (their time or arrivals will be positioned at 0 us). This does not modify the waveforms in any way. It is simply a display preference.

Display Threshold Lines Button: When this button is checked, red horizontal lines appear on the waveform graphs showing the threshold. When this button is unchecked these lines are hidden.

Display Power Spectrum(s) Button: The power spectrums can be viewed by clicking the 'Display Power Spectrum(s)' toolbar button. This will replace the waveform graphs with corresponding power spectrum graphs. Click it again to return to viewing waveforms.

Show Location Graph Button: The location graph can be displayed by clicking the 'Show Location Graph' toolbar button. Clicking it again will hide the location graph. Note that this graph is only usable in event linking and only when the event being viewed has more than one hit. You can customize this graph by clicking the 'Location Graph Setup' toolbar button.

Force Recalculation of Event Button: After changing the time of arrival of one or more waveforms in an event, click this button to recalculate the event location. If successful, the new event will be shown in the mini line display and the location graph will be modified to show an X at the location of the modified event. If a new event location could not be created using the new time of arrivals then the mini line display will show a message indicating the problem. Use the 'Reset Unexported User Changes' button to undo these changes (but only BEFORE you export them). Note that certain actions can automatically force a recalculation of the event. These actions include: Changing the location timing strategy and exporting the event.

Export As (New File) Button: Clicking the 'Export As (New File)' button allows you to create a new datafile and export the currently selected hit/wave or event to that datafile. See section 3.9.3 for details.

Export Button: Clicking the 'Export' button allows you to export the currently selected hit/wave or event to a datafile. If you have already specified a datafile for export then it will use that datafile. If not then you will be prompted to specify a new datafile name as if you had clicked the 'Export As (New File)' button. See section 3.9.3 for details.

Exit Button: Clicking this button will close the Data Lookup Display dialog box without clearing the cursors (crosshairs) placed on other graphs to denote the current hit (or first hit of the current event).

Exit and Clear Cursors Button: Clicking this button will close the Data Lookup Display dialog box and clear any cursors (crosshairs) placed on other graphs to denote the current hit (or first hit of the current event).

Loc Timing Strategy Radio Buttons: The bandpass filters and timing strategies defined in the 'Timing' page of the 'Location Setup' dialog box can be applied to the waveforms/events. Use the radio buttons in the 'Loc. Timing Strategy' section to select a strategy. The waveform/power spectrum graph(s) and the mini-line display will be refreshed to show it after filtering. The default 'Primary' strategy does not apply a filter. Note that the selected strategy must have an enabled filter for a filter to be applied. See section 3.9.7 for information regarding the 'Custom Arrival' timing feature, which is particularly useful in this dialog box.

Scroll Buttons: The 'Scroll' buttons (next to each graph) are used to change which waveform is displayed in a particular graph. They allow you to 'scroll' through the list of waveforms up or down as desired. This is particularly useful when the selected event has more than 3 waveforms.

Exporting Hits, Waves and Events

One feature available in the Data Lookup Display dialog box is the ability to 'export' the current hit/wave or event to a new datafile. This can be used as a powerful and flexible way of filtering data, as you can decide on a case by case basis what to export to the new file. If you have made any changes to the time of arrival(s), then these changes are exported as well (see section 3.9.4). To export the currently selected hit/wave or event simply click the 'Export As (New File)' button and enter a filename for the new datafile when prompted. After you do this the title of the dialog box will indicate that the currently selected hit/wave or event has been exported. Subsequent exports to the same file can be accomplished by selecting a new hit/wave or event (prev/next/select buttons) and clicking the 'Export' button.

For Systems that have the Moment Tensor Option: When you perform an export on a system that has the Moment Tensor Option you will be asked if you want to export to PT files as well. This is used for moment tensor (Sigma3D). If you aren't using this then click 'no'. See the file 'Waveform Event Synchronization.doc' for details.

Changing the Time of Arrival of a Waveform and Recalculating an Event

Another powerful feature of the Data Lookup Display dialog box is the ability to 'Modify' a hit/waveform's time of arrival, and (if event-linking) recalculate the event location. To do this, simply use the 'Select Cursor' mode and use the mouse to drag the (red) time of arrival cursor on a waveform graph to the desired point in time. As you do this you will see the new time of arrival updated on a table in the mini line display. If you are in event linking then you can recalculate the event location by clicking the 'Force Recalculation of Event' button. This will recalculate the event using the new time of arrivals. If your event has more than 3 hit/waves then you might have to use the scroll controls to access some of the waveforms to modify.

If you export a hit/wave or event that has been modified in this manner then the changes will be saved to the new datafile and accessible using the 'Custom Arrival' timing feature in location (see section 3.9.7).

Other Features

1) All settings in the Data Lookup Display dialog box are saved when the layout file is saved. This includes the various on/off preferences ('display threshold lines', 'show location graph', etc), location graph setup and custom power spectrum width.
2) If the hit/event linking dialog box is open when a 'cluster/area history' operation is performed then the dialog box will be updated with its data.

Notes

1) Try loading the 'HE-LINK.LAY' layout file and replaying 'HE-LINK.DTA'. Hit/Event linking works well on this example dataset.
2) "2D Binned Scatter" and "2D Colored Scatter" graph types store their data in a 'binned' grid structure much like a Histogram (though in 2 axes). This means that each 'point' on these graphs represents an area rather than an absolute point. The cursor display (near the bottom right corner of the screen) shows the extent of the area

represented by the point. Hit/Event linking will locate all hit/events that fall into that area (lower bound inclusive, upper bound exclusive).

3) Selected hit/events are not highlighted with crosshairs on "3D Scatter" and "Spherical Location" graph types, though hit/event linking can be initiated from them.
4) It is possible to perform hit linking on TRA datafiles even though they lack hits. In this case only the waveforms will be displayed.
5) Event Linking functionality can be disabled by unchecking the "Enable Event Linking" checkbox in the Acquire/Replay menu (this setting is NOT saved in the layout). Doing this may result in a modest performance improvement in high data rate tests with large numbers of events. Once disabled it cannot be re-enabled until a new test or replay is performed. Disabling Event Linking will not disable Hit-Linking functionality.

Custom Arrival Timing Feature

When a user-modified waveform is exported it is saved in its unmodified state to the new datafile. The user-modified TOA (time of arrival) is also saved for future use. This can be accessed through the use of the 'Custom Arrival' timing feature.

Before starting the replay, enter location setup. Set one of the timing strategies in the timing page of the location setup dialog box to the 'Custom Arrival' feature. If you want location plots to use the modified TOAs then select this strategy in your location group's 'Timing Strategies'. Or, you can simply set the Primary strategy to 'Custom Arrival' so all location groups will use it. Since 'Custom Arrival' only works in replay, this will default to 'FTC' (First Threshold Crossing) during acquisition.

<u>If you want to see the modified TOAs in hit/event linking:</u>
When the Data Lookup Display dialog box is opened it will plot the locations and waveforms based on the currently selected 'Loc Timing Strategy'. Note that the dialog will remember your previous setting and use it by default the next time it is loaded. If this strategy uses a 'Custom Arrival' timing feature then all the locations will be plotted using the modified TOAs. If this strategy doesn't use a 'Custom Arrival' timing feature then all locations will be plotted using the original TOAs.

Once the Data Lookup Display dialog box is loaded you can switch strategies through the use of the 'Loc Timing Strategy' radio buttons. Doing so will NOT re-plot all locations but it will re-plot the current one AND reposition the TOA cursors on all current waveforms. You can switch back and forth between a timing strategy that uses 'Custom Arrival' and one that doesn't to view the difference between the two.

Cluster/Area History

Cluster History and Area History provide the ability to view detailed graphical information about AE activity in a specified area on a 2D Scatter, Binned Scatter or Colored Scatter graph (called the 'source' graph). This area can be defined by selecting a cluster (Cluster History) or by designating an area manually (Area History). This information is displayed on any number of graphs defined and designated by the user as 'History Graphs'. These graphs remain inactive until the user starts a Cluster or Area History process. When the user does so the software 'replays' the data contained in the selected cluster or area onto all history graphs. The history graphs then remain active for the remainder of the test or replay, accepting/displaying any incoming data that would fall within the original bounds of the cluster or area at the time the process was started. This process can be repeated any number of times during a test or replay.

Setup

1) To use Cluster History you need to create at least one cluster graph before starting the process, usually before starting the test or replay. Area History does not require this.

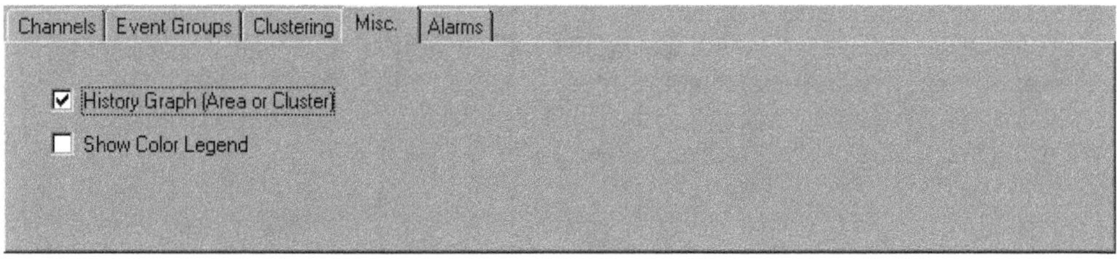

*Figure 73. History **Graph** Checkbox*

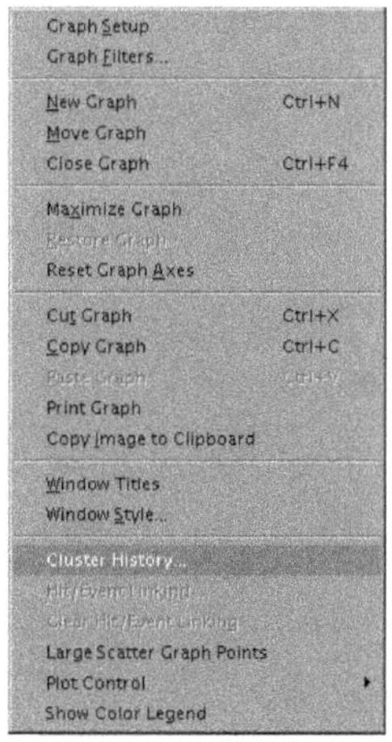

*Figure 74. **Context** Menu*

2) Create as many history graphs as desired. This can be done at any time before you start the history process. Any graph can be designated as a 'History Graph' by checking the 'History Graph (Area or Cluster)' checkbox on the 'Misc' page of graph setup. The graph's title will then include the '(Hst)' tag.
3) Start an acquisition test (with autodump on) or a replay.

Starting a Cluster History Process

Cluster History is started by right clicking on a cluster graph and selecting 'cluster history' from the menu that pops up. If this option is grayed out then it means one of the following:

1) There are no clusters on the specified graph.
2) You disabled 'Event Linking and Cluster/Area History ' from the Acquire/Replay dialog when you started the current test or replay.
3) You started a test without autodump on.
4) You started a history process or Hit/Event Linking process that hasn't completed yet.

If there are more than one cluster available on the selected cluster graph then the program will prompt the user to select one from a list.

Starting an Area History Process

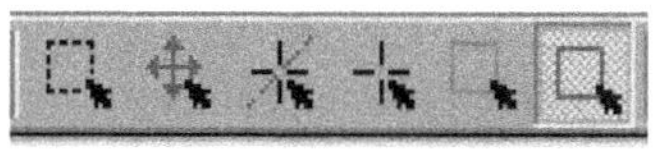

*Figure 75. **Area** History*

Area History involves manually designating a rectangular area on the graph. Activate 'Area History Mode' (blue square) from the toolbar.

If this option is grayed out then it means one of the following:
1) The active graph is of an invalid type (i.e.: histogram, waveform, 3D graph, etc).
2) You disabled 'Event Linking and Cluster/Area History' from the Acquire/Replay dialog when you started the current test or replay.
3) You started a test without autodump on.
4) You started a history process or Hit/Event Linking process that hasn't completed yet.

If you are working with a 2D Scatter or 2D Binned Scatter graph that has multiple Y axes defined it will prompt you to select which Y Axis to work with. Then use the mouse to select an area on the graph (as if you were zooming).

Execution

The Cluster/Area History process is detailed below:
1) All history graphs are re-initialized. Any data on them (from a previous history process) is cleared.
2) The setup of all history graphs are modified to reflect the setup of the source graph.

- Channel sets, location group sets, and the data source (hits/events) are made identical to the source graph.
- The min/max limits of any graph axes which share the same graph quantity as on the source graph are changed to reflect the limits of the cluster or area.

3) Special graph filters are set on all history graphs to reject any future incoming data that does not fall within the (original) cluster or area from being displayed. These filters persist until changed by another history process, the user turns off the 'History Graph' checkbox on the graph or until the end of the test/replay.
4) All history graphs are enabled. This allows them accept and display historical and future incoming data.
5) Historical data (the data within the cluster or area) is displayed on all history graphs.

Notes

1) A Cluster/Area History process can be started at any time during or after a test that has autodump ON or during/after a replay. A data file MUST be present for this feature to be available.
2) It is important to note that the bounds of a history graph reflect the bounds of the cluster or area at the time when the process was started. These bounds do not change if the cluster itself grows.
3) Cluster/Area History on event based source graphs is disabled if the 'Enable Event-Linking and Cluster/Area History' checkbox on the Acquire/Replay dialog is disabled before the current test or replay begins.

Graph Alarms

'Graph Alarms' is a feature that enables certain graph types to trigger an alarm whenever data is plotted on them above one or two pre-set thresholds ('warning' and 'trip'). They are available on 2D Histograms, 2D Scatter graphs, Binned Scatter graphs and Colored Scatter graphs during data acquisition.

How to setup a graph alarm:

An 'Alarms' tab is available in graph setup (for graph types that can use them).

1) Check the 'Enable Graph Alarm' checkbox to turn on graph alarm functionality for this graph.
2) Check the 'Enable Warning' checkbox to enable the alarm 'warning' level. Check the 'Enable Trip' checkbox to enable the alarm 'trip' level. Either or both may be used at the same time.
3) Set the vertical axis thresholds for the alarm levels you have enabled. These thresholds are inclusive, so a threshold of 80 will be triggered by a signal of 80 or above.
4) Select which vertical axis controls the graph alarm (Primary, Secondary-1, etc).
5) If desired, you may enable the 'Alarm Hold-Off' and specify its '#Signals'. See the section on 'Alarm Hold Off' below for information on this feature.

How an alarm is detected/triggered:

Graphs that have an enabled graph alarm will display horizontal lines at the levels of the thresholds set in graph setup (Green for warning, Red for trip).

A graph alarm 'Warning' will trigger when all of the following occurs.

- The 'warning' level is enabled for that graph.
- New data is displayed on this graph at or above the green threshold line.
  - This needs to occur a number of times consecutively equal to the specified '#Signals' if used.
- No graph alarm (Warning or Trip) is currently active (on this graph).

An alarm 'Trip' will trigger when all of the following occurs.

- The 'trip' level is enabled for that graph.
- New data is displayed on this graph at or above the red threshold line.
  - This needs to occur a number of times consecutively equal to the specified '#Signals' if used.
- No graph alarm trip is currently active (on this graph).

'New' data is defined as any data that is plotted after the alarm was last reset. Note that the alarm is automatically reset at the start of the test.

What happens when an alarm is triggered:

When a graph alarm is triggered several things happen:

- The audio monitor (if installed) sounds an audible beeping sound. This continues until the alarm is cleared or the test ends.
- The alarm output (WARNING or TRIP as appropriate for the level) is set high. This stays high until the alarm is cleared or the test ends.
- An alarm message (16) appears in the Line Display in red, and is saved to the datafile if autodump is on.

A small dialog box appears, displaying the conditions of the alarm. The alarm can be cleared by clicking the 'Clear Alarm' button. You can exit this dialog without clearing the alarm if you wish. It can be accessed later by selecting 'Display Last Alarm…' from the Acquire/Replay menu.

Alarm Hold Off

Normally (when the "Alarm Hold Off" is disabled) an alarm is triggered when a single data point ("signal") is plotted above threshold.

By enabling 'Alarm Hold-off' you can delay alarm triggering until a specified number of consecutive data points ("signals") are plotted above threshold. This is useful when you wish to avoid having data spikes trigger alarms. It is important to remember that the string of data points must be consecutive for an alarm to be triggered. Any data points below threshold will 'reset' the internal counters.

For example, if you set '# Signals' to 5 then an alarm would only be generated until 5 consecutive data points are plotted above threshold. Let's say your threshold was 70. The following sequence:

- 70, 80, 75, 80, 70 would trigger an alarm.
- 70, 80, 60, 80, 70, 80 would NOT trigger an alarm, because even though 5 signals passed threshold, they were not consecutive.

Alarm 'Warning' and 'Trip' both maintain their own separate internal counters (though both use the same '#Signals'), so a particular sequence might trigger an alarm 'Warning' without triggering an alarm 'Trip'.

Tips:

1) Graph alarms are checked when the graphs are refreshed. If you want to increase the rate at which they are checked then you should increase the rate at which the graphs are updated. See the settings in 'Display Mode…' under the 'Test Setup' menu.
2) Note that data does not have to be in the visible portion of the graph to trigger an alarm. Panning/Zooming of the graph will not hinder graph alarm operation.
3) Setting #Signals to 1 is identical to disabling 'Alarm Hold Off'. In fact, if you do so, the system will automatically disable 'Alarm Hold Off' for you.
4) Caution is recommended when using Compression on the horizontal axis of histograms, particularly when using "Sum" Bin Display mode. In this scenario the bins will become taller as the graph compresses horizontally, an effect of the bins being combined as part of the compression process. Be aware that data is only checked against alarm criteria only when it is first plotted, not when the graph is refreshed during compression. A bin that suddenly grows above threshold due to compression will not trigger an alarm.
5) Caution is recommended when using 'Alarm Hold Off' on graphs that have all the following settings selected:
    - Graph is plotting RMS, ASL, Threshold or Absolute Energy.
    - The 'Include Time Driven' checkbox is checked in the 'Input Data' section of Graph Setup.
    - Graph is plotting more than one channel.

   In this case you are plotting channel-based 'time driven data', which can plot multiple points from a single time driven data message. 'Alarm Hold Off' will advance its internal counter at most once per message, so this can be tricky if your graph is plotting more than one channel. If you want to use 'Alarm Hold Off' with Graph Alarms on the type of graph listed above it is advised that you create multiple single channel graphs instead.

6) It is recommended that 2D Scatter (not Binned/Colored) graphs that have a graph alarm enabled use only Manual scaling on the Y (vertical) axis. Using compression or sliding scale won't hinder graph alarm operation but it may slide the threshold line off the visible portion of the screen.

List of Shortcut Keys

The following is a list and short description of all the shortcut keys available in the software. Many of the shortcut keys follow in the tradition of 's DAQ and LOC software programs and should be familiar. Others follow standard WINDOWS conventions. Menu items usually do list the shortcut key, but the full list is provided below for convenience.

Key	Function
F1	Help Menu
Shift + F1	Context Sensitive Help
F2	Hardware Setup
F3	Graph Setup
F4	
Ctrl + F4	Close (Delete) Graph
Shift + F4	Delete Page (Entire Graph Screen)
F5	
F6	View Cluster Table (Toggle)
Ctrl + F6	Controls Cluster Table Sorting
Shift + F6	Shifts Cluster Table Position
F7	View Line Dump (Toggle)
F8	Location Setup
F9	Acquisition
Shift + F9	Acquire to Existing File
F10	Replay
F11	
F12	
Ctrl + C	Copy Graph
Ctrl + D	File New
Ctrl + H	Window Tile Horizontal
Ctrl + N	Add New Graph
Ctrl + O	Open Layout File
Ctrl + P	Print Page
Ctrl + S	Save Layout
Ctrl + T	Window Tile Vertical
Ctrl + V	Paste Graph
Ctrl + X	Cut Graph
Ctrl + W	Waveform Collection (Toggle)
Ctrl + Shift + C	Copy to Clipboard
Ctrl + Insert	Copy Graph (same as Ctrl + C)
Shift + Insert	Paste Graph (same as Ctrl + V)
Ctrl + Alt + O	Options Dialog

LOCATION

The purpose of this section is to familiarize you with setting up and using the software location modes. Location is the process of collecting incoming hits into events & analyzing the arrival times of the hits in an event to produce a source location. Individual location modes have several key differences which are summarized in the table below.

Location Mode	Hits Used	Description	Analysis Type	Required Option
Zonal	1	Trivial solution to location. Identifies the sensor closest to the source using the relative arrival times of the hits. Returns coordinates of the first hit sensor as source location.	Single	Standard
Linear	2	Designed for 1 dimensional location on a line. Performs linear interpolation between 2 sensors' coordinates based on differences in the arrival times of the first 2 hits in the event.	Single	Standard
LinearXY	2	Linear without the assumption of sensors in a straight line & events allowed between any sensor pair. For example, sensors could be in a square with sources locating on the sides or diagonals.	Single	Standard
Tank Bottom	3	Designed for analyzing evenly sed 2D circular arrays of sensors.	Single	Tank Bottom
2D Planar	3-8	Designed for an arbitrary array of sensors on a 2D plane.	Regression	Full
2D Planar (xy) 2D Planar (yz) 2D Planar (xz)	3-8	2D Planar designed to complement 3D Location. Locates using just 2 of the 3D coordinates & ignores the 3rd. Useful for some events that fail normal 3D Location.	Regression	3D
2D Anisotropic	3-8	2D Planar allowing for differing velocities in the x & y directions.	Regression	Full
Cylinder	3-8	2D Planar analysis wrapped around a 3D cylindrical shell. The ends of the cylinder may be open or have flat, spherical or elliptical endcaps. Assumes sound travel through cylinder shell rather than its interior.	Regression	Full
Anisotropic Cylinder	3-8	Cylinder allowing for differing velocities in the x & y directions. Only supports location on cylinder body, not on the endcaps.	Regression	Full
Conical	3-8	2D Planar wrapped around a 3D conical shell.	Regression	Full
Spherical	3-8	2D Planar wrapped around a 3D spherical shell again assuming sound travel via the shell only.	Regression	Spherical
3D Location	4-8	2D Planar extended to 3D coordinates assuming sound travel directly through se rather than via an external shell.	Regression	3D Location
3D T0 Location	5-8	3D Location with time zero input. Assumes all events are signaled via one channel that indicates the absolute start time of an event.	Regression	3D T0 Location

The most important difference in the location modes is the geometry they are designed for. The geometry in turn affects the number of hits used per event to calculate the location. The type of analysis also affects the number of hits used in an event. The single analysis modes use just enough hits to create one set of simultaneous equations which are then solved to extract the source location coordinates. A more complex approach uses more hits & multiple regression analysis to produce the source location that best fits all the available data. By using data from extra sensors, the regression modes yield a more accurate source location with improved immunity to errors in the individual arrival times. For more details, see the section 4.1, Theoretical Background.

Two location modes, zonal & linear, are standard features. Others are location enhancements that require additional licensing options. If you are not sure what locations options are enabled in your software, look for the Options dialog available on the Help pulldown menu. It will show a checkmark next to any purchased options. If you wish to purchase additional license options for other location modes, contact , Physical Acoustics Corp. at 1-609-716-4000.

Theoretical Background

The following section provides a brief overview of the math behind location. It is definitely not required that the average user to comprehend this material. Its purpose is to educate the inquisitive & allow the user additional insight into what location is doing behind the scenes.

The fundamental basis for the location calculation is just the simple time-distance relationship implied by the velocity of the sound wave. The absolute arrival time, t, of a hit in an event can combine with the velocity, v, of the sound wave to yield the distance, d, from the sensor to the source.

$$d = v * t \tag{1}$$

The distance between 2 points depends on the geometry of the problem. The majority of the location modes are a variation of 2 dimensional source location in a plane, although in many cases the 2D plane will wrap around a 3 dimensional object. For 2 points in a flat plane, the distance equation is just the Pythagorean theorem expressed in Cartesian coordinates:

$$d = \sqrt{(x_2 - x_1)^2 + (y_2 - y_1)^2} \tag{2}$$

This calculation is complicated by the lack of knowledge of the exact time the event originated. To get around that problem, all the times are considered relative to the first hit in the event. Each arrival time difference implies a difference in distance to the sensor relative to the distance to the first hit sensor. For the second hit sensor relative to the first hit sensor, a difference equation can be written as:

$$t_2 - t_1 = (d_2 - d_1)/v \tag{3}$$

The distance equation (2) can be combined with the difference equation (3) to yield:

$$t_2 - t_1 = [\sqrt{(x_2 - x_S)^2 + (y_2 - y_S)^2} - \sqrt{(x_1 - x_S)^2 + (y_1 - y_S)^2}]/v \tag{4}$$

where x_S & y_S are the unknown coordinates of the source. This equation contains 2 unknowns & cannot be solved by itself. To get a second equation with the same 2 unknowns, a 3rd hit is added to the event producing analogous equation:

$$t_3 - t_1 = [\sqrt{(x_3 - x_S)^2 + (y_3 - y_S)^2} - \sqrt{(x_1 - x_S)^2 + (y_1 - y_S)^2}]/v \tag{5}$$

These simultaneous equations can then be solved for x_S & y_S, although the algebra is fairly complex. Tank Bottom location uses this particular approach to produce a source location from 3 hits as did the original version of 2D Planar location before it was extended to include multiple regression analysis.

If there are additional hits added to the event, there is a question of how to use the extra information. In the case of Tank Bottom location, the algorithm simply takes the extra hits & tries them 3 at a time to produce extra source locations. It iterates through all the 3 hit combinations of the hits in the event until it exhausts them or runs into a user specified maximum. The problem with this approach is that it produces more than one source location per event & if there is any error in the timing values, the source location can be wildly incorrect. A better approach would be to average the data somehow to produce a single location.

Multiple regression analysis does just that, although it does not actually average the results of multiple 3 hit calculations directly. Rather it searches for the location that best fits all the all the available data. Each additional hit basically adds an extra equation to the set of simultaneous equations given above as (4) & (5). This can be generalized as:

$$t_i - t_1 = [\sqrt{(x_i - x_S)^2 + (y_i - y_S)^2} - \sqrt{(x_1 - x_S)^2 + (y_1 - y_S)^2}]/v \tag{6}$$

If one defines

$$\Delta t_i = t_i - t_1 \tag{7}$$

one can write:

$$\Delta t_i = [\sqrt{(x_i - x_S)^2 + (y_i - y_S)^2} - \sqrt{(x_1 - x_S)^2 + (y_1 - y_S)^2}]/v \tag{8}$$

Equations (7) & (8) give 2 ways of calculating the Δt of the ith sensor. Using the known arrival times of the hits, equation (7) calculates the observed time difference, $\Delta t_{i,obs}$,. For a given set of source location coordinates, equation (8) defines the calculated time difference, $\Delta t_{i,calc}$. Multiple regression analysis is a general purpose algorithm that minimizes the difference between 2 quantities which in this case are the observed & calculated Δt values. To do that a quantity called χ^2 (chi square) is calculated. In our case, it is defined as a sum across all sensors in the event:

$$\chi^2 = \Sigma(\Delta t_{i,obs} - \Delta t_{i,calc})^2 \tag{9}$$

This quantity is also called the fit value and is dependent on the x,y coordinates of the source location . The sum is recalculated for each potential source location. Assuming no error in the data, χ^2 will have a value of 0 at the source location. The location code searches for the values of x_S & y_S that minimize the value of χ^2. The process is an iterative search, because it is not possible to directly write down simple equations for the values of x_S & y_S that minimize χ^2.

The search is performed using one of 2 algorithms, either a Simplex search (3D or 2D Planar modes) or Powell's Method (Spherical, Cylinder & Conical Location). Both search methods are based in general on code & algorithms described in Numerical Recipes in C, 2nd edition (NRC). The code has obviously been heavily adapted for the problem of acoustic emission source location, so there are significant variations from the NRC originals. The interested user may, however, find more information about Simplex & Powell fitting & searching in NRC in chapter 10 & more information on regression analysis in chapter 15.

The math gets more complicated when extended to 3 dimensions or involves wrapping around a 3D object, but the approach is still basically the same. An important point is that the regression approach is no longer over determined in the case of the event containing the minimum number of hits, i.e. 3 hits for 2D Planar. In that case, however, regression analysis turns out to be mathematically equivalent to solving the complicated algebra of equations (4) & (5) & yields the same location coordinates.

Location Setup Dialog

When setting up location groups, you use the Location Setup dialog box (F8 shortcut key). It has 2 tabs labeled General and Timing. Source locations are produced by analyzing events. Events are just collections of hits & their associated time of arrival. The settings on the General page control how hits are combined into events & how events are analyzed by the location modes to produce source locations. The settings on the Timing page control how the time of arrival of each individual hit is determined and even whether or not a hit is considered significant or excluded from the location analysis.

There are several features to make setting up location groups easier. First you should be aware of the F5 function key. It allows the user to quickly set all rows in a column of the grid to the value of the current cell. It works on both the General & the Timing tabs. In the example setup shown below, pressing F5 would set all rows of the Overcal column to the current cell's value of 200. Alternatively, you can right click a cell & select Set Column from the context menu that appears. The context menu for the General tab, also has Copy Group & Paste Group options. Right click & select Copy Group on source row, then right click the destination row & select Paste Group. All of the group settings & the associated sensor & structure settings will be copied to the new row. By default, Paste Group will use the first group as the source, if you did not use the Copy Group option to specify a source row.

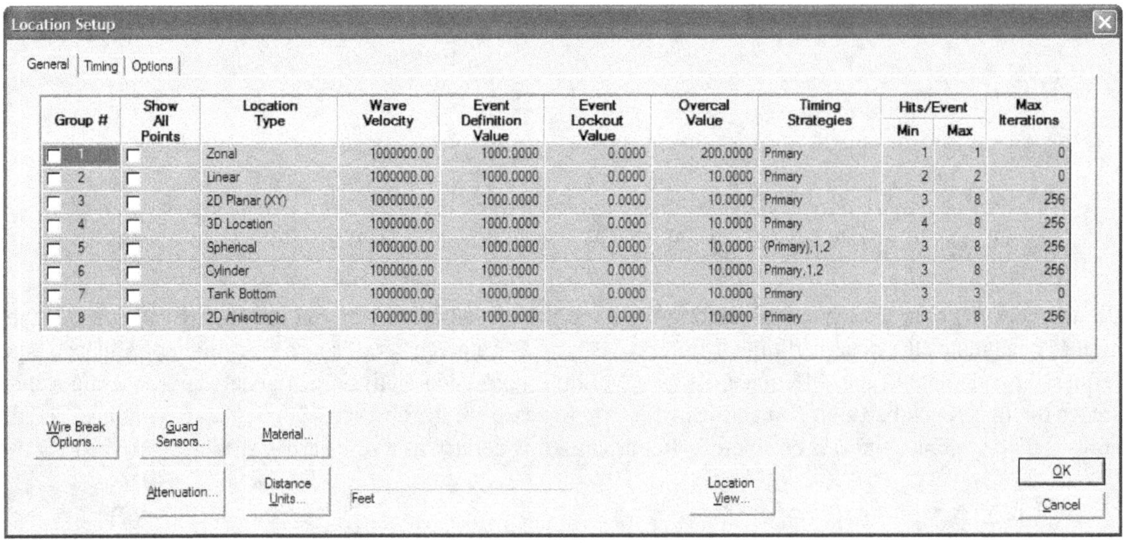

Figure 76. ***Location*** *Setup – General Page*

General Tab Settings:

The layout for the General tab is shown above. The grid on the General page of Location Setup contains a list of location groups & their type. If you click an entry in the Location Type column, an arrow appears allowing you to change the type. As mentioned above, many location types (3D, Spherical, Cylinder) appear only if you have purchased the appropriate option.

Group #: This column acts as an On/Off setting. Unchecked groups are ignored when doing a replay.

Show All Points: This column allows the user to show all source results for events that this group can successfully locate. Independent of which groups have Show All Points checked, the software always allows all enabled groups to analyze a given event to get the best results for each group and then compares the results from all successful groups to come up with a best result. For example, linear location is always presumed to give a better result that zonal location, if there are enough hits(2) to produce a location. If no groups have Show All Points checked, the software will report one source location per event & it is the best result that is shown on graphs & in the line display. There are times when a side-by-side comparison of the results of 2 locations groups are useful. In that case you can mark the groups as Show All Points & then whatever source locations they can calculate will be reported. The software will appear to sometimes produce more than one source location per event & if you look at the line display, you will see extra event listings with the same first hit times. The extra location results will also show up as extra points on your graphs too, but they only count as 1 event in the Statistics bar. In all cases you will see the software's best result as one of the points. There will be times when an event cannot be located by more than 1 group, so having Show All Points does not guarantee a result from a given location group. It is only a guarantee that you will see it, if it can produce a useful result.

Location Type: The Location Type column lets you specify which location mode to use with this particular location group. If you click on a cell in this column, a down arrow appears on the right hand side of the cell. If you click the down arrow, a drop down list of the available location choices will appear. The available choices do depend on the software licensing options as mentioned in the introduction, so you may not see all the location modes that are discussed in this manual.

Wave Velocity: The Wave Velocity is the expected rate of propagation of the acoustic emission signal you wish to monitor. It is entered in the user's Distance Units per sec. For example, if the Distance Units are in Inches, then the Wave Velocity takes units of Inches/sec. If the user's Distance Units are time based, this column is ignored by the location algorithm. The Materials button will allow you to access a database of known velocities for various materials.

See the section on the Material Properties dialog for more information. It is always recommended that you determine the appropriate value by measuring the velocity on your test structure using lead breaks, for example.

Event Definition Value: The Event Definition Value (EDV) is the length of an event in the user's Distance units. If the user's Distance Units are time based, then the EDV is directly the maximum time allowed between first & last hits in an event. If the EDV is distance based, then it is the physical length between the first and the last hit sensor of the event. The goal is to select an Event Definition value that will give enough hits to perform source location with the desired accuracy. For example, in the figure below, the 2D location requires a minimum of 3 hit sensors to locate, however more sensors can provide a more accurate source location, therefore, the inner circle radius defines the Event Definition Value and in the case shown, it will detect channels 9, 10, 15, 16, 17, 21, and 22, so up to 7 channels (hits) will be able to form the event if the EDV value as shown is entered.

Since the location calculations use time information, any distance units will be later converted to time units using the velocity entered in the Wave Velocity column. The velocity units are the user's Distance Units per second. For example, if the Distance Units is set to meters, then the velocity units are meters/sec.

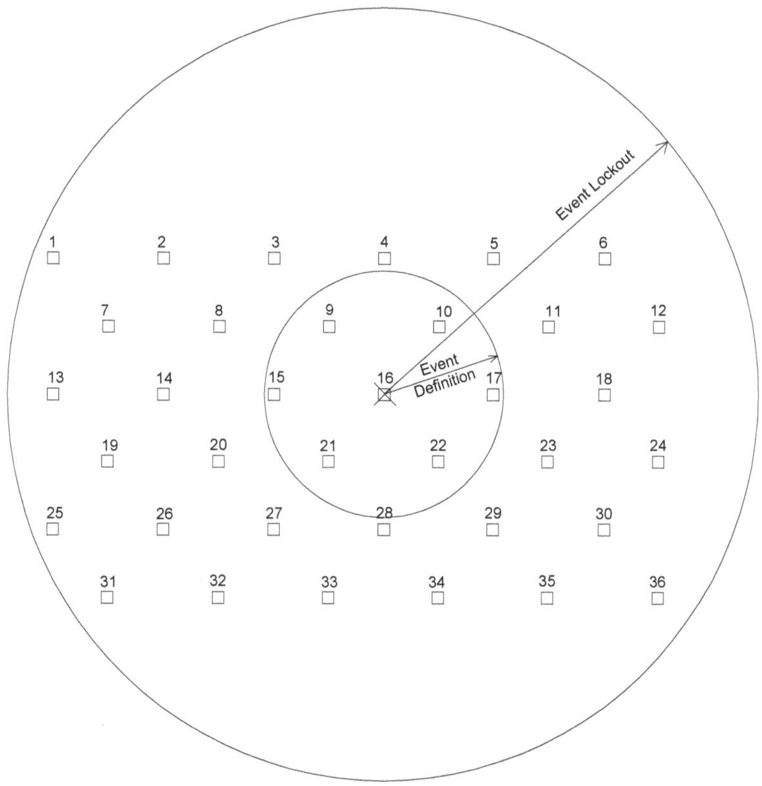

Figure 77. Relationship between Event Definition and Event Lockout.

Event Lockout Value: The Event Lockout Value is also a length of time or distance entered in user units, but it controls the interval between consecutive events. The Event Lockout of the last located event blocks hits in all location groups until it has expired. The Event Definition Value & the Event Lockout Value are both specified relative to the start of the event (time of test of the first hit). An Event Lockout Value of 0 allows overlapping events in 2 *different* location groups. Overlapping events are not permitted in the same location group no matter what the Event Lockout Value is. An Event Lockout Value larger than the Event Definition Value allows hits to be ignored after an event, for example, if reflections could generate extraneous acoustic signals. In the example shown in the figure above, all AE hits between the Event Definition Value and the Event Lockout will be ignored from the point of view of being part of the event in progress or the next event.

Overcal Value: The Overcal Value or more fully the over-calibration value is also a length that is specified in the user's Distance Units. Before data acquisition or a replay begins, the software calculates a table of time differences between sensors using their known coordinates & the user entered wave velocity. The Overcal Value is also converted to a time value if needed. After the first hit in an event, these calibration times are compared to the time differences of later hits relative to the first hit sensor. In order for a hit to be included (or accepted) as part of the event, its time difference must be less than this calibration time, but the Overcal value provides an allowance for a margin of error. The time difference can be higher than the calibration time by the Overcal Value & still be included in the event.

Timing Strategies: The Timing Strategies column is used to specify which strategies listed on the Timing page will be applied to a particular location group. You set up the strategy on the Timing page, but you control whether it is used with a particular location group using the Timing Strategies column. See the section on the Timing page for information on the settings for strategies & use of the Timing Strategies column.

Hits/Event Min & Max: The Hits/Event (Hits per Event) columns are related to the regression modes. In theory, the multiple regression can sum across any number of sensors (hits) in a given event when calculating a fit, but there are limits in the code. First the absolute minimum number is determined by the mathematical requirements specific to the location mode. For Spherical, Cylinder Location and all 2D Planar regression modes, the event must contain at least 3 hits to produce a location. For 3D Location a minimum of 4 hits are required to analyze an event & produce a location. All regression modes can make use of up to 8 hits in an event.

The Hits/Event Min & Max settings allow the user control over the analysis between these limits. For example, in 2D Planar, a minimum of 3 hits are required to produce a location result, but then this is not an over determined result. If there is an error in the timing data of any one of the 3 hits, the location calculated will be in error. With 4 or more hits, the regression analysis produces a best fit location that averages out errors in the data. Thus, the user might want to adjust the Hits/Event Min setting to 4 for 2D Planar to eliminate analysis of 3-hit events. This will reduce the number of events observed & generally reduce scatter in the results. In some cases, the user might want to only look at events with a specific number of hits. Setting the Hits/Event Min & Max settings to 4 will limit the analysis to events with 4 or more hits using only the first 4 hits to form the event.

Max Iterations: The Max Iterations column is also related to the regression modes. In this case, the setting controls the maximum number of locations evaluated in the search for the best fit location. It is not strictly speaking the same as the number of times through a search loop for 2 reasons. First some searches can test a variable number of locations during one pass though the loop. Second other searches use loops within loops. Due to the variable nature of the search algorithms, the actual number of locations tested can be slightly higher (0-3) than the Max Iterations setting. A lower Max Iterations setting will terminate long location analyses sooner, but the higher speed results in lower accuracy in the best fit location produced.

When the Location Type is not set to an regression mode, the Max Iteration column & the Hits/Event Min & Max columns will be disabled. In that case the Max Iterations column displays 0 as iterations are not used for the other modes. The Hits/Event Min & Max settings will be updated with the number of hits required for the location mode. For example, the Linear location mode requires 2 hits to produce a location and if Location Type is set to Linear, Hits/Event Min & Max will both display 2 & the user will not be able to adjust the setting.

Timing Tab Settings:

Timing strategies processes hits and their accompanying waveform to produce a time of arrival. A list of arrival times a series of hits is grouped together to form an event. The location modes analyze the arrival time differences (delta t's) in the event to calculate the source location. For an enabled location mode, the software always analyzes the event at least once. The first analysis is performed using only the arrival times from the Primary Strategy. If you use additional strategies with one of the regression modes, the software performs a second event analysis to refine the source location by adding the arrival times produced by the secondary strategies (1-4) to the event.

The settings for the strategy are on the Timing page, but you control which location groups use them with the Timing Strategies column on the General page. When you click on a cell in the Timing Strategies column, the Select Strategies dialog appears as shown below.

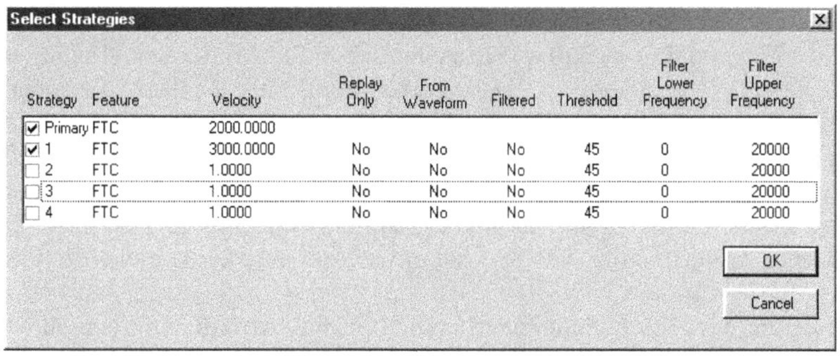

Select Strategies

Strategy	Feature	Velocity	Replay Only	From Waveform	Filtered	Threshold	Filter Lower Frequency	Filter Upper Frequency
☑ Primary	FTC	2000.0000						
☑ 1	FTC	3000.0000	No	No	No	45	0	20000
☐ 2	FTC	1.0000	No	No	No	45	0	20000
☐ 3	FTC	1.0000	No	No	No	45	0	20000
☐ 4	FTC	1.0000	No	No	No	45	0	20000

OK

Cancel

Figure 78. Select Strategies Dialog

Simply check or uncheck the strategies you want to use. The dialog summarizes the current strategy settings from the Timing page. After you click on OK, the cell in the Timing Strategies column will update to list the enabled strategies. A notable exception is the Primary Strategy which always appears in the list as either Primary or (Primary).

With additional strategies enabled, unchecking the Primary Strategy causes the arrival times from the Primary Strategy to be omitted from the second event analysis, but they are still used to perform the first analysis. If the location mode cannot produce a useful source location after the first analysis, the software will not attempt to refine the location with the secondary data. In that respect the Primary Strategy acts like a filter for the second analysis even when the Primary results are not being used in the second analysis. To reflect this implied filtering, the Timing Strategies column displays (Primary) whenever the Primary Strategy is not to be included in the secondary analysis. In the figure showing the General settings page above, Location group 5 has Strategies 1 & 2 turned on with the Primary turned off and displays "(Primary),1,2". Unchecking all the strategies is equivalent to checking only the Primary Strategy. In both cases the event is analyzed just once with the Primary Strategy and the Timing Strategies column just displays "Primary"

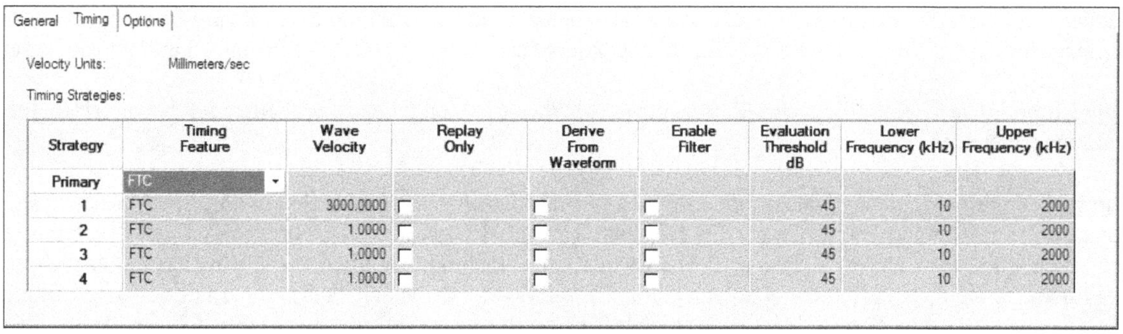

General | Timing | Options

Velocity Units: Millimeters/sec

Timing Strategies:

Strategy	Timing Feature	Wave Velocity	Replay Only	Derive From Waveform	Enable Filter	Evaluation Threshold dB	Lower Frequency (kHz)	Upper Frequency (kHz)
Primary	FTC							
1	FTC	3000.0000	☐	☐	☐	45	10	2000
2	FTC	1.0000	☐	☐	☐	45	10	2000
3	FTC	1.0000	☐	☐	☐	45	10	2000
4	FTC	1.0000	☐	☐	☐	45	10	2000

Figure 79. Location Setup – Timing Page

The current layout for the Timing page of the Location Setup dialog is shown above. For each strategy there are a number of options available. Those to the right of Derive From Waveform are all related to waveform processing.

For the Primary Strategy the only available option is the Timing Feature option. The other columns are not used because they are unneeded or not allowed for the Primary Strategy. For example, the Primary Strategy always uses the velocity associated with the location that is listed in the Wave Velocity column on the General page. The Primary Strategy is always applied during both acquisition & replay. The Primary Strategy relies on the results reported by your hardware in a single hit message, so waveform processing options are not available. The software later searches & pairs each hit message with an accompanying waveform message making the waveform processing options available for the secondary strategies.

Timing Feature: The Timing Feature column controls how the arrival times for a hit are determined. There are 2 basic choices: FTC (First Threshold Crossing) and Peak Timing (PT). All the arrival times are based on the time of test associated with the hit and a variable offset. During data acquisition, the hardware actually sets the time of test of a hit based on a threshold crossing where the Threshold level is controlled by the user in the Hardware Setup dialog. The FTC setting actually just defaults the arrival time to the time of test for the hit with no offset. The Peak Timing parameter adds the risetime (RT) as determined by the hardware to the hit time of test. Risetime is defined as the delay between the first threshold crossing and the highest amplitude peak during the Peak Definition Time (PDT). The risetime feature must be enabled in the Data Sets tab of the Hardware Setup dialog for this option to work.

In addition to FTC and PT there is a third timing feature called Custom Arrival. This is used when replaying a datafile that has had the time-of-arrival of some or all of its waveforms modified through hit/event linking. If this timing feature is used when replaying such a datafile then the modified time-of-arrival's will be used instead of the original ones. For any other timing feature, the original time-of-arrivals will be used. Custom Arrival is designed for use in replay. If it is used during acquisition it will default to FTC behavior.

Wave Velocity: The Wave Velocity column contains the velocity used during analysis by the location modes. The regression modes use a velocity to convert arrival time differences into differences in distances to the sensors. There is a velocity associated with each hit arrival time. For the Primary Strategy that velocity always comes from the location group Wave Velocity setting. During the second analysis, the velocity from the secondary strategies are used when the arrival times are added to the list of hits in the event.

Replay Only: The Replay Only checkbox controls whether the strategy is applied during both data acquisition and replay or only in replay mode. Since secondary strategies cause events to be analyzed twice & the arrival times may be extracted from a digitally filtered waveform, secondary analysis can slow data acquisition rates. In cases where high data acquisition rates are important, you may prefer to postpone the secondary analysis until after data acquisition is over. The Replay Only flag just makes it easier to disable the strategy during acquisition without having to continually access the Location Setup menu to toggle the strategy on or off.

Derive from Waveform: The Derive From Waveform checkbox controls whether the strategy attempts to extract the arrival time from the waveform. If it is disabled (unchecked), the arrival time can still vary from that produced by the Primary Strategy if the Timing Feature for the secondary strategy varies from the Primary Timing Feature setting.

Enable Filter: The Enable Filter checkbox controls whether a digital filter is applied to the waveform before the arrival time is determined.

Evaluation Threshold: The Evaluation Threshold is similar to the Hardware Setup Threshold used during data acquisition. If the Timing Feature is set to FTC (First Threshold Crossing), the Threshold directly imts the arrival time or even whether an arrival is detected. The Evaluation Threshold is applied to the waveform after the optional digital filter is applied. Note that the Hardware Setup Threshold affects what data is recorded during acquisition. The user can set the Evaluation Threshold to values higher than the Hardware Setup Threshold used during data acquisition & simulate the results that would have been recorded during acquisition with a higher Threshold. Beware, though, setting it to lower values cannot reveal data that was not recorded in the first place.

Upper & Lower Frequency: Lower Frequency & Upper Frequency columns control the bandpass cutoff frequencies. Turning on the digital filter & adjusting the filter cutoff frequencies can help isolate individual acoustic emission modes with differing wave velocities. Note that you should be aware of the sampling rate used during acquisition and avoid setting the Lower Frequency cutoff above ½ the lowest sampling rate on any channel. You should also avoid targeting frequency ranges that your sensors do not respond to or that have been removed by your hardware's analog filters. Otherwise you could be analyzing meaningless waveforms.

Options Tab Settings:

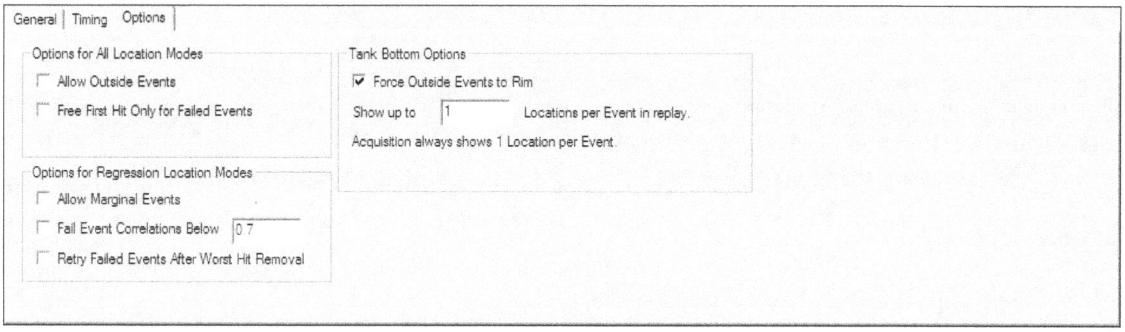

Figure 80. ***Location*** *Setup – Options Page*

There are a number of options available on the Options Page of Location Setup that control how location processing occurs. There are 3 different groups of options. Some apply to all location modes, some to only the regression location modes & some are specific to Tank Bottom Location. Here is a description of each group of options.

Options for All Location Modes

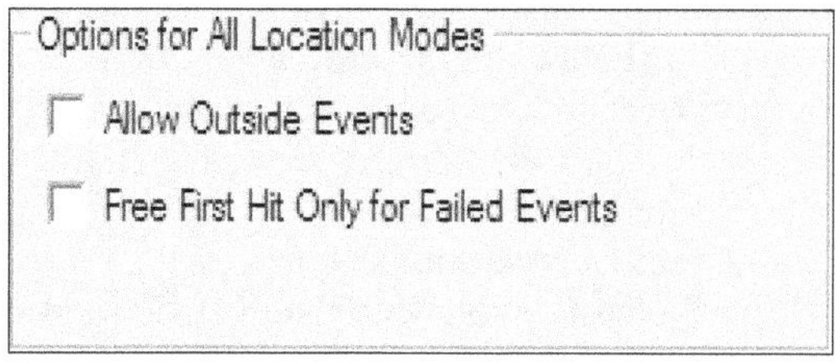

Figure 81. General Location Options

Allow Outside Events

The Allow Outside Events checkbox controls what happens to events that produce source locations outside the structure. Usually this is due to timing errors from an odd combination of hits like the 2nd hit sensor being further away from the 1st hit sensor than the 3rd hit sensor. The default behavior is to reject them as failed events & not display the calculated source locations, but the user can optionally display them. This option does not apply to the Zonal or Linear location modes, but it does apply to Tank Bottom location & all the regression location modes. For example, if 2D Planar location is paired with the Plate structure, events that locate outside the Plate limits will be rejected. The option has no effect for 2D Planar or 3D Location when they are set to use the Free structure. The software does allow a margin of error at the structure limits, so that events that locate very close to the limit still show. In general that margin of error is an extra 2% of the structure dimension. For the Plate structure, that is 2% of the length and 2% of the height. For the Tank Bottom structure, it is 2% of the radius. When you have Allow Outside Events enabled, you may wish to use manual axis limits on any 2D location graphs as outside events can sometimes have coordinates with very large values.

Free First Hit Only for Failed Events

The software assembles groups of hits into event candidates which are then analyzed by whatever location groups the user has set up. When these potential events failed to produce a valid location for whatever reason, the software has always freed up all hits in the candidate event in the past. This meant it would not reuse those hits in other candidate events. A single stray hit due to noise can easily obscure a true event that occurs right after the noise hit. Now there is an option to only free the first hit of a failed event. The other hits will be allowed to continue to be considered for use in other candidate events. Checking *Free First Hit Only for Failed Events* will allow the software to find more potential events at the expense of more processing time slowing down the replay of the user's data. This option is ignored during acquisition. During acquisition all hits in a failed event are always freed to minimize the imt of location processing on the data acquisition speed.

Options for Regression Location Modes

There are 3 options specific to the regression based location modes. These are 2D Planar (& XY, YZ & XZ variants), 3D Location, Spherical Location, Cylinder Location & Conical Location, 2D Anisotropic, Anisotropic Cylinder.

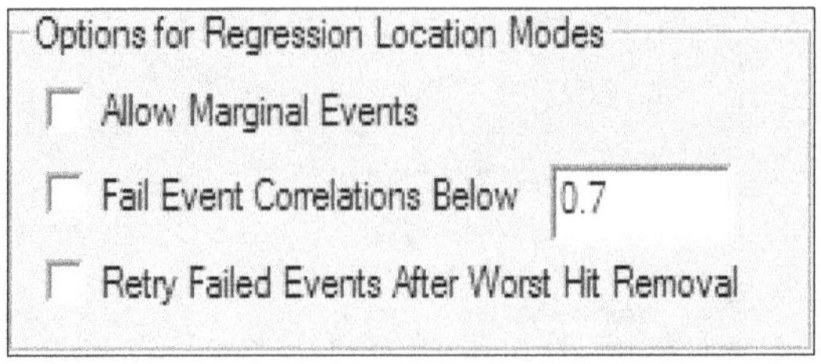

Figure 82. Regression Location Options

Allow Marginal Events

The Allow Marginal Events checkbox has the largest imt on 2D Planar & 3D Location. By default this checkbox is not enabled. Marginal events are events that the software expects to have low accuracy. These are linear events for 2D Planar & planar events for 3D Location. See the Notes sections on those location modes for a more detailed discussion of what linear & planar events are & why they are problematic.

This option is useful for comparing with the results of 's older DOS kages that did not automatically filter out these kinds of events. If the user wishes, he can enable this checkbox & see the results of these events, but be aware that it is not recommended as the calculated results have a 50% or greater chance of being incorrect. The incorrect locations can sometimes be useful, though. For example, if a 2D linear event show a result that is +5 from the line of sensors, then -5 is an equally valid answer according to the mathematics. This is also true for distance from the plane for 3D events. In both cases the software has no way of deciding between the 2 locations, but you could always do a visual inspection at both locations. Note that marginal events often produce locations outside the structure, so you may want to enable Allow Outside Events, if you are using Allow Marginal Events.

Fail Event Correlations Below
Retry Failed Events After Worst Hit Removal

All of the regression location modes now calculate a correlation coefficient. These values are reported as "q" (= quality) in the Line Display:

```
* Gp# 1[2,3,1,4,5,6] x = 11.45, y = 20, q = 0.9902 dT[ 68, 71, 302, 316, 323]
```

Correlation is also available for use on graph axes. For events with 4 or more hits (5 or more for 3D Location) the reported correlation values range from 0 to 1 with 1 being a very good event & 0 being a bad event. For events with 3 hits (4 for 3D Location) the correlation coefficient should be very close to 1.0. Note that for these events, a value of 1 does not guarantee a high quality event as timing errors cannot be detected with only 3 hits. Other location modes like Linear & Zonal will always report correlation coefficients of 0 as they do not calculate them.

There are 3 location settings relevant to correlation coefficients. There is a Fail Events Correlations Below checkbox along with an edit field to set the failure level. The regression location modes can optionally reject events with low correlation coefficients which are less likely to produce useful location information. The failure level can be adjusted based on the user's personal preference.

The third new setting is a checkbox labeled Retry Failed Events After Worst Hit Removal. This allows failed events to be reanalyzed after the worst hit of the event has been removed. Worst hit removal & reanalysis can be applied to events that fail due to low correlation values & can significantly improve the correlation coefficient which implies improved location accuracy as well. Here are a few examples:

Worst Hit Removal & Reanalysis on Events with Low Correlations

Time of Test						
* 0 00:00:23.3043055						
– Gp# 1[5,6,4,7,3,8,2,9,1,11]	x =	7.897	y =	-4.841	q =	0.483
+ Gp# 1[5,6,4,7,3,8,2,9,1]	x =	4.4,	y =	-3.246	q =	0.9967
* 0 00:02:11.5468993						
– Gp# 1[2,1,3,12,4,11,5,6,10,9,7,8]	x =	4.129	y =	7.312	q =	0. 4556
+ Gp# 1[2,1,3,12,4,11,5,6,10,9,7]	x =	2.835	y =	5.095	q =	0.9822
* 0 00:02:29.9029040						
– Gp# 1[5,6,4,7,3,8,2,9,1,11]	x =	9.131	y =	-6.009	q =	0.413
+ Gp# 1[5,6,4,7,3,8,2,9,1]	x =	4.521	y =	-3.38	q =	0.9968
* 0 00:03:22.0759543						
– Gp# 1[8,9,7,10,6,11,5,12,4,1,3,2]	x =	-6.047	y =	-7.491	q =	0.3598
+ Gp# 1[8,9,7,10,6,11,5,12,4,1,3]	x =	-3.915	y =	-4.642	q =	0.987

– = No Correlation Filtering, + = Correlation Filtering + Worst Hit Removal + Reanalysis

Worst hit removal & reanalysis can also be applied to failed outside events when Allow Outside Events is not checked. Reanalysis after worst hit removal can shift the results of the event analysis from outside the structure to inside the structure sometimes with the removal of just 1 hit. Here are some results for events that originally occurred way outside the user's structure:

Worst Hit Removal & Reanalysis on Outside Events

Time of Test						
* 0 01:05:04.1100330						
– Gp# 1 [5,6,4,7,3,10,12]	x =	71490	y =	-50270	q =	0.9988
+ Gp# 1 [5,6,4,7,3,12]	x =	15.55	y =	-13.91	q =	0.8813
* 0 01:09:39.3894840						
– Gp# 1 [8,9,10,5,1]	x =	-31600	y =	-31380	q =	0.9899
+ Gp# 1 [8,9,5,1]	x =	-15.99	y =	- 17.58	q =	0.978

– = Outside Events Allowed, + = No Outside Events + Worst Hit Removal + Reanalysis

Worst hit removal & reanalysis is never applied to marginal events. Events are considered marginal due to a problem with the geometry of the sensors included in the event. For example, for 2D Planar, an event where all the hit sensors are in a straight line is considered marginal. Removal of any hit, still leaves all the hits in a straight line, so the event is still marginal. For that reason marginal events are never reanalyzed.

Options for Tank Bottom Location

There are 2 options specific to Tank Bottom location. One allows the software to display more than 1 source location for a given event by varying which 3 hits it uses in the calculation. The Force Outside Events to Rim option works in conjunction with the Allow Outside Events to control where outside events are plotted. If Allow Outside Events is disabled, the Force Outside Events to Rim does nothing. If Force to Rim is disabled, the events are plotted at their calculated coordinates. If both Allow Outside Events & Force Outside Events to Rim are enabled, then all the outside events are plotted on the rim of the tank bottom structure. *For examples of how the options affect your Tank Bottom location results, see the notes for the Tank Bottom location mode in section 4.4.9.*

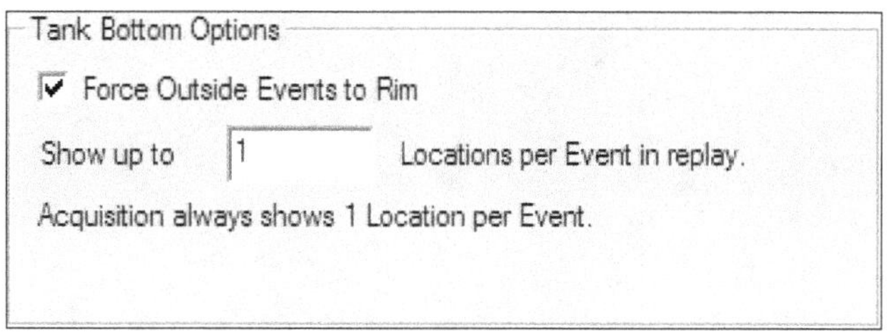

Figure 83. Tank Bottom Location Options

Distance Units Dialog

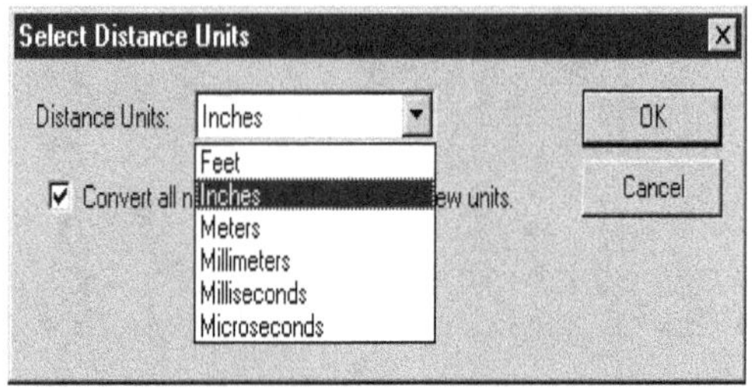

Figure 84. Distance Units

Pressing the Distance Units button opens the Distance Units dialog box. The user can select his desired units from the dropdown list. The user is given an option of converting old numerical values to the new units by simply checking the "Convert all numerical values based on new units." option. This will cause the event definition value, event lockout value, the overcal settings, the wave velocities & sensor positions to be updated as appropriate.

Material Properties Dialog

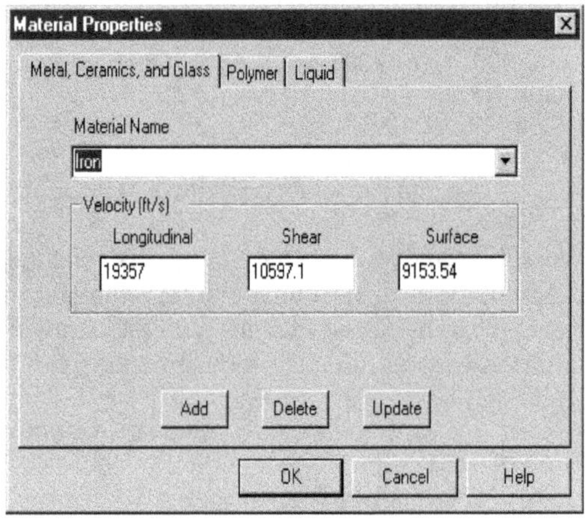

Figure 85. Material Properties

Pressing the Material... button opens the Material Properties dialog box. This dialog box is just a simple database of velocities for acoustic emission in various materials. When you find a useful velocity, simply highlight it & press Ctrl-C, Ctrl-Ins or right click & select Copy. You can then insert it into any field in the Location Setup dialog with Shift-Ins or Ctrl-V.

If you wish to add some of your own numbers, you can simply type a new entry for the Material Name, fill in the velocities & press the Add button. The Delete button removes an entry from the database, while the Update button lets you replace bad values.

Attenuation Dialog

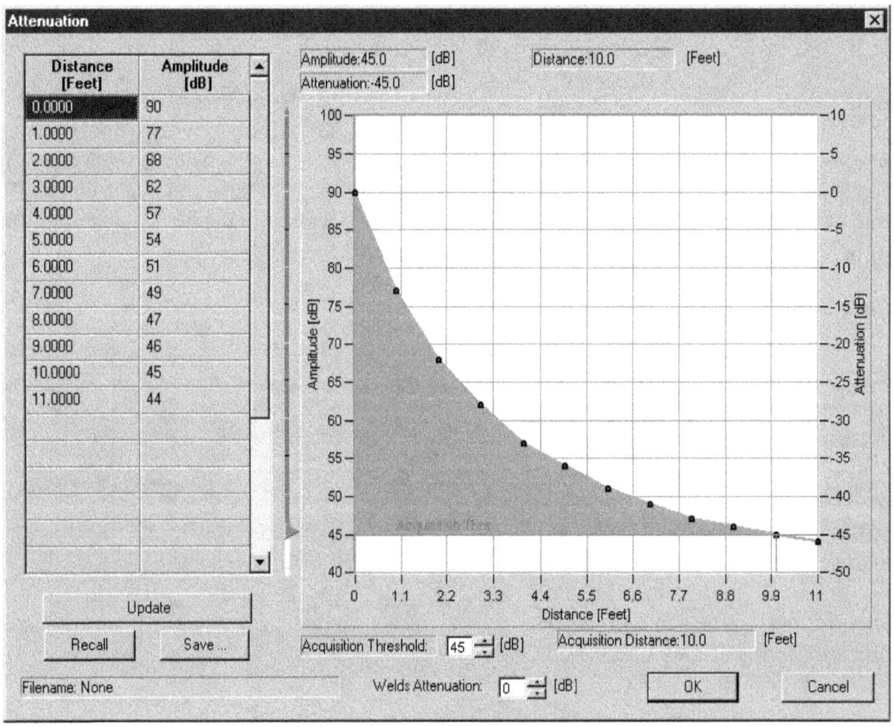

Figure 86. Attenuation Dialog

Pressing the Attenuation… button opens the Attenuation dialog box shown below. The sole purpose of this dialog is to allow the software to account for the decay of the acoustic emmission as a function of distance travelled & to estimate of the initial amplitude at the source. It uses the amplitude observed at the first hit sensor, the distance to the source for the event and the attenuation table entered by the user in this dialog. The attenuation table documents amplitude as a function of distance. The units of distance are controlled by the Location Setup Distance Units setting. The Update button will causes the graph to display the current values in the table. The user may use the Save & Recall buttons to save or recall an attenuation table to/from the disk

The basic procedure uses the table to calculate the difference between an amplitude at the source & at the observed distance from the source. If the source to first hit sensor distance falls within the range of values in the table, an amplitude at that distance is calculated using linear interpolation between the 2 nearest distances in the table. Using the table above as an example, if the source was determined to be 2.5 feet from the first hit sensor, the table has the 2 closest entries as 68 at 2.0 feet & 62 at 3.0. The amplitude predicted at 2.5 feet would be 65 (= 68 – (68-62)*0.5). The correction applied to the amplitude at the first hit sensor would be 25 dB (= 90 - 65). If the first hit amplitude was 60, the line display would report a Source Amplitude of 85 for this example. The amplitudes are not extrapolated beyond the maximum distance. If the distance is beyond the maximum distance in the table, only the maximum correction of 46 (= 90 – 44) is applied. If the user never accesses this dialog to fill in the table, no correction is applied & the source amplitude reported is identical to the amplitude of the first hit sensor.

Guard Sensor Dialog

Pressing this button opens the 'Guard Sensors' dialog box. Guard Sensors can be used to lockout out events on selected location groups whenever a hit is first detected on a guard channel. It is designed for situations in which one wishes to ignore events that originate outside of the location group.

In the example setup shown, sensors 1, 3 & 5 are protecting Location Group 1 from outside events. If there is a hit on sensors 1, 3 or 5, then Location Group 1 will not process any events for the next 1000 us. If the Event Only flag were checked, only the next event on Location Group 1 would be locked out instead of being locked out for the next 1000us. The Lockout Time field would also be disabled as its value would no longer be used.

The Guard Data Filter Setup controls what hits get saved into the DTA file during acquisition, if guard sensors are enabled. The default setting is to just save all hits, but you can restrict the data saved to just those hits that make up events & the guard sensor hits. This can be useful in situations where external noise creates a large number of hits that end up being useless for the purpose of location. Not only do the extra hits take up se in the data file, but they also slow down replays of the data file too.

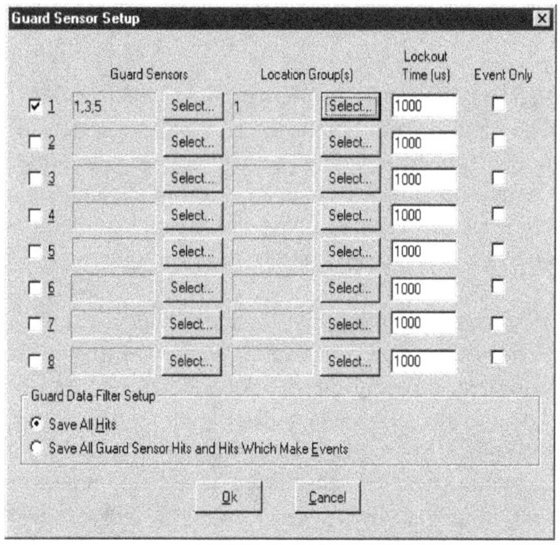

Figure 87. Guard Sensor Setup

Location View & Sensor Placement Dialog

Before discussing the Sensor Placement Dialog, there is one important point about the physical placement of your sensors that you should understand and it applies to all of the regression location modes. *You can place sensors anywhere, but you will get better results with the regression location modes if your sensor layout is made up of an evenly sed triangular array of sensors. The optimal arrangement would be a triangular grid made up of equilateral triangles, but the regression modes do not require exact equality of the triangle legs or even regular sing between the sensors.* ***Avoid creating triangles with one very long or very short side, however, as they do not work well for locating sources.***

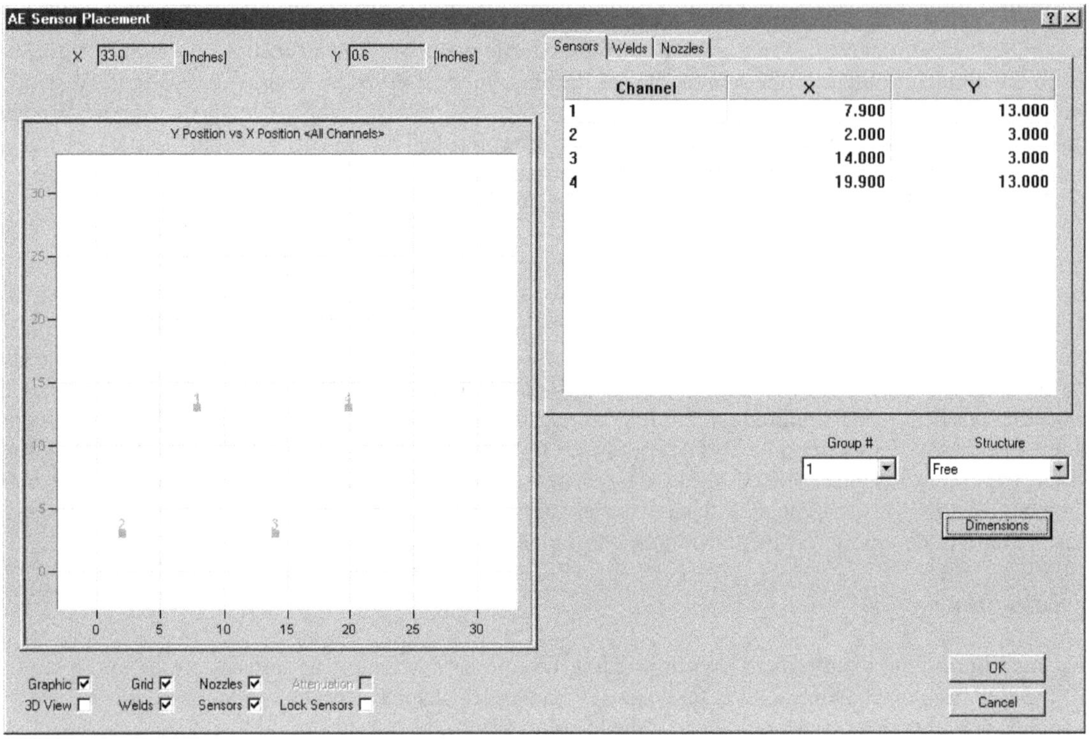

Figure 88. Sensor Placement Dialog

Pressing the Location View button in the Location Setup dialog opens the Sensor Placement dialog box. shown above. The 2 main areas are the actual view of the sensors in the left-hand side of the dialog & a listing of their coordinates in the right-hand side. The grid area for the sensor coordinates also has tabs that will display the coordinates of any welds and nozzles created. You can change the color of the sensors using the Graphing menu to access Sensor Color on the Colors submenu.

When using the Sensor Placement dialog, there are 2 ways to add new sensors. When working in the grid area, pressing the INS key adds a new sensor. The F2 key allows you to edit its coordinates & move it to the correct location. Alternatively you can right click the sensor view area to get a context menu like the one shown to the right. Place Sensor & Next Sensor allow you to put a new sensor at the current position. Next Sensor numbers it automatically using the next available sensor number. Place Sensor will ask for a sensor number. You can also use AutoPlace Sensors to create a grid of sensors with regularly varying coordinates. You work with welds & nozzles in a similar manner. The exact menu you see may differ from that shown here as it depends on the Structure.

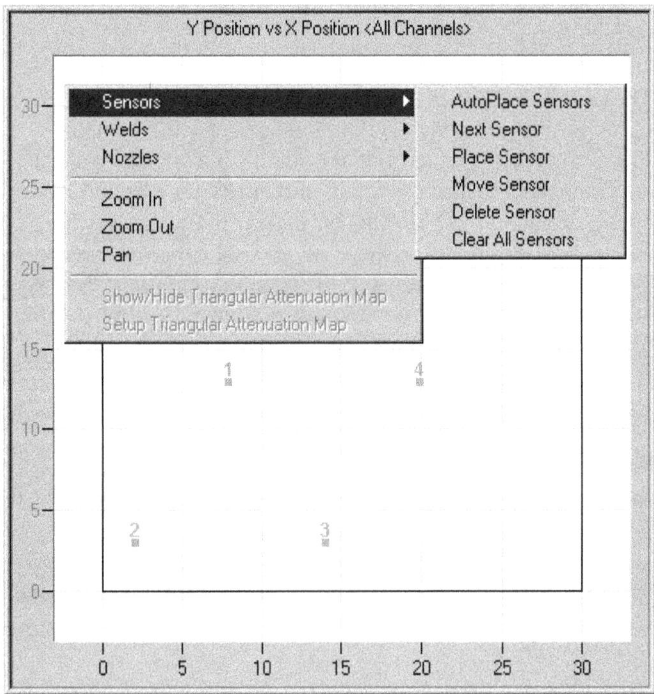

Figure 89. Sensor Menu

The Structure list is just below the grid area. Support for welds, nozzles, manual placement of sensors & autoplacement of sensors varies with the Structure setting. When creating a sensor layout for a new location group, the first thing you should do in the Sensor Placement dialog is to set the Structure type. The Group # setting allows you to quickly switch to editing the sensors for another location group without having to return to the Location Setup dialog. If you made changes to the current settings, you are prompted to save them.

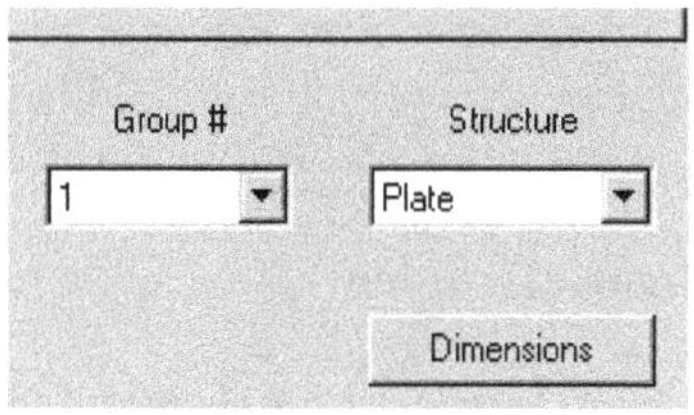

Figure 90. Structure List

Below the Structure setting is the Dimension button. If you click it, one of the dimensions dialogs will appear. They are also specific to the type of Structure. The one shown to the right is for the Plate structure. When working with a new location group, the second thing you should do in Sensor Placement is to fill in the dimensions for the structure you are working with. The dimensions affect the scaling of the sensor view area. If you have incorrect values entered, some sensors may not be visible in the sensor view.

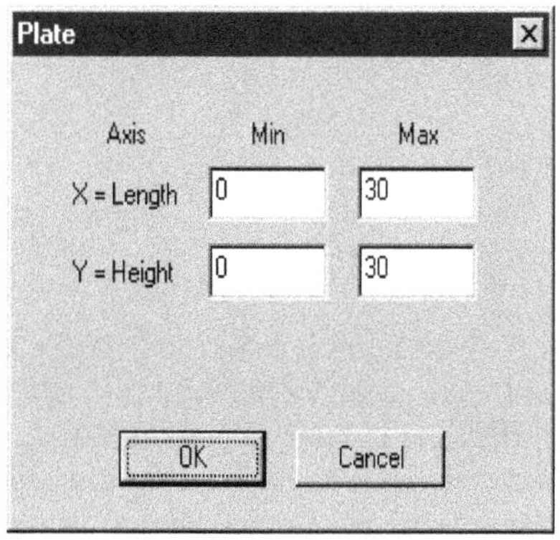

Figure 91. Plate

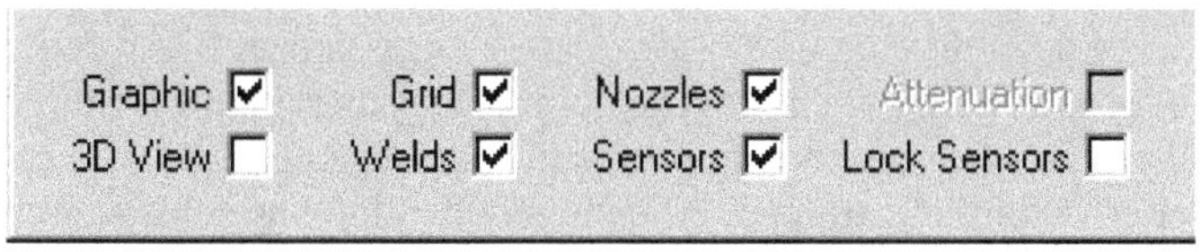

Figure 92. Sensor Placement Checkboxes

Below the sensor view area, there are a number of checkboxes that control the sensor view. The Nozzles, Sensors & Welds checkboxes control whether or not those objects are displayed. Likewise the Grid checkbox controls whether or not the sensor view shows a Cartesian grid to facilitate placing of sensors. The Graphic checkbox controls whether or not a representation of the structure is drawn on the graph. The Lock Sensors checkbox disables moving of sensors with the mouse. The 3D View checkbox will toggle the display between 2D & 3D views of the sensor layout, if that structure type supports both 2D & 3D views. Finally the Attenuation checkbox enables if you have used the Attenuation dialog to enable the calculation of source amplitudes. Checking Attenuation in the Sensor Placement dialog causes the sensor view to display an attenuation map as a series of concentric rings around each sensor as shown to the right.

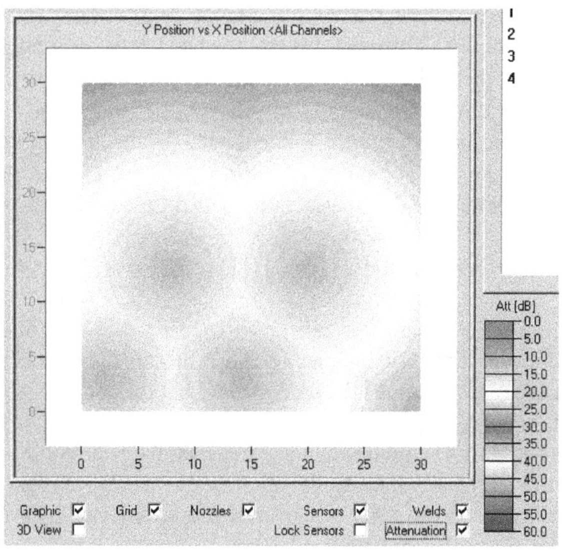

Figure 93. Attenuation

Notes on Individual Location Modes

As mentioned in the section describing the Sensor Placement Dialog, there is a point that all users should understand about the regression location modes & it is important enough to bear repeating. *You can place sensors anywhere, but you will get better results with the regression modes if your sensor layout is made up of an evenly sed triangular array of sensors. The optimal arrangement would be a triangular grid made up of equilateral triangles, but the regression modes do not require exact equality of the triangle legs or even regular sing between the sensors.* ***Avoid creating triangles with one very long or very short side, however, as they do not work well for locating sources.***

In general, you can just read the section on the location mode you are interested in using. All the regression location modes are a variation on 2D Planar, however, so you might also want to read it as well. If you'd like more information on setting up a Zonal, Linear or Tank Bottom location group, you can also see the Examples section.

Zonal Location

Zonal location is the simplest location mode there is. It just returns the coordinates of the first sensor in the list of hits that make up the event. You can place sensors absolutely anywhere & you can get better resolution on source locations by just using more sensors. A better way to improve your source locations would be to switch to a more sophisticated location technique like Linear or 2D Planar. See section 4.5.1 for a detailed example of setting up a Zonal Location group.

Linear Location Modes

Linear Location

Linear location is a step up from Zonal location in that it actually attempts to locate of the source from the timing data in the event rather than just restricting it to a region of se around the first hit sensor. Linear location uses 2 hits & the difference in their arrival times to calculate the source location using linear interpolation of the coordinates for the 2 sensors. Its drawback is that in a 2-dimensional area it restricts the source to points on the line segments that connect the sensors when real sources do not face that restriction. Still that can be good enough when looking for a source along a long series of sensors on, for example, a pipeline.

Linear location requires the sensors to be laid out in a line, although the line does not have to be perfectly straight. You could lay them out in an arc. It is not possible, however, to have the line double back on itself, so laying out the sensors in a square or circular array would not work. Linear location relies on a list of the sensors sorted by the x coordinate when deciding which 2 hits to use in an event. You can also use a vertical line. If the x coordinate of the first & last sensor in a Linear location group are equal, the list of sensors gets sorted by the y coordinates. Linear location only forms events between adjacent sensors, so there are only 2 possible 2nd hit sensors for any 1st hit sensor. When there are hits on both, it decides which hit to use based on the ratio of the Δt value of the hit divided by the maximum Δt calculated from the sensor sing & the user entered velocity. The hit with the lowest relative Δt value becomes the 2nd hit of the event used for calculating the source location. The logic is that the 2 sensors on either side of the source will have the shortest relative Δt value, while the Δt value between the 1st hit sensor & other sensors will always be higher. See section 4.5.1 for a detailed example of setting up a Linear Location group.

LinearXY Location

LinearXY location is related to Linear location in that the source location is still calculated by linear interpolation between the arrival times of the 2 hits in the events. The difference between LinearXY & Linear location is that LinearXY removes the assumption that the sensors will be laid out in a line & the sorting based on coordinates. With LinearXY location you can put the sensors literally anywhere in 2-dimensional se, so it is perfectly acceptable to have your sensors laid out in a square, a circle or a rectangular array. The advantage & at the same time the drawback to this method is that it allows events to form between any 2 sensors in the array. For a square, not only are events allowed on the sides of the square, but they can also occur on the diagonals. LinearXY does still look for the pair of sensors with the lowest relative Δt values.

Wrapped Linear Location

A common question from users of 's DOS software is how to turn on wrapping in Linear location. The software doesn't have a checkbox to enable wrapping in Linear location like the old DOS software, but there are 2 methods you can use to set it up. The first method is an older long mentioned in the HMSGHAT2000DDI software manual & it works in any version of HMSGHAT2000DDI independent of when it was released or how the user's licensing options are configured. It requires the use of 2 linear location groups. The second method only works in the newest releases of HMSGHAT2000DDI licensed for Full Location. It makes use of a newer Wrapped Linear location mode.

Method #1 — Wrapped Linear Location Using 2 Location Groups

Below are a series of graphs from the sample layout, WrappedLinear.lay, showing how it is done with 2 Linear location groups.

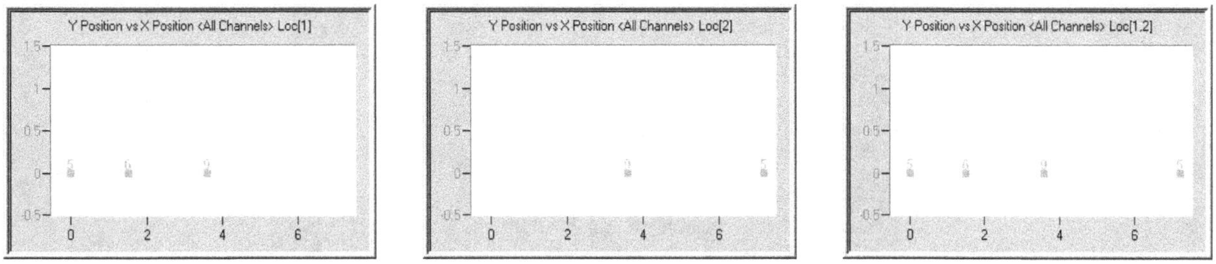

*Figure 94. Wrapped **Linear** Location using 2 Groups*

The graph on the top left is showing location group 1 which consists of sensors 5, 6 & 9. Linear location allows events between sensors pairs 5 & 6 or 6 & 9 in this location group. Turning on wrapping requires allowing events between sensors 5 & 9 and it requires a second set of coordinates for the wrapped location of sensor 5. This is accomplished by creating location group 2 shown in the graph on the top right above. Sensor 9 is common to both location groups at the same coordinates. Sensor 5 in location group 2 is at its wrapped location which is just the position of sensor 5 in group 1 offset by the structure circumference. The graph on the bottom just illustrates that you can superimpose both location groups 1 & 2 on a single graph using Graph Setup. You also need to turn on both groups in Location Setup. Do not enable Show All Points for both groups, though, if you want the software to decide whether events between sensors 6 & 9 for group 1 or events between sensors 5 & 9 for group 2 make the most sense based on relative Δt values. If you were to enable Show All Points, then you could end up with a single set of hits producing an event between sensors 6 & 9 as well as between 9 & the wrapped location of sensor 5.

Method #2 — Wrapped Linear Location Mode

Method #2 is illustrated in the WrappedLinear2.lay example layout. In more recent versions of HMSGHAT2000DDI a Wrapped Linear location mode has been added to the software.

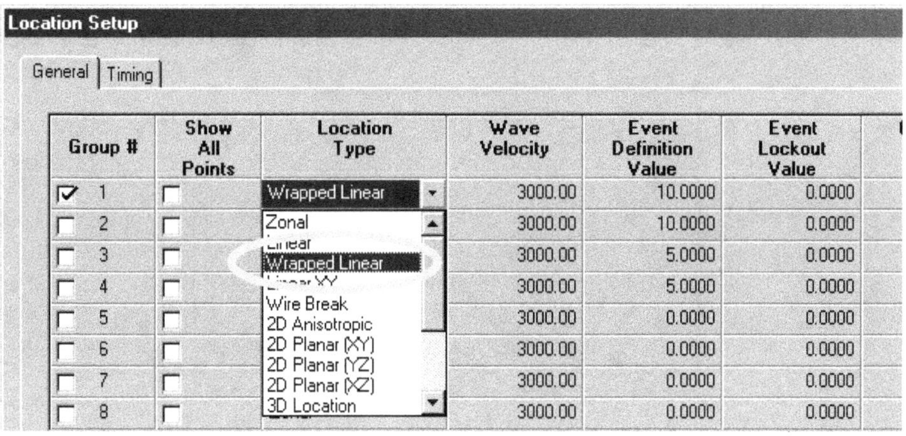

Figure 95. Wrapped Linear in Location Setup

If you can't see Wrapped Linear in the list of available location modes in the Location Setup dialog, check your licensed options by selecting Options under the Help pulldown menu. The Options dialog will appear: The Wrapped Linear location mode requires the Full Location licensing option.

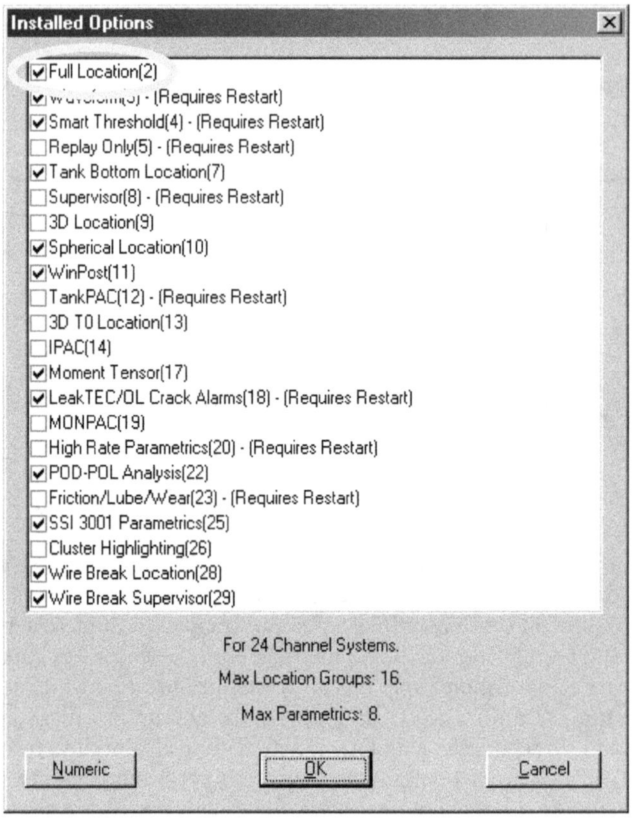

Figure 96. Full ***Location*** *Option in Installed Options Dialog Box*

If you do have an enabled Full Location license option & still can't see a Wrapped Linear location mode in Location Setup, you will have to either upgrade to the latest version of HMSGHAT2000DDI for your hardware or use Method #1 for doing wrapped linear location. Contact Group Inc, Products & Systems Division, Customer Service (customer.service@group.com) for information any needed upgrades or licensing options.

The Wrapped Linear location mode consolidates the 2 location group method into 1 location group. Here is the location graph from WrappedLinear2.lay which demonstrates the newer Wrapped Linear location mode.

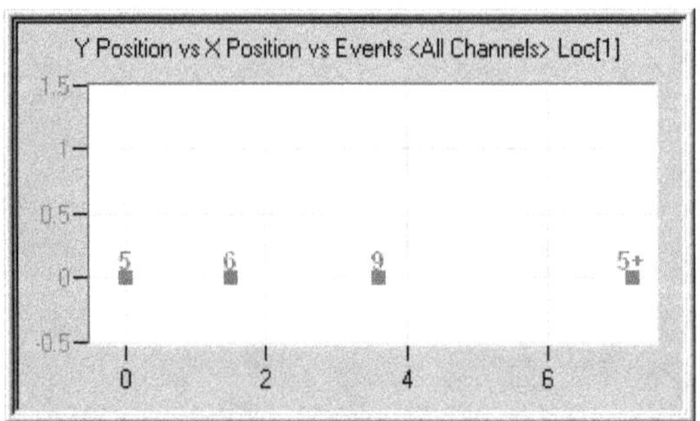

Figure 97. Graph ***Showing*** *Wrapped Linear Location Group*

Here the wrapped location for sensor 5 is marked on the graph as "5+". Putting the same sensor in 2 locations is a bit of a complication for HMSGHAT2000DDI, because the Sensor Placement dialog is only designed to have 1 set of coordinates for each sensor. The way around this is to invent a new sensor in Sensor Placement that is used for the wrapped sensor location. Here's the Sensor Placement dialog showing the sensors for this location group:

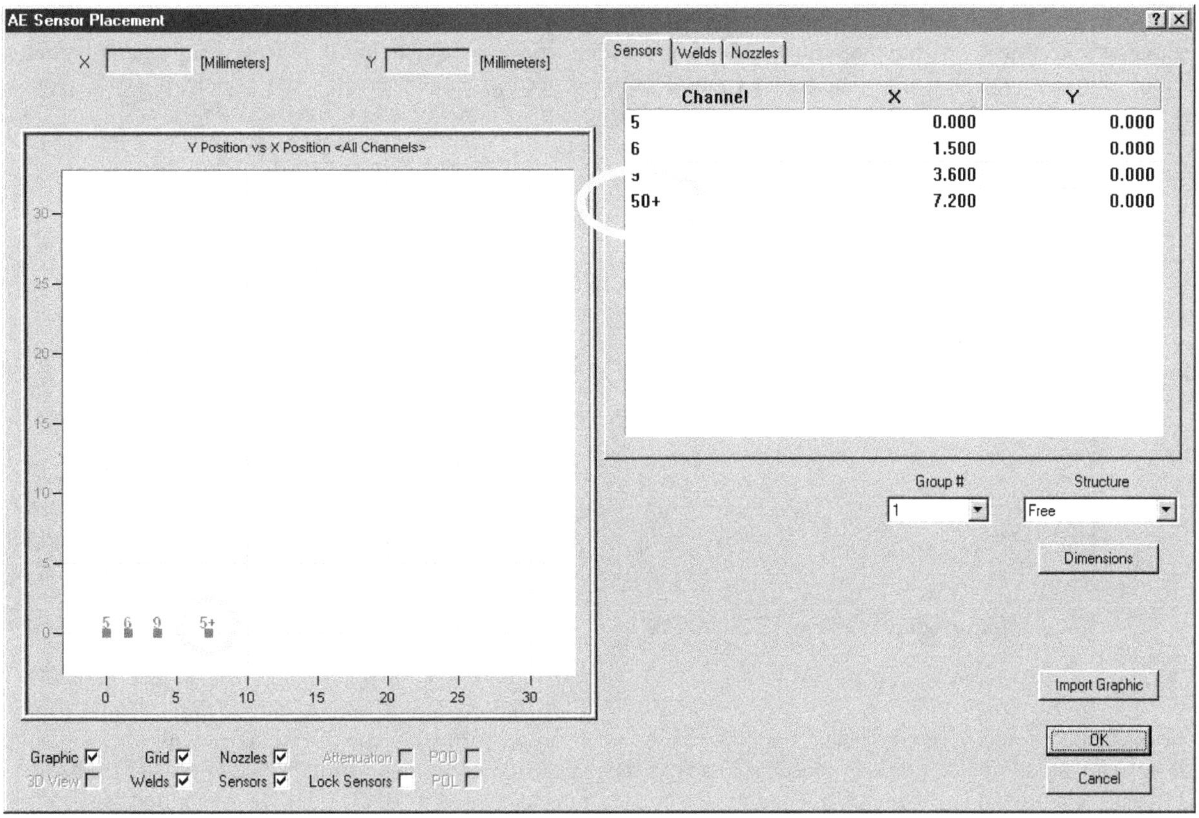

Figure 98. Wrapped Linear Location Group in Sensor Placement

Notice that the table of sensors lists sensor "50+", but the sensor diagram shows sensor "5+" at the location of "50+". These are the same wrapped sensor as indicated by the "+" signs. The only location mode that relabels a sensor on the graph is Wrapped Linear. If this location group is changed to a regular Linear location group, then the "+" signs will disappear in the grid & on the graph & sensor "5+" on the graph would become sensor 50. There is nothing magical about sensor number 50 used in the table. It could have just as easily been labelled sensor 51, 52 or any other unused channel number.

Wrapped Linear location always replaces the label of the last sensor with the number of the first sensor & a plus sign to indicate that it is the wrapped sensor location. It picks the first & last sensors based on their coordinate values, not their sensor channel ID's. Internally it sorts the list of sensors based on the X coordinate to find the sensors with the lowest & highest X coordinates & then labels the highest X, rightmost sensor on the graph with the ID of the lowest X, leftmost sensor & the "+" sign. It also adds the "+" sign in the table to let you know which sensor it picked as the last sensor. If you input a vertical array of sensors where all the sensors have the same X coordinate, Wrapped Linear will sort the sensors based on their Y coordinate to decide which sensors are the first & last sensors to identify the wrapped sensor.

2D Planar & 2D Planar (XY)

There is no fundamental difference between 2D Planar & 2D Planar (XY) location except on the menus. Enabling 3D Location also enables 2D Planar (YZ) & 2D Planar (XZ) location. To distinguish the standard 2D Planar from those 2 modes, it is relabeled 2D Planar (XY) location whenever the 3D Location option is enabled. 2D Planar (YZ) & 2D Planar (XZ) location are discussed in the section on 3D Location.

For 2D Planar location, a particularly difficult problem is locating the source of an acoustic emission event, if all the hits are from sensors in a straight line (linear events). For linear events, even if a distance from the line can be determined, there is no way to distinguish positive & negative directions away from the line. Random error in the data can also make determining the distance from the line highly suspect. By default, the software filters out linear events and does not attempt 2D Planar location on them. You could always enable Allow Marginal Events (See discussion in section 4.2.4), but then you could end up flooding your graphs with lots of low accuracy results. A better solution is to use a layout that minimizes the chances of linear events being observed. An example of a bad layout is shown below. It is bad because of the horizontal sensor sing is much smaller than the vertical sing.

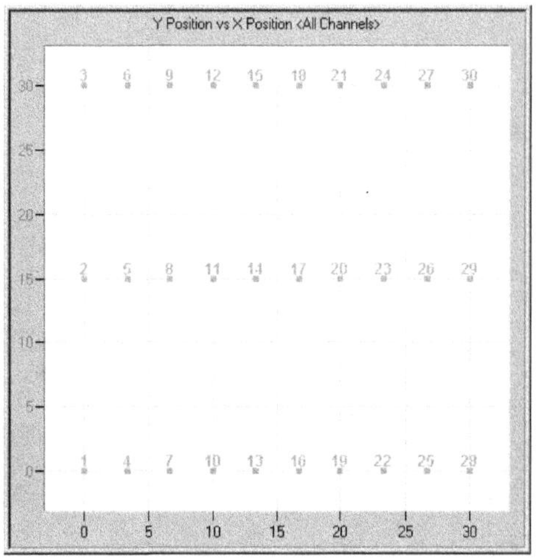

Figure 99. Bad 2D Planar Layout

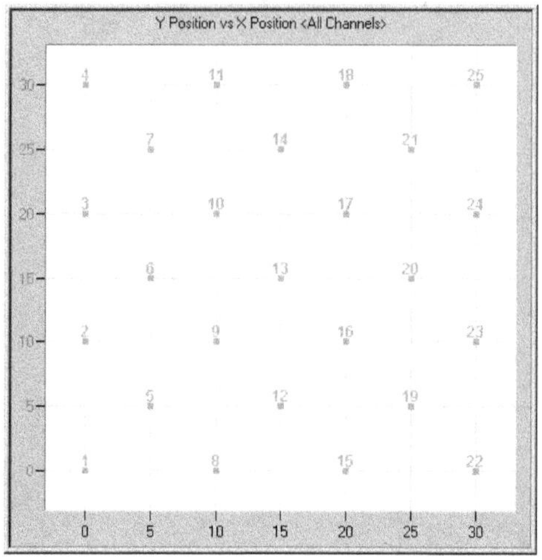

Figure 100. Good 2D Planar Layout

A better layout would have all distances in between the sensors approximately equal. Triangles with one very long side or one very short side can also cause problems as they approach becoming a straight line of sensors. The ideal 2D Planar layout would be made up of equilateral triangles, but exact equality of all sides is not required as shown in the good layout above . As mentioned in the introduction, you should avoid creating triangles with one very long side or one very short side. Events based on triangles with one very long or very short side become linear events when one side goes to infinity or to 0. Short of those limits, though, they can still act similar to the linear events, and won't accurately locate any sources outside of the triangle.

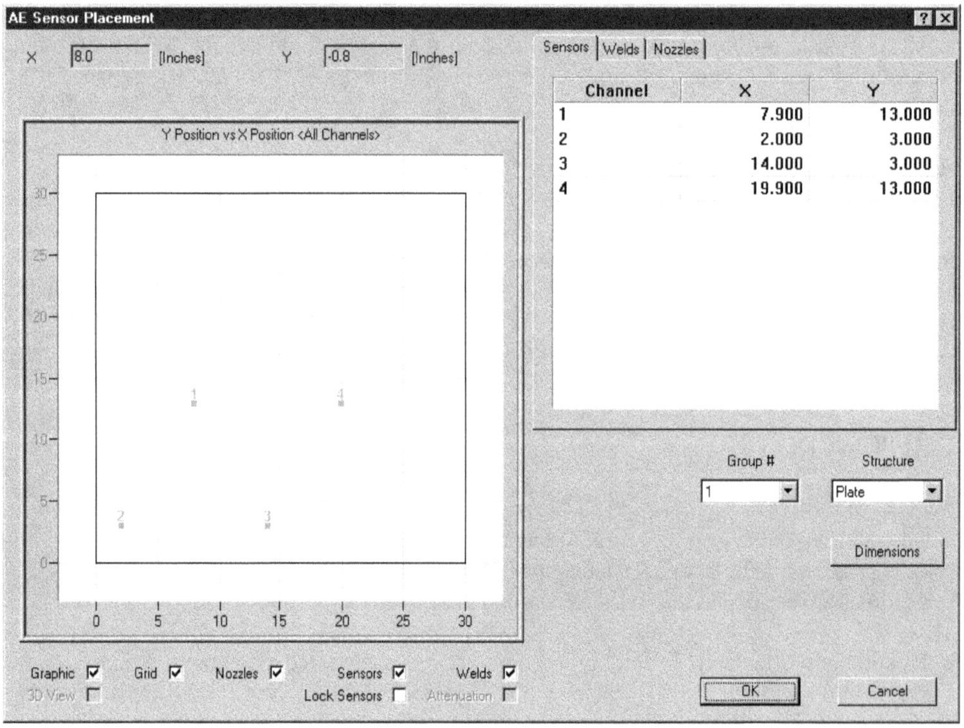

Figure 101. 2D Planar Sensor Placement

Here is Sensor Placement set up for a 2D Planar mode with the Plate structure. 2D Planar location is designed to work with either the Free or Plate structures. Although there isn't any difference in 2D Planar & 2D Planar (XY) location during analysis, there is one difference in Sensor Placement. A 2D Planar (XY) location group has the option of displaying the sensors in 3D graph, if the user wishes it. For a 2D Planar (XY) location group with the structure type set to Free, the 3D View checkbox enables. The user may use it to toggle between 2D & 3D views of the sensors. The checkbox disables for the Plate structure, though.

The Dimensions dialog for both the Free & the Plate structure contain the max/min limit settings for both the X & Y axes. These settings control the scaling of the graph in the sensor view of Sensor Placement. For the Plate structure, they have added significance. When 2D Planar is analyzing events, it will restrict results location groups using the Plate structure to points on the plate. To allow locations close to the limits to report, the software uses a tolerance of 2% of the axis length when deciding whether it is an acceptable result. For example, if x has a minimum value of 0 and a maximum value of 100, then results with an x coordinate greater than 102 or less than -2 are thrown out as invalid. If you don't want to limit your results, you can either set the Structure type to Free or stick with Plate & enable the Allow Outside Events checkbox in Location Setup. See section 4.2.3 for more details.

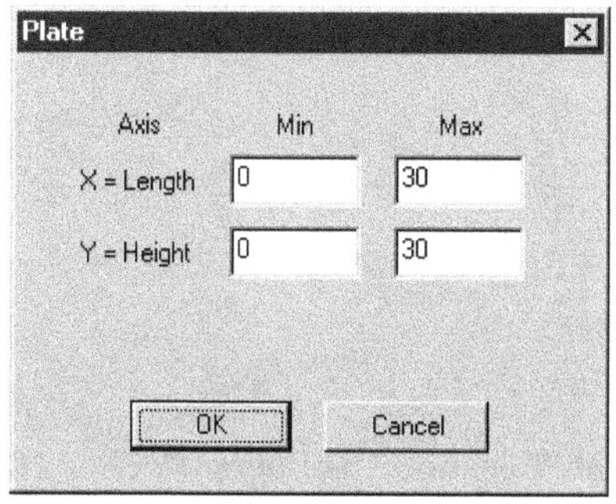

Figure 102. Plate

After setting up the sensors & a 2D Planar location groups, you will want to create a location graph. Use the Graphing pulldown menu's New Graph option & then configure the graph using the Graph Setup dialog box (F3 shortcut key). To the right is a 2D X-Y location graph based on the 2D Binned Scatter graph type where the location group is using a Plate structure. If you create a 2D Binned Scatter, 2D Colored Scatter or 2D Scatter graph that uses X & Y Position, the graph will automatically draw the sensors on the plot for any location groups selected on the Event Groups tab.

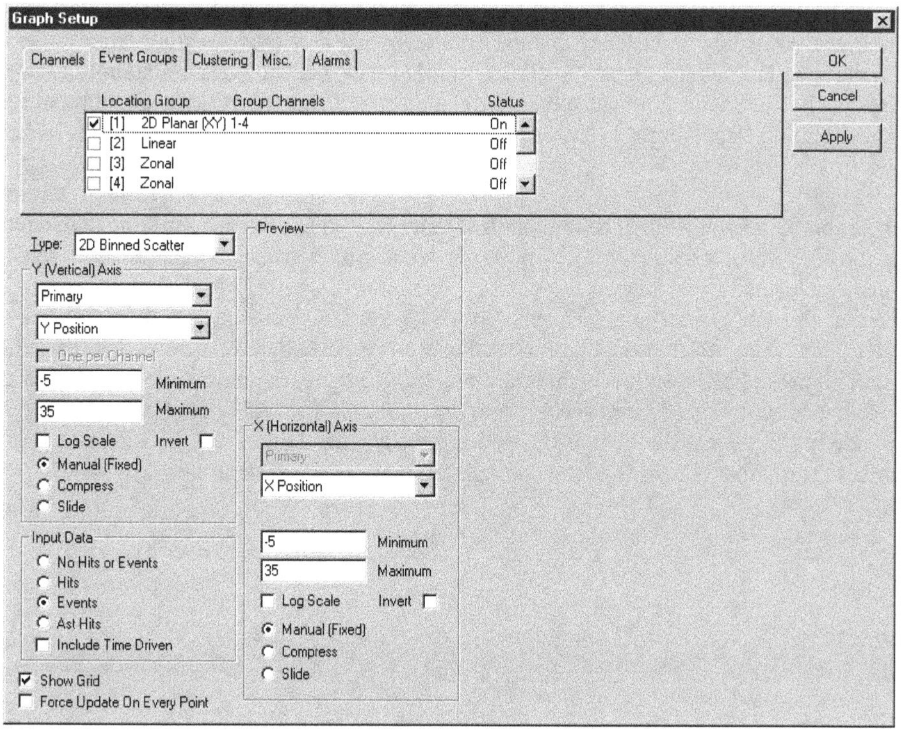

Figure 104. 2D Planar Graph Setup

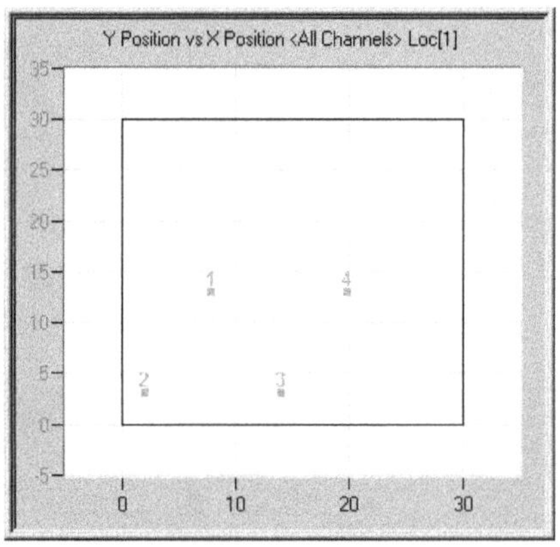

Figure 103. *2D Planar Graph*

In the Graph Setup dialog, make sure that the Input Data is set to Events and use the Event Groups tab to select which location group(s) to use for source information. Below is the corresponding Graph Setup dialog to illustrate the typical settings.

3D Location, 2D Planar (YZ) & 2D Planar (XZ)

Primarily this section focuses on 3D Location, but 2D Planar (YZ) & 2D Planar (XZ) are related. All 3 location modes are designed to use 3D coordinates. The 2D Planar (YZ) & 2D Planar (XZ) modes are similar to normal 2D Planar & 2D Planar (XY) location with the exception that you can enter 3D coordinates & have the software locate the source while ignoring either the X or Y coordinate. The reason these modes are useful relates to a problem with sensor placement & event handling for 3D Location.

The equivalent of the linear event problem for 2D Planar location is a problem with planar events for 3D Location.

A planar event is one in which all the hits are on sensors in the same plane. For planar events, it is hard to determine distance from the plane not to mention distinguishing sources in the positive direction from those in the negative direction. By default, the 3D Location algorithm filters out planar events and does not attempt to locate them as the distance from the plane is often quite simply useless. You can override this behavior by enabling the Allow Marginal Events checkbox in Location Setup, but it exists primarily for compatibility with older versions of 's DOS software. See section 4.2.4 for more details.

Because of the planar problem, you will get better results if you set up your sensors to minimize the number of sensors that line up across the body of the test structure. Avoid using a rectangular grid layout that lines up sensors on opposing faces of your structure. Instead try to stagger the sensors, so that no 2 sensors are directly across from one another. The bad sample layout above uses a rectangular approach with 3 rows of 3 sensors directly opposite one another. The good layout uses a triangular approach by staggering each row on one side by half the sensor-sensor distance and then it inverts the pattern on the opposite wall. The diagram may not show it well, but sensor 4 is below the midpoint of sensors 1&2, above the midpoint of 7&8 and across from the midpoint of 12&13. Both sides have 3 rows of sensors, but the left-hand side has 3-2-3 pattern of sensors, while the right-hand side has a 2-3-2 pattern of sensors. This staggering minimizes the number of sensor planes through the body of the structure.

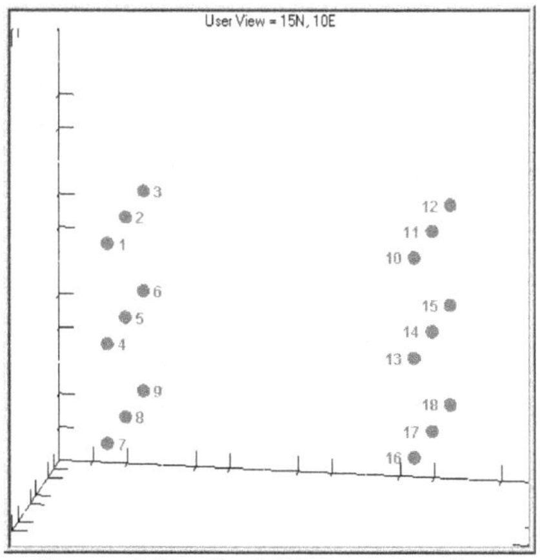

Figure 105. Bad 3D Location Layout

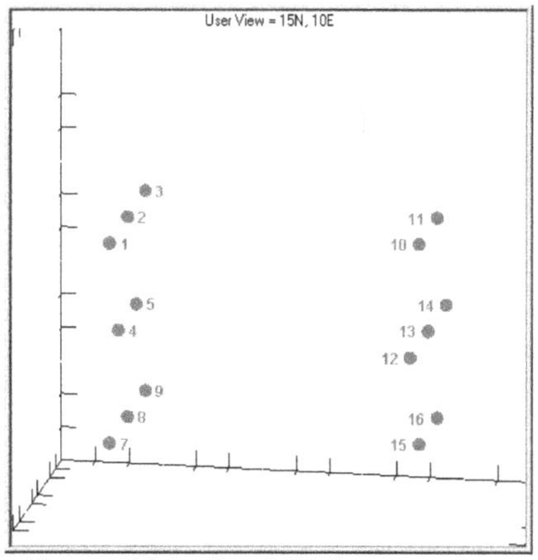

Figure 106. Good 3D Location Layout

One problem with 3D Location filtering out planar events is that pencil lead breaks will not locate with 3D Location unless the lead break is energetic enough to register with a sensor on one of the other faces. If you enable Allow Marginal Events, you may get results far outside the structure. A good alternative in this case are the 2D Planar (XZ) & (YZ) location modes. The distinction between them is what sensor coordinates they use when fitting the data. Both allow you to enter 3D coordinates, but ignore one of the values when locating. 2D Planar (XZ) use the X & Z sensor coordinates & ignores the Y values. 2D Planar (YZ) ignores the X coordinate. They are useful when doing pencil lead breaks by sensors confined to one face of your structure. With the 3D Location option enabled, 2D Planar (XY) will allow you to enter 3D coordinates & only locate using the x & y coordinates. Fundamentally that's no different than the standard 2D Planar location. Both are ignoring the z coordinate, but as a convenience one allows you to enter a non-zero value for Z.

For best performance, any 2D Planar groups using 3D coordinates should be confined to just one face of the test structure. Parallel sensors on an opposing face could appear as duplicate sensors with identical coordinates when the 3rd unused coordinate is ignored & that would lead to errors in the event location calculations. You should also note that during acquisition a result from a 2D location mode does not necessarily imply that the source was confined to the 2D location plane. Just as for planar events in 3D Location, distance from the plane is indeterminate. You can draw a perpendicular to the 2D plane at the best fit location and the source event could be anywhere on the line. Of course for pencil lead breaks, you will know in advance that the source is confined to a given face and be able to check your sensors before starting acquisition.

For 3D Location, the Sensor Placement dialog shown here allows you to directly enter X, Y, Z coordinates for any groups that are considered a 3D Location group with the right structure type. 3D Location is designed to use either the Free & the Box structure types.

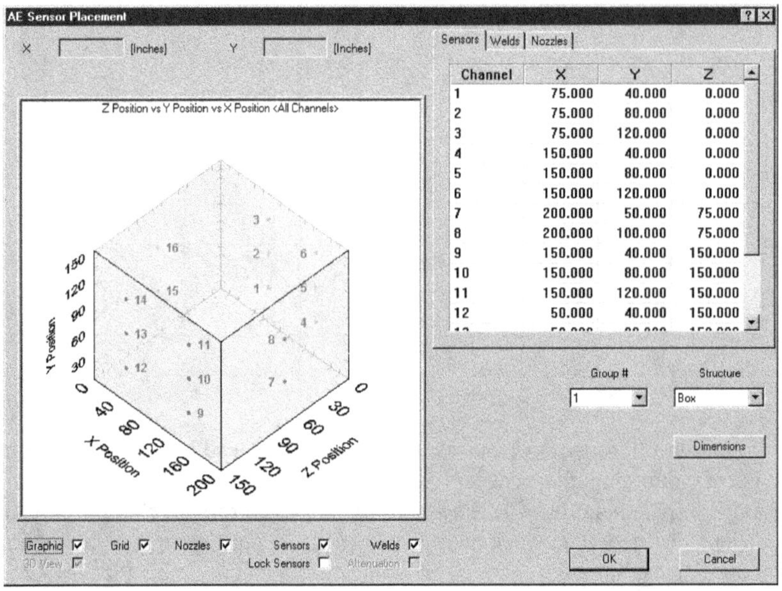

Figure 107. 3D Sensor ***Placement***

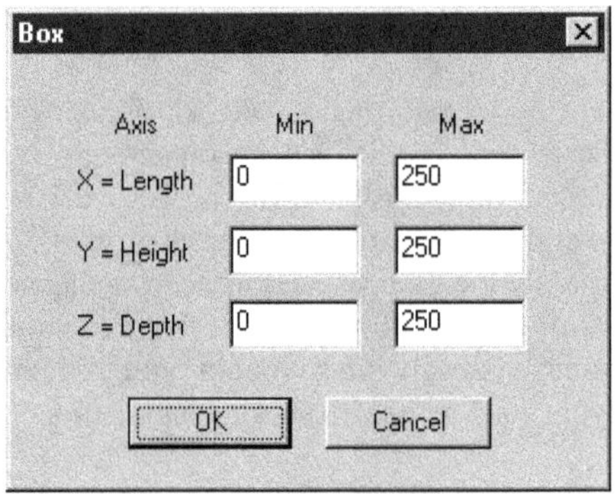

Figure 108. Box Structure Dimensions Dialog

The Dimensions dialog for the Box & Free structures is similar to the one for the 2D Free or Plate structures with the addition of max/min settings for the Z axis. Similar to the way the Plate structure limits 2D Planar results, the Box structure acts as search limits for 3D Location. If any location results are more than 2% above or below the limits, the result will be considered bad & is thrown out. For example, if Z Min is 50 & Z Max is 150, then any location results where z is above 152 or below 48 will be filtered out. As with 2D Planar, using the Free structure avoids imposing any limits on the results. Alternatively, you can enable Allow Outside Events in Location Setup. See section 4.2.3 for details.

You cannot graphically enter 3D sensor locations in Sensor Placement, but you can graphically see the sensors in the 3D view. If you do not see a grid column for Z or a 3D view of your sensors, go back to the Location Setup dialog & make sure the group type is set to 3D Location, 2D Planar (YZ) or (XZ). Also make sure you have the structure type set to Free or Box.

Use the INS key & F2 to insert a new sensor & edit its coordinates in the grid. As an alternative to creating all the sensors manually, you can also use the AutoPlace Sensors option to create a grid of sensors with varying x,y coordinates & a fixed z coordinate of 0. You would then manually edit the z coordinates of each sensor. Note that you can use AutoPlace Sensors multiple times without erasing the old sensors. That allows you to build up a 3D grid one layer at a time. Right click the graph in Sensor Placement to find AutoPlace Sensors.

To create a location graph for 3D Location, use the 3D Scatter graph & set each axis to X, Y & Z Position which correspond to the x, y & z coordinates of a located source point. The graph to the right is a 3D Scatter graph set up as a location graph. As shown below, Graph Setup has the Input Data is set to Events & the Event Groups tab has the 3D Location group enabled. If you create a graph that uses X, Y & Z Position, the graph will automatically draw the sensors and structures on the plot for enabled location groups.

For 3D Scatter plots, both the sensors & the data points are drawn as small cubes. You may use the Graphing Colors menu to change the colors. The data points use the color assigned to Plot 1 under Plot Colors and are drawn slightly smaller than the sensor points. If you would like larger points, right click your graph & turn on the Large Scatter Graph Points. Both the sensor points & the data points will get larger.

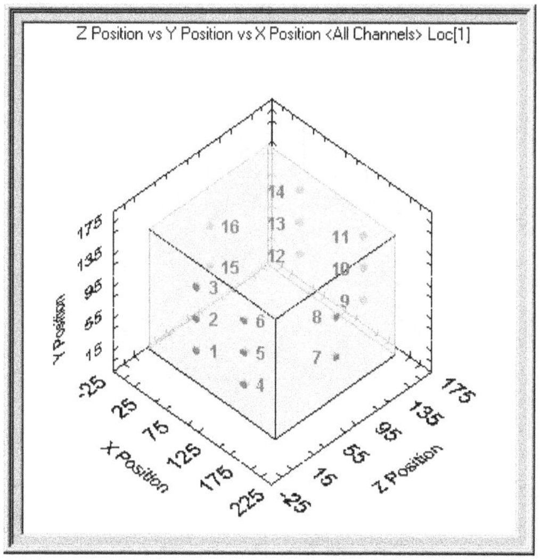

Figure 109. 3D Graph

Graph Setup for the example graph above is shown here. It sets the X (Horizontal) Axis to X Position, Y (Vertical) Axis to Y Position & Z (Depth) Axis to Z Position. You may prefer to remap the axes, if you are used to discussing z as height. For example if you are used to z being vertical, y being horizontal and x being out of the plane of the monitor(paper), set the X (Horizontal) Axis to Y Position, the Y (Vertical) Axis to Z position and the Z (Depth) Axis to X Position. You may also want to change the positive direction of the Z (Depth) Axis. The Invert checkboxes will let you specify whether the positive direction is in or out of the monitor. Similarly you can also invert the horizontal & vertical axes.

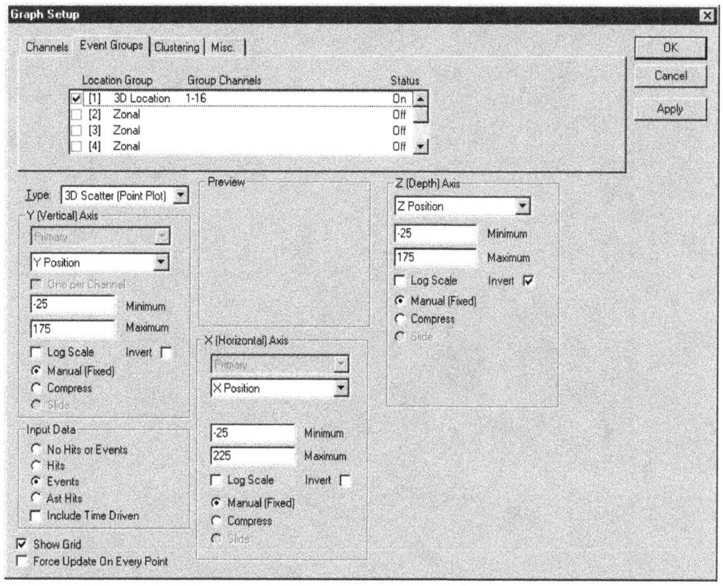

Figure 110. 3D Graph

You can also create 2D location plots using X, Y & Z Position settings on a 2D Binned Scatter graph as done in the section on 2D Planar. If you do, the sensors will also draw on the 2D plots using 2 of the 3 sensor coordinates ignoring the unused dimension. For example, an X & Z Position graph would show the sensors at their x & z coordinates & ignore their y coordinate. This allows you to create front, top and side views of your structure as shown below. These graphs also illustrate another benefit to staggering your sensor locations – legible sensor markers. If your sensors line up on opposing faces, you will end up with overlaid sensor markers on the 2D graphs. That can make them hard to read.

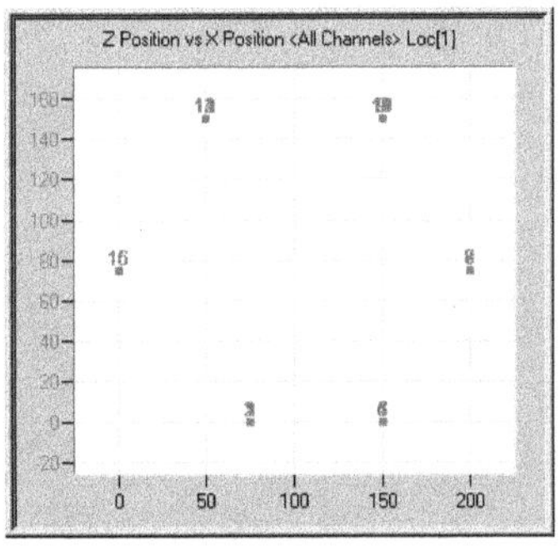

Figure 111. Top View

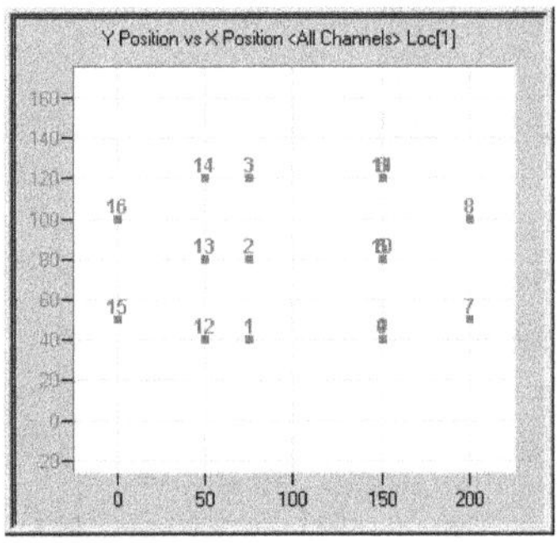

Figure 112. Front View

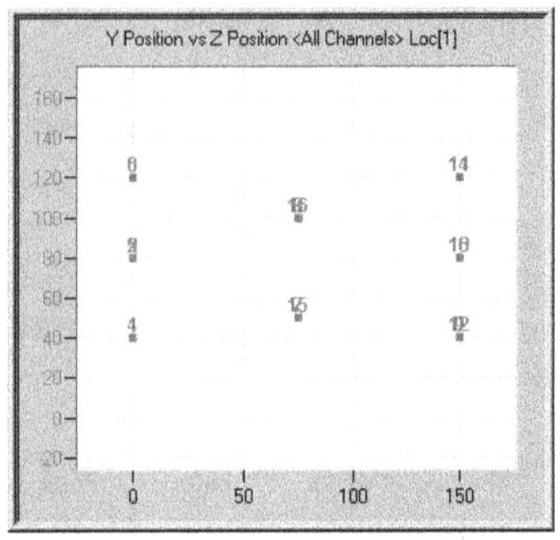

Figure 113. Side View

3D Location on Solid Cylinders

3D Location has been enhanced to allow the use of a cylinder structure to determine the limits of acceptable 3D source locations. The 3dLocCylinderDemo.lay file is a sample layout file included with HMSGHAT2000DDI that demonstrates how to set up & use a cylinder structure with 3D Location. Location Setup is configured as for any other 3D Location group (see 3D Location section above). Sensor Placement controls the user's choice of structure. The user simply selects Horiz. or Vert. Vessel as the Structure setting:

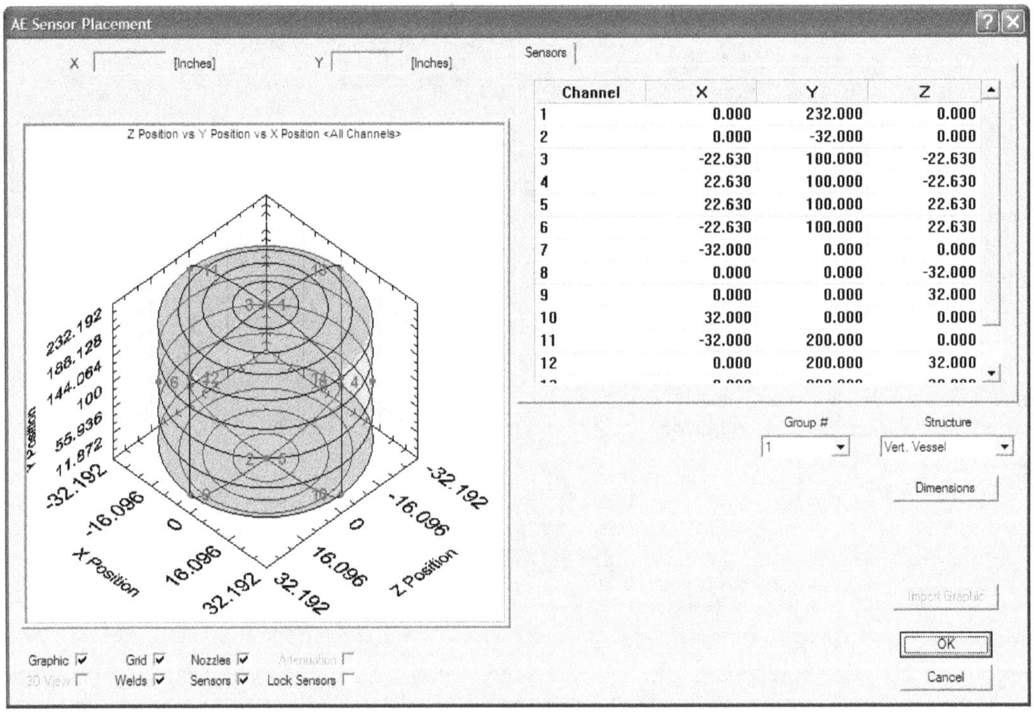

*Figure 114. Sensor **Placement** for 3D Location on a Solid Cylinder*

The user then uses the Dimensions dialog to configure the size of cylinder:

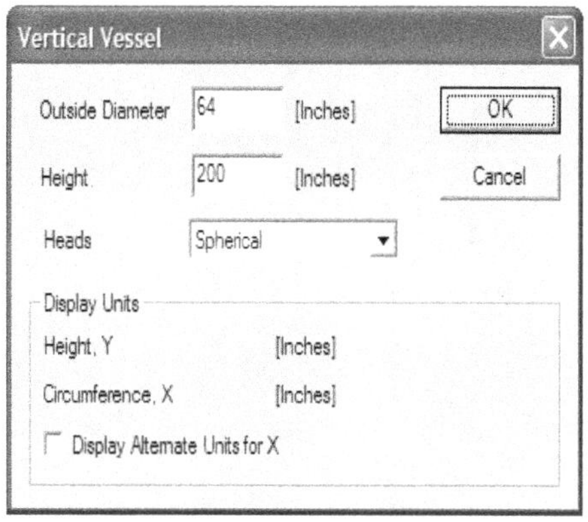

Figure 115. Vertical Dimensions

Sensors are placed on the cylinder using the grid in the Sensor tab of the Sensor Placement dialog box just as for the Box & Free structures. Simply click on the grid & press the Insert key to add a new sensor. Edit the coordinates to values needed. You can highlight a sensor in the grid & press the Delete key to remove an unwanted sensor.

The results of 3D Location on a solid cylinder can be displayed just like any other 3D Location result in a 3D Scatter graph that is set up to display the X, Y & Z Location results on each of the axes:

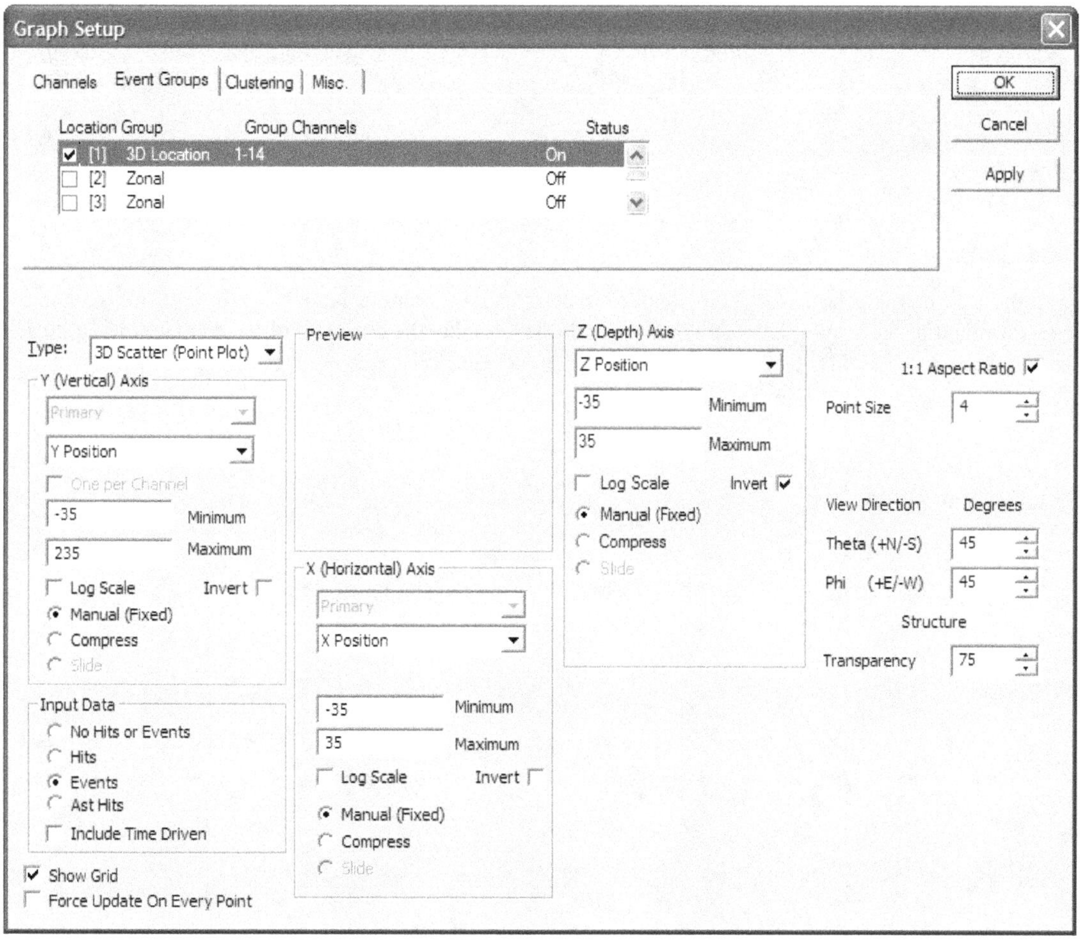

*Figure 116. Sensor **Placement** for 3D Location on a Solid Cylinder*

Note that the user checks the desired location group on the Event Groups tab to determine which location group generates the location results. The user also can control the transparency level of the cylinder and the viewing angle in Graph Setup as shown above.The results display on the 3D Scatter graph after a replay of the data:

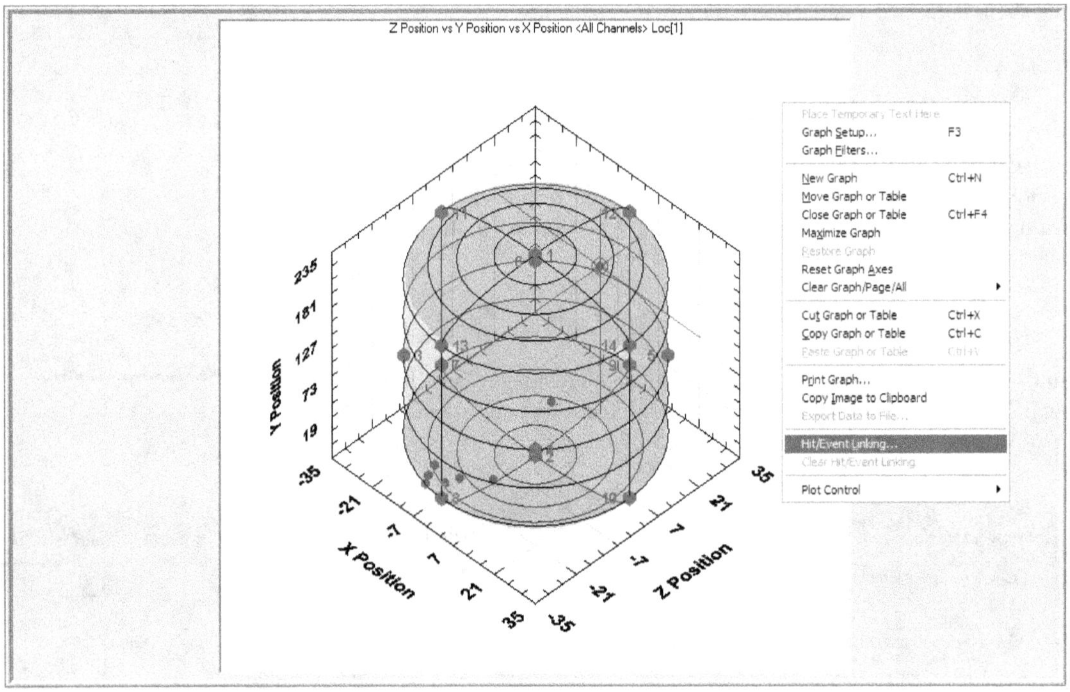

Figure 117. *Sensor Placement for 3D **Location** on a Solid Cylinder*

If desired the user can use the Data Cursor feature to select a data point & launch Hit/Event Linking using the right click context menu of the graph. Hit/Event linking will then display the corresponding event results in its dialog box:

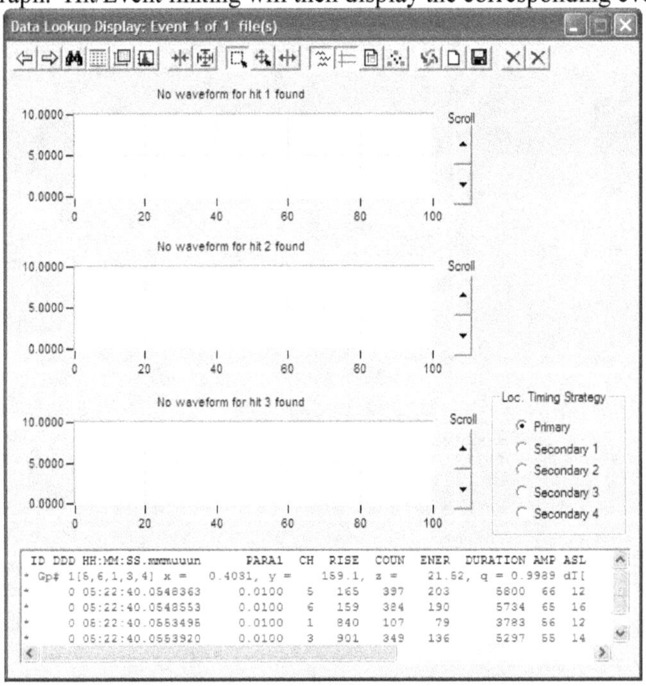

Figure 118. *Hit/Event Linking on 3D Solid Cylinder Results*

Spherical Location

Using Spherical location is similar to using the 2D Planar location mode, although there are differences in the coordinate system used & & how a typical location plot is set up. Mathematically, Spherical location behaves like 2D Planar where both axes wrap back around on themselves. There aren't any new pitfalls in sensor placement, but the linear event problem that affects 2D Planar also affects Spherical. If you attempt location using just 4 sensors on the equator of a sphere, Spherical location would always fail as all events would be due to sensors in a straight line. You could override that behavior by enabling Allow Marginal Events in Location Setup at the risk of having many poorly located events. See section 4.2.4 for details. For better results, use at least 6 sensors (north & south poles & 4 evenly sed on the equator). The maximum number of sensors is limited only by the number of available hardware channels. When using more than 6 sensors, it is recommended that you stagger alternate rows of the sensors to yield a triangular grid and produce more locatable events.

The Sensor Placement dialog has been enhanced to allow you to directly enter theta & phi coordinates anytime the structure type is set to Sphere as shown here. Sensor coordinates are entered in degrees. Theta is defined as the north/south angle (latitude) and should be between +90 and –90. +90 is 90 degrees north (top of the sphere), while –90 is 90 degrees south (bottom of the sphere). Phi is the east/west angle (longitude) and ranges between +180 & –180. A phi of +90 is 90 east (right), while –90 is 90 west (left) relative to the origin. +180 (east) and –180 (west) are actually the same location due to the wrapping of the sphere. A white + symbol marks the theta, phi origin of 0,0. You can quickly generate sensor layouts with the AutoPlace Sensors dialog. Just right click the graph in the Sensor Placement dialog.

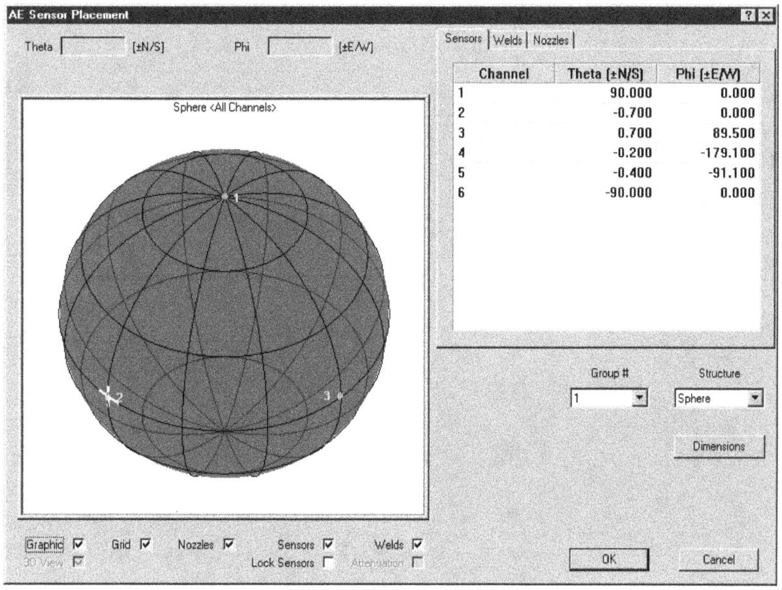

Figure 119. Spherical Sensor Placement

Alternatively you can use the INS key to manually insert a new sensor & then edit the sensor coordinates on the sensor grid.

Spherical location is designed to work only with the Sphere structure type. You will receive a warning if you attempt acquisition or replay with a Spherical location group that attempts to use any other kind of structure. The Dimensions button allows you to set the diameter of the sphere. The drawing of the sphere in Sensor Placement does not change in size with the diameter, but the diameter is a key parameter required when making use of the sensor coordinates in the location calculations. If your diameter is set incorrectly, location will often fail all spherical events.

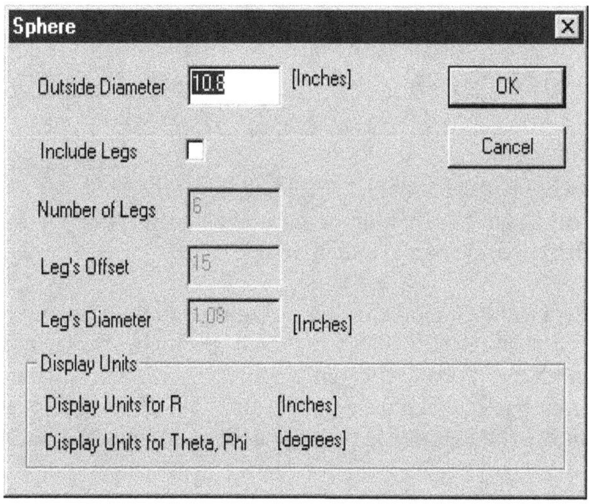

Figure 120. Sphere Dimension Dialog

There is an Include Legs option in the Dimensions dialog that allows the drawing of the sphere to show support columns if desired. Spherical location does not directly support location on the legs, but you could use Cylinder location on the legs, if location on the legs is required. You can specify the number of legs, their diameter and the offset in degrees of the first leg. The rest of the legs are drawn evenly sed around the sphere.

After setting up the sensors & each of the location groups, you will want to create location graphs to display the results. It is possible to use any of the standard 2D graph types and display theta or phi on one of the axes. Just pick Theta Position or Phi Position from the list of available features. The easiest way, however, is to use a Spherical Location graph like the one shown at the right.

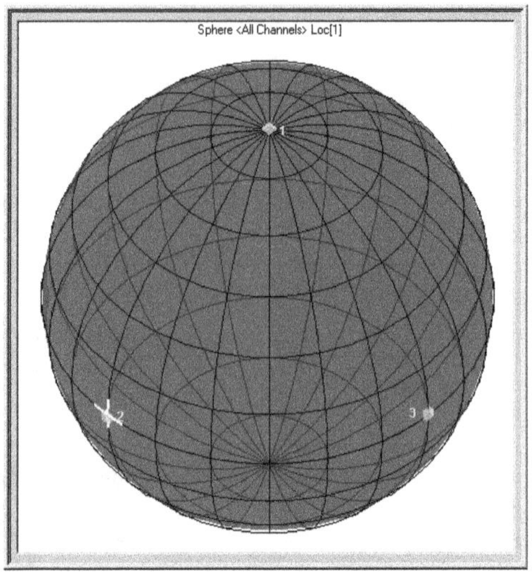

*Figure 121. Spherical **Location** Graph*

In use, a Spherical Location graph can contain a maximum of 4095 data points. During acquisition or replay, the newest data points replace the oldest ones on the graph once the 4095 point limit is reached.

When viewing the Spherical Location plot, you can use the mouse to rotate the sphere by clicking & dragging with the left mouse button. If you use the wheel on a wheel mouse or hold down the Alt key while clicking & dragging with the left mouse button, you can make the

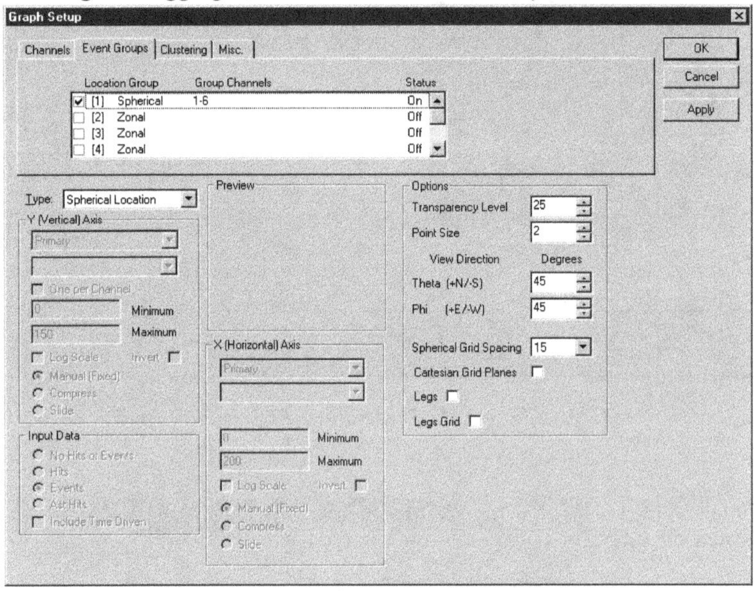

*Figure 122. **Spherical** Graph Setup*

sphere larger or smaller. Holding down the Shift key while clicking & dragging pans the sphere up/down/left/right. The Reset Axes button on the toolbar will cause the sphere to revert to the view originally specified by the View Direction settings in Graph Setup.

Here is the Graph Setup dialog configuration for the graph shown above. Unlike the normal 2D & 3D graphs, for the Spherical Location plot you do not specify the axis quantities or the data type (always plots events), so the Input Data and X & Y Axis settings are disabled. The graph will automatically draw sensors on the plot for any spherical location groups selected on the Event Groups tab. If you do not see a sphere on the graph after you exit Graph Setup, go back into Graph Setup & make sure at least one spherical location group is selected in Event Groups.

There is an Options section specific to the Spherical Location plot. It contains Transparency Level, Point Size, View Direction, Spherical Grid Sing & Cartesian Grid Planes. Here is brief description of what they do.

Transparency Level: The Transparency Level can vary between 0 and 100%. When set to 0%, the sphere becomes opaque & points that plot on the back side of the sphere are not visible. If you use 25 or 50%, the points on the back side become visible, but are a darker shade than those on the front side.

Point Size: Point Size directly affects the size of the data points & the sensor markers. The Spherical Location plot draws both the sensors & the data points as small diamonds. A bigger Point Size yields bigger data points & bigger sensor markers. The sensor markers are always slightly larger than the data points. You may also use the Graphing Colors menu to change the colors of the sensors, data points and the sphere (structure color). The data points use the Plot 1 color in Plot Colors.

View Direction: The View Direction setting in Graph Setup controls what angle the sphere is viewed from using the same conventions as the sensor coordinates. This allows you to create top, bottom and side views of your sphere. For example, if you enter Theta = -90 and Phi = 0, the sphere will draw as if you are looking directly at the south pole. Below is a page of sample graphs from the SphereDemo.lay example file.

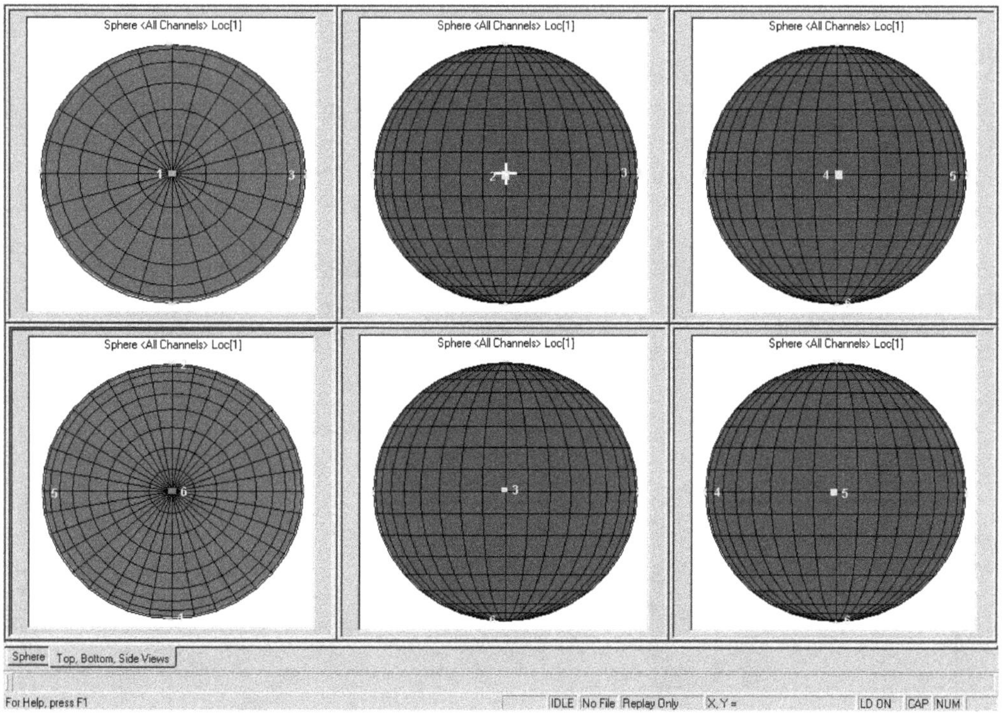

Figure 123. Spherical Location Graphs From Differing Viewpoints.

Spherical Grid Sing: The Spherical Grid Sing parameter controls how a grid is overlaid on the sphere. Available choices are 10, 20, 30, 45, 90 degrees and None.

Cartesian Grid Planes: The Cartesian Grid Planes checkbox toggles an X, Y, Z frame of reference on/off.

Legs: The Legs checkbox controls the display on the graph of any support columns defined for the structure.

Legs Grid: This option turns on an additional grid for the legs.

Cylinder Location

Cylinder Location is designed for locating source emissions on a hollow cylinder with either flat, elliptical, or hemispherical endcaps or no endcaps at all. (If you are interested in location on solid cylinders, see section 4.4.5 3D Location on Solid Cylinders.) It is similar to 2D Planar Location with the x, y coordinates being the unwrapped cylinder distances, but one or both of the axes wraps around the body of the cylinder.

When setting up your sensors on your cylinder, there are several guidelines to keep in mind. As with 2D Planar, the software cannot successfully locate any source when all the hits are from sensors in a single straight line. You can force it to do so with the Allow Marginal Events checkbox in Location Setup, but the advice on good vs. bad sensor layouts for 2D Planar applies here too. Avoid rectangular sensor layouts where the distance between rows differs greatly from the distance between columns as shown in the Bad 2D Planar Layout diagram. As stated before, the optimal arrangement would be a triangular grid made up of equilateral triangles. For Cylinder location, also be sure your sensor triangles on the heads are not significantly different in size than the triangles on the body especially if the cylinder has flat heads. If not, you may get spurious event locations clustering along the head-body border.

Sensors that are 180 degrees apart can also cause mathematically invalid events too. Geometrically, it is similar to problem with a straight line of sensors. If an event consists of a sensor on one side of the cylinder & 2 others exactly 180 degrees around the body, there is no way to distinguish source locations in the positive direction from the negative . As with linear events, you can force it to do so with the Allow Marginal Events checkbox in Location Setup. See section 4.2.4 for details. Generally it is recommended that you use 3 sensors around the circumference of the cylinder in a given row. If you must use 2 sensors per ring around the cylinder, definitely stagger the alternate rows or else you will not see any locatable events.

Use the Sensor Placement dialog shown below to set up the cylinder as well as place your sensors. Cylinder location is only designed to be used with the Horizontal or Vertical Vessel structure types. For Cylinder location, the Structure list in Sensor Placement *must be* set to either Horizontal or Vertical Vessel (cylinder) as shown above. A warning displays in acquisition or replay with a Cylinder location group using any other kind of structure.

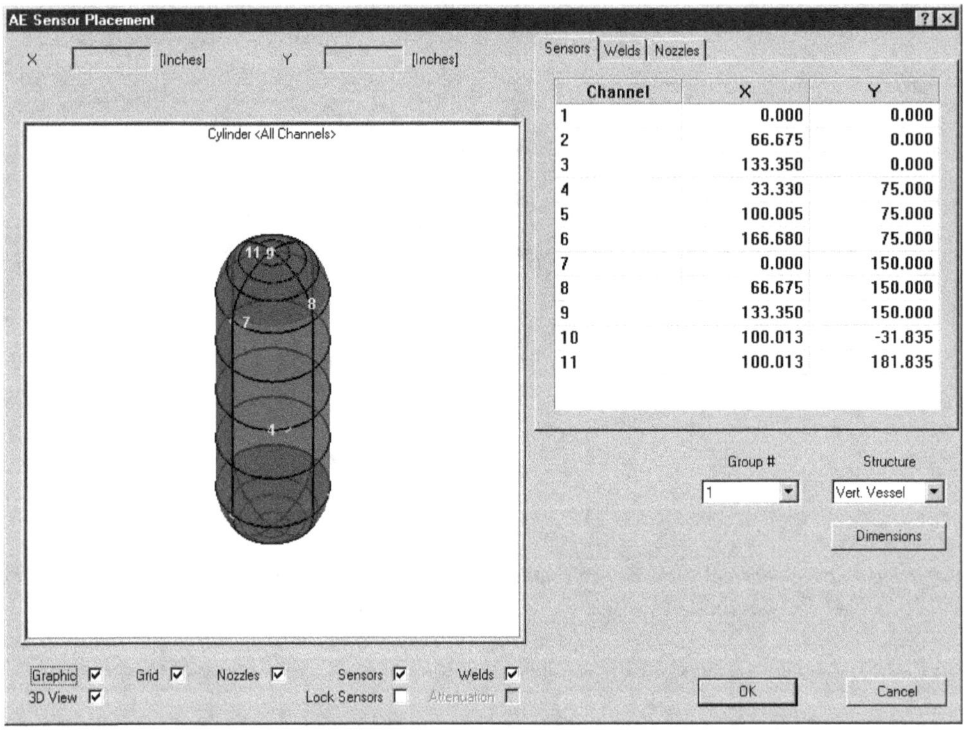

Figure 124. Cylinder Sensor Placement (3D View)

By default, Sensor Placement shows a 3D cylinder view. If you prefer a 2D view, simply uncheck the 3D View option at the bottom of the dialog & you will be presented with an unwrapped 2D view of the cylinder shown below.

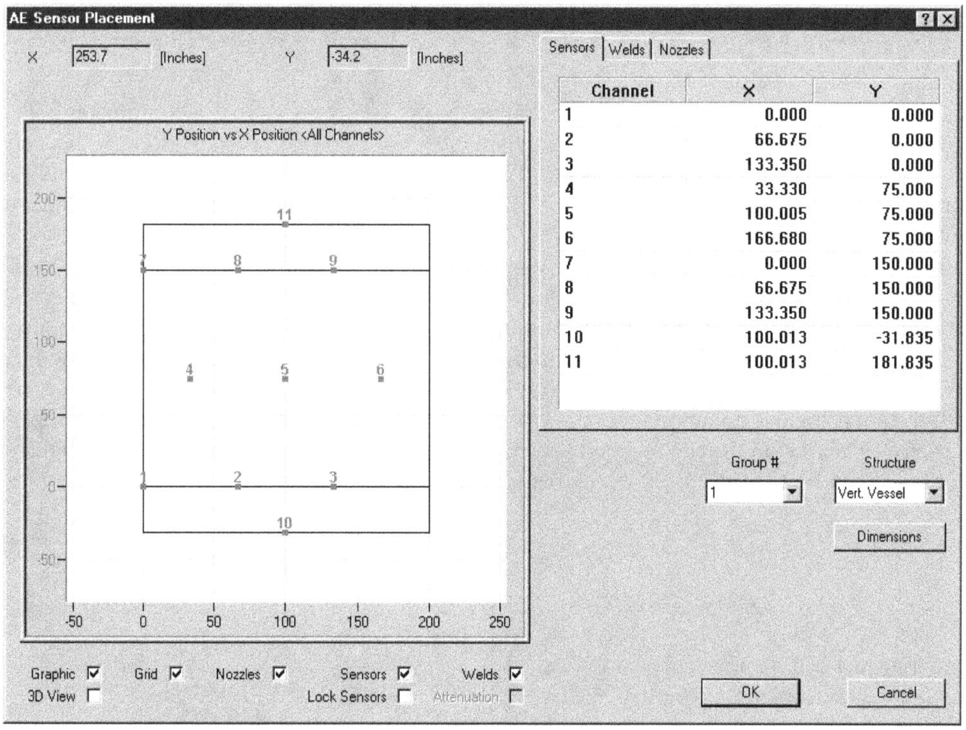

Figure 125. Cylinder Sensor Placement

The Sensor Placement dialog in 2D view & 2D location plots draws a rectangle to represent the body of the cylinder based on the user entered dimensions. You use the Dimensions dialog shown to the right to enter an Outside Diameter and the Height (vertical) or Length (horizontal). For a Horizontal Vessel, the rectangle extends along the x axis from 0 to Length and along the y axis from 0 to the circumference calculated from the user entered Outside Diameter. For a Vertical Vessel, the rectangle extends along the x axis from 0 to the circumference and along the y axis from 0 to Height.

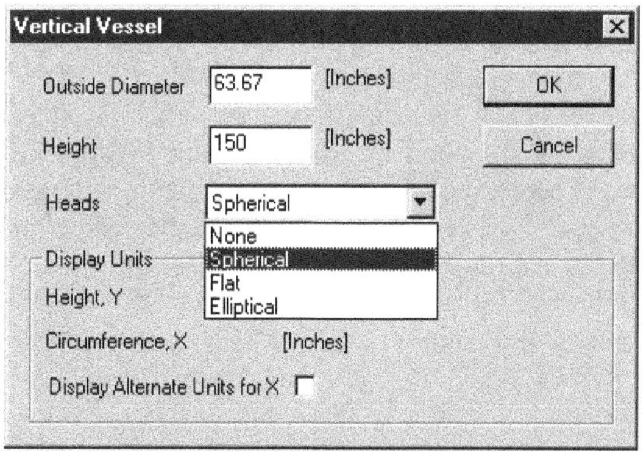

Figure 126. Vertical Vessel

Both dimensions are also vital to accurate source calculations. The Diameter is used to allow for the wrapping around the cylinder circumference. The Height or Length in combination with the Diameter is used in eliminating 2D locations off of the cylinder body caps itself.

There is also a dropdown list to select the type of Head on the cylinder. Available choices are Spherical, Flat, Elliptical & None. If a head type other than None is selected, the body rectangle drawn in Sensor Placement will have 2 additional rectangles on the top & bottom (Vertical Vessel) or left & right ends (Horizontal Vessel). The head rectangles extended the body rectangle by a distance determined by the head type. For Spherical or Flat, it is the length of the cap radius which is also the body radius and half the user entered diameter. In the horizontal case, the rectangle will extend along the x axis from –radius to Length+radius, while in the vertical case the y axis will run from –radius to Height+radius. For Elliptical heads, a 2:1 ratio is assumed between the body radius & the perpendicular head radius, so the distance is ½ the body radius or ¼ the user entered diameter.

You can quickly generate rectangular & triangular sensor layouts with the AutoPlace Sensors dialog available by right clicking on the sensor view. If you are using the 2D view of the cylinder, you can also manually place sensors on the cylinder by right clicking. After using either option, you can edit the coordinates of the sensors if structural features (valves, nozzles, welds) require shifting sensors around. Just enter the x,y coordinates in the user's preferred distance units.

When you enter sensor coordinates & you are not including the caps, express your x, y coordinates with a lower limit of 0 and an upper limit of the circumference or the length/height. For example, when working with a vertical cylinder of height 150 and diameter 63.66 (circumference = 200), express x as a value from 0 to +200 and y as value from 0 to +150. Basically, just make sure your sensor is visible in the body rectangle drawn in the Sensor Placement dialog.

If you are using Spherical or Flat heads, these limits are widened for the y axis by the radius to –31.83 to +181.83. These extensions allow for the entry of sensors on the heads. For this vertical cylinder, a pole sensor on the bottom cap would have a y coordinate of -31.83. The x coordinate for a cap on a vertical cylinder corresponds to a rotation angle, but is entered as a fraction of the body circumference. In this case the limits are 0 to +200 and sensors placed 120 (=360/3) degrees apart are separated by 66.67 (=200/3). For a cap sensor placed directly at the pole of the hemisphere, the rotation angle is unimportant in the source location calculations, but a value of half the circumference will place the sensor halfway up the cap rectangle border. In the case of the vertical cylinder shown above, the pole sensor on the bottom cap is at -31.83, +100 (any x really) and +181.83, +100 for the top cap.

When you are finished setting up your sensors & ready to actually do data acquisition, you will need a location graph to display the results of the source location calculations. You can either use a standard 2D graph as described in the section on 2D Planar location or you can use a Cylinder Location graph. Like the Spherical Location graph, a Cylinder Location graph is limited to 4095 data points with newer points replacing older ones once the limit is reached. A Cylinder Location graph is shown at right and its corresponding Graph Setup dialog settings are below. It has an Options section that is very similar to the one for Spherical Location. See the section on Spherical Location for a description of the options common to both.

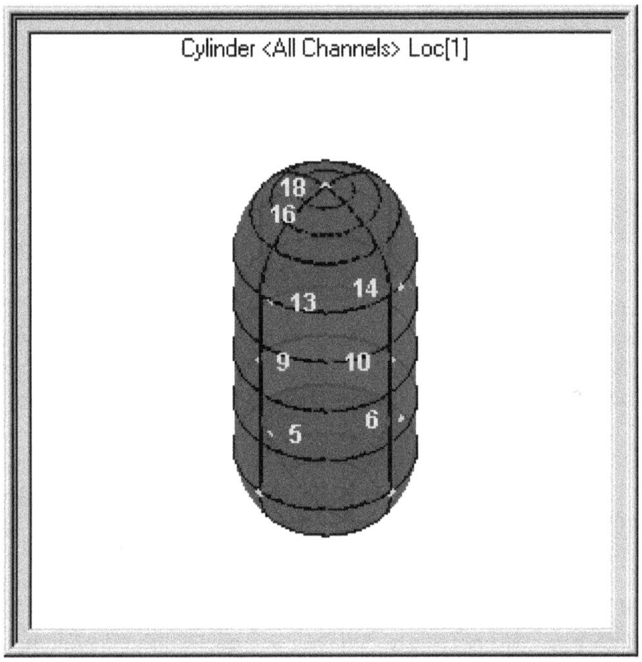

Figure 127. Cylinder Graph

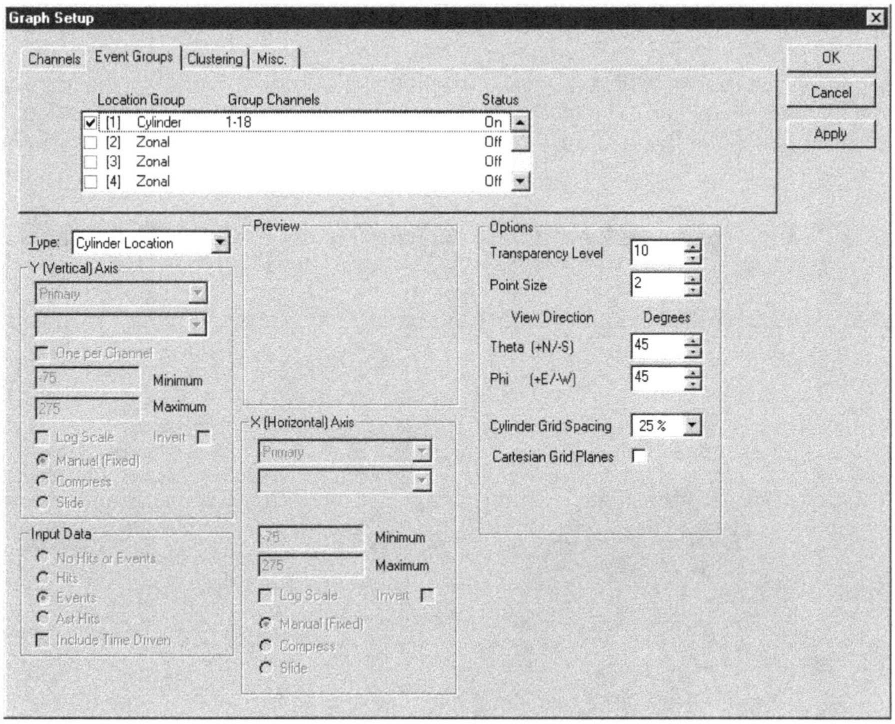

Figure 128. Cylinder *Graph Setup*

The Spherical Grid Sing option is replaced by the Cylinder Grid Sing option, although it behaves in a similar fashion. You may set the Cylinder Grid Sing to None, 10%, 20%, 25%, 50% or 100% to control the appearance of a grid overlaid onto the surface of the cylinder. The percentages are applied to the heads & the body separately as the dimensions for the 2 regions can vary greatly. If the user selects a grid sing of 25%, the body will be divided up into 4 equal regions and the head will also be divided up into 4 regions. If the user selects 100%, only the boundary between the heads & the body will be marked. Cylinder Location graphs always use Events as the Input Data, but as for the 2D graphs, make sure you set the Events Groups to the appropriate location group or your graph will not show any data & will not draw the cylinder associated with the group.

Conical Location

Conical location is set up almost identically to Cylinder location mode. In Location Setup, the user selects Conical for the Location Type on the General tab. The other settings in the table are described in the section on the Location Setup dialog. In Sensor Placement, set the Structure list to Vert. Cone or Horiz. Cone as shown on below.

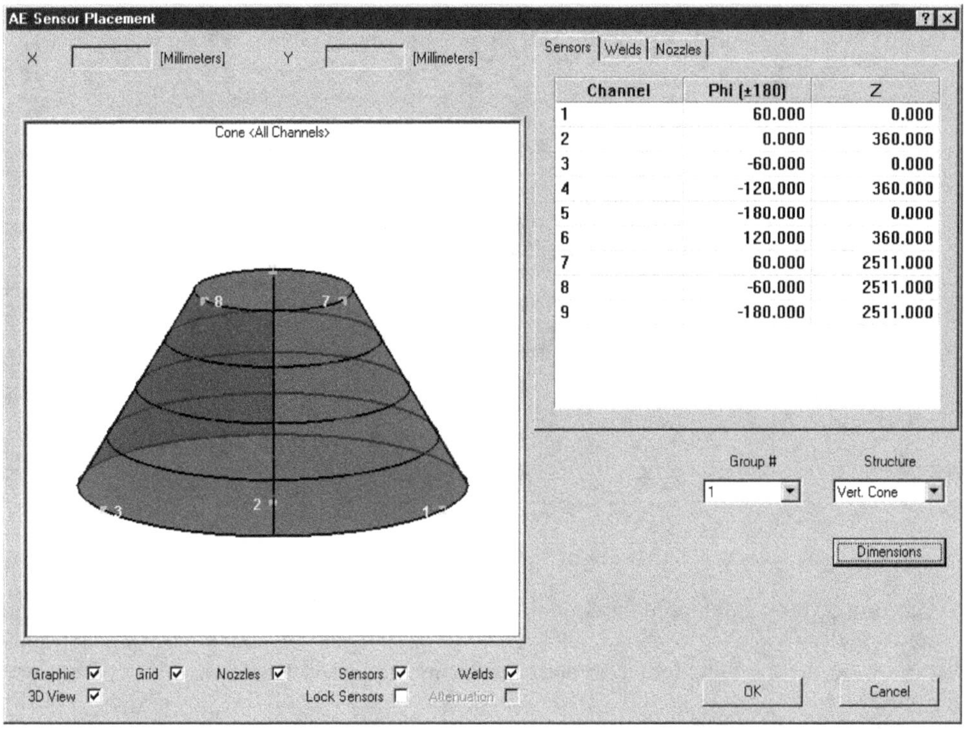

*Figure 129. Conical Sensor **Placement** (3D View)*

The Dimensions button calls up the Cone dimensions dialog allowing the user to enter the upper & lower diameters as well as the height (or the left & right diameters & length, if horizontal). The height or length has added significance when conical location is analyzing data. Conical results are limited to source locations to within +/- 2% of the height. In other words, points located wildly off the cone body are thrown out as bad results.

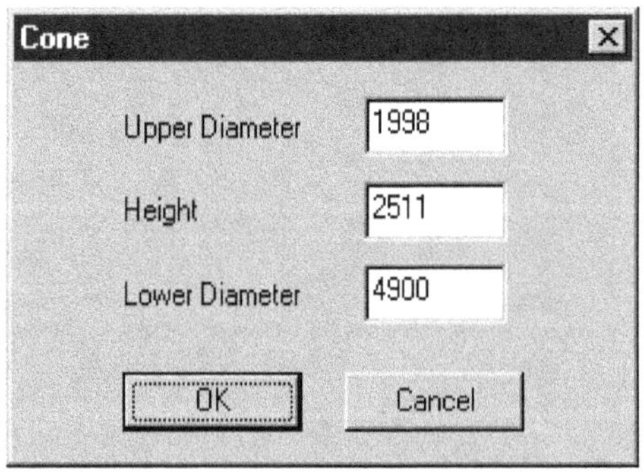

Figure 130. Cone

Sensor coordinates are entered as Z & Phi. The Z axis is the cone axis independent of horizontal vs. vertical configuration. Z starts at 0 at the bottom of the cone & goes up to the height of the cone (or 0 at the left & up to the length). Phi should be entered between -180 to +180 degrees to keep the sensors within the 2D drawing of the unrolled cone, although location will work with values from 0-360 as well.

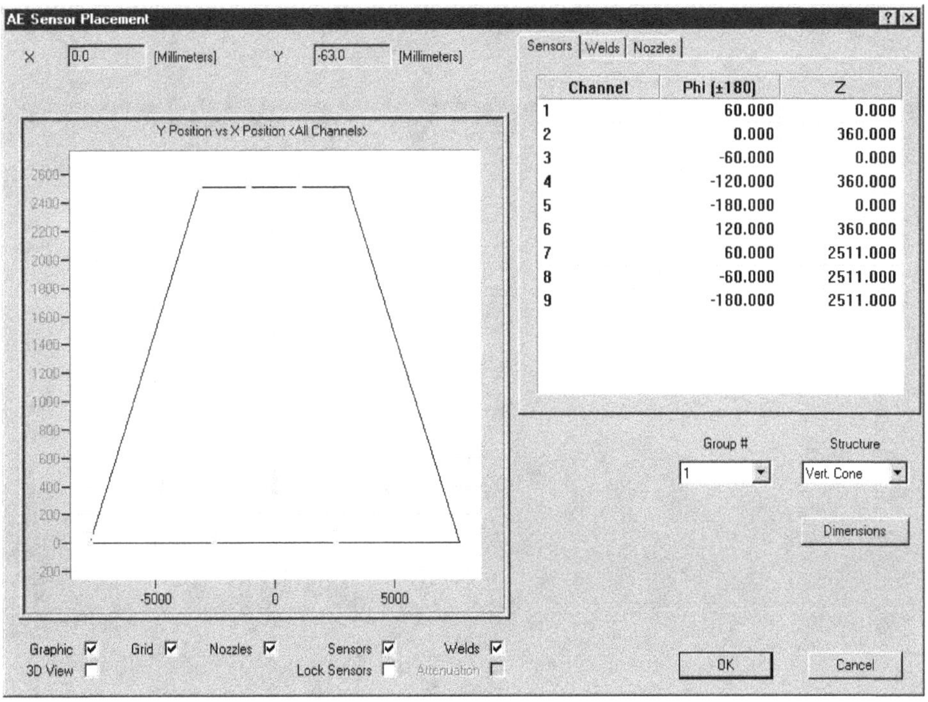

Figure 131. Conical Sensor Placement (2D View)

You can toggle between 2D & 3D graphs in Sensor Placement with the 3D View checkbox. The 2D view translates the phi value into a fraction of the circumference. The trapezoid is drawn from +circumference/2 to -circumference/2 where the diameter depends upon the z value in the cone.

There is also a 3D Conical Location graph similar to the Cylinder Location graph as another graph type in Graph Setup. It has options similar to those for the Cylinder Location graph & the Spherical Location graph. See their sections for a description of these options.

In use the 3D Conical Location graph looks like the 3D graph in Sensor Placement. The sample layout, ConicalDemo.lay, gives an example on the 3D Cone page. A Conical Location graph can contain a maximum of 4095 data points. During acquisition or replay, the newest data points replace the oldest ones on the graph once the 4095 point limit is reached.

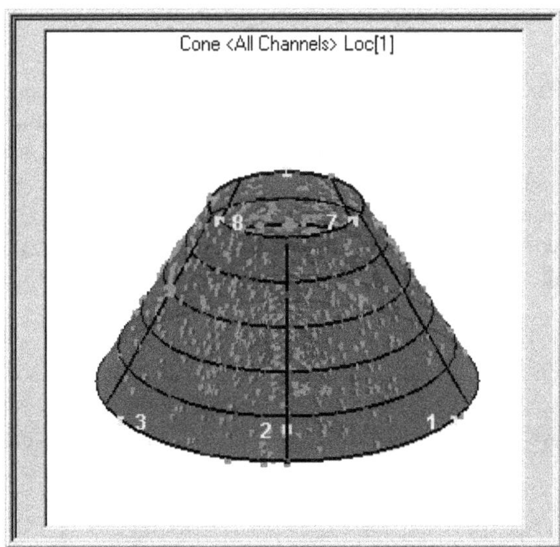

Figure 132. Conical Graph

Anisotropic Location Modes — 2D Anisotropic & Anistropic Cylinder Location

These location modes allow for differing velocities in the x & the y directions. They are almost identical to their corresponding isotropic modes, 2D Planar & Cylinder Location. The velocity in the Wave Velocity column of the General tab corresponds to the velocity along the X axis for these modes. When one of these location modes is selected in Location Setup, another field will appear in Sensor Placement as show below that allows the user to enter the Y axis wave velocity, Vy.

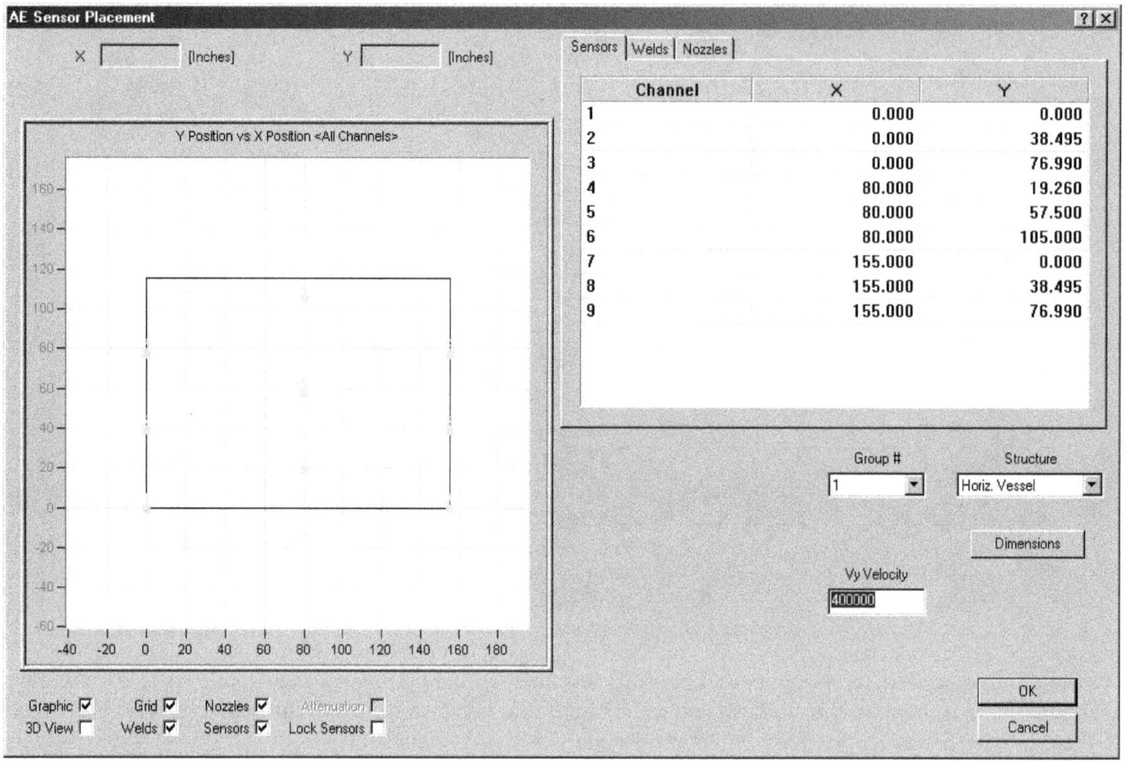

Figure 133. Sensor Placement for Anisotropic Location Modes.

These location modes do not support the use of Timing Strategies other than Primary & the software will not let you select other strategies if you try. The reason is that the alternate strategies only have one velocity associated with them, but the anisotropic modes require 2. The anisotropic location modes rely on a coordinate transformation to enable the standard location modes, 2D Planar or Cylinder location, to analyze the data. That coordinate transformation is based on the ratio of the X & Y axis velocities being constant, but allowing alternate timing strategies with varying velocities would violate this assumption. Anisotropic cylinder location also does not support location of the cylinder endcaps for much the same reason. One might reasonably expect that if the velocities along the X & Y axis differ substantially, then the velocities on the endcaps might also differ. For example, flat endcaps might use the velocity down the length of the cylinder, while spherical endcaps might need the velocity observed around the circumference of the body of the cylinder. The coordinate transformation cannot allow for those kinds of variations, so location on the endcaps is not supported. Except for those limitations, the anisotropic location modes, 2D Anisotropic & Anisotropic Cylinder, function just like the their isotropic counterparts, 2D Planar & Cylinder.

Tank Bottom Location

Tank Bottom location is designed for use with evenly sed 2-dimensional circular arrays of sensors. You must use it with the Tank Bottom structure. A direct consequence of requiring evenly sed arrays of sensors is that the user is not allowed to manually place sensors on the Tank Bottom structure. Instead one uses the Dimensions button in Sensor Placement to bring up the Tank Bottom dimensions dialog shown to the right. Unlike other dimensions dialog boxes, this one adds a section for automatically generating the evenly sed circular array of sensors. By default the sequence starts with sensor 1 placed at 0 degrees or 12:00 on the circle, but the user can start with other channel numbers and specify the rotational offset to the first sensor in degrees. You set the diameter of the tank bottom circle in this dialog box as well. After you click on OK, you will be asked if you want to delete the previously placed sensors. You must answer Yes to have the changes take effect. If you answer no, the changes will be cancelled. Once you have things configured correctly, Sensor Placement will reappear with the display updated based on your settings as shown below.

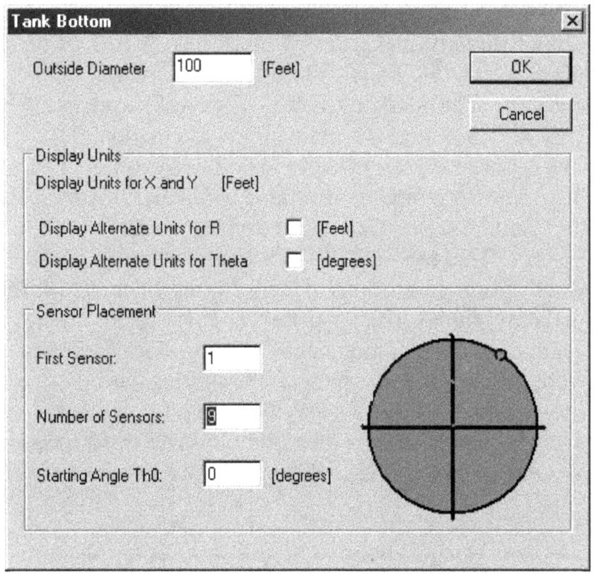

*Figure 134. Tank Bottom **Dimensions** Dialog*

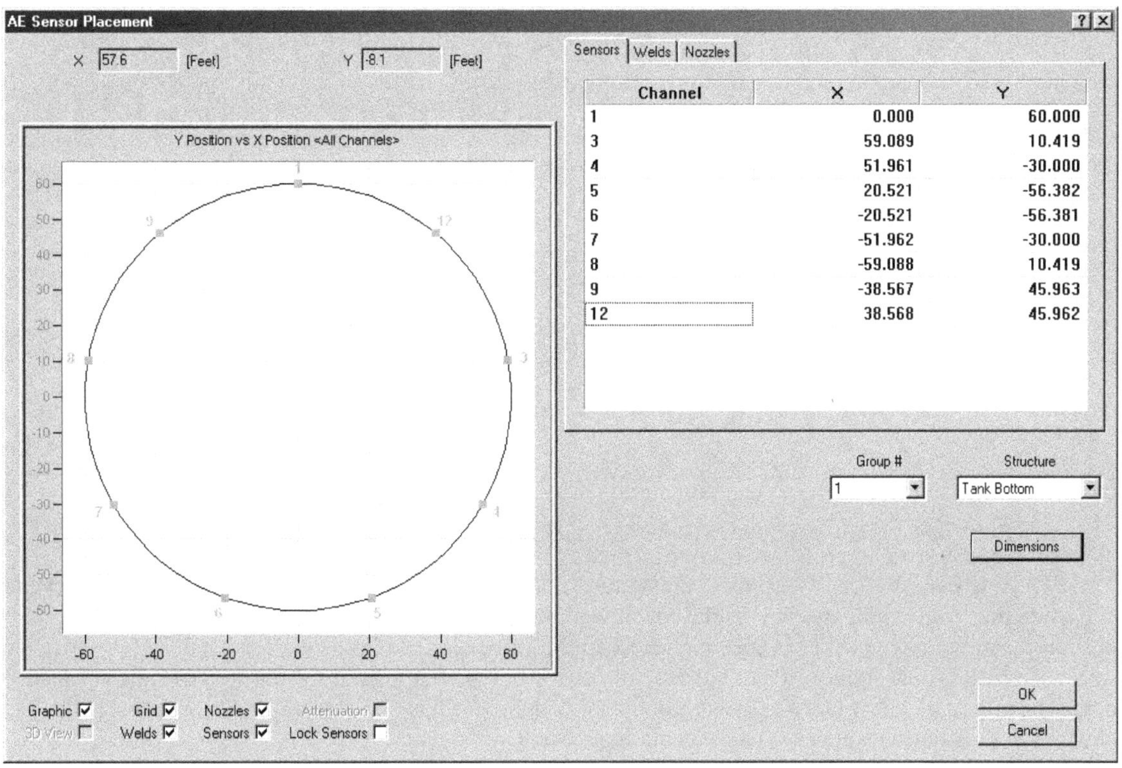

Figure 135. Sensor Placement for Tank Bottom Location.

Note that you cannot edit the X & Y coordinates of the sensors in the sensor grid array when using a Tank Bottom structure. You can edit the channel numbers, though, to allow non-sequential ordering of the sensors on the tank. In the example above sensor 2 was renamed 12, even though, the X & Y coordinates have been altered.

Tank Bottom location determines the source location by directly solving the time-distance equations generated from an event with 3 hits. It cannot directly make use of more than 3 hits when calculating a source location. When there are more than 3 hits available, however, it can vary which 3 hits are used in the calculation. The user can tell the software to do this by using the Tank Bottom Options dialog available thru the Tank Bottom Options button in Location Setup. The top field in the dialog controls whether or not variations of the 3 hits are calculated & the results displayed on location plots. By default, the software will only show 1 location per event & variations are never shown in acquisition mode. When limited to 1 location per event & there are more than 3 hits, the algorithm uses the first 3 available hits on the assumption that those sensors were the closest to the source. This setting has no effect on events that only have 3 hits available. Below are 2 graphs showing how the results vary for a data file with about 600 events. The graph on the left is limited to 1 location per event, while the one on the right allows up to 10 locations per events. Note that the software counts these variations as extra events, so the total events for the graph on the right is reported to be about 2500 events.

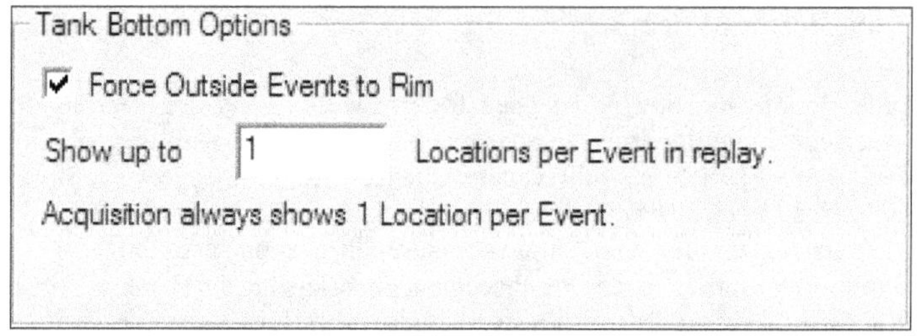

Figure 136. *Tank **Bottom** Options*

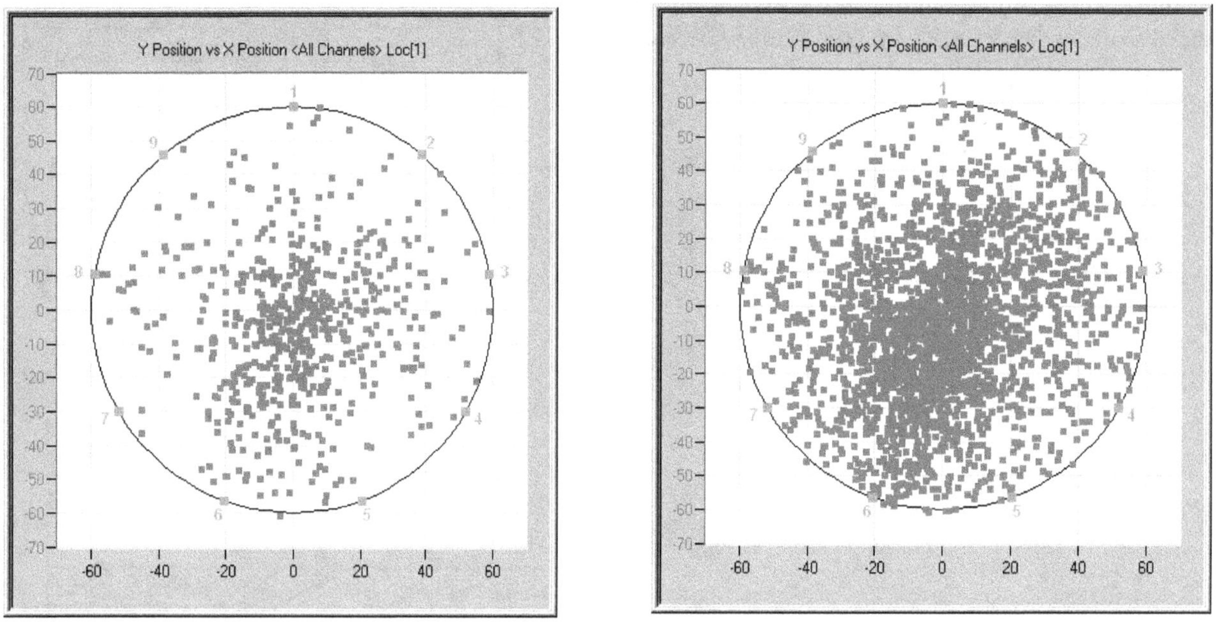

Figure 137. *Tank **Bottom** Location Results with 1 (left) vs. 10 (right) Locations per Event.*

The Force Outside Event to Rim checkbox at the bottom of the Tank Bottom Options dialog controls how events that locate outside the tank bottom area are displayed. It works in conjunction with the Allow Outside Events checkbox in Location Setup. Allow Outside Events applies to all the regression modes as well as Tank Bottom location, but Force Outside Events to Rim only affect Tank Bottom location. Allow Marginal Events does not affect Tank Bottom location. The effects of Allow Outside Events combined with Force Outside Events to Rim are shown in the 3 graphs below which are restricted to showing just 1 location per event.

As discussed in the section on the Location Setup dialog box, "Allow Outside Events" controls whether or not the software reports events that locate outside the structure. If it is off, "outside events" are considered failing events and in general are not reported. Some events may occur very close to or just outside the tank rim & still be considered inside events, because the algorithm uses a tolerance of ±2% beyond the radius when deciding whether an event is inside or outside the tank. That 2% tolerance allows for rounding errors that can push events at the rim just outside the tank bottom. The results of Tank Bottom location with Allow Outside Events disabled are shown below in the graph on the right. If Allow Outside Events is disabled, the Force Outside Events to the Rim setting has no effect on the results. If Allow Outside Events is enabled & Force Outside Events to Rim is disabled, all events plot at the calculated source locations as shown below in the middle graph. If both Allow Outside Events & Force Outside Events to Rim are enabled, then events that would plot outside the tank bottom area are redirected to the rim of the tank as shown below in the graph on the right. The software draws a line from the center of the tank to the calculated source location, finds the point this line segment crosses the tank rim & reports it as the source location. Although this location is not an accurate source location, it does allow the user to see that an event occurred. A drawback to enabling Force Outside Events to Rim is that it can obscure events that actually do locate at the edge of the tank. It is provided primarily for backwards compatibility when comparing with results observed in 's older DOS applications.

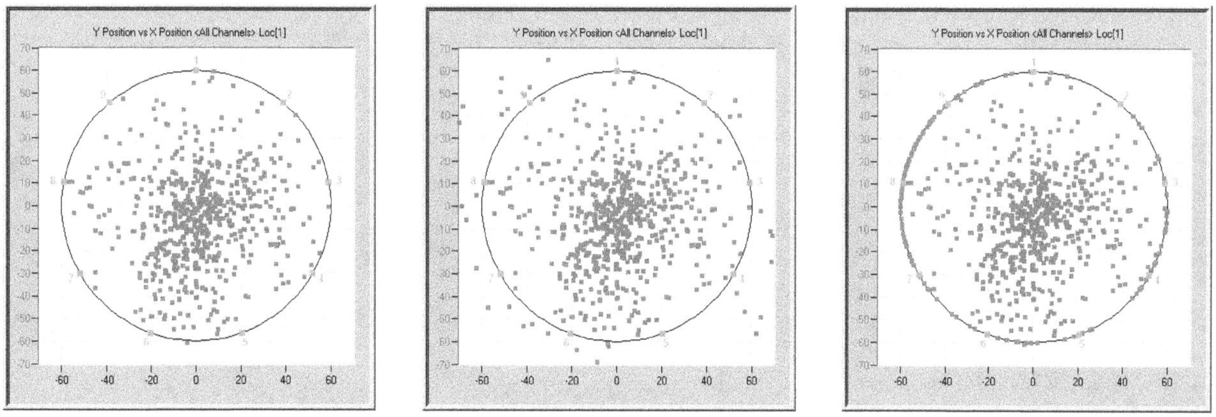

*Figure 138. Tank Bottom Location Results with Allow Outside Events disabled (left) & enabled with Force Outside Events to Rim **disabled** (middle) & enabled (right).*

Location Examples

It is not feasible to provide detailed examples of all location modes in this manual. There are, however, example layout files & sample data files for each location mode. When you are starting out with a new location mode, you can simple recall the sample layout & inspect the location related menus (Location Setup, Sensor Placement & Graph Setup dialogs). After you learn how the example is set up, you can make changes to better reflect your test setup & save the modified layout under a new name. Below is a table summarizing the example layouts & data files for the various location modes.

Location Mode	Layout File	Data File
Zonal	Zonal_Linear.lay	Zonal_Linear.dta
Linear	Zonal_Linear.lay	Zonal_Linear.dta
(Wrapped) Linear	WrappedLinear.lay	WrappedLinear.dta
Tank Bottom	TankBot.lay	TankBot.dta
2D Planar	2D-Planar.lay	HE-Link.dta
Cylinder	OpenCylinderDemo.lay	LongVessel.dta
Conical	ConicalDemo.lay	ConicalDemo.dta
Spherical	SphereDemo.lay	SphereDemo.dta
3D Location	3dLocDemo.lay	3dLocDemo.dta

In addition to the example files, there are detailed examples given here to illustrate how to use Zonal, Linear, 2D Planar & Tank Bottom location.

The steps in setting up and running a location test are pretty similar no matter what location mode you are using. Here is an overview of those steps:

1. Create a layout file with the hardware set up appropriately (e.g. the correct number of sensors turned on, detection threshold set properly).
2. Bring up the Location Setup dialog (=F8 key) and start by setting the distance units.
3. Next fill in the location group information. Enable the group & select the Location Type. Enter the Wave Velocity, Event Definition Value & Overcal Value. If this is your first time working with location, leave the Event Lockout Value, Timing Strategies, Max Iterations, Min & Max Hits/Event at the default values. That applies to all the settings on the Timing page as well. These parameters are more fully described in section on the Location Setup dialog under the headings of General Tab Settings & Timing Tab Settings.
4. Next define the sensor locations on the structure using the Sensor Placement dialog. Use the Location View button in Location Setup to display it.
5. In Sensor Placement, select the Structure type, then click on Dimensions & enter the structure information.
6. Now create your sensors by right clicking the sensor view area & or using INS & F2 in the sensor grid.
7. When finished, click OK in both the Sensor Placement dialog & in the Location Setup dialog.
8. Next create a suitable location graph using the Graphing menu (New Graph) & configure it using Graph Setup (=F3 key). See the notes on individual location modes for information on the appropriate location graphs.
9. Now we are ready to run a trial AE test using pencil lead breaks to verify that everything is set up properly. If everything works, you're ready for acquisition.
10. If you end up staring at an empty graph, here is a list of things to double-check. Some are obvious, but personal experience shows them to be easy mistakes.

a. Is the Input Data set to Events in Graph Setup?
b. Is this location group checked on in the Event Groups tab of Graph Setup?
c. Is this location group checked on in the Group # column of Location Setup?
d. Are there any events from the group in the Line Display?
e. If there are events, are the axis limits in Graph Setup wide enough to see the positions?
f. If no events, are the Location Setup settings reasonable? Did you look up a velocity like 10,000 ft/sec & enter 10,000 when it should be 120,000 because the Distance Units are Inches? Verify the Event Definition & Lockout Value are reasonable given your sensor sing & structure type. They all use the same units, so the values should be related to sensor-sensor sing.

Set Up and Testing of a Zonal or Linear Location Group

As an easy to set up layout, we will be using two sensors located 30 inches apart. This sensor layout will be used to illustrate Zonal location in this section & Linear location in the next section. We will be using example layout and data files called Zonal_Linear.lay and Zonal_Linear.dta which are installed in the AEData directory under your software program directory, for example "C:\Program Files\Physical Acoustics\HMSGHAT2000DDI for DiSP\AEData". If you do not have these sample files, we will be showing the setup screens so that you can duplicate the setup and conduct your own acquisition test.

In this section we will be setting up a Zonal Location Test following the steps identified above. For the example, we are assuming that we have two sensors in Channels 1 & 2 of the AE system.

Step 1. Start by using the File pulldown menu to access the Open Layout menu option or press Ctrl-O. Select the layout file named, "Zonal_Linear.lay" & recall it.

Step 2. Enter the location setup menu by either pressing the F8 function key, clicking on the Toolbar icon that looks like a "bullseye" target called "Location Setup" or select Location on the Acquisition Setup pulldown menu.

The Location Setup dialog will be displayed as shown in this section. This is the first menu that needs to be set up in our step by step procedure. By default, the General page displays. Notice the General tab near the top left of the dialog box. The General page of Location Setup allows you to control the setting of the individual location groups. Up to 8 location groups can be set up, although an additional licensing option provides for up to 32 groups.

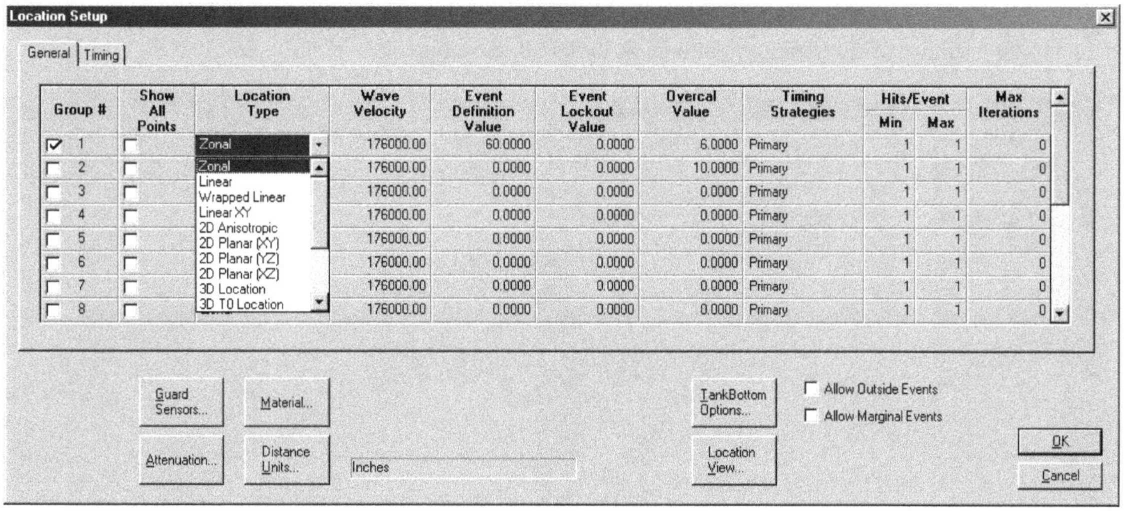

Figure 139. *Setting up Location* ***Group*** *Information*

In addition, near the bottom center of the dialog, the user can set up the location distance units. Various selections including Metric, English and time units can be selected for distance units. Selecting a real distance value is most appropriate since the values read out can be directly measured when trying to locate a source on a structure. Use the Distance Units button to select inches as in this example.

Step 3. In this example, we will select a Zonal location group as group #1. Follow the Group #1 line and enter the information as shown. First select the zonal group. This is done by left-clicking the mouse on the down arrow on the right side of the Group #1, "Type" box. A dropdown list will appear to allow the user to select from the various alternatives (Zonal, Linear, etc.). Select Zonal.

For the Wave Velocity, the velocity entered will be related to your Distance Units setting. In this case, we've selected inches as our units, so the velocity is in inches/second. You can look up a velocity using the Material Properties dialog by clicking on the Material button. The database contains values for Longitudinal, Shear & Surface velocities. It is often better to directly measure the velocity using the AE system. Simply use the Line Display mode to determine the actual transit time (Δt) to traverse between two sensors. You can then divide the distance between the two sensors by the Δt to arrive at the estimated velocity. In order to determine this correctly you really need to simulate the type of defect you are trying to detect. Enter the desired velocity in the edit box. In our example, we used an aluminum bar and a pencil lead break as a source. Although the Longitudinal velocity is 249,000 inches/second and the Shear velocity is 123,000 inches/second, we experimentally determined a test velocity of 176,000 inches/second.

For the Event Definition Value, we want to select a generous Event Definition using the same units as the Distance Units selected. This should be set to make sure that from any point in the location array, a minimum desired number of sensors will be hit, forming an event. In our example, two sensors 30 inches apart, the Event definition should be at least 30 inches to make sure that the two sensors get hit within a single event. We have entered 60 in the example above.

For Event lockout Value, we enter a value (in the selected distance units) to assure that the there is enough time (or distance) for all additional hits and reflections related to the event have had a chance to die down. The amount of time or distance selected will be the "dead" time as we wait for the event to die down. In this example, we will leave it at the default of 0 as reflections are not expected to be a problem for this setup.

The Overcal value is also set in the selected distance units and is usually set at an amount of approximately 10% to assure that minor variations in wavepath velocity are not ignored by the location algorithm. When you are just learning to use location, it is best to leave the settings on the Timing page & the Timing Strategies column at the defaults. The Max Iterations, Min & Max Hits/Event cannot be changed for Zonal as they are not used by it. They only apply to the regression location modes. All of these parameters are more fully described in section on the Location Setup dialog under the headings of General Tab Settings & Timing Tab Settings.

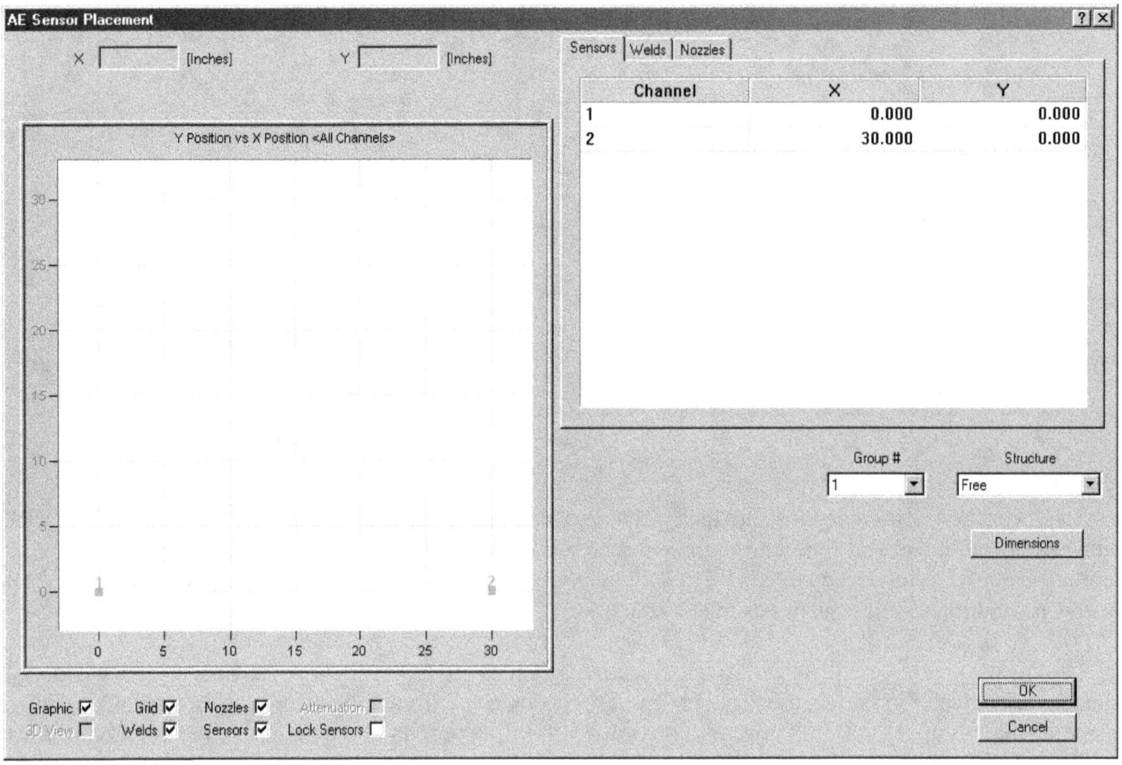

Figure 140. Sensor Placement Dialog

Step 4. Once the location group settings have been entered, we can define the sensor setup using the Sensor Placement dialog. Click on the Location View button on the bottom right side of Location Setup to display it. The figure above shows the Sensor Placement dialog box. There are 2 main area to the dialog box. The sensor layout view on the left-hand side and the sensor grid with the coordinates of individual sensors on the right hand side.

When we click Location View, the sensor layout for that particular location group is displayed. If we had group 1 highlighted, but wanted to display the sensors for group 5, we could change the group displayed with the Group dropdown listbox just below the sensor grid. Select Group 1, if it is not already selected.

Step 5. Next, we need to select the "Structure". The Structure setting is also just below the sensor grid & just to the right of the Group setting. This is currently set to Free. Free allows sensors to be placed anywhere without worrying about the specific structure geometry such as a plate, box or sphere.

Next, we set the Structure dimensions. For the Free structure, this isn't required for location, but the dimensions control the axis limits of the graph in the sensor view area. To change them we click the "Dimensions" button and the Free dimensions dialog shown at the right appears. Here we've set the X and Y minimum and maximum values to 0 & 30.

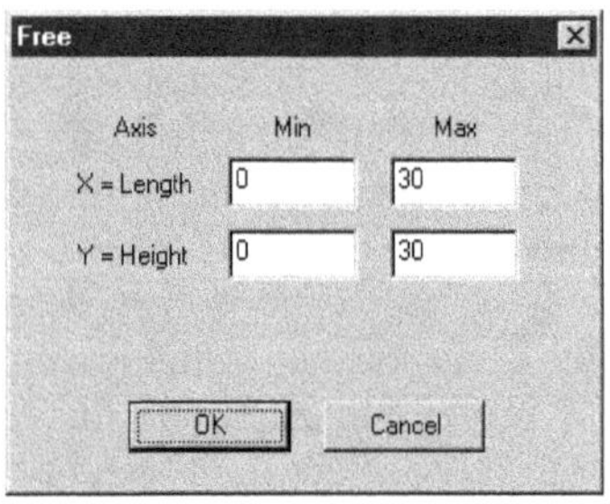

Figure 141. Free Dialog

Step 6. We are now ready to create our sensor layout. In Sensor Placement example above, you can see that the sensors have already been placed at the bottom of the graph area, at locations of (0, 0) and (30, 0) using an (X, Y) arrangement. The software makes it easy to place and move sensors on the screen and to get them in an exact position.

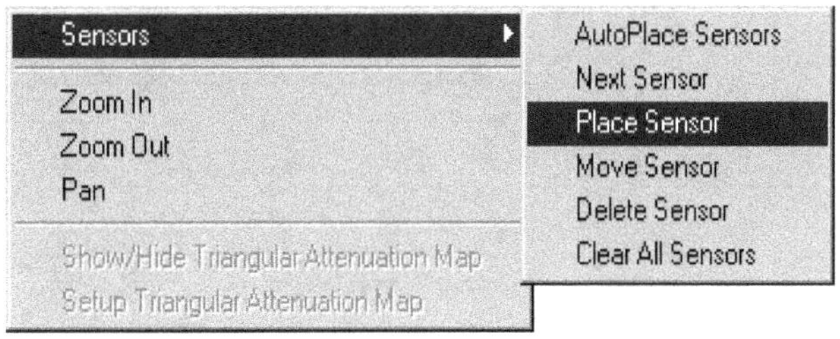

Figure 142. Place Sensor

To place a new sensor, we right click inside the graph region. The context menu to the right appears. The menu options we want to concentrate on are on the Sensors submenu allowing you to perform various functions related to sensor placement.

Since we want to place a sensor, simply click on the "Place Sensor" selection. Upon doing so a small menu box comes up asking you to verify the number of the sensor to be placed or to change it. When you're done, click OK and the mouse cursor will change to a red sensor icon with a crosshair cursor. Each time you click you will generate another sensor in your layout. Notice the x,y coordinates of the mouse cursor are shown above the sensor view area. When you are finished placing sensors, right clicking will return the mouse to its normal arrow shape. Simply follow the right click with a left click off the menu, if you don't need any of the context menu options. Whatever you do, don't press the ESC key. That is equivalent to pressing the Cancel button for the dialog & you'll lose all your changes to the sensor layout.

There are 2 ways to move sensors around to correct bad placement. First, the right click Sensors menu has a Move Sensor option. When you select it, the mouse cursor changes to a red sensor icon with a 4-headed crosshair cursor. To use it, click on the sensor you want to move. The sensor will turn red. Now click a second time at the spot where you want to move the sensor & it will shift to the new location. Right clicking ends the Move Sensor mode.

If after graphically placing or moving a sensor, its placement is not accurate enough, you can alter the position to an exact value by editing the X or Y value in the sensor grid. Left-click the mouse to highlight the desired box to be changed and enter the exact value, followed by the Enter key. Alternatively, press the F2 key to edit the current value.

Note the other sensor placement and edit capabilities provided by the submenu. You can automatically place the "Next" sensor by number by pressing the "Next Sensor" button. In this case the sensor channel number selection box does not come up. It simply enters the Place Sensor mode using the next highest number sensor. Please note and experiment with these sensor placement management selections.

By now you may have added several sensors to your setup that you want to delete. Do this by right clicking over the sensor you want to delete and then select "Delete Sensor". The mouse cursor with change to a red sensor with a crosshair labeled "del". When you click on a sensor, a "Delete this sensor?" message box asks for confirmation. Click "Yes", to delete the sensor from the graph and from the table. Again right clicking puts the cursor back to normal. Return to the original sensor layout as shown above when you are finished exploring the Sensors menu options.

Step 7. In the previous steps we have fully defined the location setup. The next step in the process is to set up graphs to see our location in acquisition or replay. We do this by first saving our sensor placement & location setup. Click the OK button in Sensor Placement & then in Location Setup. DO NOT CLICK Cancel in either one as you will lose any changes you made while in these dialog boxes. This is a very good time to save your layout on disk as well using the File pulldown Save Layout As menu option. When you don't need to rename the layout, you can just use Ctrl-S to update the current layout on disk.

Step 8. Next we need to create a suitable location graph. If you have been following along with our example layout file called Zonal_Linear.lay, you will be seeing a screen like that shown below. In this screen you can see two graphs. The first or top graph is set up already as a location graph. Since this is being set up for zonal location, we will only see the activity in a histogram or bar chart type of arrangement.

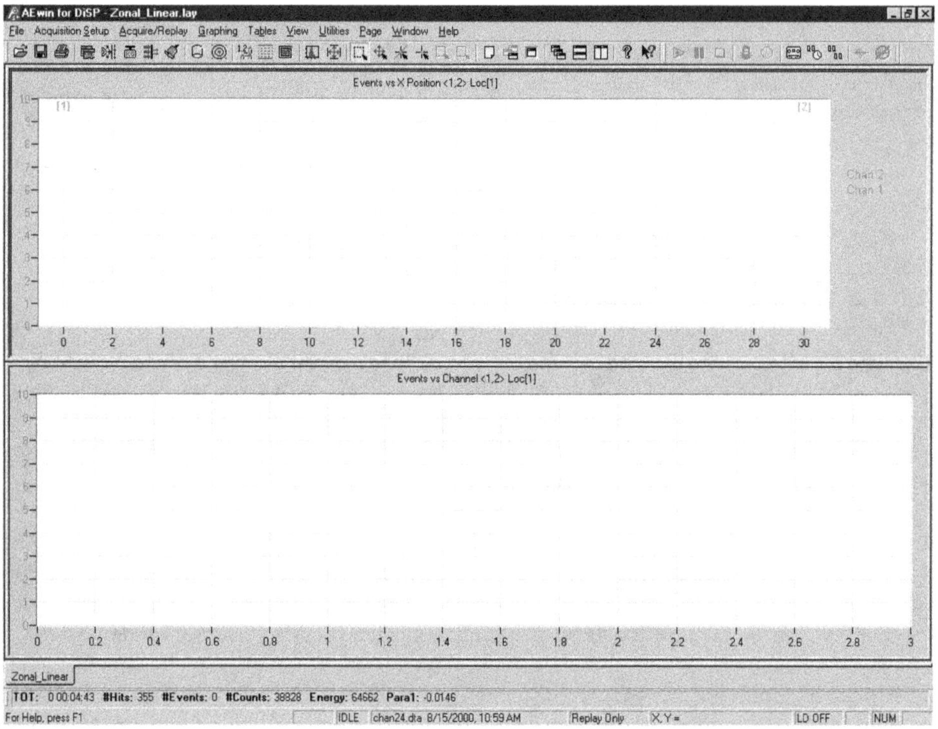

*Figure 143. Zonal – **Linear** Location Graph*

Assuming that this graph was not already set up in the desired location mode, let's go through the steps of setting up the graph for a Linear or Zonal location. First, we select the graph that we want to change to the location graph. We do this by right clicking the mouse when it is inside the graph region of the graph that we want to work on. Once we do that a menu appears. This menu gives us many choices but we will select the top one called "Graph Setup". Upon selecting this a graph setup menu appears like that shown below.

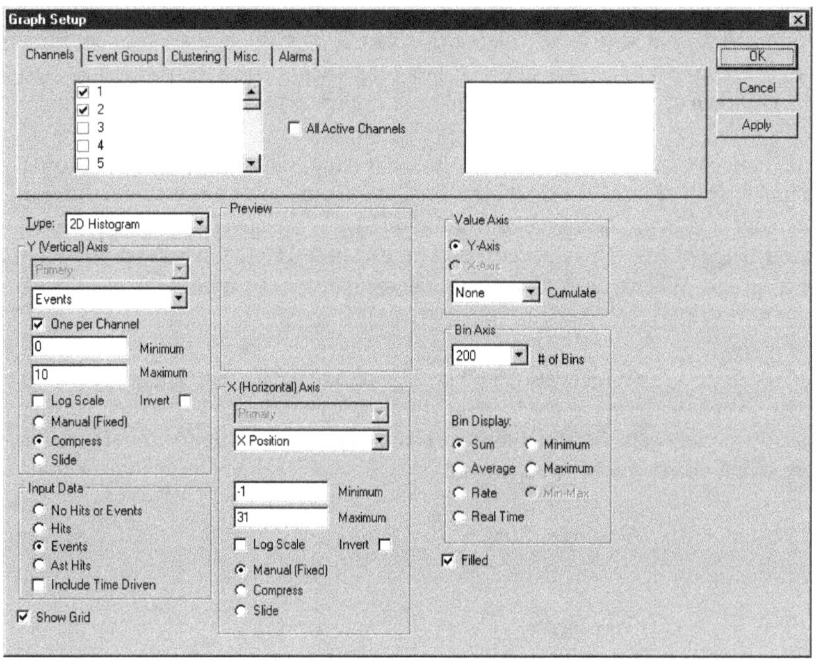

*Figure 144. **Graph** Setup Dialog*

This screen capture shows us how we have set up the Graph setup menu for a linear location. This gives us a lot of insight into how the graph setup menu works and how to set it up. Lets note first the layout of the menu. We see at the top of the menu, a selection area for Channels, Event Groups and Clustering. These settings allow us to select exactly those items. With the Channels Tab, we can select which channels will be displayed on the graph that we are setting up. In this case we are only selecting channels 1 and 2 for viewing in the graph. You can see that we have set group #1 in Event Groups if you select that tab. Clustering allows you to set up parameters that link closely sed hits/events together.

The next menu area is the Graph Type selection box. This is located on the left just below the Channel, Event Group and Clustering selection area. Various graph types can be selected, but for this application we will use a 2D Histogram.

Below the Graph Type selection, is the Y (Vertical) Axis setup menu area. Going down one item at a time in this menu is the plot # selection within the graph. This selection normally allows you to display more than one plot on a graph, but it is disabled for this 2D Histogram type, because we have checked the One per Channel checkbox (see below). Basically the selection (Primary, Secondary-1, 2 etc.) controls which plot the Y Axis settings apply to.

Next is the Y-Axis parameter to be plotted on the graph. By clicking this, you can see many selections. However, we have selected Events. This will display the number of Events detected on the Y-Axis.

The checkbox One per Channel allows us to select a unique per channel plotting color that will give us a multiple color by channel plot. In our case for events and 2 channels, we will get a different color for the channel that was hit first if this is checked. If not we will only get one color plot.

The next two edit boxes allow us to set the starting Y-axis Minimum and Maximum settings.

The next checkbox Log Scale allows us to set a log or linear graph scale. Beside it is an Invert checkbox which allows the user to reverse the positive to negative direction on the axis.

The next 3 radio buttons are related, allowing you to select one type of graph scaling. Manual(Fixed) will force the graph limits as set above to remain fixed. Compress will compress the graph so that the minimum graph value remains but the maximum keeps changing as the graph limit is exceeded. And the Slide selection will allow both the minimum and maximum limits to change or slide as data is received that is beyond the current limits.

The next box allows us to select what type of Input Data we are plotting. In our case since we are plotting a location graph, we want to choose the Event data for plotting.

Below this is a Show Grid checkbox for displaying a grid on the graph background.

The next column to the right, shows the X (Horizontal) Axis settings. The X-axis settings have similar setup boxes as were described for the Y-Axis and will not be described here. Just note how they have been set up to display the X-Position for location and that we have set up the minimum and maximum graph values to include the sensor positions of 0 and 30 inches. Also note that we have selected fixed scaling on the X-axis because we do not expect the X axis position to change. If this were a time axis we would be using a "compression", or sliding graph scaling.

The last column on the right changes according to the graph type being set up. If a 3-D graph were being set up, this would be the column for the Z-axis setting. However, since we have selected a graph type of 2-D Histogram, settings for the Value Axis and the Bin Axis appears. Again, for now just note the settings and you can get familiar with what they do by making a change and seeing how it affects the graph later.

At the bottom of the third column is a checkbox that allows you to decided whether the histogram shows unfilled, outlined bars or filled solid bars.

Please note and try to make sense of all the selected menu settings. Although it might look complex at the start, if you have gone through these settings, one at a time you will see that all is quite simple. Exit the graph menu by selecting the OK button at the top right of the Graph setup menu.

By looking at the top of the graph we see its graph title of "Events vs. X Position <1,2> Loc(1)". This summarizes what is being displayed. First is the graph type "Events vs. X-Position". Next we are displaying data from channels 1 and 2. Finally, we are displaying results from location group #1.

Now that we are finished, we see that we have set up a Linear location graph and are ready to test it. However, before we do, we have a second graph located below the first one. Let's take a look at it. By the title, we see that it is an Events vs. Channel graph, plotting channel 1 and 2 data and also working with data from location group #1. We can look further to see how this graph is set up by entering the graph set up menu (right clicking within the graph area and selecting the "graph menu" selection at the top of the submenu).

Take a look to see how this has been set up. The main difference is in the X-Axis graph parameter and the Bin Axis setting # of bins. This should help give a little insight into how you can set up different graphs.

Now we are ready to proceed with a test or a replay. As a reminder, as a precaution always save your Layout file before trying it out. It is not fun to have all your work disappear because power went down or something happened to lose the layout.

Step 9. The next step after setting up the graphs is to test the layout & verify that it functions properly during acquisition. We can do this by carrying out a trial acquisition and location test if we have set up our sensors or we can do this by replaying our test file called **Zonal_Linear.DTA** . To do this, refer to the sections on going into acquisition or replay (section 3.4 and 3.5). In our example we will replay the DTA file by going to the Replay function, selecting the file "Zonal_Linear.DTA and clicking Start. The data file eplays and the results should look like the screen below.

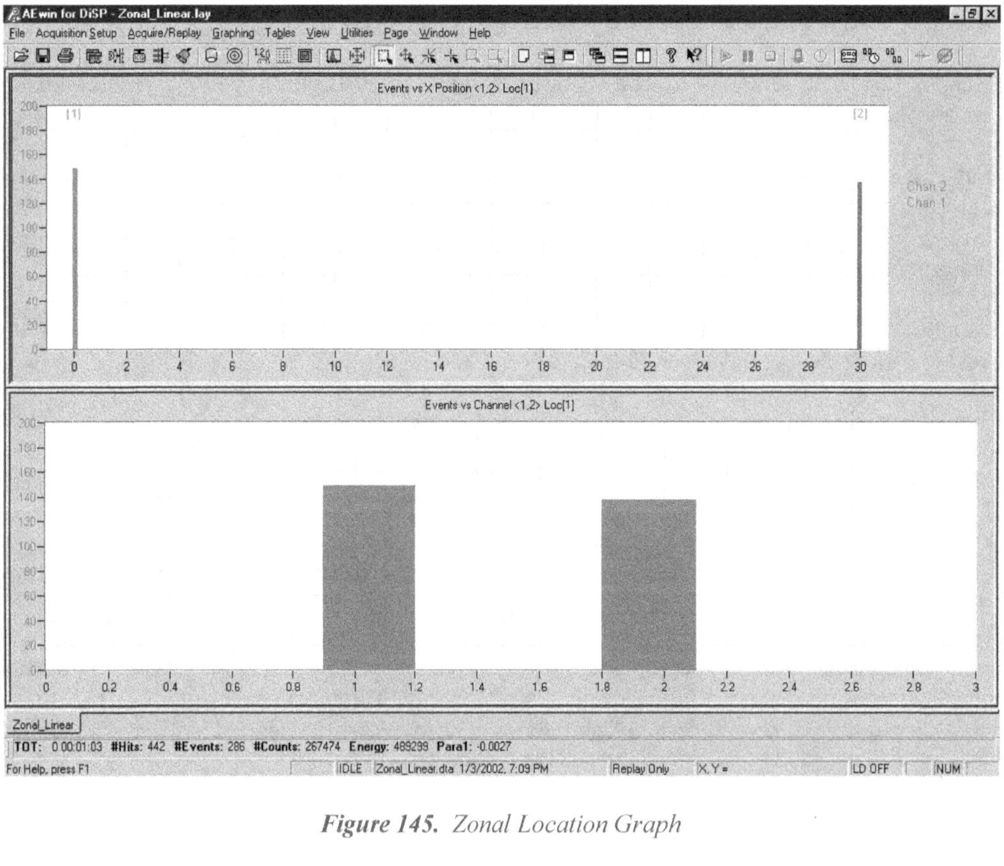

Figure 145. *Zonal Location Graph*

The top Events versus position graph shows that all the locations occurred at each of the sensor positions, even though in reality the events were distributed along the length between the two sensors. You can also see that there are fewer events on the right side, or from sensor #2. This is confirmed in the Events versus channel graph below the location graph. The reason for the events showing up at the channel of a Zonal location group is that again the zonal group provides information on which sensor was hit first.

From the above display it seems that our setup is good and we could proceed on to test with a setup like this. However, you can modify some of the settings in the graph setup menu & repeat the replay of the data to see how the graphs are affected by changes and to get familiar with the graphing capabilities.

To illustrate the ease with which changes can be made, we'll now modify the location setup & transform the Zonal location group into a Linear location group. Simple use F8 to bring up the Location Setup dialog & change the Location Type of group #1 from Zonal to Linear, Next use F10 to replay Zonal_Linear.DTA again. Note how the results on the location graph below compare with the results for the zonal location shown above using the same Zonal_Linear.dta data file. There are now locations along the length of the top graph indicating that some events have located in between the 2 sensors..

This example shows one of the very nice capabilities built into the software, the ability of changing a setup and visualizing a different result upon replay. Remember that simply by changing the Zonal setup to a Linear location setup we got two different graphical results. This is very useful when analyzing data.

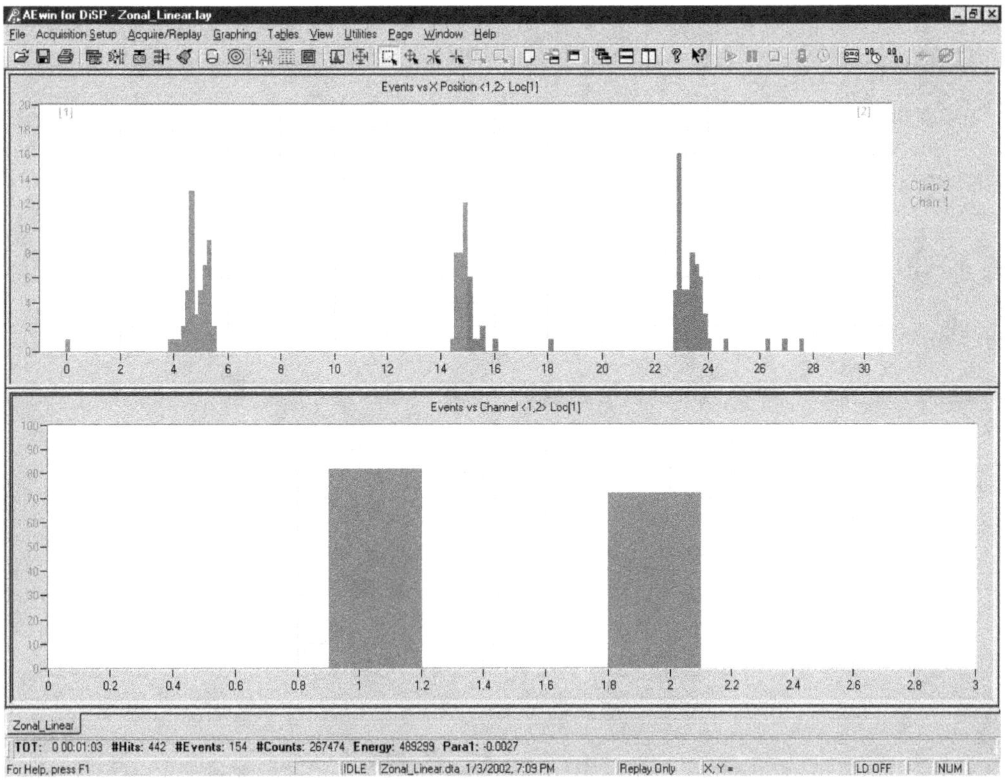

Figure 146. ***Linear*** *Location Graph*

Setup and Testing of a 2D Planar Location Group

As a detailed example, we will be setting up a 2D Planar location group with 4 sensors. We will be using the 2D-Planar.lay and chan24.dta example layout and data file and which are in the AEData directory by default. If you do not have these examples, we will be showing the setup screens so that you can duplicate the setup and conduct your own acquisition test.

As explained at the beginning of this chapter, 2D Planar location is part of the full location option. You can check for the full location option using the Options dialog available on the Help pulldown menu. Look for a checkmark next to Full Location. Alternatively you can just look for 2D Planar in the list of available location modes under the Location Type column in Location Setup.

2D Planar location lets you place multiple sensors anywhere on your test structure without restricting you to a fixed grid arrangement. This is convenient as many structures have pipes or valves that may make a fixed triangular or rectangular arrays of sensors impossible to set up. The position of the source event is calculated from the differences in arrival times between the first 3 valid hits, ≥ 2 Δt values, & can use as many as 8 hits in the calculation. The positions of the sensors for first 3 hits always determines a triangle. The ideal arrangement for 2D Planar would be a layout made up of equilateral triangles, but exact equality of the triangles & sides of the triangles is not an absolute requirement. You will get better results if you avoid creating sensor triangles where one side of the triangle is much larger or smaller than the other 2.

Setting up for 2D Planar location is very easy in the software and is similar to setting up Zonal or Linear location. The steps are identical to those outlined above, only some of the menu selections are changed for the 2D Planar location. Let's provide a quick run-through of setting up 2D Planar location using the same steps outlined above for the other location setups. We will be using the example layout file, 2D-Planar.lay.

Step 1. Use the File pulldown menu's Open Layout option (Ctrl-O) to recall 2D-Planar.lay.

Step 2. Enter the Location Setup dialog (F8 key) & use the Distance Units button to select Inches.

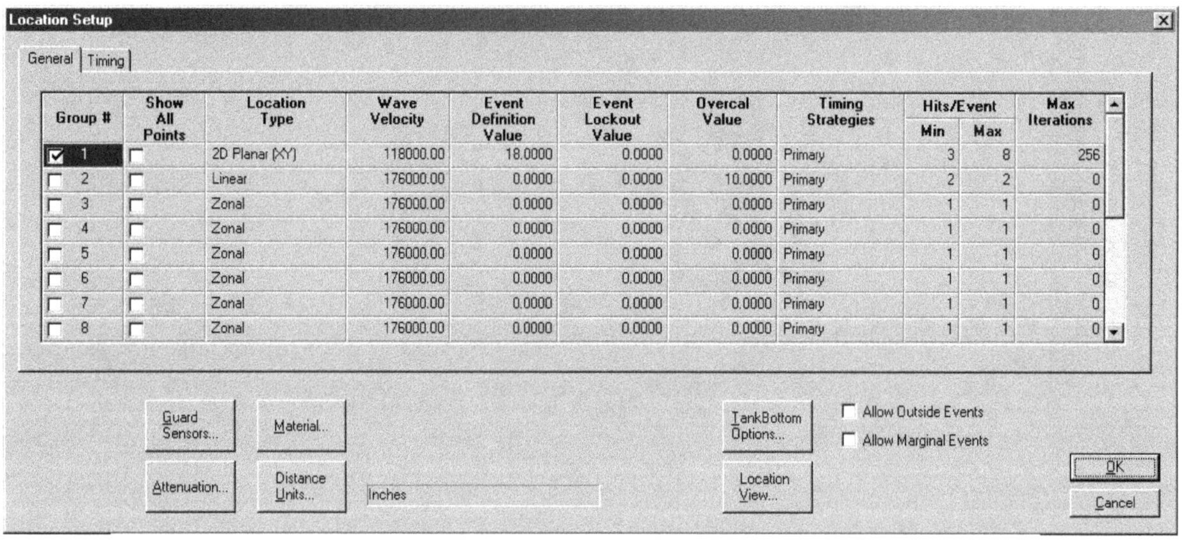

Figure 147. 2D Planar Location Setup

Step 3. Now configure the settings for location Group #1 as shown above. Enable the group by putting a check next to #1. Set the Location Type to 2D Planar. Enter 118,000 for Wave Velocity. Set the Event Definition Value to 18, Event Lockout to 0 & Overcal to 0. Leave the Timing Strategies, Min & Max Hits/Event, Max Iterations & all settings on the Timing page at the defaults.

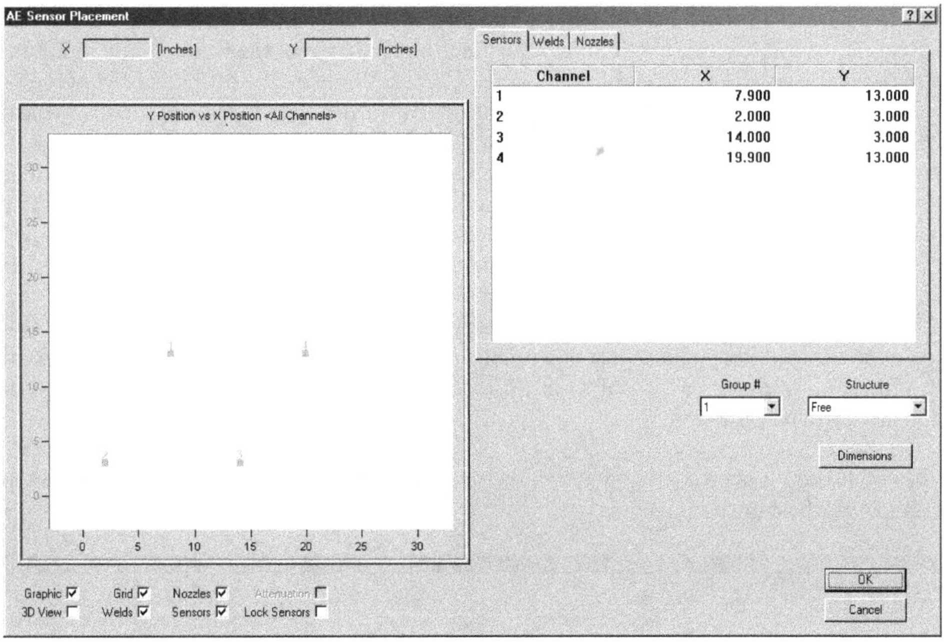

*Figure 148. Sensor Placement for 2D-Planar.**lay** Location Setup*

<u>Steps 4-5.</u> Click on Location View to bring up the Sensor Placement dialog. Set the Structure type is to Free & use Dimensions to set 0 & 30 for the Min & Max values for both x & y.

<u>Steps 6-7.</u> Set up the sensor locations on the screen using the same procedure described in the previous section for zonal location. The only difference now is that you will set up 4 sensors in a 2 dimensional layout. An example is shown above along with the coordinates of the 4 sensors. Even though we are showing a triangular arrangement, 2D Planar is not restricted to a specific sensor pattern. When you are done placing sensors, click OK in Sensor Placement & Location Setup to return to the main menu.

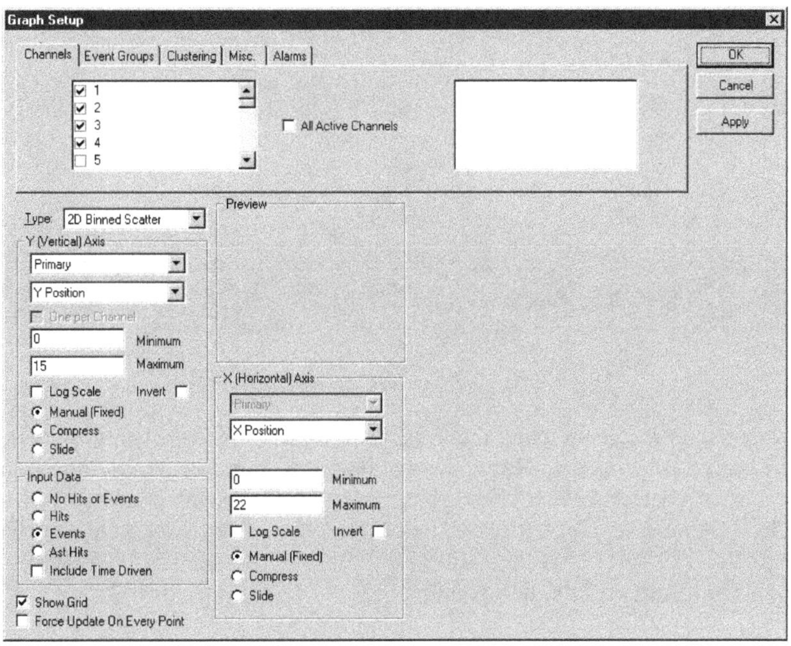

*Figure 149. Graph Setup for **2D**-Planar Location Plot*

Step 8. Next, define a 2D X-Y location plot using the graphing menu following the procedure described in the section for zonal location . The screen capture above shows the graph setup menu for the location plot in 2D-Planar.lay while the one below shows the graph (top graph) for that layout. You can compare the differences between the setup above and the graph setup for the zonal and linear location graph shown earlier. First, you can see that we have enabled all 4 sensors for display that have been setup in the location view menu. Next, you can see that we have selected a 2D Binned Scatter (point plot) for the graph type. Next, you can see that we have changed the Y-Axis to plot "Y-Position" instead of Events. You can also see that we have changed the axis coordinates in both the X and Y axes. Finally, you can see that there is no "bin or value axis settings" to the right of the X axis setup as there was for the 2D Histogram graph setup. 2D Binned Scatter graphs don't require any user choices about the binning strategy. All 2D Binned Scatter graphs use a fixed 200x200 binned grid when plotting the data. The point of using a 2D Binned Scatter graph rather than a 2D Scatter graph is speed. Its binning strategy decreases the number of data points actually plotted, so that multiple point at one location graph as just one point. The 2D Colored Scatter graph is similar to the 2D Binned Scatter, but it will allow you to plot the points in different colors that reflect how many original data points are superimposed at any particular location.

The result of the graph setup is shown in top graph of the screen below. In this graph you can see the sensor locations, although your graph will be empty at this point.

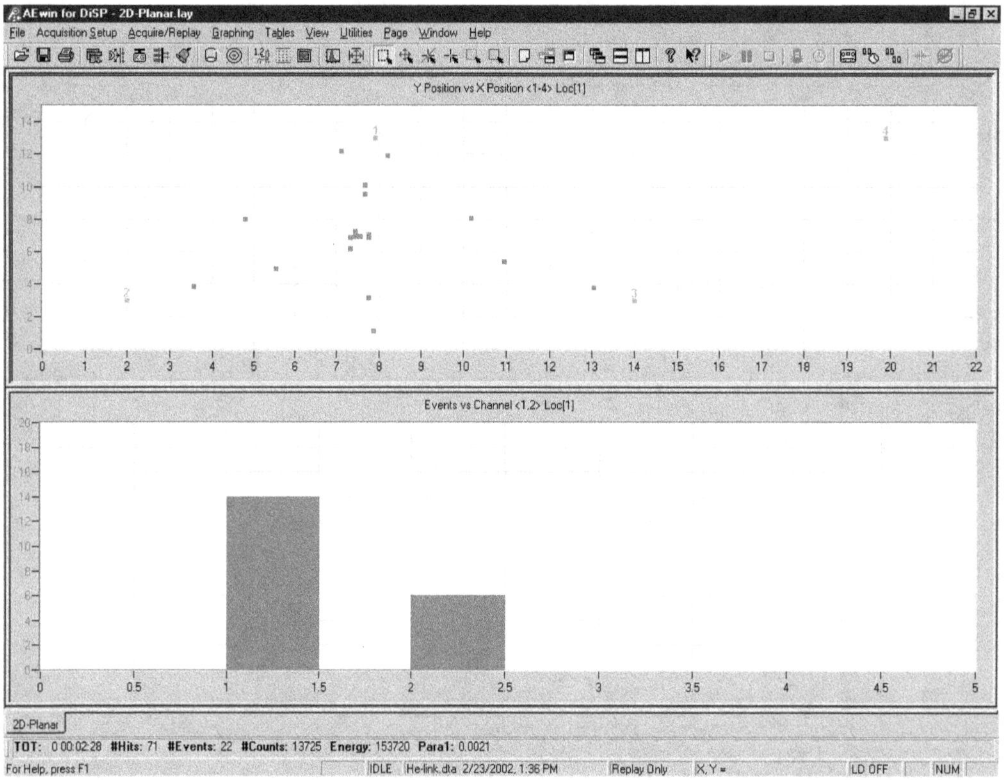

Figure 150. 2D-Planar Location Point Plot

Step 9. Now perform a test of the setup using the same procedure as described in the previous section for zonal location. Replay the data file HE-Link.dta. Results in a point plot will be plotted as dots on the screen where each dot represents the location of a detected event. Now your graphs should resemble those shown above. You can right click on the graph & look for the Large Scatter Graph Points for better visibility of the data points, if you prefer.

Setup and Testing of a Tank Bottom Location Group

Tank Bottom location is a specialized AE source location mode, developed specifically for monitoring the floors of large cylindrical vessels. The software assumes a tank with a circular bottom with sensors equally sed around the circumference of the tank. Tank Bottom location should not be confused with TANK. The difference between them is that Tank Bottom software performs a location calculation and projects the result of that location determination onto the tank floor graphic. In contrast, TANK© is a knowledge & experienced based procedure for assessing and grading the integrity of tank floors.

If you have purchased the Tank Bottom location software option, this section will help you to setup for your first Tank Bottom test. Please first verify that you have the Tank Bottom software option by selecting the Options dialog from the Help pulldown menu. If you have this option, it will be checked. If the option is not present and you desire it, please contact the Customer Service Department at customerservice@ndt.com.

Tank Bottom location follows along the same location setup, menus and setup procedure as do the other location modes. Setting up Tank-Bottom location is a simple 3 step process including;

1. Accessing the Location Setup menu (F8), from the "Acquisition Setup" main toolbar menu, and setting up a Tank Bottom location group.
2. Accessing the Location View menu to set up the Tank bottom structure, dimensions and sensor placement.
3. Setting up a Tank Bottom location graph in the Graph setup menu.
4. Acquiring and/or Replaying data.

Assuming one is familiar with the software, the following is a step by step procedure for setting up and running the Tank Bottom Location software.

Configuration of Location Setup

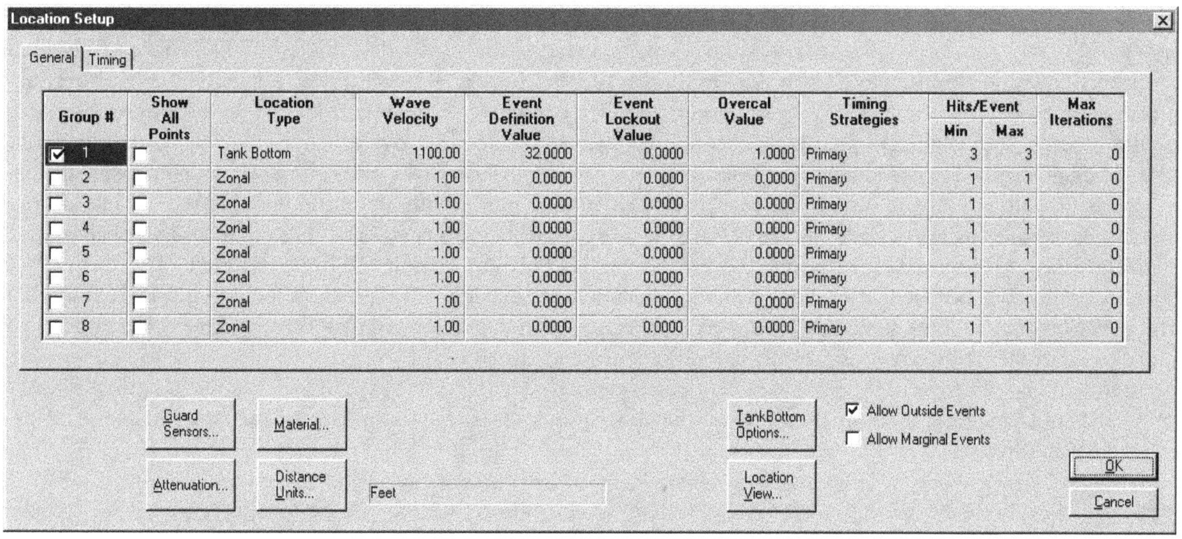

Figure 151. Location Setup Menu

The first step in setting up for a Tank Bottom location application is to access the Location setup menu. This is done by selecting Location from the Acquisition Setup pulldown menu (or by pressing the F8 key). The Location Setup menu appears as shown above. As an example we have shown a Tank Bottom location group as group number 1. We give an overview of setting up this location group below, but you may also wish to read section 4.2 on the Location Setup dialog box for more detailed information on the individual settings.

The first step in setting up a location group is to first select the distance units that you want to use for your work. You will know what distance units you are working in by viewing the textbox at the bottom center of the Location setup menu. In the example above, the current measurements are in feet. To change from the default distance units, left-click the "Distance Units" selection box at the bottom of the Location setup menu. Upon clicking this button, you will see the dialog shown to the right appear. Select whichever units you desire to work with and click "OK". Upon clicking OK, you will see the units textbox at the bottom of the location setup menu show these new values.

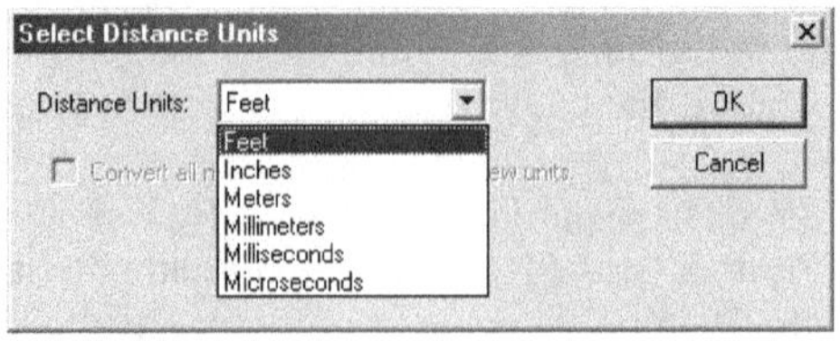

Figure 152. Distance Units Dialog

It is important to setup your distance units as early as possible into the location setup because many the location setup entries including "Wave Velocity", "Event Definition Value", "Event Lockout value", and "Overcal value" all use the units selected or in the case of velocity it will be units/second. By dealing in direct units, the user can think in terms of distance values when setting up these entries.

Returning to setting up the Group #1 (as shown by the first row of the Location setup menu), we will address each column entry below.

Group#: The first column shows that we have checkmarked the Group #1 Location group to be an active location group. Putting a check in this box activates that location group and tells the software to perform a location calculation on based on the setup data in this menu, the Location setup "Timing "menu (see the upper tabs) and the "Location View" and "Tank Bottom Options".

Show All Points: This allows the user to show all source results for events that this group can successfully locate. Independent of which groups have Show All Points checked, the software always allows all enabled groups to analyze a given event to get the best results for each group and then compares the results from all successful groups to come up with a best result. For example, linear location is always presumed to give a better result that zonal location, if there are enough hits(2) to produce a location. If no groups have Show All Points checked, the software will report one source location per event & it is the best result that is shown on graphs & in the line display. There are times when a side-by-side comparison of the results of 2 or more location groups are useful. In that case you can mark the groups as Show All Points & then whatever source locations they can calculate will be reported. The software will appear to sometimes produce more than one source location per event & if you look at the line display, you will see extra event listings with the same first hit times. The extra location results will also show up as extra points on your graphs too, but they only count as 1 event in the Statistics bar. In all cases you will see the software's best result as one of the points. There will be times when an event cannot be located by more than 1 group, so having Show All Points does not guarantee a result from a given location group. It is only a guarantee that you will see it, if it can produce a useful result.

Location Type: In the next column, we have selected "Tank Bottom location" as the location type. By left clicking your mouse in the Location type area on this first row, a selection box will appear with all the available location option choices that are enabled within the software. "Tank Bottom" will be one of the choices, select this.

Wave Velocity: For Wave Velocity, we suggest that you use the velocity of the fluid that is in the tank as a starting point. It is recommended you perform some tests to try to empirically determine what that actual velocity is once you get on the tank. This can be done by creating an imt signal near one sensor, that travels from one side of the tank to the other and monitoring the time it takes for the signal to travel the liquid path to the sensor on the opposite side. You can then measure the time it has taken to arrive and use that with the knowledge of the diameter to arrive at a velocity. However, you must be careful not to mistake the faster metal path arrival, traveling around the tank.

If you do not know the velocity of the fluid in the tank & cannot easily measure it, you can access a built in "Material Properties" database within the software. To access the "Material Properties" database, simply click the "Material" button at the bottom of the Location Setup menu. The Material Properties dialog box will appear as shown below. Along the top of the menu, you can select what type of materials group that you want. In our case, the "liquid" database has been selected. By clicking on the down-arrow in the Material Name text entry box, a list of materials (liquids) in our case will be shown. Simply scroll down and select the desired material (or liquid). In our case, we have selected Motor oil. You will note that the velocity will automatically read out in the distance units that you have selected in the Location setup menu. If the material selection you want is not in the database, you will be able to enter this into the material database by using the "Add" "Delete" and "Update" keys near the bottom of the Materials database menu. You will note that for the example of the Motor oil in the figure the velocity is 5708. We can make use of this velocity by entering it into the "Wave Velocity" column of the location group setup row. The direct way is to simply left-click the cell in Location Setup, type the appropriate value into it and hit the "Enter" key. Remember to enter the value in the same units as that listed at the bottom of the Location setup menu. As an alternative, you can Copy & Paste the value by highlighting it in the Material Properties dialog box & pressing Ctrl-C. In Location Setup, highlight the cell, press F2 to edit its contents & then press Shift-Ins

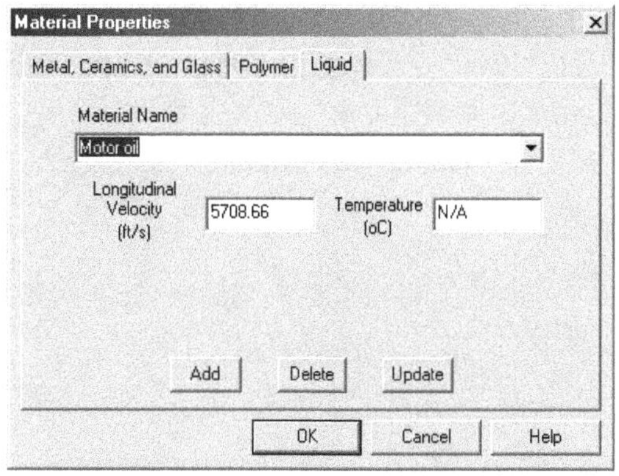

Figure 153. Material Properties Dialog

Event Definition Value: The Event Definition Value is the length of an event in the user's distance units. The Event Definition Value should be set to make sure that an AE source, from any point in the location array, will intersect a minimum desired number of sensors (3 is the minimum), forming an event. In our example of a 30 foot diameter tank, we have selected an Event Definition value of 32 which assures that all the sensors in the tank will have the capability of detecting an event regardless where it originates.

Event Lockout Value: The Event Lockout Value is also entered in the user's distance units. The Event Lockout Value controls the interval between consecutive events. Enter an Event lockout Value starting from the beginning of the Event and long enough to assure that the there is enough time (or distance) for all additional hits and reflections related to that event, to have had a chance to die down. In order for the Event Lockout Value to be effective, it needs to be set longer than the Event Definition Value, since they both start at the same time. There are no rules of thumb for setting Event Lockout Value, since the event related activity is dependent on the structure and sensitivity settings of the system. It is best to experiment to find the best selection.

Overcal Value: The Overcal value is also set in the selected distance units and is usually set equal to approximately 10% of the Event Definition Value. This is to assure that the location algorithm does not ignore minor variations in the wavepath velocity.

Timing Strategies: The user can select from some different timing strategies as set up in the "Timing" tab located at the top of the Location Setup menu. Please note that the Tank Bottom location algorithm does not support multiple timing strategies like the advanced regression location modes can.

Hits/Event (min/max): This entry is used with the regression location modes which are able to use more than 3 hits when calculating the source location for an event. Tank Bottom location is not, however, a regression based location mode. It only uses 3 arrivals to determine location, so these fields are fixed at 3 & are not adjustable for Tank Bottom location.

Max Iterations: Again this setting only applies to regression location modes not Tank Bottom location, so you cannot modify this setting for Tank Bottom.

Once you fill out all this information, you should proceed to the "Location View" menu where you can setup up the Tank Bottom "Structure" and dimensions and set sensors around the tank. An important point with all these settings is that they do not affect the data that is stored in the DTA file during acquisition, because the location results are not stored in the DTA file. They are recalculated each time you do a replay of the data, so you can change any of the above entries at any time & reanalyze the data to try to improve the location accuracy.

Configuration of Sensor Placement

The Sensor Placement dialog is entered from the Location Setup by clicking on the "Location View" button. Upon selecting this option, the dialog box shown below will appear. This menu is the base starting point for setting up the

Tank bottom and other location based structures. The Tank Bottom structure is one of the easiest location setups to create since the sensors can be placed and automatically. Please perform the following steps to set up all the aspects required in the Sensor Placement dialog shown below.

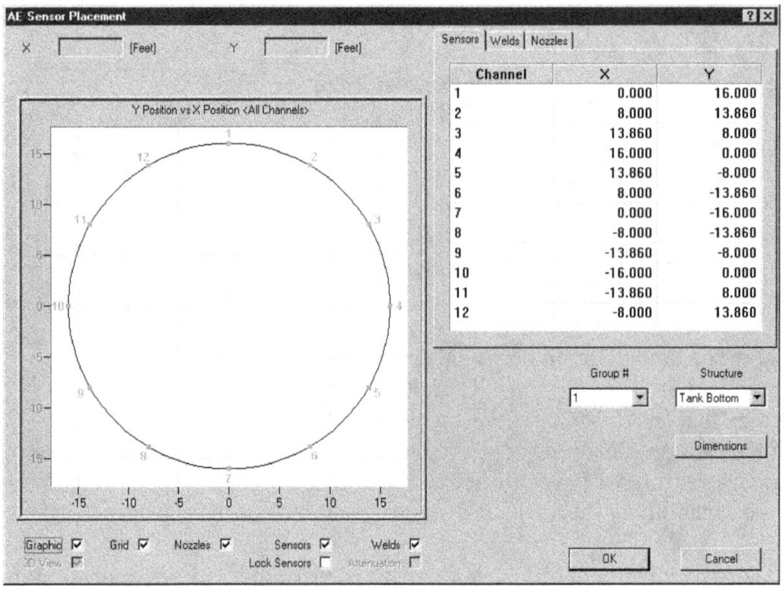

Figure 154. Tank Bottom Sensor Placement

1. First, make sure that you have selected the location group that you wish to work with. In our case, we had set up location group #1 in the Location Setup menu, so make sure Group #1 is selected in the Group selection text entry box.

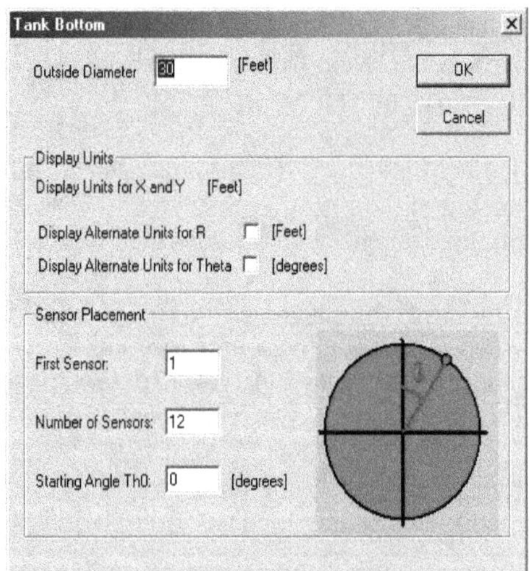

Figure 155. Tank Bottom Dimensions

2. Next, we go to the "Structure" menu and select the "Tank Bottom" structure as the desired structure. This gives us a chance to talk about our "structure" based location capability. Since the user is setting up his application on a particular type of structure (such as a plate, a pipe, a vessel, a sphere, a tank bottom, etc., etc., it is important for the location to think in terms of structures as well. This is what has been implemented in the software and is an extremely advanced concept that only is doing.
3. Once you have selected the Tank Bottom as your Structure type, you should then click the "Dimensions" button. The dialog box shown to the right will appear. This menu is a customized user configuration menu for the type of structure you have selected and allows the customer to set up certain aspects of that structure. In the case of a Tank Bottom, the menu allows the user to enter the Tank diameter (which we have set to 30 feet in our example, allows you to set alternate units and also allows you to enter information about the automatic sensor placement. First it lets you select the starting sensor number as well as the total number of sensors that will be equally sed around the tank and the starting phase, assuming that 0 degrees is that the top of the tank bottom (or North). In our example, we have set up for our sensors to start at sensor #1, starting at the top or 0 degrees and the placement of 12 sensors. When accepted (by clicking the OK button), the sensors will be automatically placed in a clockwise fashion at equal angular sing, on a circular tank bottom in the Location View area. This allows the user to view the sing. Please note the table

of sensor positions that is automatically generated. This table can be exported to the clipboard (by right clicking on the grid and selecting 'copy to clipboard') for insertion into a report file.

4. After the sensor placement step, the user should then go and set up the sensors on the tank in the locations indicated by the Location view menu. Once done and the sensors have been connected to the appropriate AE channel number, you are ready to proceed to the setting up the Location graph and begin acquisition.

Setting Up a Tank Bottom Location Graph

The next step in the process of getting ready for carrying out a Tank Bottom location test is to set up the Tank Bottom Location screen. This is a straightforward process which is carried out in conjunction with the Graph Setup menu. The following are the steps in setting up a Tank Bottom Location Graph.

1. First, exit the Sensor Placement dialog and the Location Setup menu by clicking the OK button. Please note that if you click "Cancel" that you will lose all the setup changes that you have entered. Make sure you click the OK button before proceeding. Upon getting out of the Location setup menu, you will be back to the main screen showing some graph setups.

2. First, let's add a graph page by either right-clicking on one of the graph tabs at the bottom of the screen or by clicking on the "New Page" icon on the icon toolbar (the icon looks like a sheet of paper). This action will add a new tab to the bottom of the graph selection tabs and will create a blank screen. Next, lets name the new Graph Tab, "Tank Bottom" by right-clicking on the tab and selecting the "Rename Page" selection.

3. Next, lets add a graph to the page by either right-clicking in the new "Tank Bottom" page or selecting the "New Graph" icon on the icon toolbar menu at the top of the screen (its next to the "new page" icon). Once you do this, a graph window will pop up. Maximize the graph to be full screen by clicking on the "Tile Horizontal" or "Tile Vertically" icons on the top icon toolbar. You are now ready to create your Tank Bottom Location graph.

4. To setup for a Tank Bottom Location graph, first you need to enter the Graph Setup menu by right-clicking in the Graph area of the graph you want to setup. This is the topmost selection in the context menu that comes up when you right-click. Please note the Graph setup dialog box to the right.

5. Next, select the "Event Groups" Tab at the top of the Graph setup menu and Check the location group #1 which is identified as a Tank Bottom location group (if you have properly followed all previous steps).

6. Next, select the type of Graph. In the case of a location graph, you want to select a 2D Binned Scatter Plot. This text entry box is just above the Y-vertical axis settings.

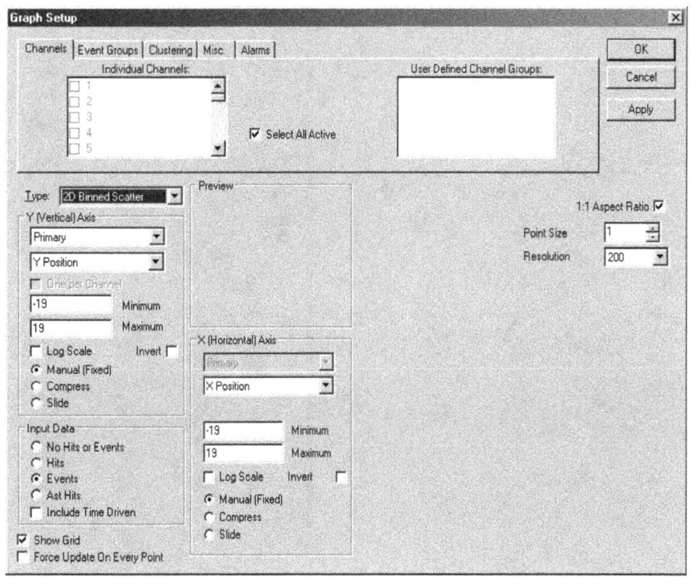

Figure 156. Tank Bottom Graph Setup

7. Next, setup the Y-(vertical axis) and X-(horizontal axis) for display. First set the Y vertical axis to be a Primary graph (this means that you will only be plotting one graph on this screen plot. The next scrolling textbox allows you to select the feature or value that you want to plot. Since this is a location plot you want to plot the Y position versus the X position, so select these for your Y and X settings.

8. Next, you need to select the graph scale. Tank Bottom location assumes that the center of the tank is the 0,0 position, so the axis limits should be a little larger than the tank radius,. In our case the radius is 15, so we'll use ±19 as the limits for both X & Y. Since we want our Tank Bottom outline to look like a circle rather than an ellipse, we've checked the 1:1 Aspect Ratio. This checkbox controls whether the software is allowed to stretched the graph to fill the available se or whether it is required to keep the X & Y axis the same length.

9. The last important setting is to select Events as your Input data to the graph. Events will show the result of the location determination on the incoming AE data.

10. With all these settings done, we can exit the graph setup menu by clicking the OK button. If you did everything correctly, you will get a nice looking tank bottom screen like that shown above. You can have other graphs on this same screen, you are not limited to just the one graph, if you want to just show a portion of the tank in different regions.

11. Note that the structure and the sensors were automatically plotted on the screen. This is because the software was instructed that the structure of the Location group #1 was a Tank Bottom location and was automatically able to draw this structure as soon as it saw that the graph which was being set up was an X Position versus Y Position location graph. If you do not see the Tank Bottom outline in the graph, go back to step 5 & make sure Event Groups has location group #1 checked & that it says Tank Bottom next to it.

You have now totally completed the setup of a Tank Bottom location test and are ready to acquire or replay data.

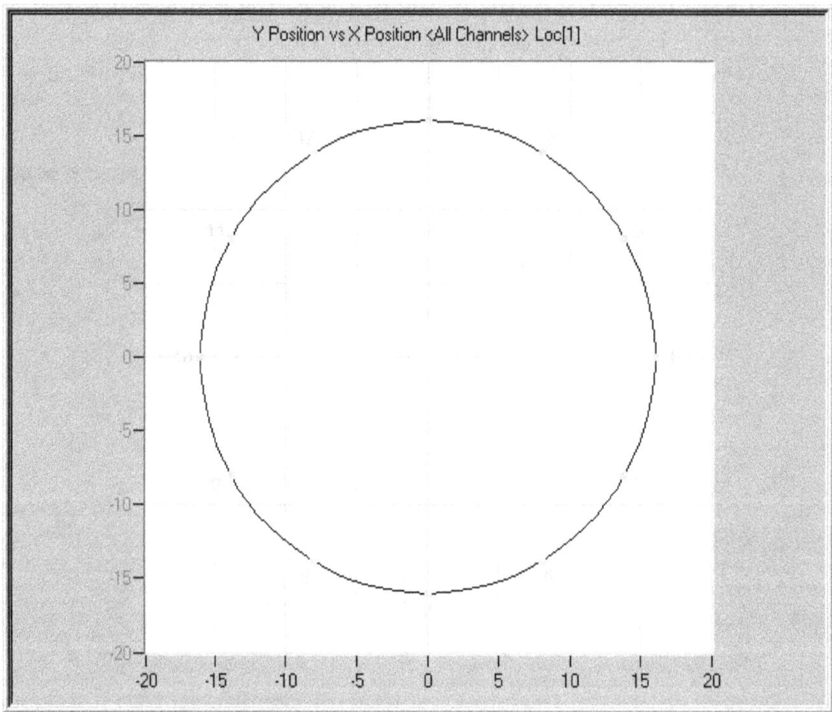

Figure 157. Tank Bottom Graph

Acquiring and Replaying and Displaying Data

Now that the entire setup is complete, you are ready to acquire or replay data from your Tank bottom test. To acquire data, simply enter data acquisition and begin recording data. You should always perform a test to determine if the location algorithm is operating properly before actually performing the true acquisition test. Verify that the location is performing properly and the AE hardware and software settings are all correct to minimize noise and maximize the types of signals you want to record. Once satisfied, start acquiring test data.

Once acquisition is completed you can start replaying and analyzing data. There is a Tank Bottom data file named TankBot.dta in your AEdata directory for you to experiment with. You can use it to test Tank Bottom location. If you do a replay of TankBot.dta using the TankBot.lay layout file, you should see results like that shown here.

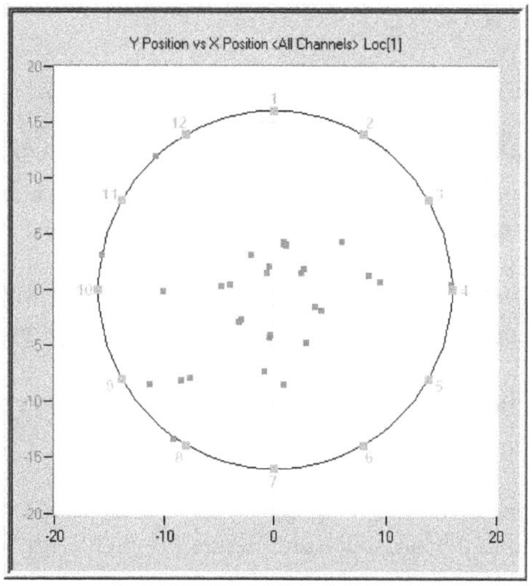

Figure 158. Tank Bottom Results

All standard analysis and filtering tools can be used on the data to eliminate noise and assure relevant data.

There is another option that has been built into the Tank Bottom Software that allows multiple location calculations and plots to be displayed on the Tank Bottom location screen. To access this processing feature, look for the Tank Bottom Options button in Location Setup (F8 or select Location on the Acquistion Setup pulldown menu). This button is just above the "Location View" button as shown on figure 139. The resulting Tank Bottom Options dialog box is shown here.

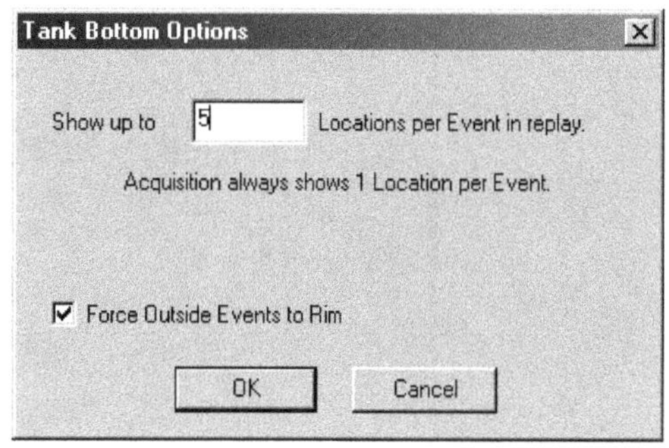

Figure 159. Tank Bottom Options Dialog

This option allows you to choose to display more than 1 location per event. Normally, the number is set to 1 as a default but we have set it to 5 to allow up to 5 location calculations per event to process and display on the screen. This uses other hits in the event in order to calculate location in case the first hits of the events have some problems. Note that this option only affects replay mode to keep acquisition at its highest speeds. Once this is selected and the file is replayed, you can see the difference by comparing the graph at the right with the one above. As can be seen, there are more events displayed on the graph to the right, as there should be. What we learn by this is that the location is pretty good and matches pretty closely telling us that the original location data in the previous graph is valid.

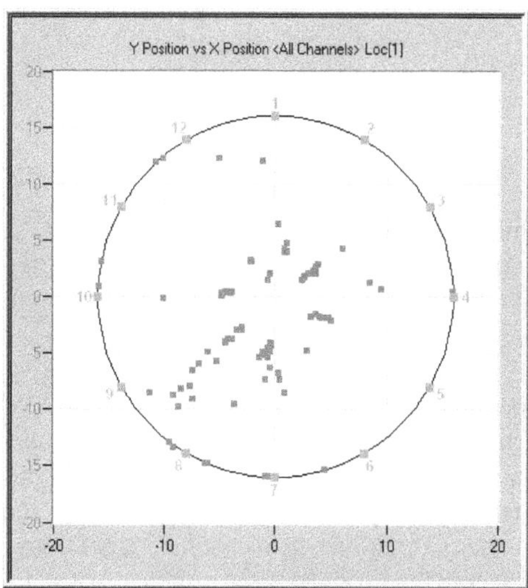

Figure 160. Tank Bottom Results

For a discussion of the Force Outside Events to Rim checkbox , its interaction with Allow Outside Events in Location Setup & how that can affect your results, see the Tank Bottom Location notes in section 4.4.9.

VIRTUAL PARAMETRICS

Virtual parametric data is data that the software collects from hardware usually manufactured by another company & treats it similar to the parametric data normally collected with the hardware manufactured by Physical Acoustics from the SAMOS & PCI2 boards & the Sensor Highway Smart Monitor system. The DiSP board does not support use of virtual parametrics. This readme file documents how virtual parametrics are set up & used in HMSGHAT2000DDI in general along with specific details on using the individual instruments. The instruments supported included the Vaisala Weather Station, the Corr Instruments' NanoCorr, Measurement Computing's PCI-DAS6033 & USB-2533 64 channel boards & the Group 9380-2114 Strain Gauge Module. Note that the Strain Gauge Module is only supported by the Sensor Highway Smart Monitor system. It is not available for the SAMOS or PCI2 systems.

Configure Virtual Parametrics Dialog Box

To make use of virtual parametrics in HMSGHAT2000DDI it is necessary to tell the software what kind of virtual parametric hardware you will be using before acquisition starts. This is done with the Configure Virtual Parametrics menu option under the Acquisition Setup dropdown menu.

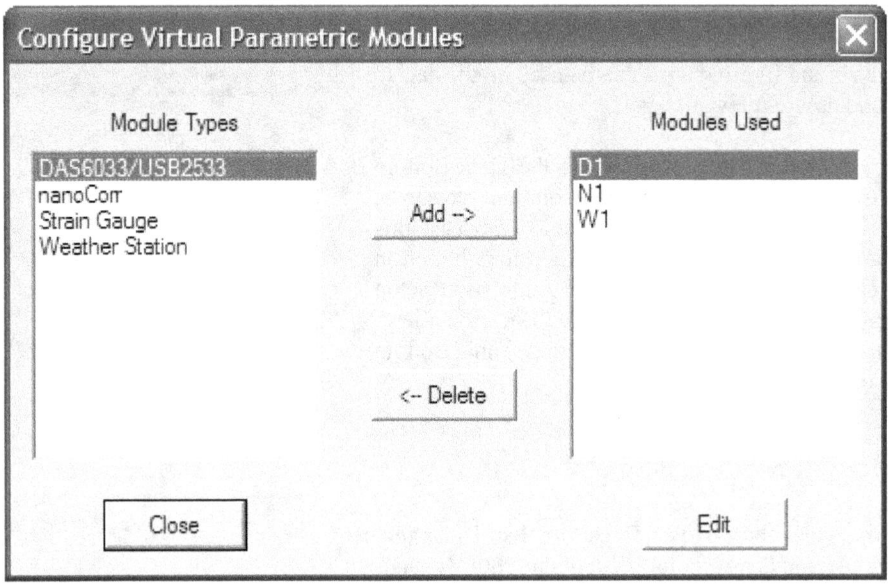

Figure 161. Configure Virtual Parametrics Dialog Box.

This dialog box can be accessed several ways. First it is available directly on the Acquisition Setup dropdown menu. You can also use the hotkey combination Ctrl+F2 to bring it up when you are at the main menu. It can also be accessed thru a button on the Virtual Parametrics tab in the Hardware Setup dialog box. More on the Virtual Parametrics tab in Hardware Setup below.

The column on the lefthand side of this dialog box displays the Module Types list. This is just a list of the types of virtual parametric hardware that is currently supported by HMSGHAT2000DDI. Simply highlight the module you want to use & click on the "Add –>" button. Alternatively, you can doubleclick the module type to add it as well. Each module type will lead to a different Virtual Parametric Module Setup dialog box specific to that hardware. More on each dialog box below. Once the module has been configured & the user clicks on OK, the module name will appear in the Modules Used column. If you no longer wish to use a module, you may highlight it in the Modules Used column & click on on the "<– Delete" button.

Modules listed in the Modules Used column can be doubleclicked to edit their settings. Alternatively, you can highlight the module name & click Edit to launch the appropriate Virtual Parametric Module Setup dialog box. For some virtual parametric types, you may wish to work with more than one instrument. For example, you might want to make use of 2 NanoCorr instruments for collecting data. If you wish, you may copy the setup for an existing module by highlighting it in the Modules Used column & clicking on the Copy button. A new setup dialog box will open with an exact duplicate of the settings.

Before you can successful save the new modules settings, you must change the Module Name & for many modules you need to change the Module ID. See the Virtual Parametric Module Setup dialog box for the DAS6033/USB2533 for an example. The Module ID is often used as an address or board identifier when communicating with the device. For example, Weather Station instruments can communicate via a shared usb port & come configured by default to address 0. If you wished to use more than one, the second Weather Station would have to be reconfigured to appear at another address like 1 or 2. The Module ID is how the user tells the software which instrument to access. It is not a problem if a NanoCorr module & a Weather Station module both use an ID of 0, but if you were using 2 Weather Stations, they cannot both have ID values of 0.

You can have up to 4 modules of a given type like a NanoCorr instrument. There is also a maximum limit of 256 total channels of virtual parametric data that are allowed. Weather Station modules have 22 channels of data, NanoCorrs have 46, Measurement Computing DAS6033 & USB2533 boards have 64 channels & Strain Gauge modules have 4 channels. You may mix & match modules as needed, but you are limited to no more than 4 modules of a given type & no more than 256 channels total from all modules in use.

There are many settings that are common to all the Virtual Parametric Setup dialog boxes. Even if you are not using a Measurement Computing board to collect virtual parametric data, it is suggested that you read that section as well as the sections appropriate for the hardware you are using, because the common settings are described in this first section.

Measurement Computing's PCI-DAS-6033 & USB-2533 Boards

The Measurement Computing boards are designed for 64 channels of analog input. The DAS-6033 uses a PCI slot in your computer, while the USB-2533 is an external board that connects to one of your computer's USB ports. They can monitor any instrument or sensor that outputs a voltage between +/- 10v making them very versatile for virtual parametric input. Here is a description of the available options in the setup dialog box shown below.

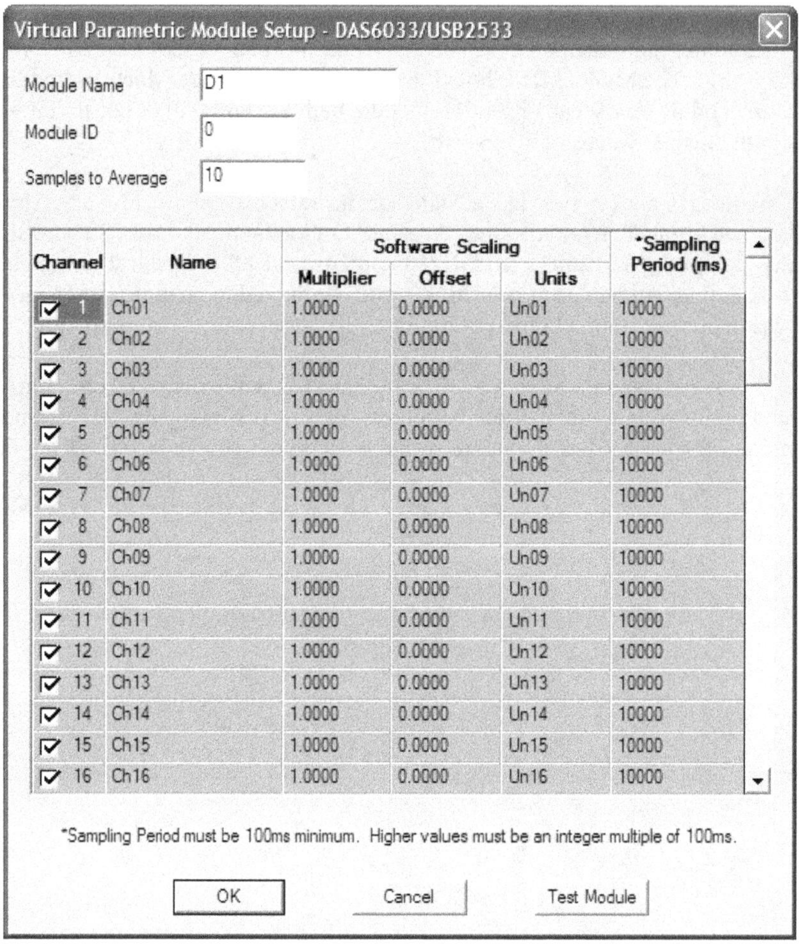

Figure 162. Configure Virtual Parametrics Dialog Box.

Common to All Modules

Module Name – Module Name is just a string used to identify the source of virtual parametric data in the Line Display. Because of the potential for confusion, you can not use the same name for 2 modules. It is also recommended that you keep the Module Name short as it does affect how much data can be displayed on one line in the Line Display.

Module ID – Module ID is an integer used to uniquely identify the module in the software. You may not have 2 modules of the same type with the same ID. In the case of the PCI-DAS6033 & the USB-2533 boards, this number corresponds to the board number used in Measurement Computing's InstaCal program. This defaults to 0 for the first board. If you have multiple boards, the second board number is usually 1. Enter the same number here that is used in InstaCal to access the hardware.

Channel On/Off Checkbox – This is just a simple On/Off toggle that specifies whether or not to collect data from this channel. If you wish to turn all the channels on or off, simply check the box for the first channel & then press the F5 key. F5 sets all values in the column to the value of the selected cell.

Channel Name – This is a string that can help you identify what data a channel is collecting. For example, channel 9 could be a voltage reading that represents the current temperature reading. In that case you might prefer to replace the default label of Ch09 with Temp.

Multiplier, Offset & Units – These 3 settings are identical to those for regular parametrics in HMSGHAT2000DDI. The Multiplier & Offset allow the user to rescale the reading from the hardware to units more to the user's liking. Multiplier & Offset are floating point numbers. The raw reading is multiplied by the Multiplier & then the Offset value is added to it. The Units setting holds a string to represent the new units. For example, if the original reading was in $^{\circ}$C & the user want the reading in $^{\circ}$F, then the Multiplier would be 1.8 (= 9/5) & the Offset would be set to +32. Units could then be set to F. For the DAS6033 & USB2533 all raw reading are actually expressed in volts, but the same principle applies if you are convertings volts to loading pressure or a rotation speed.

Sampling Period – All modules have a Sampling Period setting, but some modules have only one setting for all channels. The DAS6033 & USB2533 boards can, however, vary the sampling rate for each channel independent of the others, so there is a column in their setup dialog box to allow a value for each channel. The values are entered in milliseconds & the minimum value is 100 ms. Higher values must be an integer multiple of 100 ms. If you wish to set all channels to the same sampling rate, type the value desired into one cell in the column & then press F5.

Test Module – This is not an option per se, but this button is useful for testing whether or not HMSGHAT2000DDI can successfully communicate with your Measurement Computing board. It will either tell you "Could not detect the board" or "Board successfully detected". If it was not detected, verify that the Module ID setting is set to the same value as the board number assigned in the Measurement Computing's InstaCal program. If InstaCal can't see the hardware, HMSGHAT2000DDI won't be able to either. If it's a USB-2533, you can doublecheck that the USB cable hasn't come loose. If it's a DAS6033, make sure the board is fully seated in its PCI socket.

Specific to PCI-DAS6033 & USB-2533 Boards

Samples to Average – The other modules do not have a Samples to Average setting. It is specific to the DAS6033 & USB2533 boards. You can enter a number from 1 to 256. This is the number of samples that the board acquires when making a reading. The samples are collected as fast as possible for the DAS6033 & USB2533 (about 640ms/reading) & then averaged to give the reading reported for that sampling period.

Vaisala Weather Station

The Vaisala Weather Station is designed to collect information on current weather conditions such as temperature, rainfall amounts & windspeed.

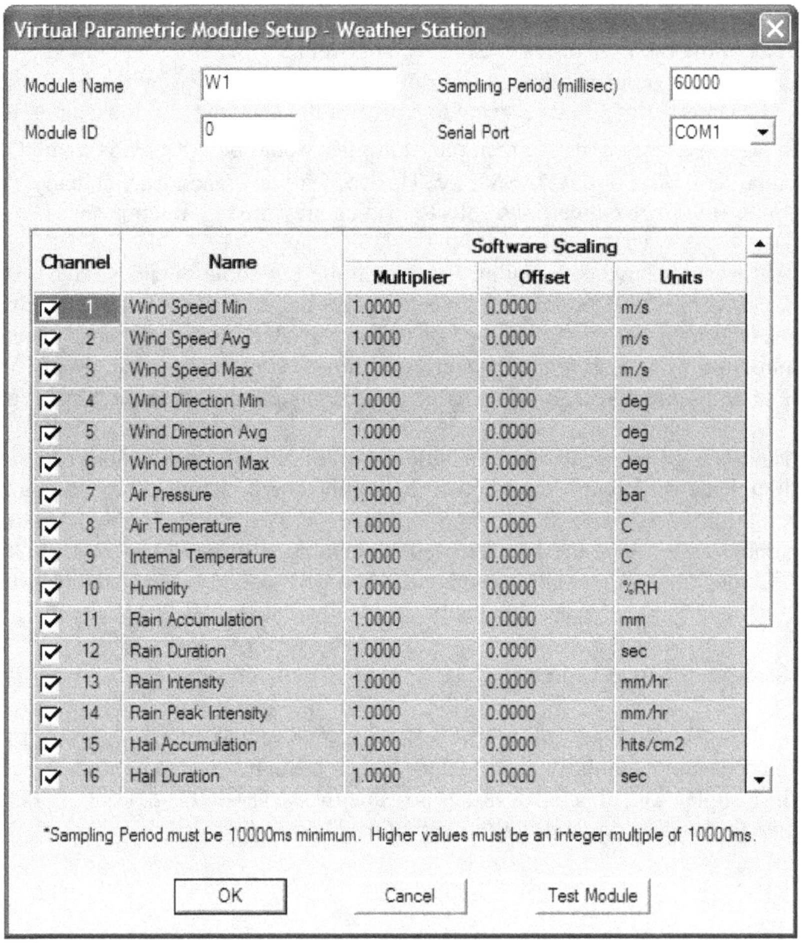

*Figure 163...Weather **Station** Setup Dialog Box.*

Common to All Modules

Module Name, Channel On/Off Toggle, Channel Name, Multiplier & Offset – These are described in the section on the Measurement Computing boards. Please refer back to that section for detailed descriptions.

Module ID – Module ID is an integer used to uniquely identify the module in the software. You may not have 2 modules of the same type with the same ID. In the case of the Vaisala Weather Station, this number corresponds to the device address on the shared serial connection. This defaults to 0 for the Weather Station. If you were using multiple Weather Stations, the second one would have to be configured to use another address like 1. You can use the Vaisala Configuration Tool to change the device address using its Device Settings dialog box. Enter the same number here that is used in the Vaisala Configuration Tool to access the hardware.

Units – The Units string is common to all virtual parametric modules, but a discussion of the Weather Station is not complete without a few comments on the units. The default units listed in the Weather Station setup dialog are the units the data is transmitted in from the instrument over a serial port connection. HMSGHAT2000DDI specifies what units the Weather Station should use when it initializes the instrument at the start of data acquisition. The user can use the Vaisala Configuration Tool to modify the instruments default units, BUT HMSGHAT2000DDI will override any changes made in the configuration program. Do not try to change the units this way. For example, if you desired wind speed expressed in km/hr instead of m/s & you enter km/hr here & try to use the Vaisala Configuration Tool to do switch to readings in km/hr, you will get a reading in m/s & it would be misleadingly labelled with km/hr. Instead set the Multiplier & Offset settings here to 3.6 (3600s/hr * 1km/1000m) & the Offset to 0 to scale the incoming m/s readings to km/hr. Don't forget to update the Units string to km/hr as well.

Sampling Period – For the Weather Station, the Sampling Period must be at least 10,000 ms. You can set a higher value, but not a lower one. Higher values should be integer multiples of 10s (10000ms), so valid numbers are 10000, 20000, 30000, but not 15000, 25000.

Specific to Vaisala Weather Station

Serial Port – The Weather Station is designed to communicate over a serial connection. It can be configured to communicate over an RS-232, RS-422, RS-485 or via USB connection. HMSGHAT2000DDI expects the instrument to communicate at the defaults of 19200 baud, 8 data bits, no parity, 1 stop bit. The only user configurable parameter in HMSGHAT2000DDI is what serial port is connected to the Weather Station.

Test Module – You may use the Test Module button to verify that HMSGHAT2000DDI can properly communicate with the Weather Station. You may receive the cryptic Windows error message "The system cannot find the file specified.". Windows treats serial ports like a file for reading & writing information which is why it says it cannot find the file. In this case the file is the serial port & what Windows is saying is that it cannot open that port. Most likely you've used a serial port that does not exist on your computer. If you receive a "Weather Station communication failed." error message, the port exists, but the Weather Station did not respond to the test message.

Troubleshooting Problems with an RS-485 Connection

If you happen to be using an RS-485 connection, there is one more instrument setting that may affect your ability to reliably communicate with the Weather Station. If you start up the Vaisala Configuration Tool & it cannot maintain a connection with your Weather Station, HMSGHAT2000DDI will not be able to maintain a connection with it either. In that case, you may need to adjust the RS-485 line delay parameter. It cannot be adjusted by HMSGHAT2000DDI, but it can be adjusted with the Vaisala Configuration Tool, if the connection is somewhat reliable. Look for it in the Device Settings dialog box. If the Configuration Tool cannot maintain a connection, you will need to use Windows HyperTerminal to adjust the parameter.

Configure HyperTerminal to use the same serial port that the Vaisala Configuration Tool is using. In HyperTerminal this is done using the File Properties menu option. On the Connect To tab, set the serial port to use 19200 baud, 8 data bits, no parity, 1 stop bit. Next on the Settings tab, click the ASCII Setup button. Check "Send line ends with line feeds", "Echo typed characters locally" & "Append line feeds to incoming line ends." Click on OK to save your changes & then again to exit the properties dialog box.

When you exit the properties dialog box, HyperTerminal will attempt to open the port. Assuming the port opened successfully, you should be able to type a command to the Weather Station instrument & have it respond. Simply type the device address, 0 by default & press Enter. The instrument should respond with the same value. If you don't know the device address, type ? and then press Enter. The response should be the device address. All commands to the Weather Station start with its device address & commands are case sensitive. Assuming this instrument is using the default address of 0, you can check the line delay parameter by typing:

```
0XU<enter>
```

The instrument should respond with something like:

```
0XU,A=0,M=P,T=0,C=2,I=0,B=19200,D=8,P=N,S=1,L=0,N=WXT520,V=1.00<enter>
```

Look for the L=# entry. This is the line delay parameter in millisecs. You can update the value to 100 ms by typing:

```
0XU,L=100<enter>
```

which the instrument should respond to by echoing the command back to you:

```
0XU,L=100
```

You can doublecheck that the setting was stored by using the 0XU command again. Once you have adjusted the line delay, disconnect from the port in HyperTerminal & try the Vaisala Configuration Tool. If it still cannot maintain the connection with the instrument, close the Configuration Tool, return to HyperTerminal & readjust the line delay. Valid line delay values range from 0 to 10000 ms. Start low & slowly make it larger by 50 or 100 ms at a time. Otherwise you may jump over the range of values that work. Most likely you will only need a delay between 0 and 250 ms, but it does depend on the length of your RS-485 cable so the exact value is hard to predict.

Corr Instruments' NanoCorr

Corr Instruments' NanoCorr is designed to measure corrosion rates of metal probes. It is unique in that it is the only instrument that the software cannot actively control. All the software can do is monitor it's output on a serial port & import the data readings. The user must configure it with NanoCorr front panel & then make the HMSGHAT2000DDI software settings match the NanoCorr settings.

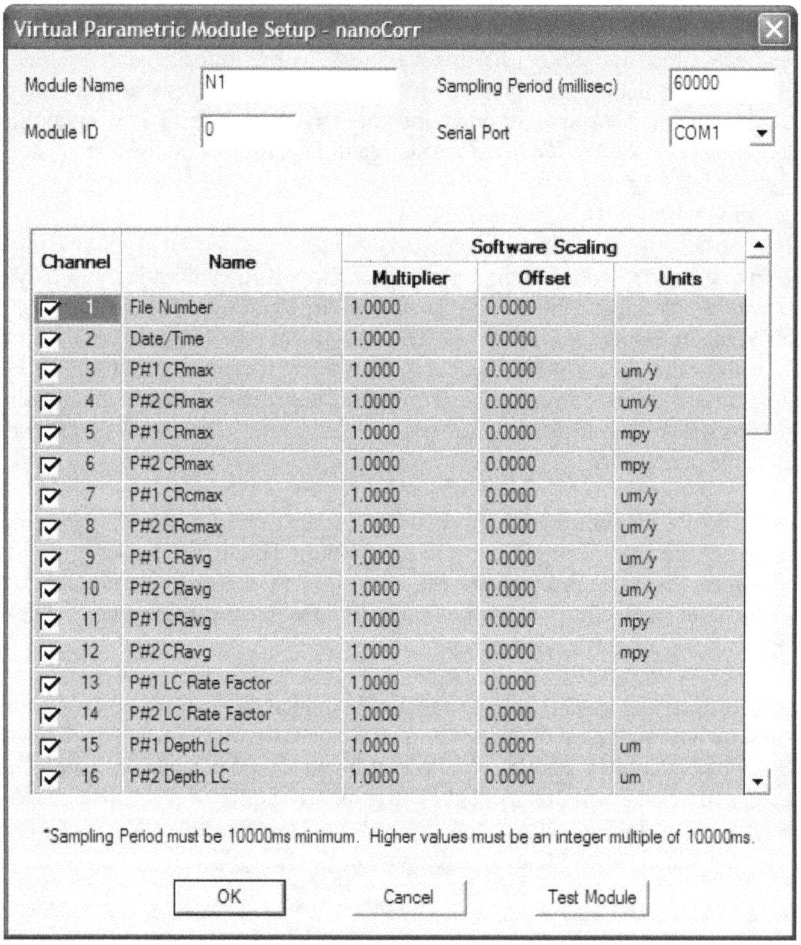

Figure 164...NanoCorr Setup Dialog Box.

Common to All Modules

Module Name, Module ID, Channel On/Off Toggle, Channel Name, Multiplier & Offset – These are described in the section on the Measurement Computing boards. Please refer back to that section for detailed descriptions.

Sampling Period – For NanoCorr as for the Weather Station, the Sampling Period must be at least 10000 ms. You can set a higher value, but not a lower one. Higher values should be integer multiples of 10s (10000ms), so valid entries are 10000, 20000, 30000, but not 15000, 25000.

Specific to NanoCorr

Serial Port – The NanoCorr is designed to communicate over an RS-232 or RS-485 serial connection. It can also be configured to communicate over an Ethernet connection masquerading as a serial connection. Select the port number of the serial port that the NanoCorr is hooked up to. If using an Ethernet connection, refer to Chapter 5 of the NanoCorr manual for details on how to set up the connection & determine the serial port number to use here. HMSGHAT2000DDI expects the instrument to communicate at the defaults of 115200 baud, 8 data bits, no parity, 1 stop bit.

Test Module – You may use the Test Module button to verify that HMSGHAT2000DDI can properly communicate with the NanoCorr. As for the Weather Station, you may receive the cryptic Windows error message "The system cannot find the file specified.". Again Windows treats serial ports like a file for reading & writing information which is why it says it cannot find the file. In this case the file is the serial port & what Windows is saying is that it cannot open that port. Most likely you've used a serial port that does not exist on your computer. Assuming the port exists, you will receive the following message:

> COM Port opened. Must wait for nanoCorr to transmit a new reading.
> Click on OK when ready to check for a data string.

The NanoCorr test capabilty is very primitive due to the limitation that HMSGHAT2000DDI can only monitor the output of the NanoCorr over its serial port connection. It cannot send a command to the NanoCorr & have it respond. The Test Module button just checks for incoming data the chosen serial port. At this point, you will have to wait until you feel the NanoCorr has transmitted data to the PC. If you used the front panel of the NanoCorr to tell it to collect new readings every 2 minutes, you will have to wait 2 minutes before pressing the OK button. If you press the button, before data has been transmitted to the PC, you will receive a see a "No data received from nanoCorr." message. If data has been received, you will see a message that shows you the text received. It should be a series of numbers that looks something similar to this NanoCorr data record:

```
NanoCorr data string read =

"186 2008-07-23 15:22:54    591.0344848     23.269074       591.0344848     122.956886
     4.840822        4.806843        0.0962117       0.0202236       4.757399        -0.554344
     -109.181907     6.709134        0.0209924       -       0       722047  -27     -89     -17531
     602829  -220275 25      -73     49      -293030 -315945 -399937 -40374  -43860  -82     47
```

If you don't end up seeing any data, you might want to try using HyperTerminal to monitor the serial port that the NanoCorr is connected to. Set up HyperTerminal as described in the "**Troubleshooting Problems with an RS-485 Connection**" section for the Weather Station, but use 115200 as the baud rate instead of 19200. The NanoCorr by default communicates at 115200 baud rate, 8 data bits, no parity, 1 stop bit. You still want to turn on "Append line feeds to incoming line ends." Just leave HyperTerminal open till you see data strings come in.

If you leave the window open long enough to see 2 data records received, you can tell how long the time is between data readings. Note in the example data string above, the time of the data reading is the 3rd number in the string. Just compare the 2 readings to figure out the time interval. That's how long you need to leave the message box open when it is waiting on data from the NanoCorr. If HyperTerminal nevers receives any data, neither will HMSGHAT2000DDI. Either you are using the wrong serial port or the NanoCorr is not actively collecting data.

Mistas Group 9380-2214 4-Channel Strain Gauge Module

As noted above, the Strain Gauge Module is only supported by the Sensor Highway Smart Monitor system. It is not available for the SAMOS, PCI2 or DiSP systems.

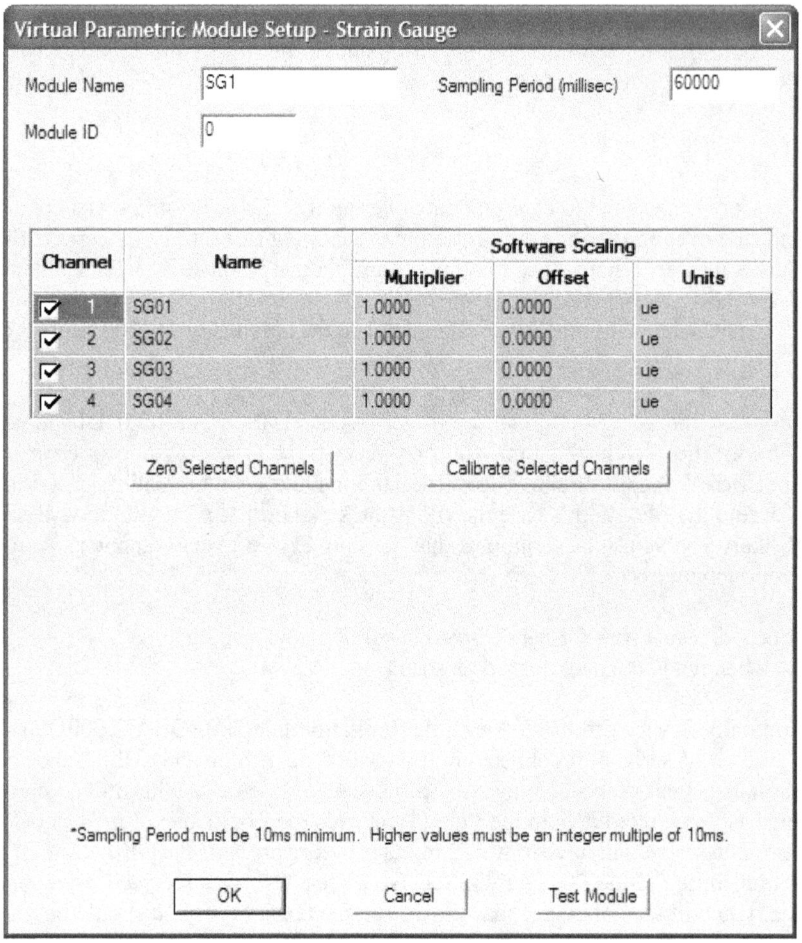

Figure 165...Strain Gauge Module Setup Dialog Box.

Common to All Modules

Module Name, Channel On/Off Toggle, Channel Name, Multiplier & Offset – These are described in the section on the Measurement Computing boards. Please refer back to that section for detailed descriptions.

Sampling Period – For the Strain Gauge module, the Sampling Period must be at least 10 ms. You set a higher value, but not a lower one. Higher values should be integer multiples of 10s (10000ms), so valid entries are 10000, 20000, 30000, but not 15000, 25000.

Specific to Strain Gauge Module

Test Module – You may use the Test Module button to verify that HMSGHAT2000DDI can properly find the Strain Gauge hardware. If there is a problem, you will get a "Strain Gauge 0 could not be detected." If all goes well, you will get a "Strain Gauge 0 successfully detected." message & it will give you information on the firmware & PCB versions of the hardware.

Zero Selected Channels and *Calibrate Selected Channels* – These buttons work with the channels that are checked on in the grid. They just run thru the routines responsible for zeroing & recalibrating the channel readings.

Virtual Parametrics Page in Hardware Setup

The Configure Virtual Parametric Modules dialog box can also be accessed thru the Virtual Parametrics tab in Hardware Setup. This tab in Hardware Setup (shown below) will let you adjust the setup of all the virtual parametric modules without having to jump in & out of the individual setup dialog boxes for each module. You cannot adjust all the settings for all modules, but you can control the main ones. You can turn channels on & off, revise the channel Name, Multiplier, Offset & Units string. The Module Name is displayed, but it is a read only column. To change the Module Name, you will have to display the Configure Virtual Parametric Modules dialog box & use it to bring up the setup dialog box for the module you want to rename. The Sampling Period also requires accessing the individual setup dialog boxes as some modules have 1 setting for the whole module & others allow individual settings for each channel.

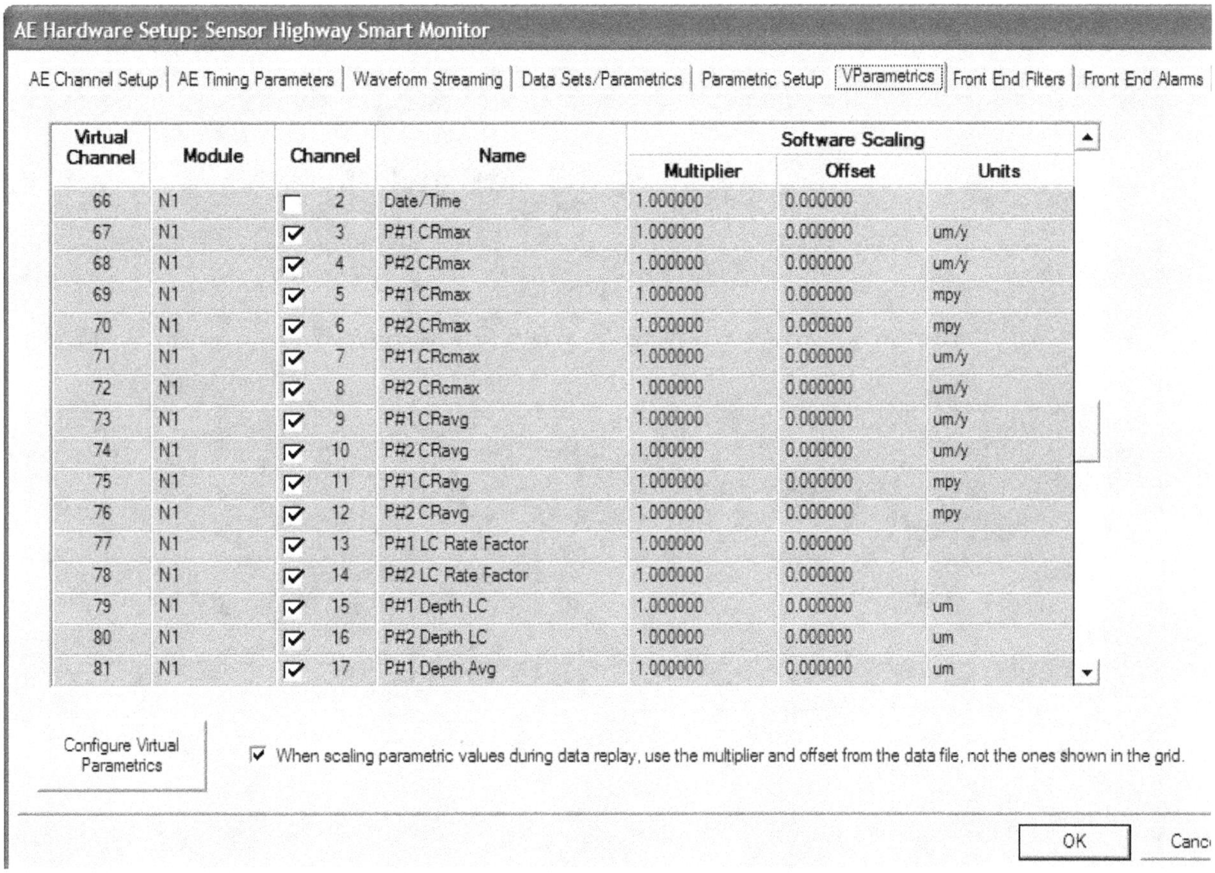

Virtual Channel	Module	Channel	Name	Software Scaling		
				Multiplier	Offset	Units
66	N1	☐ 2	Date/Time	1.000000	0.000000	
67	N1	☑ 3	P#1 CRmax	1.000000	0.000000	um/y
68	N1	☑ 4	P#2 CRmax	1.000000	0.000000	um/y
69	N1	☑ 5	P#1 CRmax	1.000000	0.000000	mpy
70	N1	☑ 6	P#2 CRmax	1.000000	0.000000	mpy
71	N1	☑ 7	P#1 CRcmax	1.000000	0.000000	um/y
72	N1	☑ 8	P#2 CRcmax	1.000000	0.000000	um/y
73	N1	☑ 9	P#1 CRavg	1.000000	0.000000	um/y
74	N1	☑ 10	P#2 CRavg	1.000000	0.000000	um/y
75	N1	☑ 11	P#1 CRavg	1.000000	0.000000	mpy
76	N1	☑ 12	P#2 CRavg	1.000000	0.000000	mpy
77	N1	☑ 13	P#1 LC Rate Factor	1.000000	0.000000	
78	N1	☑ 14	P#2 LC Rate Factor	1.000000	0.000000	
79	N1	☑ 15	P#1 Depth LC	1.000000	0.000000	um
80	N1	☑ 16	P#2 Depth LC	1.000000	0.000000	um
81	N1	☑ 17	P#1 Depth Avg	1.000000	0.000000	um

Figure 166... *Virtual Parametrics Page of* ***Hardware*** *Setup.*

There is one setting on this tab that you cannot adjust anywhere else. At the bottom of the tab is a checkbox labelled:

> "When scaling parametric values during data replay, use the multiplier and offset from the data file, not the ones shown in the grid."

During data acquisition, the virtual parametric settings that are in use are stored in the DTA file. This checkbox controls whether you use the settings in your layout or from the DTA file when replaying the data. This allows you flexibility in the units that are displayed during a replay. If you use the settings from the DTA file, you will see the data in the same format as it was collected. If, however you want to changes some units, you would want to use the settings from your layout file, so that you can rescale the data using the Multiplier & Offset values & display the correct units string. If you don't need to change any units, you should normally stick with using the settings in the DTA file, so that your data displays properly.

If you use the settings from your layout, you should start with the layout that you used to collect the data, because it will have the correct modules already defined. When you create a layout & tell HMSGHAT2000DDI which modules you are using, it keeps track of the data using a virtual channel number. These are shown in the lefmost column in the grid. They range from 1 to the total number of virtual parametric channels available for the modules in use. If you are only using 1 module, the virtual channel numbers are identical to the module channel numbers. If you are using 2 or more modules, though, the virtual channel numbers for the 2nd module start where the 1st modules channels end.

For example, the Measurement Computing DAS6033 & USB2533 have 64 channels of data available, while the Weather Station has 22. When they are both being used, the virtual channel numbers for the Weather Station start at the end of the 64 from the Measurement Computing module, so channel 1 for the Weather Station corresponds to virtual channel number 65. These are important, because you need them to graph virtual parametric data. In this case, if you want to graph channel 1 for the Weather Station, you need to look for Virtual Parametric 65 in Graph Setup. These numbers can also show up in the Line Display. More on this in the next section.

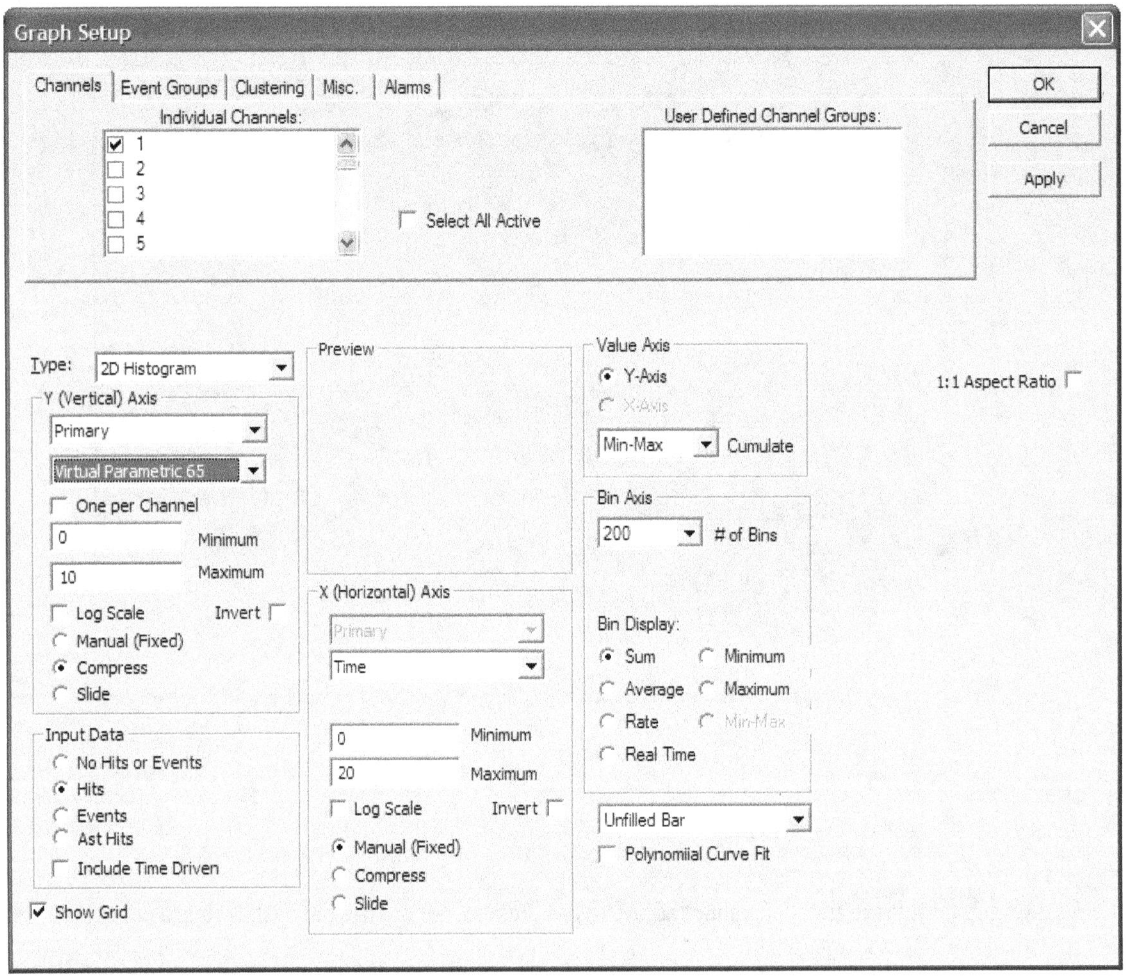

Figure 167...Accessing Virtual Parametric in Graph Setup.

Line Display & Virtual Parametric Data

To get virtual parametric data to show up in the Line Display, you will need to turn on both the Line Display & make sure virtual parametric is being output. Use the Line Display Setup dialog box shown below. Check "Enable Line Display" in the upper lefthand corner to turn on the line display.

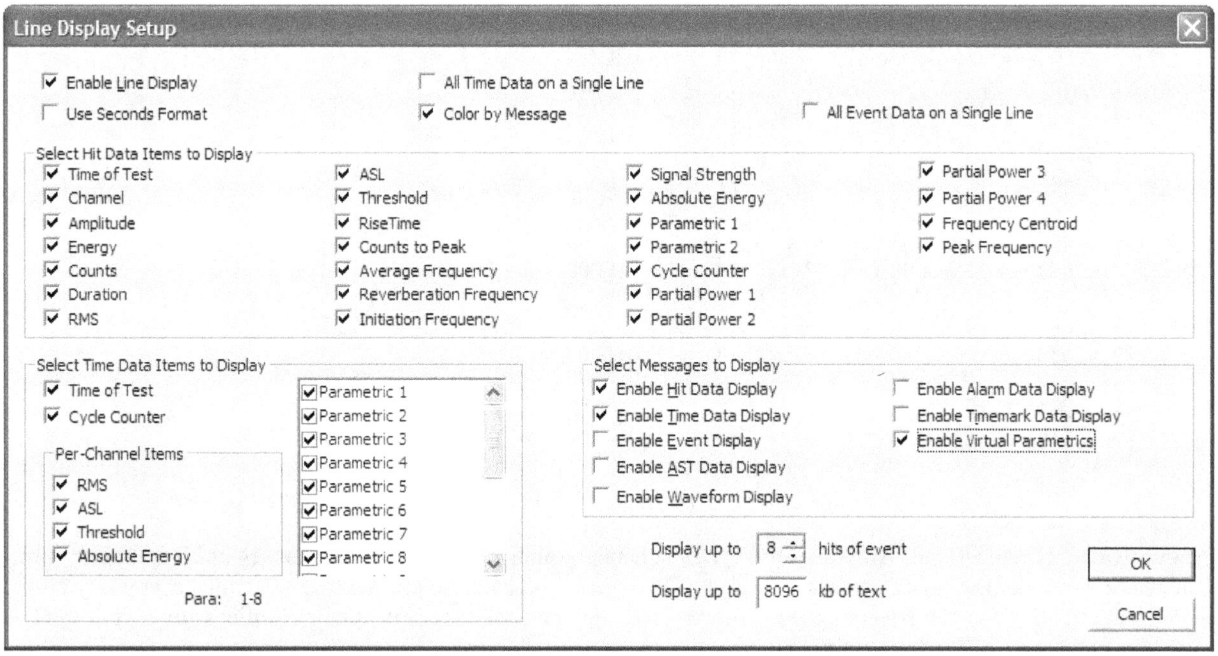

*Figure 168...Enabling Virtual Parametrics in Line **Display** Setup.*

Check "Enable Virtual Parametrics" in the "Select Messages to Display" section in the lower right hand side of the dialog box. Note there is a difference between the line display being on & being visible. You can have it on, but not have it visible. You can use the F7 key to show or hide the Line Display window itself. When you replay a DTA file with virtual parametric data in it, it will show up in the Line Display as message ID 9. Here is a sample for data collected from a NanoCorr instrument:

```
9    0 00:00:18.2061840
     N1:P#1 CRmax          =   0.3219 um/y     N1:P#2 CRmax          =   0.1372 um/y
     N1:P#1 CRmax          =   0.0127 mpy      N1:P#2 CRmax          =   0.0054 mpy
     N1:P#1 CRcmax         =   0.3219 um/y     N1:P#2 CRcmax         =   0.1372 um/y
     N1:P#1 CRavg          =   0.0522 um/y     N1:P#2 CRavg          =   0.0207 um/y
     N1:P#1 CRavg          =   0.0021 mpy      N1:P#2 CRavg          =   0.0008 mpy
     N1:P#1 LC Rate Factor =   6.1660          N1:P#2 LC Rate Factor =   6.6439
     N1:P#1 Depth LC       =   0.0000 um       N1:P#2 Depth LC       =   0.0000 um
     N1:P#1 Depth Avg      =   0.0000 um       N1:P#2 Depth Avg      =   0.0000 um
     N1:P#1 LC Depth Factor =  1.8404          N1:P#2 LC Depth Factor =  1.2792
     N1:P#1E               =   0.0013 V        N1:P#1E               =   0.0013 V
     N1:Temperature        =  24.8901 oC       N1:pH                 =   7.0001
     N1:ORP                =   0.0072 V        N1:Remarks            =   0.0000
     N1:P#1I Offset        =  18.0000 pA       N1:P#2I Offset        =-170.0000 pA
     N1:P#1E#1             = -75.0000 pA       N1:P#1E#2             =-217.0000 pA
     N1:P#1E#3             = -10.0000 pA       N1:P#1E#4             =  61.0000 pA
     N1:P#1E#5             =  16.0000 pA       N1:P#1E#6             = -13.0000 pA
     N1:P#1E#7             = 130.0000 pA       N1:P#1E#8             =  83.0000 pA
     N1:P#1E#9             =  26.0000 pA       N1:P#2E#1             =  30.0000 pA
     N1:P#2E#2             = -99.0000 pA       N1:P#2E#3             =  30.0000 pA
     N1:P#2E#4             =  -7.0000 pA       N1:P#2E#5             = -27.0000 pA
     N1:P#2E#6             =  10.0000 pA       N1:P#2E#7             =   2.0000 pA
     N1:P#2E#8             =  50.0000 pA       N1:P#2E#9             =  10.0000 pA
```

Figure 169...Sample Line Display Output for the NanoCorr.

If you replay a DTA file with virtual parametric data in it & choose to scale the data using the settings in the layout file, but aren't using the same virtual parametric modules as were used to collect the data, the software may not be able to correctly display the virtual parametric data. Here is the same NanoCorr data as above replayed with a layout that did not have any virtual parametric modules defined to illustrate how that looks:

```
9    0 00:00:18.2061840
     ??:VCh1   = 202.0000      ??:VCh2   =   0.0000
     ??:VCh3   =   0.3219      ??:VCh4   =   0.1372
```

```
??:VCh5   =   0.0127      ??:VCh6   =   0.0054
??:VCh7   =   0.3219      ??:VCh8   =   0.1372
??:VCh9   =   0.0522      ??:VCh10  =   0.0207
??:VCh11  =   0.0021      ??:VCh12  =   0.0008
??:VCh13  =   6.1660      ??:VCh14  =   6.6439
??:VCh15  =   0.0000      ??:VCh16  =   0.0000
??:VCh17  =   0.0000      ??:VCh18  =   0.0000
??:VCh19  =   1.8404      ??:VCh20  =   1.2792
??:VCh21  =   0.0013      ??:VCh22  =   0.0013
??:VCh23  =  24.8901      ??:VCh24  =   7.0001
??:VCh25  =   0.0072      ??:VCh26  =   0.0000
??:VCh27  =  18.0000      ??:VCh28  =-170.0000
??:VCh29  = -75.0000      ??:VCh30  =-217.0000
??:VCh31  = -10.0000      ??:VCh32  =  61.0000
??:VCh33  =  16.0000      ??:VCh34  = -13.0000
??:VCh35  = 130.0000      ??:VCh36  =  83.0000
??:VCh37  =  26.0000      ??:VCh38  =  30.0000
??:VCh39  = -99.0000      ??:VCh40  =  30.0000
??:VCh41  =  -7.0000      ??:VCh42  = -27.0000
??:VCh43  =  10.0000      ??:VCh44  =   2.0000
??:VCh45  =  50.0000      ??:VCh46  =  10.0000
```

Figure 170...Line Display Output Without Any Virtual Parametrics Modules Defined.

Notice that the Module Name is displayed as ?? & the Channel Name is VCh#n where #n is the virtual channel number. The readings are the raw unscaled values from the DTA file. You can also see a display like this, if you rely on the settings in the DTA file & it's an older DTA file that does not have the virtual parametric settings stored in it. In this case you just need to define your virtual parametric modules & change to scaling the data using the settings in the layout.

The number of channels displayed will allow you to guess at the virtual parametric modules in use when the data was collected, because the total number of channels differs for each type of virtual parametric data. The Measurement Computing boards have 64 channels of data, the NanoCorr has 46, the Weather Station 22 & the Strain Gauge Module only 4 channels. If you were to mistakenly use a Weather Station module & replay NanoCorr data, you would get a display like the following one where some channels are defined & some aren't:

```
9   0 00:00:18.2061840
    W1:Wind Speed Min          = 202.0000 m/s          W1:Wind Speed Avg          =    0.0000 m/s
    W1:Wind Speed Max          =   0.3219 m/s          W1:Wind Direction Min      =    0.1372 deg
    W1:Wind Direction Avg      =   0.0127 deg          W1:Wind Direction Max      =    0.0054 deg
    W1:Air Pressure            =   0.3219 bar          W1:Air Temperature         =    0.1372 C
    W1:Internal Temperature    =   0.0522 C            W1:Humidity                =    0.0207 %RH
    W1:Rain Accumulation       =   0.0021 mm           W1:Rain Duration           =    0.0008 sec
    W1:Rain Intensity          =   6.1660 mm/hr        W1:Rain Peak Intensity     =    6.6439 mm/hr
    W1:Hail Accumulation       =   0.0000 hits/cm2     W1:Hail Duration           =    0.0000 sec
    W1:Hail Intensity          =   0.0000 hits/cm2h    W1:Hail Peak Intensity     =    0.0000 hits/cm2h
    W1:Heating Temperature     =   1.8404 C            W1:Heating Voltage         =    1.2792 V
    W1:Supply Voltage          =   0.0013 V            W1:Reference Voltage       =    0.0013 V
    ??:VCh23                   =  24.8901              ??:VCh24                   =    7.0001
    ??:VCh25                   =   0.0072              ??:VCh26                   =    0.0000
    ??:VCh27                   =  18.0000              ??:VCh28                   =-170.0000
    ??:VCh29                   = -75.0000              ??:VCh30                   =-217.0000
    ??:VCh31                   = -10.0000              ??:VCh32                   =  61.0000
    ??:VCh33                   =  16.0000              ??:VCh34                   = -13.0000
    ??:VCh35                   = 130.0000              ??:VCh36                   =  83.0000
    ??:VCh37                   =  26.0000              ??:VCh38                   =  30.0000
    ??:VCh39                   = -99.0000              ??:VCh40                   =  30.0000
    ??:VCh41                   =  -7.0000              ??:VCh42                   = -27.0000
    ??:VCh43                   =  10.0000              ??:VCh44                   =   2.0000
    ??:VCh45                   =  50.0000              ??:VCh46                   =  10.0000
```

Figure 171...Line Display Output Without All Virtual Parametrics Modules Defined.

Anytime you see virtual parametric data in the Line Display showing ?? for the Module Name & virtual channel numbers, you know that your virtual parametric setup is incorrect. Just modify the modules in use in your layout & replay the data again till you get the number of channels defined in the layout to match the data in the DTA file. Most often the simplest solutions to these problems is to display the data using the settings stored in the DTA file. As stated earlier, though, this does not work for early version files that do not contain the virtual parametric settings used to collect the data.

OPTIONS FOR HMSGHAT2000DDI

Supervisor Option

The 'Supervisor Option' allows the user to restrict access to the software by the use of a password. With this option installed the software will start up in ***Operator Mode*** by default.

In ***Operator Mode*** the user has limited access to the system. He can:

- Load a layout file.
- Print/Export graphs or copy them to the clipboard.
- Change the Test Title.
- Edit the Logistics Information.
- Setup the Audio Monitor (if present).
- Perform a test or AST.
- Replay an existing data file.
- Toggle the Line Listing Display on/off.

The user can switch to supervisor mode by clicking 'Switch to Supervisor Mode' from the File menu and entering the password.

In ***Supervisor Mode*** the user has full access to the system. He can perform all the functions possible in operator mode plus he can:

- Change the password.
- Create and save layout files.
- Access the various Test Setup menus.
- Create, modify and delete graphs.
- Modify the arrangement of the graphs, create and delete pages, etc.

How to set the password the first time. The first time the user tries to switch to supervisor mode the software will ask for an initial password. Enter a password of your choice which must be 4-8 alphanumeric characters in length. It is not case sensitive. From this point on the user must enter this password to switch to supervisor mode.

Note: If your system came with HMSGHAT2000DDI pre-installed and the program doesn't request an initial password, this means that the initial password was set at the factory. The password in this case is 'HMSGHAT2000DDI'.

How to change the password: The user can change the password by switching to supervisor mode and clicking 'Change Password' from the file menu.

What to do if you forget the password: If you forget the password the only way to recover is to uninstall the software and reinstall it. This will allow you to enter a new password.

LeakTEC Alarms

LeakTEC Alarms (part of the 'LeakTEC/On-Line-Crack Alarm' Option) are used to monitor ASL or RMS levels (using Time Data) during acquisition and alert the user if they stray beyond certain pre-set lower and upper limits. If ASL or RMS levels stray beyond these limits for a certain period of time an Alarm is generated.

Setup

If you have the 'LeakTEC/On-Line-Crack Alarm' Option installed then you can enable and customize these alarms by using the LeakTEC Alarm Setup dialog box located on the Acquisition Setup menu. Its controls are described below.

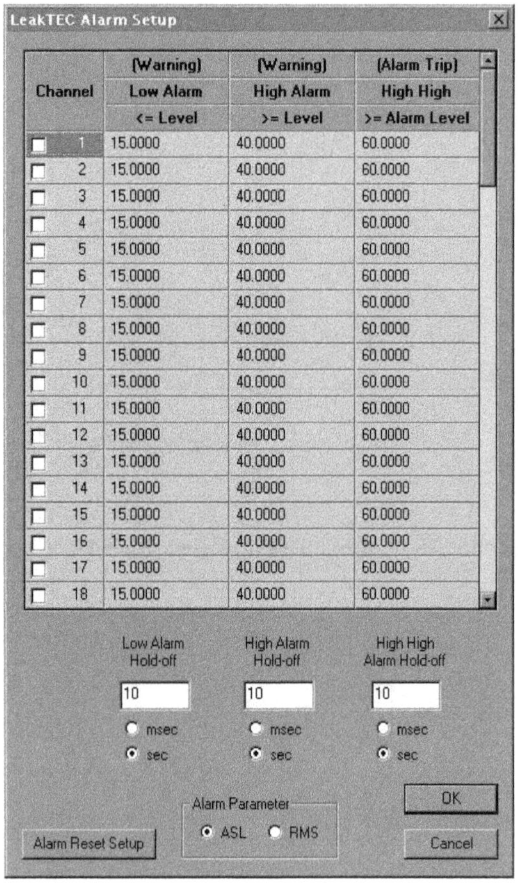

Figure 172. LeakTEC Alarm Setup

Channel – Each channel can be enabled/disabled and customized independently. Click a channel's checkbox to enable that channel.
Low Alarm – Set the Low Alarm Level for a channel in this cell. An alarm 'warning' is triggered if ASL/RMS levels drop to this level or below for a duration of time equal to the 'Low Alarm Hold-off' period.
High Alarm – Set the High Alarm Level for a channel in this cell. An alarm 'trip' is triggered if ASL/RMS levels equal or exceed this level for a duration of time equal to the 'High Alarm Hold-off' period.
High High Alarm – Set the High High Alarm Level for a channel in this cell. An alarm 'trip' is triggered if ASL/RMS levels equal or exceed this level for a duration of time equal to the 'High High Alarm Hold-off' period.
Low Alarm Hold-off – This is the period of time that a channel's ASL/RMS must remain at or below the Low Alarm Level before a Low alarm 'warning' is triggered.
High Alarm Hold-off – This is the period of time that a channel's ASL/RMS must remain at or above the High Alarm Level before a High alarm 'trip' is triggered.
High High Alarm Hold-off – This is the period of time that a channel's ASL/RMS must remain at or above the High High Alarm Level before a High High alarm 'trip' is triggered.
ASL/RMS – Select which Time Dataset feature to monitor using this control.
Alarm Reset Setup – Click this button to enter the Auto Alarm-Reset dialog box. This can be used to set the system to automatically clear the alarms periodically.

Operation

If the ASL/RMS for a particular channel violates an alarm level (low, high or high high) and remains continuously in violation of that level for a duration of time equal to the appropriate 'Hold-Off' then an alarm is triggered.

What happens when an alarm is triggered

1) An alarm message (16) is generated. If autodump is on then this message is saved to the datafile. If the line display is active then it will be displayed there with the time was triggered and the conditions that triggered it.
2) The Alarm Display dialog box is opened (or updated if already open). The details of the alarm and the conditions that triggered it are shown. The alarm can be cleared by clicking the 'Clear Alarms' button.
3) If the alarm is an alarm 'warning' then the Alarm Warning output line is set high. If the alarm is an alarm 'trip' then both the Alarm Warning and Alarm Trip lines are set high.
4) An audible alarm is triggered if you have an audio board installed and enabled.

On-Line Crack Alarms

On-Line Crack Alarms (part of the 'LeakTEC/On-Line-Crack Alarm' Option) are used to monitor hits or events during acquisition and alert the user if the hit/event or energy rates exceed certain pre-set levels. If these rates exceed these limits within a certain period of time an Alarm is generated.

Setup

If you have the 'LeakTEC/On-Line Crack Alarm' Option installed then you can enable and customize these alarms by using the On-Line Crack Alarm Setup dialog box located on the Acquisition Setup menu. Its controls are described below.

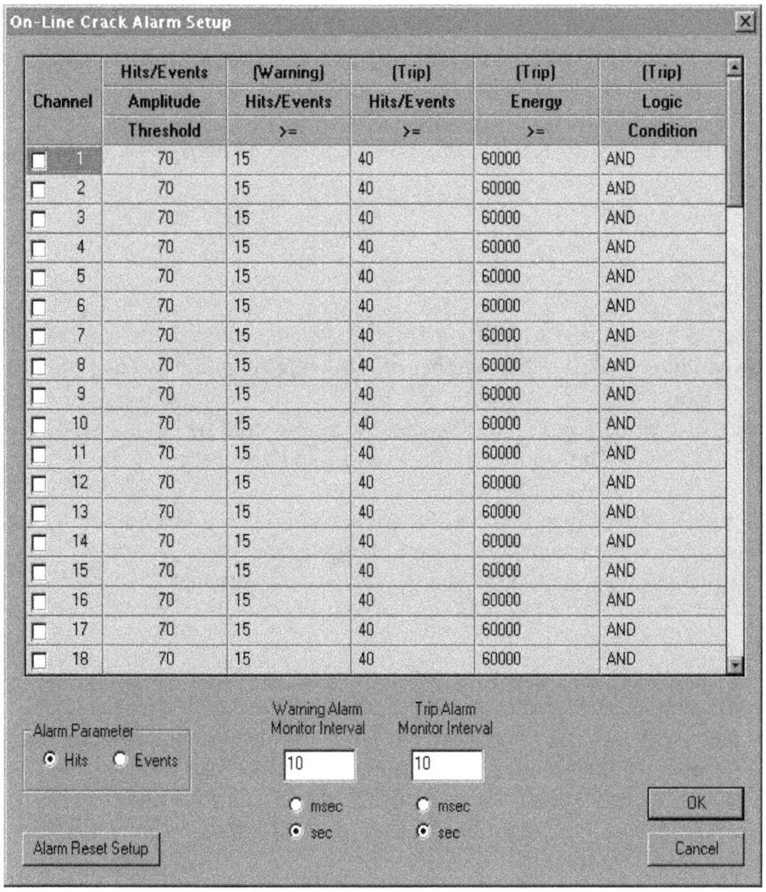

Figure 173. On-Line Crack Alarm Setup

Channel – Each channel can be enabled/disabled and customized independently. Click a channel's checkbox to enable that channel.

Hits/Events Amplitude Threshold – This is the amplitude that a hit (or first hit of an event) must equal or exceed in order to be counted towards a 'Warning Hit/Events' or 'Trip Hit/Events' alarm. This does not affect energy alarms.

[Warning] Hits/Events – A 'Hits/Events Warning' alarm will be triggered if at least this many hits/events that meet the Amplitude Threshold are acquired within a period of time not exceeding the Warning Alarm Monitor Interval.

[Trip] Hits/Events – A 'Trip Hits/Events' alarm will be triggered if at least this many hits/events that meet the amplitude threshold are acquired within a period of time not exceeding the Trip Alarm Monitor Interval. If the logic condition is 'ENERGY' then this alarm type is disabled and will not trigger. If the logic condition is 'AND' then the Trip Energy condition must be met as well before an alarm will be triggered.

[Trip] Energy – A 'Trip Energy' alarm will be triggered if the total energy of any number of hits/events (with any amplitude) are acquired within a period of time not exceeding the Trip Alarm Monitor Interval meets or exceeds this value. If the logic condition is 'Hits/Events' then this alarm type is disabled and will not trigger. If the logic condition is 'AND' then the 'Trip Hits/Events' condition must be met as well before an alarm will be triggered.

[Trip] Logic Condition – The logic condition controls how the Trip alarms are calculated. If it is set to 'Hits/Events' then the 'Trip Energy' condition is ignored and no alarms will be triggered using it. If it is set to 'Energy' then the 'Trip Hits/Events' condition is ignored and no alarms will be triggered using it. If it is set to AND then both the 'Trip Hits/Events' and 'Trip Energy' conditions must be met before an alarm is triggered. If it is set to OR then meeting either condition will trigger a Trip alarm.
Warning Alarm Monitor Interval – This is the period of time in which hits/events (passing the threshold) are accumulated towards a Warning alarm.
Trip Alarm Monitor Interval – This is the period of time in which hits/events (passing the threshold) and energy are accumulated towards a Trip alarm.
Alarm Parameter – Select whether to use hits or the first hit of events for alarm determination.
Alarm Reset Setup – Click this button to enter the Auto Alarm-Reset dialog box. This can be used to set the system to automatically clear the alarms periodically.

Operation

Four kinds of alarms can be generated by this option.

- Warning Hits/Event – A number of hits/events that met the Amplitude Threshold were acquired within a period of time not exceeding the Warning Alarm Monitor Interval. The number of hits/events met or exceeded the [Warning] Hits/Events value.
- Trip Hits/Event – A number of hits/events that met the Amplitude Threshold were acquired within a period of time not exceeding the Trip Alarm Monitor Interval. The number of hits/events met or exceeded the [Trip] Hits/Events value.
- Trip Energy – The total energy of all hits/events (with any amplitude) that were acquired within a period of time not exceeding the Trip Alarm Monitor Interval met or exceeded the [Trip] Energy value.
- Trip Hits/Event and Energy – Both the 'Trip Hits/Event' and 'Trip Energy' conditions were met.

What happens when an alarm is triggered

1) An alarm message (16) is generated. If autodump is on then this message is saved to the datafile. If the line display is active then it will be displayed there with the time was triggered and the conditions that triggered it.
2) The Alarm Display dialog box is opened (or updated if already open). The details of the alarm and the conditions that triggered it are shown. The alarm can be cleared by clicking the 'Clear Alarms' button.
3) If the alarm is an alarm 'warning' then the Alarm Warning output line is set high. If the alarm is an alarm 'trip' then both the Alarm Warning and Alarm Trip lines are set high.
4) An audible alarm is triggered if you have an audio board installed and enabled.

APPENDIX I

UTILITIES MENU HMSGHAT2000DDI™ SOFTWARE

APPENDIX - II

UTILITIES MENU HMSGHAT2000DDI™ SOFTWARE

DIAGNOSTIC DEVICES INC.

94 Wolcott Road, Simsbury, CT 06070

Advancing Signal Processing Beyond The Year 2005

www.Diagnostic-Devices.com

HMSGHAT2000DDI™ SOFTWARE

Part No. 6301-7001
Version E5.20

Installation, Operation and User's Reference Manual

INTRODUCTION

HMSGHAT2000DDI is a 32 bit WINDOWS, Data Acquisition and Replay program capable of running 's DiSP, SAMOS, PCI-2, , SENSOR HIGHWAY or SPARTAN AT/2000 based products. HMSGHAT2000DDI uses full WINDOWS resources including: Setting of any WINDOWS available screen resolutions, Printing, Networking, Multi-Tasking, Multi-Threading, etc. It is capable of operating in WINDOWS 2000/XP/Vista operating systems. HMSGHAT2000DDI is fully compatible with 's standard (DTA) data files. This allows you to replay and analyze all your previously collected AE files. HMSGHAT2000DDI is easy to learn, operate and use. The software has all the acquisition, graphing and analysis capabilities that you have come to expect in your AE system, plus many more new and enhanced features to ease your data analysis and visualization tasks. Multiple copies of HMSGHAT2000DDI can be run at once, one for acquisition with full display and analysis capabilities and one or more as Replay for additional analysis purposes.

Some of the general AE display and analysis capabilities found in HMSGHAT2000DDI includes the following: HMSGHAT2000DDI provides a framework for easily adding graphs and graph screens to the main system WINDOW. HMSGHAT2000DDI has a very flexible line listing (dump) capability, which can be called at any time, in any screen, which can be scrolled up or down. HMSGHAT2000DDI has various user selectable tool bars including; Setup icon toolbar, acquisition control, line dump, status toolbar and statistics tool bar. HMSGHAT2000DDI has many built in, enhanced features including; Hit-Event Linking, Graph zooming and panning, more flexible graph setup, more graphs, and built-in filtering functions (graphical filters and Post Filtering). Printing capabilities include: Print a graph, a screen or multiple screens to a WINDOWS or network printer, to the clipboard or to a JPEG file.

Some of HMSGHAT2000DDI graphing capabilities include:

- Exceptional 2D and 3D graphing capabilities. Multiple graphs can be displayed on a screen, limited only by the resolution of the screen itself.
- Graphs are individually sizable on a screen, making for a very flexible arrangement. The user can set up one (or more) large graphs for visualization with multiple supporting small graphs alongside or around the main ones.
- Multiple Graphs are arranged on a screen, however the user desires. Multiple screens can be setup, each accessible by selecting a user labeled tab. Therefore, a user can set themes for screen layout (e.g. Location & Cluster analysis, waveform analysis, AE activity analysis, AE Feature correlation analysis, Alarm Analysis, themes to name a few)
- Many different types of graphs can be set up including; histograms, point plots, 3-D graphs, waveforms, FFT's, multiple plots on a single graph with coloring options, etc.
- All graphs can easily be enlarged to full screen (maximized) by the touch of a button.
- All graphs have full Cursor Readout capability, either continuous or point by point movement.
- All graphs (2-D and 3-D included) can be infinitely zoomed and panned for close-up analysis.

In terms of Location determination, the standard version of HMSGHAT2000DDI has zonal and linear location modes, while the optional "Full Location" software adds 2D Planar Location capability to the list. In addition, there is also a 3D Location option available and a Spherical 3D location option. Key features regarding location include the following:

- Multiple (8) location groups standard, (32 optional), each capable of using all the AE channels in the AE system. Mouse oriented sensor setup includes; Set sensors manually (click and drop), automated sensor setup for triangular setup, and easy sensor editing drag and drop, or table entry based sensor position.
- Multiple 1 dimensional, 2 dimensional and 3 dimensional location modes.
- Flexible Clustering, cluster reporting and Cluster statistics available on any point plot graph screen (not just limited to location).
- User selectable location structure visualization to ease setup. Options include; Plate, vertical vessel, horizontal vessel, sphere, and Free, each with specific setup menu's geared for easy setup of that structure, including automated and manual placement of sensors, weld lines, nozzles, etc.
- Visualization overlay planes to ease setup and viewing of location setup and results. Overlay planes include; Sensors, grid, attenuation map, welds and nozzles.

- Advanced Location setup capabilities provides improved event detection classification, and accurate source location techniques.

The HMSGHAT2000DDI program runs under WINDOWS 2000/XP/Vista operating systems.

System Requirements

HMSGHAT2000DDI software requires a DiSP, SPARTAN AT-2000, PCI-2, , SAMOS or SENSOR HIGHWAY System with Pentium III (or equivalent). Although lesser system computer configurations will operate HMSGHAT2000DDI, it is recommended that at least a 2 GHz processor, 1Gigbyte RAM, 100 gigabyte hard disk and 17" display monitor with 1280 x 1024 resolution be utilized. These days this type of computer power is easily available at low cost. Please consult the factory if your computer is below these standards.

If HMSGHAT2000DDI is being installed explicitly for replay and analysis, it is also recommended to use a similar computer base as that described above.

Software Installation and Licensing

The HMSGHAT2000DDI program is generally pre-installed on all AE system computers from the factory. In special instances, such as software upgrades, however, you will need to perform the installation yourself from the supplied CD. Likewise, if your AE system was built at our factory it will come pre-licensed. You will generally need to perform licensing yourself only when installing HMSGHAT2000DDI on a new computer.

If you are installing HMSGHAT2000DDI yourself, you will need to follow the simple installation instructions below. Software installation and licensing is a two step process which will be discussed in the next two sections for an example of "HMSGHAT2000DDI for DiSP". Other products such as PCI-2 and SAMOS will be very similar.

Software Installation

To fully install HMSGHAT2000DDI, you must perform the following steps.

h) Exit all running applications.
i) If you have a License Key already plugged into your system please remove it now. This is a green or gray USB dongle that is used in licensing. It should not be plugged in during HMSGHAT2000DDI installation.
j) Run SETUP.EXE from the root folder of the HMSGHAT2000DDI installation CD. You can find this through Windows Explorer.
k) Upon executing SETUP.EXE you will follow a typical Windows installation process. Select all the default folders (highly recommended but not mandatory), if desired.
l) Install the PCI-DSP4 driver if you are doing acquisition. See the section below titled "Installing the Data Acquisition Board driver if you are doing acquisition".
m) Install the Audio driver if using an audio board. See the section below titled "Installing the Audio board driver if using the Audio Board".
n) If you have a License Key then plug it into a USB slot now. Windows may need to install the driver for the Key at this point. Simply follow the prompts and accept the default values.

Software Licensing

Licensing Basics

systems come pre-configured with licenses (either internal-file or USB Key based). If you received such a system then you should not have to do anything beyond plugging in the License Key (if you were issued one). If you install HMSGHAT2000DDI yourself on a new computer then you will need to license it. The section "Licensing Instructions" below gives instructions on how to do so.

HMSGHAT2000DDI can be licensed using either of two methods.

- License Keys – These are green or gray USB dongles that can be purchased either with your system or at a future time. They can be used to run the licensed program on ANY computer they are plugged into. This makes them very portable. Note that these are not the same as the black or purple "Hardware Security Keys" used by older versions of HMSGHAT2000DDI.
- License Files – This is HMSGHAT2000DDIs default licensing system. If you have not purchased a Key you can acquire a License File from Customer Service by following the steps in the section titled "Licensing Instructions" below. This will license a single copy of HMSGHAT2000DDI to the computer of your choice.

Licensing Instructions

If you have a Licensing Key (Green or Gray USB Dongle).

c) Plug the Key into a USB port. If this is the first time you have done this on this particular USB port then Windows may want to install the driver. Let it do so now. Follow the prompts and accept the defaults. It should only take a few moments.
d) Run HMSGHAT2000DDI.

If you don't have a Licensing Key.

e) Run HMSGHAT2000DDI. You will most likely see a dialog box like this one.

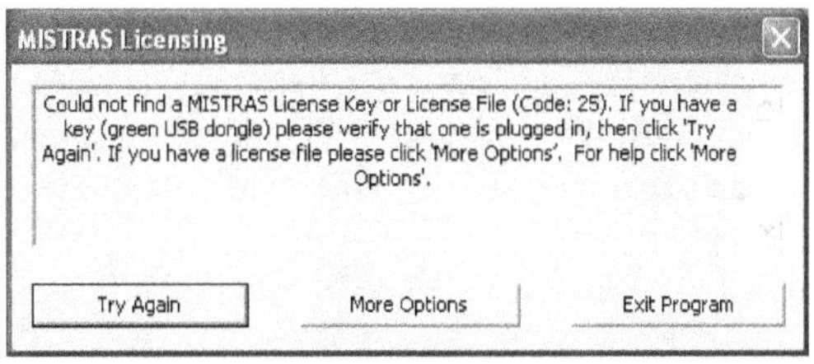

Figure 1. Initial Licensing Screen

f) Click "More Options". This will bring up the " Licensing" dialog box shown here.

Figure 2. MISTRAS Licensing Screen

g) The center section labeled " License File" guides you through the process of acquiring and installing a license file. It is a two step process as follows.
   a. Contact Customer Service – You will need to contact Customer Service to acquire a license file. The easiest ways are via email or by phone. In the sunken area of the menu is a block of information that Customer Service will need in order to create your license. You can copy/paste this into an email if you are using this method (be sure to enter your name, company and sales order#), or simply have the dialog box open while you call so the information will be at hand. Once Customer Service has this information they will create a license and email it to you. You can exit HMSGHAT2000DDI at this point if you wish, while your license is being made for you. If so, you can return to this dialog box later by following the same instructions above.
   b. Install your License File – Once you have received your license file from Customer Service click the "Install License File" button and browse to the file. You should receive a message indicating success, at which point you can click the "Start HMSGHAT2000DDI" button at the bottom of the screen (which previously was "Exit HMSGHAT2000DDI").
h) HMSGHAT2000DDI is now licensed.

Upgrading an HMSGHAT2000DDI License

If at any time you decide you would like to purchase additional options for HMSGHAT2000DDI such as "3D Location", "HMSGHAT2000DDI Post" or "LeakTEC/Online Crack Alarms" (there are many more options to choose from), you can purchase them by contacting Customer Service. You will not need to re-install HMSGHAT2000DDI or return your License Key (if you use one) to the factory. Customer Service can upgrade your license via email.

- If you use a License Key then Customer Service will send you a special file called a "Patch" file.
- If you use a License File then Customer Service will send you a new one to replace it.

In either case once you have received the file you will need to open the 'Upgrade License' dialog box. This can be accessed from one of two places.

- If HMSGHAT2000DDI is displaying a licensing error message on startup such as the one in 'figure 1' above then click "More Options" to open the " Licensing" dialog box. From here click the "Upgrade License" button found at the bottom left portion of the screen.
- If you can access HMSGHAT2000DDI normally then open the "Options" dialog box from the "Help" menu. Click the "Upgrade License" button.

This will open the "Upgrade License" dialog box shown here.

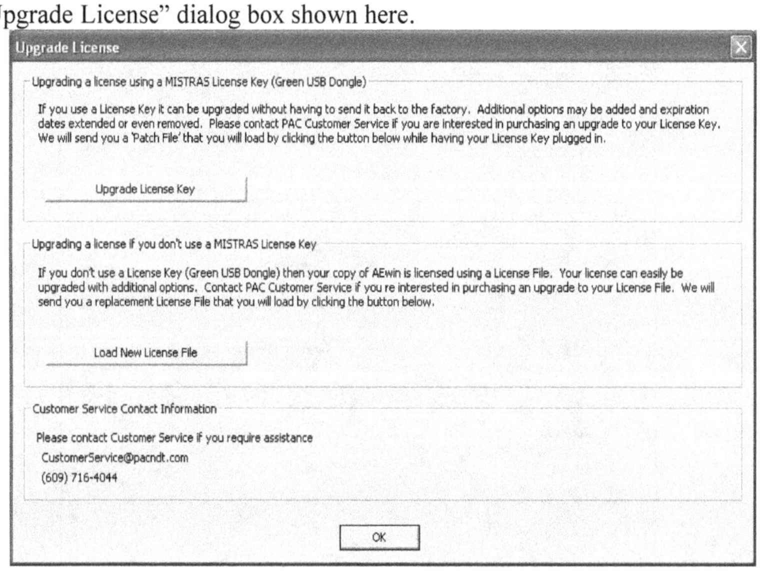

Figure 3. Upgrade License Screen

- If you are upgrading a License Key then make sure it is inserted into a USB slot. If you have more than one License Key make sure that only one is inserted at any one time. Click "Upgrade License Key" and select the "Patch File" emailed to you by Customer Service.
- If you are upgrading/replacing your License File then click "Load New License File" and select the License File emailed to you by Customer Service.

That's it. HMSGHAT2000DDI may need to close. If it does not, or when you re-start it, HMSGHAT2000DDI will use your newly upgraded license.

Licensing FAQ and Troubleshooting

Q: My license has stopped working when I upgraded HMSGHAT2000DDI. What do I do?
A: Starting in mid 2009 HMSGHAT2000DDI was upgraded to a new licensing system that is more flexible and user-friendly. The old "Hardware Security Keys" (black or purple USB dongles) have been replaced by " License Keys" (green USB dongles). Likewise the old "lservrc" license files have been replaced as well. If you have an old "Hardware Security Key" or "lservrc" license file then this will need to be replaced with a new version. Please contact Customer Service for assistance.

Q: HMSGHAT2000DDI won't install and gives the following error message: "There is a problem with this Windows Installer kage. A program required for this install to complete could not be run. Contact your support personnel or kage vendor.".
A: Make sure you run "Setup.exe" to install HMSGHAT2000DDI, not "HMSGHAT2000DDISetup.msi".

Q: After plugging in a License Key Windows gives an error message like one of these:

- "You must be a member of the Administrators group on this computer to install this hardware: USB Device".
- "You do not have sufficient security privileges to install hardware on this computer. Please contact your site administrator, or log out and log in again as an administrator and try again.".

A: The first time you plug a Key into a particular USB port Windows will attempt to install the driver for it. If you do not have sufficient privileges to do so you will get this warning. In this case you will need to log in as an administrator or contact your network or IT administrator for assistance. To avoid this problem in the future, be sure to plug the Key into all USB ports one by one (when you have admin privileges) to install the driver for all ports.

Q: Can I use my License Key in multiple computers?
A: Yes. License Keys are portable. Simply exit HMSGHAT2000DDI on your first computer, remove the Key, plug it into your other computer and start HMSGHAT2000DDI there. You will have the same options as before.

Q: Can I install my License File on multiple computers?
A: No. License Files are created for a single computer. You can purchase additional license Files or Keys for replay purposes from Customer Service.

Installing the Data Acquisition Board driver if you are doing acquisition

The various Data Acquisition Boards (PCI-DSP4, AEDSP, etc) and Audio Monitor Cards (ISA and PCI) require a driver in order to function properly in Windows.

Normally, when Windows is first booted after installing the card it will prompt for a driver for it. Point Windows to the appropriate driver folder on the CD and follow the instructions.
The driver folders are located on the CD at the following locations:

- – Drivers\AEDSP.
- **All other boards** – Drivers\Pci.

If for some reason you need to install a driver manually then read the document (drivers.doc or readme.txt) found in the appropriate driver folder. Use the procedures described there only if the above automatic driver installation does not work. It is not normally required to install the driver this way. But if problems arise, that is how to do it.

Installing the Audio board driver if using the Audio Board

The Audio Monitor Board (ISA and PCI) require a driver in order to function properly in Windows.

Normally, when Windows is first booted after installing the card it will prompt for a driver for it. Point Windows to the appropriate driver folder on the CD and follow the instructions.
The driver folders are located on the CD at the following locations:

- **ISA Audio Board** – Drivers\IsaAudio.
- **PCI Audio Board** – Drivers\Pci.

If for some reason you need to install a driver manually then read the document (drivers.doc or readmr.txt) found in the appropriate driver folder. Use the procedures described there only if the above automatic driver installation does not work. It is not normally required to install the driver this way. But if problems arise, that is how to do it.

Warning Notes

Please note the following warnings:

5. If the software is installed on a computer without AE hardware boards (e.g. PCI-DSP4 boards) installed, the program will run in replay only mode.
6. If the software is installed on a computer with AE hardware (e.g. PCI-DSP4) board(s) installed, it will not run in acquisition mode unless the Windows drivers for those boards are installed.
7. If a audio board is installed in the system, the software will not be able to control it unless its drivers are also installed.
8. If the entire system came from , the drivers and software are pre-installed at the factory.

Installing a Shortcut Icon for HMSGHAT2000DDI on the Desktop

Once the HMSGHAT2000DDI software installed as described in the previous sections, and assuming that all the default selections have been selected, the software can be started by the following mouse operated (activated by pressing the left mouse button) sequence:

Start → Programs → Physical Acoustics → HMSGHAT2000DDI for DiSP

Other forms of HMSGHAT2000DDI, such as SAMOS and PCI2 will have similar shortcuts.

An HMSGHAT2000DDI desktop Icon can easily be created by performing the same sequence above except pressing the right mouse button when pointing to HMSGHAT2000DDI. Upon doing this a menu will appear. You can either select "create shortcut" (which will create a second copy of the HMSGHAT2000DDI program title) and drag it to the desktop or you can select "Send to", then select the "desktop (create shortcut)" menu button which will place this extra copy directly on the desktop. Once created, the Icon title can be modified by left-clicking on the title. Now the program can be run directly by double clicking the HMSGHAT2000DDI Icon on the desktop.

HMSGHAT2000DDI LAYOUT (.LAY) FILES AND DATA (.DTA) FILES

This section discusses the various file types associated with HMSGHAT2000DDI. The most important files you will be working with everyday with HMSGHAT2000DDI are the Setup Files (called Layout files in HMSGHAT2000DDI) and the Data files. Your working copy of HMSGHAT2000DDI will have examples of both these layout files and test data files. You will be creating, using and modifying these files in the course of your work.

The initialization or setup files, referred to as "Layout" files, distinguished by the suffix .LAY, hold all AE test setup information for the HMSGHAT2000DDI software. A .LAY file stores the hardware setup, graph setup, acquisition setup, filters, alarms and location setup. Other program settings, such as graph item colors and various preferences are stored in the Windows registry.

A .LAY file can be loaded from disk by means of the "Open Layout" option in the "File" Menu of the HMSGHAT2000DDI program. Once loaded, all the initialization information from the layout file immediately becomes effective for system operation. A .LAY file is written to disk by means of the "Save Layout" or "Save Layout As" option in the File Setup menu. All the initialization information currently established in the system is written to the file. A .LAY file can have any name followed by the .LAY extension.

There is one special .LAY file known as the Autoload file. It is distinguished by the special name Layout.LAY. The Autoload file, when present, is automatically read from disk when the HMSGHAT2000DDI program begins to run. If the Autoload file is not present, the system is set up with default values held in the HMSGHAT2000DDI program itself. The Autoload file is a convenience feature to speed your work.

AE Test Data Files are identified by the filename extension .DTA. You may specify any name for a data file using standard Windows file naming conventions. This is done in the Start acquisition dialog box from the Acquire Menu. If a new name is not entered, the system will use the default data file name e.g. TEST0000.DTA for the first data written after bootup. The next file will be TEST0001.DTA, then TEST0002.DTA, and so on. It is convenient to use numbers at the end of the filenames (e.g. ABCD000.DTA) as the system will then increment the numbers to help keep the files in order and organized.

Layout and data files may be written to any disk drive and any specified subdirectory. A good practice is to save both .LAY files and .DTA files to the same drive and subdirectory, keeping related test files together. It is also a good practice to name the files very similar (e.g. test.LAY and test0000.DTA) as this will help keep the layout files organized with the data files.

There has been an intentional name and file-extension change from the "Initialization" (.INI) file of the "Legacy" (DOS) based programs to the "Layout" (.LAY) file for setup files utilized by HMSGHAT2000DDI. The file extension for a data (.DTA) file remains consistent. There is a good reason for this. Due to the extensive changes in migrating to Windows, and the many, many extra capabilities that have been added to the setting up of the system, it is impossible to keep consistency in the setup files between the previous legacy software and the HMSGHAT2000DDI software. Therefore, to avoid confusion, the extension file name has been changed. Now whenever a user sees an INI file, they will know that this can only be used with the previous legacy software, while the .LAY layout file can only be used with the HMSGHAT2000DDI software.

However, with the DTA file, the compatibility remains between the data file structure and AE .DTA data files can be read and displayed by either the DOS or the HMSGHAT2000DDI software.

OPERATING HMSGHAT2000DDI

The purpose of this section is to provide a quick software operational run-through, to provide a familiarization with operating and using HMSGHAT2000DDI. This is the most important part of this HMSGHAT2000DDI software user's manual as it will provide a view of the important screens and commands necessary to get HMSGHAT2000DDI setup, and running in both Replay and Acquisition modes.

Starting HMSGHAT2000DDI

The first step in the process of using HMSGHAT2000DDI is booting up the program on your AE system. As the HMSGHAT2000DDI program is a standard Windows program, it is started in the same way as any other WINDOWS program. This is accomplished either by using the "Start" command or by selecting the HMSGHAT2000DDI Icon from the WINDOWS desktop.

It is assumed that the AE computer is booted up and in its opening screen which is the desktop layout showing all the available program icons to select. If the HMSGHAT2000DDI icon (as shown on the right) is on the desktop, all you need to do is select this by double clicking the left mouse button, once the mouse is positioned over the icon. If the icon is not on the desktop you can create the shortcut by referring to section 1.2.6.

Alternatively you can start HMSGHAT2000DDI by selecting the following sequence starting with the Start menu at the bottom left side of the screen:

Start → Programs → Physical Acoustics → HMSGHAT2000DDI

The boot-up sequence for HMSGHAT2000DDI starts by loading the executable program HMSGHAT2000DDI.exe. Once HMSGHAT2000DDI is loaded, the program searches the startup "AE Data" subdirectory for the default Layout (setup file) called "Layout.LAY". Upon finding this file, HMSGHAT2000DDI loads it and becomes fully configured and ready to run in accordance with the setup information in that default Layout.LAY file. If no Layout.LAY is found, HMSGHAT2000DDI chooses a common default starting setup that is ready for use. In either case an HMSGHAT2000DDI startup screen similar to figure 4 will appear.

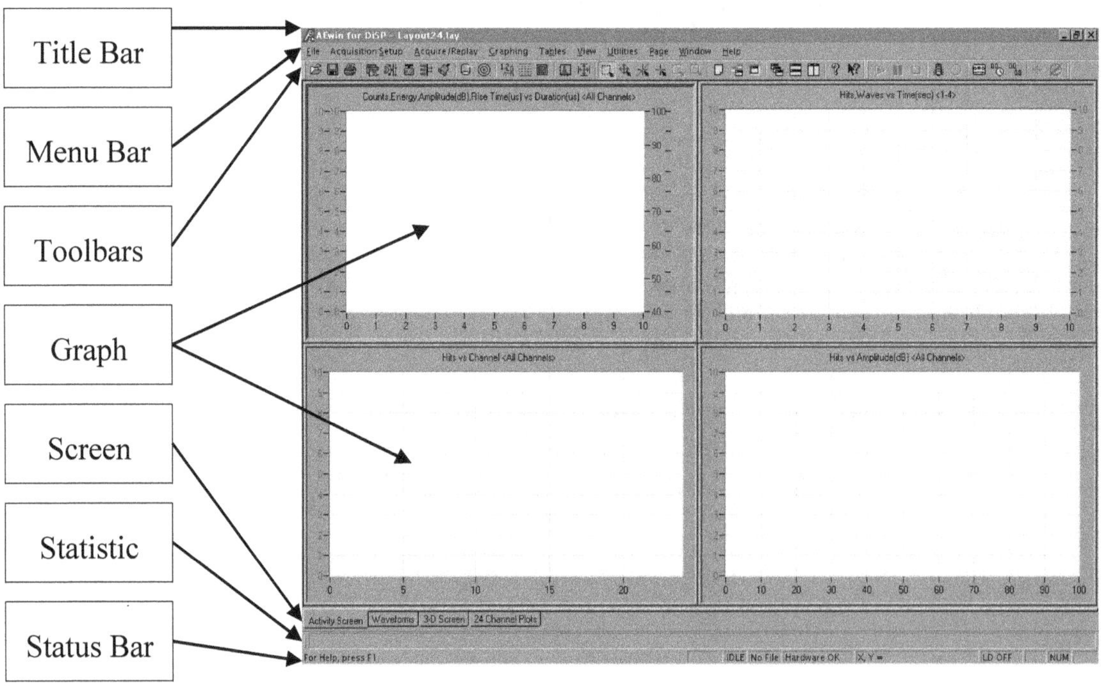

Figure 4. HMSGHAT2000DDI Opening Screen

HMSGHAT2000DDI Screen Layout Familiarization

Figure 4 shows a typical 4-screen layout for HMSGHAT2000DDI. Above and below the graph area are various informational, status and setup toolbars. Each bar is labeled with a name. Depending on the preferences selected for your program, you may have other toolbars or be missing some. The "View" menu in the main menu bar area controls which bars are on or off. The following is a short description of each area identified on the HMSGHAT2000DDI screen. This section will get you familiar with the layout and the overall function. Further chapters will provide detailed insight into the use of each of these functions.

Title Bar: This is a typical WINDOWS Title Bar with all the standard capabilities. At the leftmost position is the Physical Acoustics Logo (Logo) Icon from which you can manage the HMSGHAT2000DDI program with functions such as minimize, restore, maximize, move, size or Close the program altogether. At the far right are the application buttons for Minimize, Restore and Close. Back on the right hand side next to the Logo Icon is the program name (in this case it is HMSGHAT2000DDI for DiSP) and next to that is the name of the current layout file being used by the program. If an asterisk (*) appears next to the file name, it indicates that one or more settings have been changed since the last time the layout file was loaded or saved.

Main Menu Bar: The main menu bar provides a series of text commands running across the line, including: File, Acquisition Setup, Acquire/Replay, Graphing, Tables, View, Utilities, Page, Window, and Help. Most of these items contain pull-down sub-menus of their own, many of which in turn contain graphical dialog boxes and prompts of their own. You narrow down the range of choices until you arrive at the specific choice you want.

To display a pull-down sub-menu from a main menu entry, simply move the mouse pointer to the name of the menu item on the menu bar and click the left mouse button. This process is known as "clicking" (or "left-clicking") on an item, and will herein be referred to as such. To use the keyboard, simultaneously press the ALT key and the underlined letter of the menu item name at the same time. Once you have opened a pull-down menu item from the main menu, you can use the left and right arrow keys to move through each entry, allowing you to see the pull-down menu for each main menu entry. You can also use the up and down arrow keys to move through each selection of the pull-down menu. To select an item from a pull-down sub-menu, click on the desired action, or use the up and down arrow keys to select an action, and then press ENTER. Pressing the underlined letter of a menu item will also select that item. If an item in the pull-down menu appears "grayed out," this action is not available in the current software or at the current time. Use the <ESC> key to remove the pull-down menu display.

Most of the selections in the pull-down sub-menus will display a dialog box requesting information necessary to carry out the command you requested. A dialog box contains areas where you enter information, select options or commands, or activate controls to execute or cancel a command. There are several types of boxes and buttons available within a dialog box. These boxes and buttons are described briefly below and will be described in further detail in subsequent sections of this manual. It is important that you be familiar with dialog boxes and be comfortable using them, for they are the means by which most functions are carried out in the HMSGHAT2000DDI software.

Icon Toolbars: Many of the menu items accessible from the Main pull-down menus can also be accessed by use of the Icon Toolbar and Acquisition Control Toolbar just below it. Figure 5 shows the Icon toolbars.

Figure 5. Icon Toolbars

To determine the function of any of the Icon's simply move the mouse cursor over top of the desired icon. Upon resting there for approximately 1 second the Icon's function will appear. To select an icon as a shortcut command simply click on it with the left mouse button.

Graph Area: The graph area is the most important area of the screen. This is where the AE information in the form of graphs and line listings appear regarding the AE examination status. The graph area can be set up in many ways with a minimum of 1 graph to a maximum only limited by the graph visibility. More about the graph setup, functionality and flexibility will be discussed later in the Graph setup menu.

Screen Page Tabs: HMSGHAT2000DDI can show many different graph screens, these are called screen or graph "Pages". Accessibility to viewing any screen is provided via the Page tabs located just below the graph area. The Page tabs can be named to be synonymous with the graph theme as shown in figure 4 with the 4 Page tab names, Activity Screen, Waveform Screen, 3D Screen and 24 channel plots. Using the right mouse button and clicking on a page will display a 'context' menu from which pages can be added, inserted, deleted or renamed. The context menu is also a quick way to add new graphs and close existing graphs.

Statistics Bar: This toolbar area is reserved for display of test statistics. Test statistics allow the user to view key Acoustic emission test indicators including: Cumulative AE counts, Total AE Hits, Total AE Events, Total # Waveforms, Cumulative Counts, Cumulative Energy, Time of Test, Disk Free Se, Parametrics and others. These are very useful activity indicators during a test and very useful summary statistics after a test. The user can select any or all of these to customize his statistics display bar.

Status Bar: The Status bar provides useful information regarding HMSGHAT2000DDI and test status. On the left is a Text bar providing status and help information regarding the functions being carried out. Near the center is a status bar. In figure 4 the word "Idle" appears in this bar. Other status information such as "Test Paused", "Replay", "Test Stopped" "Test Active", "Abort Test", etc. are displayed in this text field. Next to the Test status field is a filename field informing the user the name of the file being replayed or the name under which an AE test is being saved. Next to that is a system diagnostics text field. In figure 4, this status is "Hardware OK". Next to that is a cursor position, showing the current location of the Cursor on the graph. When selected, the X, Y position will be read out in actual screen units.

Navigating the HMSGHAT2000DDI Menus

This section provides a detailed reference of HMSGHAT2000DDI's major menu items and dialog boxes. The first time through, we suggest you just look at the menus (to get familiar with them) and use the setups provided with your HMSGHAT2000DDI installation, that have been predefined by ; later, you can make your own choices within the menus to control system operation.

From the Opening Screen (Figure 4), pay attention to the Main Menu Bar near the top of the screen. This specific area is shown again below. Each main menu item will be described. This brief introduction to navigating the menus is provided in order to get ready for guiding the user in the sequence of starting and replaying a test.

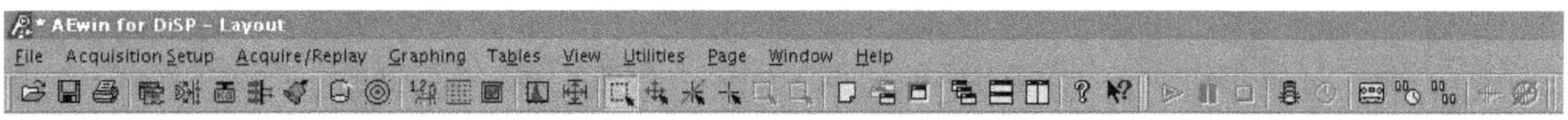

Figure 6. Main menus and toolbars

File Menu

New Layout

The 'New Layout' menu item is used to remove all graphs and pages and restore the graph layout to a default setting of one page and one simple graph. This command does not alter any other settings.

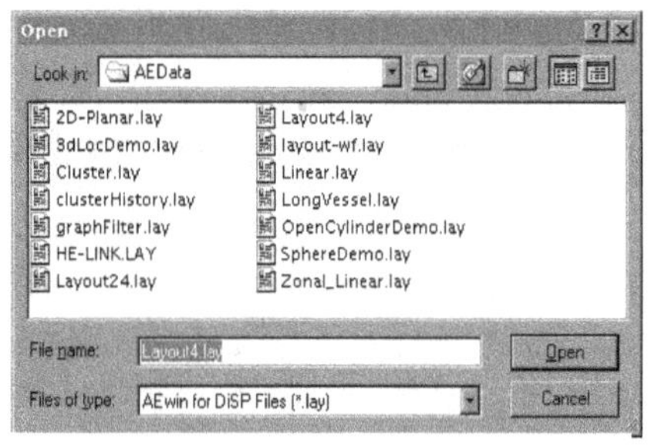

Figure 7. Open Layout...

Open Layout...

The 'Open Layout...' menu item is used to load a specified layout file's setup information into HMSGHAT2000DDI.

Save Layout/Save Layout As…

The 'Save Layout…' and 'Save Layout As…' menu items are used to save the setup information in HMSGHAT2000DDI to a specified layout file. When the 'Save Layout As…' item is clicked the 'Save As' dialog is displayed. Here you can enter a new layout filename to create or an existing one to overwrite. Once this is done you will see the layout filename you entered on the upper left corner of the screen. You can use 'Save Layout…' from then on to update the file with any changes you have made since you have last saved.

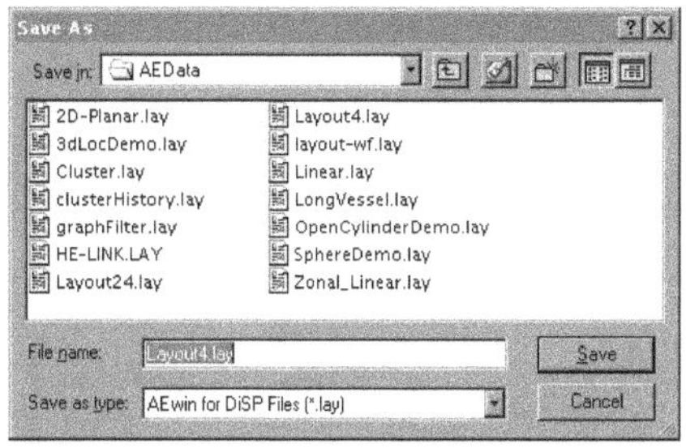

Figure 8. Save Layout…

Specify Data Folder…

The 'Specify Data Folder…' menu item opens a dialog box of the same name. It is used to select the default folders for storage of Data and Layout files. These are the folders that are displayed by default in the various Open/Save Layout and Acquire/Replay dialog boxes. The 'Layout File Folder' is also where HMSGHAT2000DDI looks for the autoload 'Layout.lay' file on startup. By default these are both set to the AEData folder found in the installation folder of HMSGHAT2000DDI.

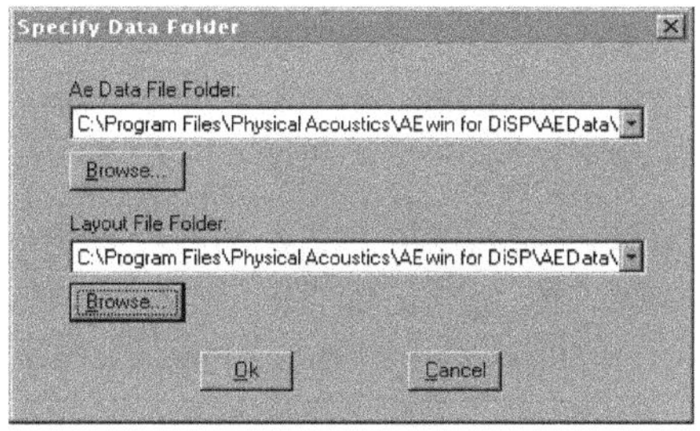

Figure 9. Specify Data Folder

Print Page(s)/Print Graph…

The 'Print Page(s)…' and 'Print Graph…' menu items are used to print a graph or multiple pages of graphs. 'Print Graph…' is used to send the currently selected graph to the printer, zoomed as much as possible while still retaining aspect ratio. 'Print Page(s)…' is used to send a selection of pages to the printer one page at a time, each page zoomed as much as possible while still retaining aspect ratio. You can select 'All pages' or a range of pages.

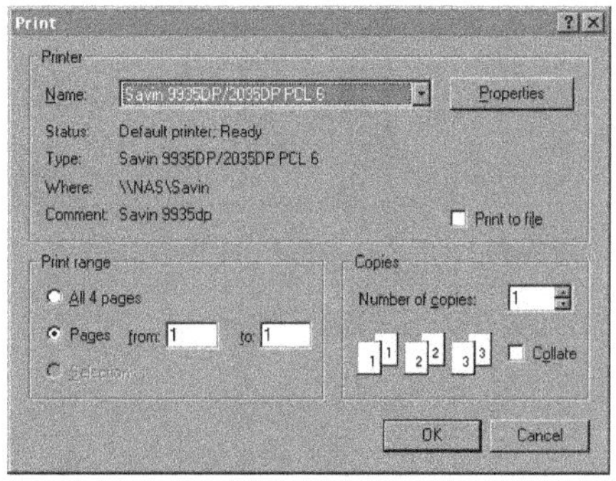

Figure 10. Print…

Export to JPG…

'Export to JPG…' is used to save pages as JPEG files in the same way that 'Print Page(s)…' sends pages to the printer. Here you enter a base filename and a destination folder. Each page will be saved as a separate JPEG file with its name formed from the combination of the base name and the page's title. e.g. If you had 2 pages named 'waveforms' and 'activity' and you set the base name to be 'TEST' then the two files will be named 'TESTwaveform.JPG' and 'TESTactivity.JPG'.

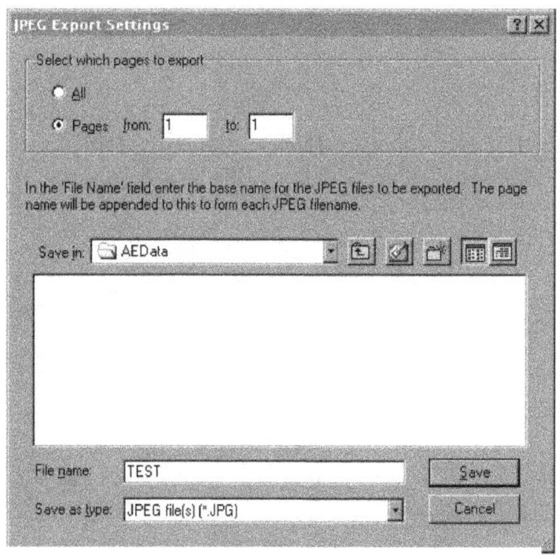

Figure 11. Export to JPG

Copy Screen to Clipboard

'Copy Screen to Clipboard' is used to save an image of the entire screen to the clipboard. This image can then be pasted in other applications such as word processors and image editors.

Exit

'Exit' is used to terminate the HMSGHAT2000DDI program. If you have made any unsaved changes to your current layout HMSGHAT2000DDI will prompt you to save your layout first, exit without saving or cancel.

Acquisition Setup Menu

Hardware...

This is the dialog box where all hardware settings are selected before the test. There are a number of property pages that each deal with a different part of hardware setup. Each is described below. Note that there are some differences in this dialog box between products. For instance 'HMSGHAT2000DDI for Spartan' has a Gain setting but no Waveform Setup whereas 'HMSGHAT2000DDI for DiSP' has no Gain setting but can have Waveform Setup if the Waveform option has been installed. Features specific to one or more types of hardware are noted by the names of the applicable hardware types. These types are briefly described below.

Name	Description	Waveforms
ICC	2 Channel card for **Spartan** systems. *	Not Available
DSP4	4 Channel card for **Spartan** systems. *	Not Available
AEDSP	2 Channel card for systems.	Standard
PCI-DSP4	4 Channel card for **DiSP** systems.	Optional
SAMOS	8 Channel card for **SAMOS** systems.	Standard
PCI-2	2 Channel card for **PCI-2** systems.	Standard

* ICC and DSP4 cards can be used simultaneously in the same system.

Standard Channel Setup

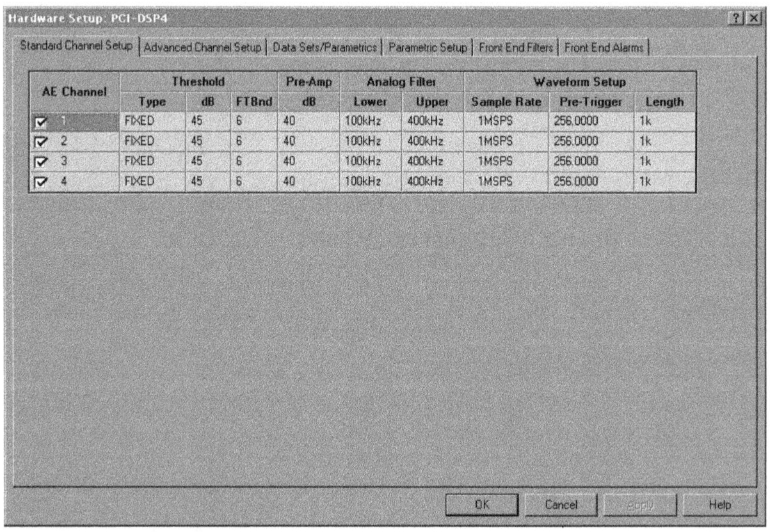

Figure 12. Standard Channel Setup

In the 'Standard Channel Setup' page you can enable or disable each channel and edit their common settings. If your system has the waveform option installed you can also edit each channel's waveform settings.

<u>AE Channel</u> – You can enable a channel by clicking its checkbox.

Threshold – 'Fixed' threshold is used when you want the threshold to remain unchanged at the 'dB' level you set (in the dB column). 'Floating' threshold is used under conditions of high, varying background noise. When floating threshold is selected, the value shown in the 'dB' column represents the threshold level at the start of the acquisition, the threshold will vary to 6dB (or FTBnd value on systems that have this column) above the current ASL reading for that channel. The threshold will not go below the initial value specified in the 'dB' column. Threshold is the prime variable that controls channel sensitivity.
Gain (dB) [ICC/DSP4, SAMOS, PCI-2] – Enter the hardware gain here.
Pre-Amp (dB) – Enter the pre-amp gain of your sensor's pre-amplifier here.
Analog Filter [not available for ICC] – The Lower and Upper drop-down list boxes allow you to select from the available analog filter values for low pass and high pass filters on each active channel. This is done by simply selecting the desired value from the drop down list boxes. Note that if waveform analysis is desired, these values should be kept below ½ of the Sample Rate. Make sure to raise your Sample Rate if this becomes an issue.
Sample Rate [only with waveform option] – This is the rate at which the data acquisition board samples waveforms on a per second basis. A sample rate of 1 MSPS (Mega Samples Per Second) means that one waveform sample is taken every μsec. A sample rate of 2 MSPS means that one waveform sample is taken every ½ μsec., etc. The sample rate can be selected by clicking on the down arrow on the drop down list box, then clicking the mouse on the desired sample rate.
Pre-Trigger [only with waveform option] – This value tells the software how long to record (in μsec.) before the trigger point (the point at which the threshold is exceeded). The user may enter a value in μsec in the pre-trigger edit box. The minimum allowable pre-trigger value is zero. The maximum allowable pre-trigger value is calculated by dividing the hit length by the sample rate in MHz. If, for example, the hit length was 1 k (1 k = 1024) and the sample rate was 4 MHz, then the maximum allowable pre-trigger value would be 1024/4 = 256 μsec.
Hit Length [only with waveform option] – This determines the size of a waveform message. Clicking the down arrow on the drop-down list box for Hit Length will display a list of allowable hit lengths from 1 k to 15 k. At a 4 MSPS sampling rate, a hit length of 1 k will allow up to 256 μsec. of data, a hit length of 2 k will allow 512 μsec of data (2 * 256), and so on.

Timing Parameters

In the 'Timing Parameters' page you can enable or disable each channel and edit Peak Definition Time (PDT), Hit Definition Time (HDT), and Hit Lockout Time (HLT).

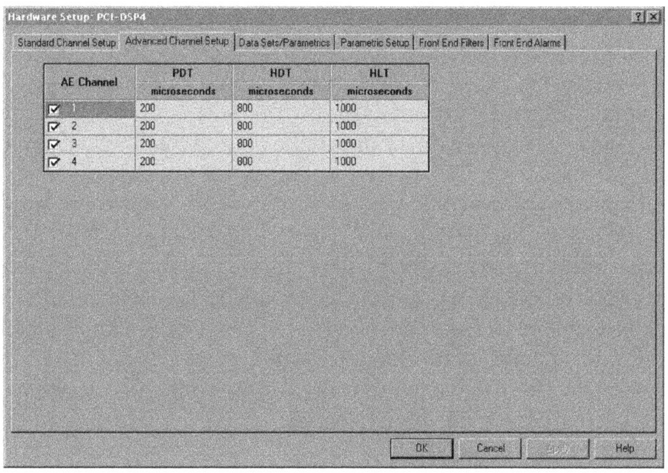

Figure 13. Timing Parameters

PDT, HDT and HLT are timing parameters of the signal measurement process. In brief, a proper setting of the PDT ensures correct identification of the signal peak for risetime and peak amplitude measurements. Proper setting of the HDT ensures that each AE signal from the structure is reported as one and only one hit. With proper setting of the HLT, spurious measurements during the signal decay are avoided and data acquisition speed can be increased. Recommended values for general-purpose testing are:

	PDT	HDT	HLT
Composites, Non-Metals	20-50	100-200	300
Small Metal Specimens	300	600	1000
Metal Structures (high damping)	300	600	1000
Metal Structures (low damping)	1000	2000	20000

Data Sets/Parametrics

HMSGHAT2000DDI features user-selectable data sets. This means that if you do not plan to use all the measurable AE parameters, you can leave some out and enjoy the compensating advantages of higher data throughput speed and more economical use of storage se. Selecting Data Sets/Parametrics from the Hardware Setup dialog box displays the Data Sets/Parametrics page that follows.
Note that the above figure is specific to the DiSP product. Your product may have more/fewer options available.

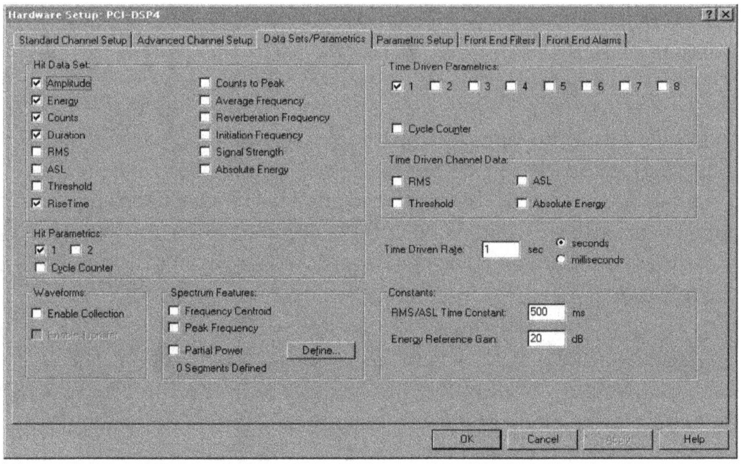

Figure 14. Data Sets/Parametrics

Hit Data Set

This box allows you to select which measured signal features are to be included in the reporting of each AE hit. The features to be included are so marked with a check in the checkbox next to the feature. Features can be included/excluded from the data set by selecting or deselecting the checkbox. "Time of test" is not shown because it is always included in the data set. A short description each hit feature in the Hit Data Set is as follows:

1) Amplitude - highest voltage in the AE waveform, expressed on the dB_{AE} amplitude scale.
2) Energy - time integral of the absolute signal voltage. The reported magnitude, depends on the value selected for Energy Reference Gain. Proportional to Signal Strength.
3) Counts - number of times the signal crosses the detection threshold.
4) Duration - time from first to last threshold crossing (µs).
5) RMS - root mean square voltage during the timeframe of the hit, referred to the input to the signal processing board.
6) ASL - RMS, converted to the dB_{AE} scale ($0dB_{AE} = 1\mu V$ at the sensor, before any amplification).
7) Threshold - detection threshold, on the dB_{AE} scale.
8) Rise Time - time from first threshold crossing to highest voltage point on the waveform (µs).
9) Counts to Peak - number of threshold crossings from first to highest voltage point on the waveform.
10) Average Frequency - Counts divided by Duration, divided by 1000 (thus, kHz). Note that this is not a spectral domain calculation, but a calculation from time domain features.
11) Reverberation Frequency - (Counts - Counts to Peak) divided by (Duration - Risetime). .
12) Initiation Frequency - Counts to Peak divided by Risetime.
13) Signal Strength - time integral of the absolute signal voltage, expressed in pVs (picovolt-seconds) referenced to the sensor, before any amplification. Proportional to Energy.
14) Absolute Energy - time integral of the square of the signal voltage at the sensor before any amplification, divided by a 10kΩ impedance and expressed in aJ (attojoules).

Hit Parametrics

This box allows you to select which parametric inputs are to be included in the description of each hit. A maximum of two parametric inputs (e.g., external load signals), or one parametric input plus the cycle counter, can be included in the Hit data set. These items, however, must also be included in the Time Driven Data Set, otherwise an error message will appear.

Waveforms
This box is available if you have the waveform option installed.
Enable Collection – This checkbox is used to signal the data acquisition board(s) to acquire waveforms. If this checkbox is unchecked then the board(s) will not be able to collect waveforms for the duration of the test.

Enable Transfer – This checkbox is used to signal the data acquisition board(s) to transfer collected waveforms to the software for processing/display and (if auto-dump is enabled) saving to a data file. If this checkbox is unchecked then the waveforms will not be visible to the user and won't be saved in a data file until the checkbox is turned on. Unlike Enable Collection this checkbox can be toggled on/off at any time during a test. When Enable Transfer is OFF the system can experience a boost in performance especially at high data rates if you are saving to a data file. Turning Enable Collection OFF can grant a greater performance boost at very high data rates as the system concentrates on hit-based data.

Spectrum Features

This box is available if you have the waveform option installed. Spectrum Features are derived from the power spectrum of the waveform associated with a hit:

4) Frequency centroid - the center of mass of the power spectrum graph.
5) Peak frequency - the point where the power spectrum is greatest.
6) Partial Powers - see 'Partial Power Features' below.

You must check Enable Collection in the Waveforms box for these features to be usable. Transfer of waveforms can still be set off so that the system can record spectrum features without recording waveforms.

Partial Power – Partial Power is a feature which is derived from the power spectrum of the waveform associated with a hit. It is a percentage and is calculated by summing the power spectrum in a specified range of frequencies (up to four of these ranges can be specified), dividing it by the total power in another range of frequencies, and multiplying the result by 100.

To use partial power you must

3) Check the partial power checkbox.
4) Define at least one partial power segment. To do this press 'Define'. This brings up the Spectrum Segment Setup dialog box described below.

Spectrum Segment Setup is used to define the partial power features. For each of the 4 definable segments the dialog box has a checkbox to enable the segment, an edit box for defining the start frequency and an edit box for defining the ending frequency.

The group box titled 'Total Power Range/Equal Sing Setup' contains two edit boxes. These edit boxes serve a dual purpose. The main reason for them is to define a range of frequencies over which the power spectrum will be summed.

The result of the summation is used in the calculation of total power for the partial powers calculation. Their other purpose is to allow the user to spread the frequency range evenly among the enabled segments. The sing will occur when the user presses the 'Apply Sing' button which is located to the right of the edit boxes.

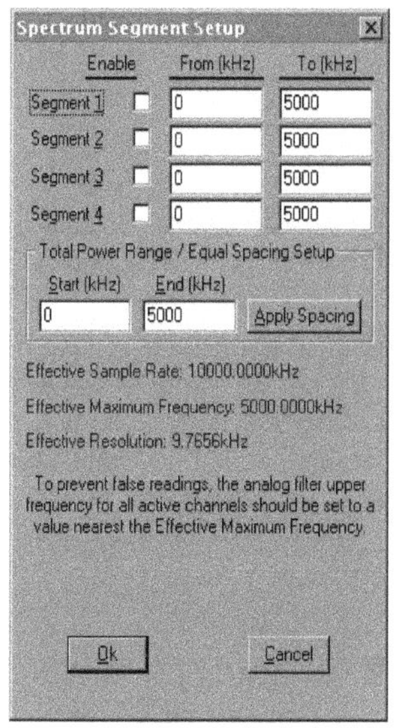

Figure 15. Spectrum Segment Setup

CAUTION: If the edit boxes are used to set up equal sing, always make sure the edit boxes contain the Total Power range values when the dialog box 'OK' button is pressed. Otherwise the partial powers will be based on the equal sing range.

The dialog box also displays information based on the sample rate and hit length of the first active AE channel:

4) Effective Sample Rate – The sample rate of the decimated waveform (the FFT calculation is always for 1k of samples). For a 1k hit length the effective sample rate is the same as the hardware sample rate. For a 2k hit length it would be half the hardware sample rate. Etc.
5) Effective Maximum Frequency – This is one half that of the Effective Sample Rate and is the maximum frequency which can be defined for any segment.
6) Effective Resolution - This is the Effective Maximum Frequency / 512 and is equivalent to the frequency range of a single FFT bin.

The system does not automatically set the input filters based on the sample rate to reduce aliasing, so the dialog box will remind the user to set the input filter to a value closest to the Effective Maximum Frequency.

The Spectrum Segments are global for all active channels, if the sample rate and/or hit length is not the same for all active channels a warning will be displayed to alert the user.

Time Driven Parametrics and **Time Driven Channel Data**

Additional to the hit descriptions, HMSGHAT2000DDI allows the recording of RMS, ASL, Threshold, Absolute Energy, Cycle Counter and Parametric inputs, at regular time intervals even in the absence of AE activity. These are called Time Driven messages. This box allows you to select a set of features for them.

Time Driven Rate – This controls the frequency of the readings and the data to be included in the records. Time driven data is useful for leak monitoring and for recording changes in load and background noise levels. Also, time driven data can be used for updating hit data graphs, involving parametric values or time, even when no hits are coming in. The interval you set must lie in the range 10 msec to 1,800 sec and must be a multiple of 10 msec. Typical values are in the range 1000-60,000 msec (1–60 seconds). For frequent parametric value updates, you may like to choose a short interval, but this leads to lengthy disk files and data listings, so it is not necessarily the best choice. When the time driven data set is used to update graphs, it is a good idea to set the rate less than or equal to the 'Seconds Between Updates' field selected in the Display Mode dialog box.

Constants

RMS/ASL Time Constant [not in DSP4/ICC] – The time constant can be set from 10ms to 1000ms in steps of 10ms. Previous versions of the program had a fixed time constant of 500ms.

Energy Reference Gain [PCI-DSP4 and AEDSP only] – In some instruments, such as SPARTAN, Energy is a gain dependent quantity. However, some systems (like DiSP and) don't have a gain setting so the Energy was designed to give a reading as if 20dB of gain were applied. The Energy Reference Gain allows you to match the value of Energy reported by system to that of a SPARTAN-2000 or other instrument which has programmable gain. The value of the Energy Reference Gain has a valid range of 0 to 60 dB.

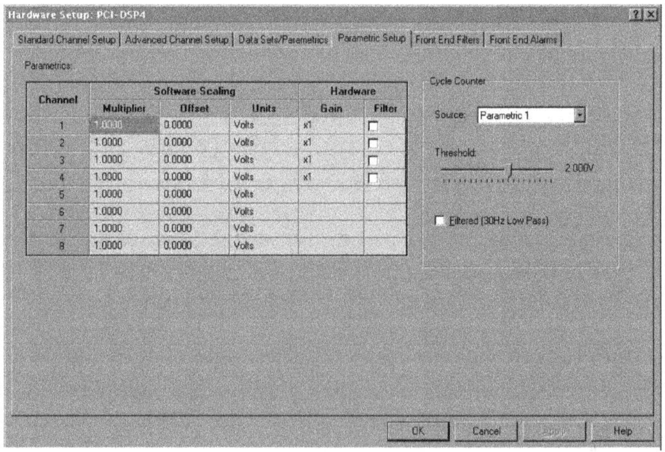

Figure 16. Parametric Setup

Parametric Setup

The Parametric Setup page allows you to scale the voltages measured at the parametric input, such that the system displays the corresponding load or pressure values. This page also allows you to tell the system what units (kN, psi) you want displayed on your graph axis and printouts, and it allows you to select from four available filters on gain.

The Parametric Scaling process used by HMSGHAT2000DDI is: Displayed Value = (Measured Voltage x Multiplier) + Offset

The default values set up in the delivered Autoload file are 1.00, 0.00 and "Volts" so that the system will display the raw voltage delivered to the Parametric Input socket on the rear panel.

To scale the output of a linear load-measuring transducer, the easiest way to find the correct Multiplier and Offset is to go into a simulated test and turn on the Line Listing Display (F7). From the display read the parametric voltages for two widely separated loads. Suppose these readings are V1 at load L1, and V2 at load L2. The correct values for Multiplier and Offset are then:

Multiplier = (L1 - L2) / (V1 -V2)
Offset = (V1*L2 - V2*L1) / (V1 - V2) = L1 - Multiplier * V1

Once you click OK to exit this dialog the scaled values of the parametric input will now appear on all graph displays and data listings. Minimum and maximum values for graph axes should be specified in the scaled values, and the entered unit will appear on the graph (e.g., 0–1000 psi). The raw data on disk will contain the original voltage readings and the scaling will be done again each time the file is replayed.

The Hardware section allows you to select from one of four available values for a hardware gain filter. Most cards (all except ICC) contain a programmable amplifier as well as a 30 Hz Low Pass filter for better resolution and less noise in the data collection process.

The **Cycle Counter** section allows you to control the source and threshold of the cycle counter [PCI DSP-4, AEDSP only].

- The Source can be selected as parametric 1-4 (on most data acquisition boards).
- The Threshold slide control allows you to set the threshold for the cycle counter. The cycle counter counts every time a signal crosses the threshold. The threshold can be set within a -10.00 volts to +10.00 volts range. This voltage should be set based on the load signal coming into the system. It should be set low enough to assure constant, accurate triggers.
- The Filtered checkbox allows you to enable an independent 30 Hz low pass filter on cycle counter.

Front End Filters

Selecting this action allows you to set up the "front end filters." These filters apply accept/reject criteria to the hit descriptions as the data is processed in the data acquisition board, so that rejected hits or waves do not even reach the main processor board.

There are two kinds of front-end filters: **AE Hit Filters** reject hits and any waveforms associated with them based on the criteria you set. **Waveform filters** work the same way but reject only the waveforms, allowing the hits to remain. The setup is identical for both.

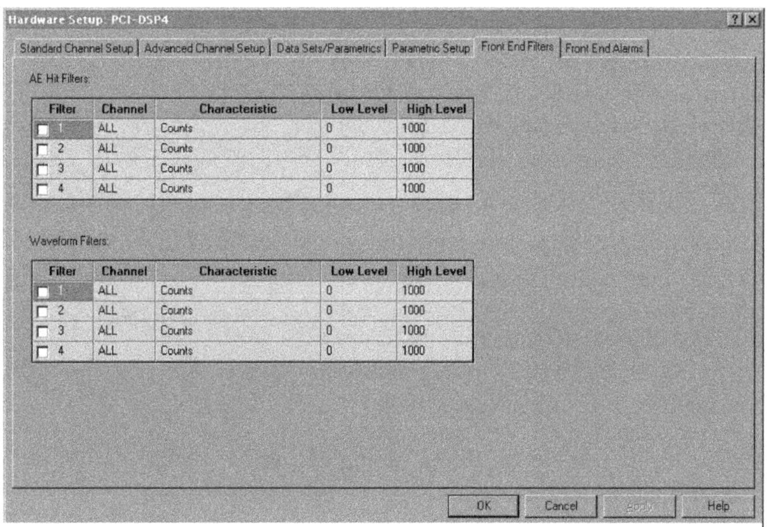

Figure 17 Front End Filters

There are four filters available at the present time. All four filters will always be displayed. To activate a particular filter simply check the 'Filter' checkbox associated with the filter you want to use.

Use the Channel drop-down list box to select which channel (or all channels) to filter on.

Use the Characteristic drop-down list box to select an AE Characteristic to filter on. Each hit will be individually screened to see whether the measured value of the specified characteristic falls in the acceptance range from LOW LEVEL to HIGH LEVEL, inclusive. For the message(s) (hit and/or wave) to be accepted, the hit must fall in the acceptance range for all active filters affecting that channel. Change the filter LOW and HIGH levels by entering new values in the text boxes. If a hit is not within the acceptance range, the message(s) will be eliminated from the front end signal processing.

Front End Alarms

There are four alarms defined in the alarms page. This page is almost identical to the front-end filter page with the exception that the user may select the type of alarm to be Single Channel Single Hit or Single Channel Cumulative. An alarm is generated if one or more of the alarm criteria is exceeded. Upon alarm, the PC computer speaker will sound, and an alarm output on the I/O connector will indicate an alarm. In addition, an alarm popup box will appear on the screen indicating if the alarm is a warning or a trigger, which alarm was exceeded and the value that triggered the alarm. To reset the alarm click the 'Clear All Alarms' button on the alarm popup box. See section 3.3.3.6 entitled 'Display Last Alarm' for more information.

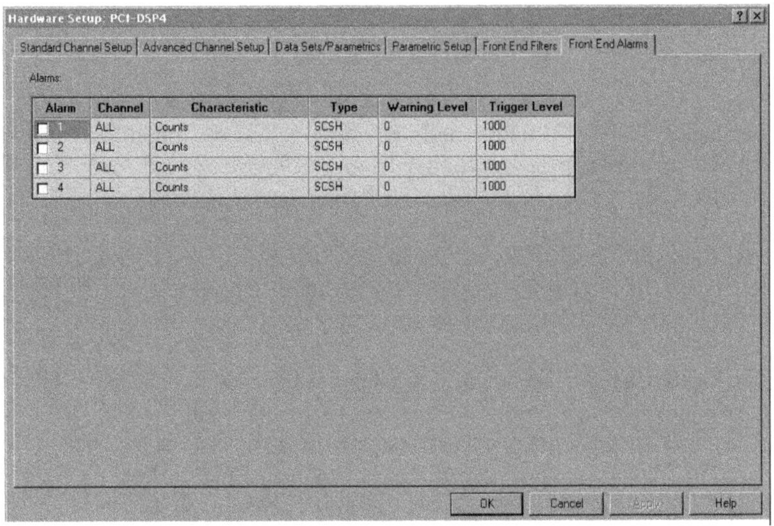

Figure 18. Front End Alarms

DeltaT Front End Filters (only)

Delta-T Front End Filters are used to eliminate "on board" AE hits which are part of the same event but outside the acceptance range selected in the DeltaT Filters Setup page.

A DeltaT Filter is located within each (AEDSP) board, so that this filter can only eliminate "on board" AE hits. This type of filter is very useful for cases where linear location is used. It will successfully filter hits that come from the outer areas of the sensors and beyond. Delta-T Front end Filters are very useful in 2 channel systems.

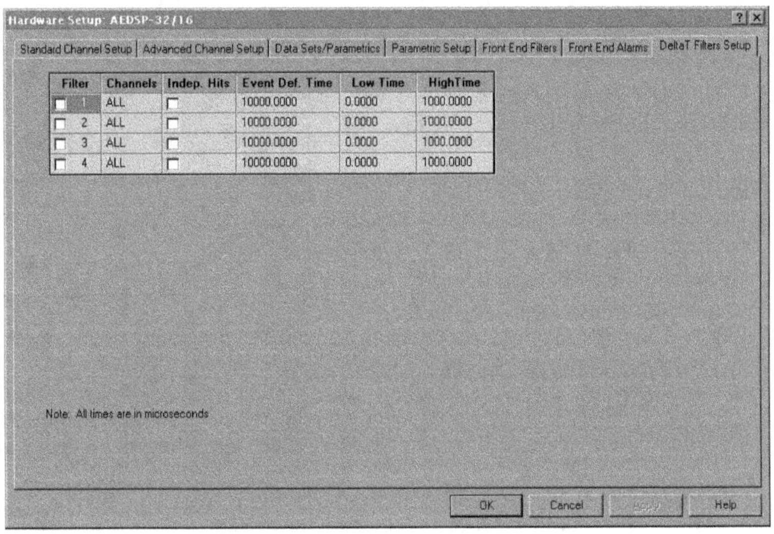

Figure 19. DeltaT Front End Filters

In cases where more than two channels are needed (such as planar location) a front end filter cannot be used since the hit information is not available from other boards. For this situation a "high end" filter is needed and Graph Filters can be used for this purpose. See the section on Graph Filters for more information.

The DeltaT Filters page is very similar to the Front-End Filters page discussed previously. Four different filters are available to allow programming of each board differently. Alternatively, the first filter can be chosen as an "ALL," which will have all boards filter using the same set-up parameters.

Each menu item and its function is discussed below.

Filter: This toggles between On and Off just like the Front End Filter page does.

Channel: This selects the channel numbers for the DeltaT filter. Because the DeltaT Filters operate on a per board basis, *pairs* of channels must be selected (e.g., 1-2, 3-4...) or ALL channels may be selected.

Indep. Hits: Allows independent hits (those hits not associated with an Event) to be accepted or rejected for transmission to the high-end, at the discretion of the user.

Event Def. Time: This is the maximum event time (in μsec) required in order for the AEDSP-32/16 to identify paired hits on the board as events. This should be set at the maximum time expected for an AE signal to propagate from the first to the second sensor plus 10% over range.

Low Time: This is the lower limit of acceptable Delta-T values in μsec. Below this is not an acceptable difference and the hits will be discarded. Negative numbers can be used. For delta-T filters, a hit occurring closer to the even channel is assigned a negative delta-T.

High Time: This is the higher limit of acceptable Delta-T values in μsec. Above this is not an acceptable difference and the hits will be discarded. As above, negative numbers indicate that the hit occurred closer to the even channel

The information set up in the DeltaT Filter page is downloaded into each specified AEDSP board with the rest of the set-up information. Upon acquisition, each AEDSP board has software event detection and filter routines implemented as follows:

As hits are received from both channels, the board has an event detection queue whereby it checks hit pairs (hits on opposite channels) against the Event Definition Time to determine if it is an event. To provide for asymmetric acceptance ranges between two sensors, the delta-T is always calculated from the lower numbered sensor to the higher numbered sensor. Hits whose delta-Ts are longer than the Event Definition time are considered independent hits. In this case, the first hit will pass to the high end as an independent hit.

In those cases where an event is detected (hits that have a delta-T within the Event Definition Time) its Delta-T is checked against the high and low levels for that board. If the Delta-T is outside of the range set in the "Low Level" and "High Level" fields (less than the Low Level and higher than the High Level), then both hits will be rejected and neither hit nor waveform data will be passed to the high end. If the Delta-T is within the "Low Level" and "High Level" values then the hits (and waveforms) are allowed to pass to the high end.

Delta-T filters are activated independently of any location processing. This allows use of Delta-T filters in tests with extremely high data rates, whereby the data is stored and analyzed later with the slower location algorithm.

PCI-2 Waveform Streaming (PCI-2 Only)

Background

The PCI-2 AE System on a Card is capable of streaming waveform data continuously into physical memory on the PC. It uses an onboard DMA controller to read samples from each channel's 4K FIFO and transfers them into a circular buffer in the computer's memory. When a stream to disk event is detected (this being a threshold crossing, an external digital trigger, a manual trigger, or a set periodic interval), the waveform is pulled out of this buffer and written to the output data file. This data file (with a WFS extension) is created separately from the normal acquisition data file (with a DTA extension) used to store collected AE data. Once the output file is created it can be viewed and analyzed with the WaveViewer program supplied on the installation disk.

Waveform Streaming Setup

Waveform streaming settings are located under the Hardware menu item under the Acquisition Setup menu (Hardware Setup Figure 20) or by pressing the F2 accelerator key on the keyboard. This will display the Hardware Settings dialog box. The waveform streaming setup is accessed by clicking on the Waveform Streaming tab near the top of the Hardware Setup dialog box.

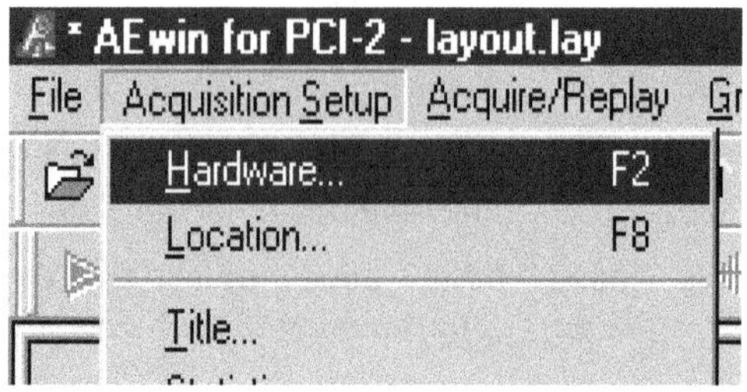

Figure 20. – Hardware Setup

Per Channel Waveform Streaming Settings

AE Channel for Streaming:
Selects a channel for streaming. When streaming is triggered, data from all selected channels will be stored to the file. Click the box to toggle this setting. A checked box indicates that a channel is enabled for streaming. A channel must be enabled for AE data collection to be able to stream. The program will enforce this by automatically enabling any channel that is enabled for streaming. If it is undesirable to have AE data from a channel that is streaming, go to the Standard Channel Setup and set the channel threshold type to 'Fixed' and raise the threshold level to its highest value (typically 99 dB).

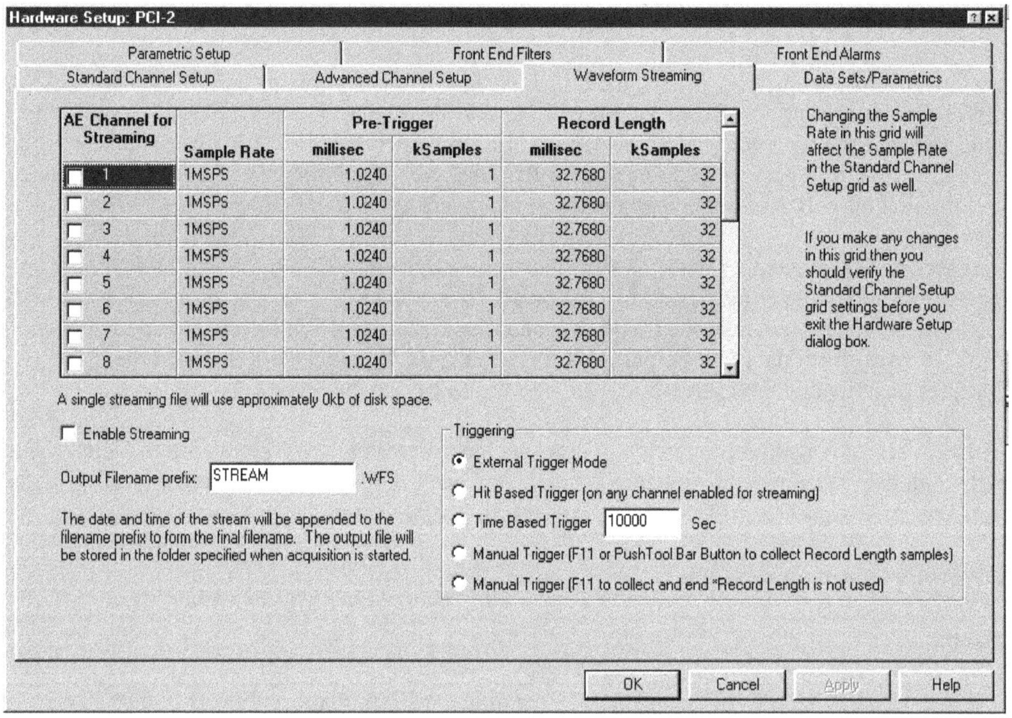

Figure 21. – Hardware Setup Dialog, the Waveform Streaming Setup Section

Sample Rate:
Selects the sample rate. This setting is the same as the one in the Standard Channel Setup. If it is changed here, it will automatically be changed in the Standard Channel Setup (and vice versa) because AE and streamed waveforms Are always collected at the same sample rate. Channels on the same board must have the same sampling rate (channel 1 = channel 2, 3=4…) but different rates are allowed between boards.
Also, the system cannot transfer more than 10MSPS total. This means 1 channel can be enabled at 10MSPS or 2 channels at 5MSPS… Always try to keep the total MSPS for all enabled channels as low as is possible. Changing the sample rate also changes the time value for pre-trigger and record length.

Pre-Trigger:
This is the amount of data that occurred prior to the actual trigger of the streaming that should be saved to the file. It cannot exceed either 32k samples or 4 seconds whichever is smaller. The pre-trigger amount cannot exceed the record length, because the record length includes the amount of pre-trigger. The value can be entered as either kilo-samples or milliseconds. When one value is entered into its field, the program will automatically calculate the other value. The program always stores the value as the number of kilo-samples, therefore it may not be possible to obtain an exact value in milliseconds.

Record Length:
This is the total amount of data that will be streamed once the stream is triggered. The value can be entered as either kilo-samples or milliseconds. When one value is entered into its field, the program will automatically calculate the other value. The program always stores the value as the number of kilo-samples, therefore it may not be possible to obtain an exact value in milliseconds.

Waveform Streaming Trigger Modes

Each time a valid trigger is detected, a new file is opened for streaming. Any trigger condition that is met while a stream is in process is considered to be invalid and is ignored.

External Trigger:
In this mode the streaming is triggered when an external pulse is applied to Pin 12 of J6 (digital input #2) of the PCI-2 board that contains channels 1 and 2. The pulse should be TTL level, normally low. It should go to a high state for at least 1 microsecond before returning to low. Theoretically it is possible to trigger this input with a pulse that is only 50 nanoseconds wide. Precautions should be taken to make sure that the input signal is clean or false triggers could occur. The signal must return to a low state for at least 10 microseconds before returning high to trigger another stream. If a stream is in process when the trigger input goes from low to high, the transition will be ignored and will not affect the current operation. The input must go from a low state to a high state while streaming is idle in order to trigger any streaming.

Hit Based Trigger:
In this mode the streaming is triggered whenever an AE hit is detected on any of the channels that are enabled for streaming. Additional hits that occur while streaming is in process do not affect the triggering.

Time Based Triggering:
In this mode streaming can be set to occur on a regular periodic basis. Make sure that the record length is less than the specified interval. Otherwise, for example, if the stream is to be 1.5 seconds long and the interval is set to every 1 second, the interval from 1.5 to 2.0 seconds will not be stored to a file. The first interval begins when acquisition is started.

Manual Trigger:
There are two types of manual triggers. Either will start when either the F11 key or the 'Trigger a Waveform Stream' acquisition toolbar button is pressed.
In the first case the stream will terminate when the number of samples specified in the record length field is reached.
In the second case either the F11 key or the 'Trigger a Waveform Stream' acquisition toolbar button needs to be pressed to terminate the stream.

General Waveform Streaming Settings

Enable Streaming:
This must be checked in order for streaming to be available during acquisition. This is a quick way to enable and disable the feature while still preserving all of the settings.

Output Filename Prefix:

This specifies the first part of the name of the output data file. It can be changed to any valid Windows filename characters. It is recommended that this be used to differentiate the data from different tests or experiments. The second part consists of the date and time that the stream was triggered.

Filename example: **`STREAM20040804-090744.WFS`**
`PREFIXYYYYMMDD-HHMMSS.WFS`

Here, STREAM is the default prefix. Following the prefix is the time and date of the stream. Four digits of year (Y), two digits of month (M), two digits of day (D), a hyphen, two digits hour (H, 0-23), two digits of minutes (M) and two digits of seconds (S). In the case of the example, the stream was started on Aug 4, 2004 at 9:07:44am.

Suggested Procedure for Configuring Waveform Streaming

- Select the channels for streaming from the first column by clicking on the check box.
- Select the sampling rate for each channel. Channels located on the same AE acquisition board must have the same sample rate. The program will enforce this.
- Select the pre-trigger. This can be a specified in milliseconds or K samples (1K samples = 1024 samples). It is easier to specify milliseconds, but if the sampling rate is subsequently changed, the milliseconds value will be recalculated and displayed.
- Select the recording length in milliseconds or K samples (1K samples = 1024 samples). There is no set limit on recording length other then the hard disk available. But the amount of physical memory on the computer does affect the ability to view the data using the WaveViewer program. For file lengths greater then 1.5Gbytes at least 256 Mbytes should be installed. As with pre-trigger, subsequent changes to the sampling rate will force the millisecond value to be recalculated and displayed.
- Select the Triggering Mode
- Specify the Output Filename Prefix
- Click the Enable Streaming Check Box. This box must be checked for streaming to be enabled.
- Press the 'Apply' button to verify the settings, fix any errors that show up.
- Go back to the Standards Channel Setup and verify that the waveform settings are acceptable.

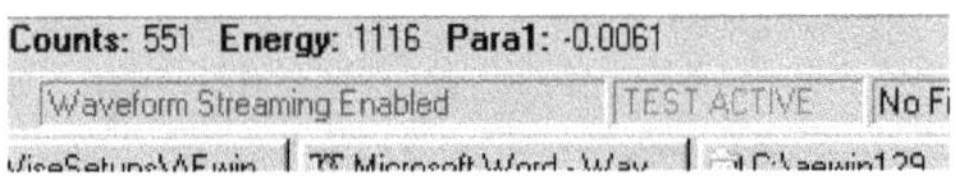

Figure 22. – Status bar indicating that streaming is enabled

After configuring all options the system can be put into acquire mode to begin data collection. If waveform streaming is enabled, the status bar at the bottom of the HMSGHAT2000DDI window will display "Waveform Streaming Enabled".
When a waveform save is triggered the text in the status bar will indicate it.

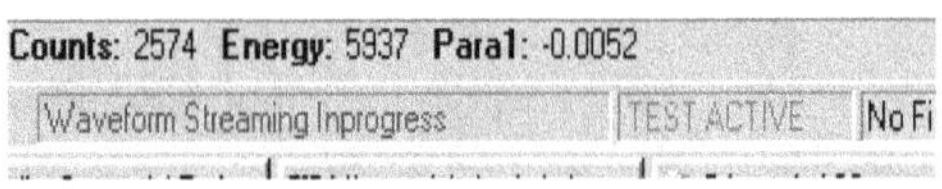

Figure 23. – Status bar indicating that streaming is in progress

When the waveform save is complete the text in the status bar indicate it.

This message will persist for 4 seconds before reverting to "Waveform Streaming Enabled" text.

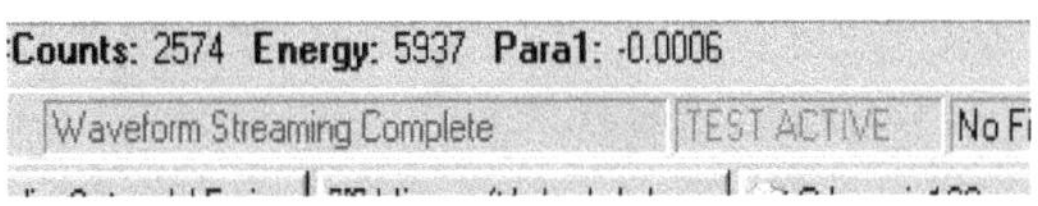

Figure 24. – Status bar indicating that streaming is complete

The file is placed in the last folder that was used to save a DTA file. If an acquisition is done and Autodump is enabled (producing a DTA file), the streaming data file will be created in the same folder. If Autodump is off, the streaming data will be placed in the last folder that was used for AE data. The streaming file is always created and there is no way to disable it. The destination folder for the streamed data cannot be specified without also saving AE data.

Viewing a Streamed File

Once a WFS file is created it can be viewed using the WaveViewer software kage. This is included on the PCI-2 installation CD. Please see the WaveViewer User's Guide for complete instructions on using the software and viewing waveforms.

Waveform Streaming Performance and Trouble Shooting

The rate at which waveforms can be saved to a data file is a function of many different variables. The card is capable of streaming up to 20 Mbytes per second. This would correspond to 10M samples per second to memory. Unfortunately the path from memory to the data file is dependent on hard disk performance and CPU utilization, so it is not possible for us to say what the highest achievable rate will be. The DMA controller requires physical non-paged memory in order to operate. If the sampling rate of the PCI-2 card is set to 100Ksamples per second and the duration of the waveform is greater than 4 seconds then the amount of memory required is (100Ksamples)*(2 bytes)*(4 seconds)*(2 channels) = 1600K. If the sampling rate of the card is set to 5M samples per second, then the physical non-paged memory requirement increases to (5Msamples)*(2 bytes)*(4 seconds)*(2 channels) = 80,000,000 bytes. So there must be enough physical memory to map the buffer. The normal AE hit and waveform processing of the card also required physical memory as well. If HMSGHAT2000DDI locks too much physical memory, then overall OS performance will decrease. Because of these difficulties HMSGHAT2000DDI will not allocate more than 20% of physical memory for its use. If there is insufficient physical memory available on your system to allocate the waveform DMA buffers the following error message will be displayed during waveform streaming setup:

Insufficient Physical Memory, Reduce Recording Length

If this happens, you can reduce the sampling rate or the waveform length until there is enough physical memory for the buffers. Another option is to increase the physical memory installed in the system.

Figure 25. – Buffer overrun error

For long duration waveform saves the possibility exists that the CPU will not be able to write the waveform data to the output file at an equal or greater rate than the waveform is being read from the PCI-2 card. If this happens then the buffer overrun error message will appear.

There are several solutions to this problem. First is to reduce the CPU utilization by turning off non-essential items. Disable the line listing display. Reduce the number of plots. Disable AE hit waveforms. Don't save AE data to disk. Other options include reducing the sampling rate of the channels or the number of channels being written to the data file. Upgrade you hard disk or add more physical memory.

Channel Group Setup... (for Non-HMSGHAT2000DDI Products)

The 'channel group setup...' dialog box is used to define and edit channel groups. A 'channel group' is simply a named list of channels. You can select this group in graph setup instead of selecting channels individually. The names of any channel groups selected for a graph are displayed on the graph's title for easy identification.

To create a group, click the Add Group button. This brings up a dialog box shown at right. From here you can name the new group and select its member channels.

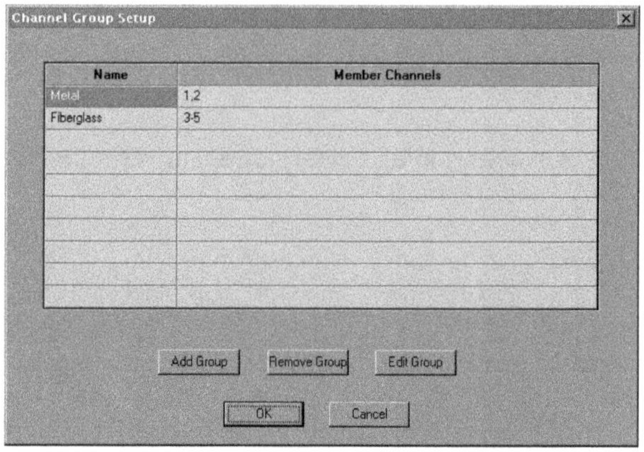

Figure 26. Channel ***Group*** *Setup*

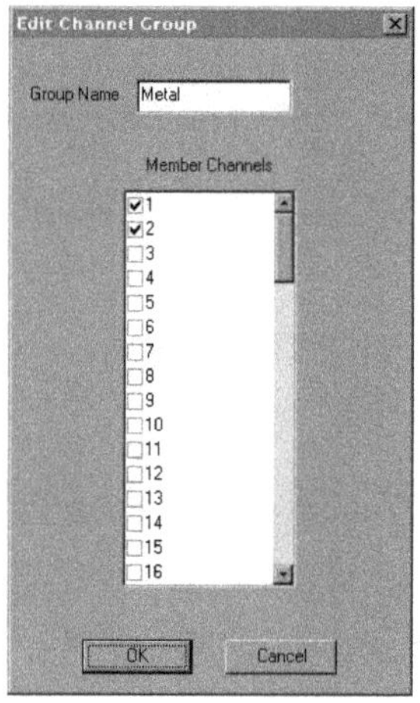

Figure 27. Edit Group

To modify an existing group, highlight it and click the Edit Group button. This brings up the same dialog box. To delete an existing group, highlight it and click the Remove Group button. Note that when you remove a channel group it is automatically removed from any graphs it was selected in.

Location...

The Location Setup dialog box is used to define and enable event groups (location groups). Use of Location involves more than just this dialog, however. For a full description of the Location dialog as well as other Location details see section 4.

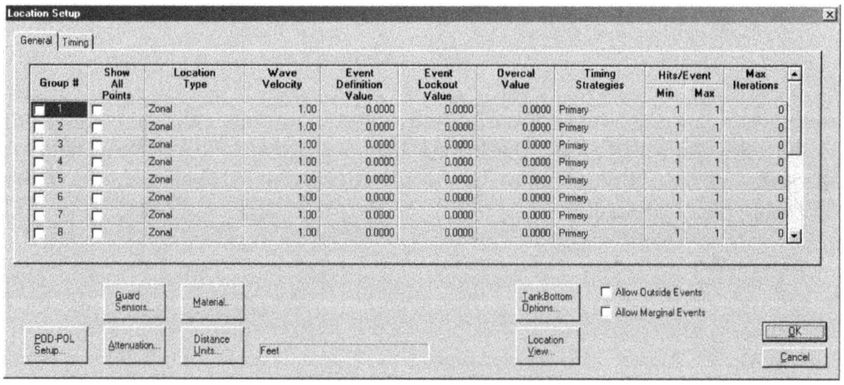

Figure 28. ***Location*** *Setup*

Title…

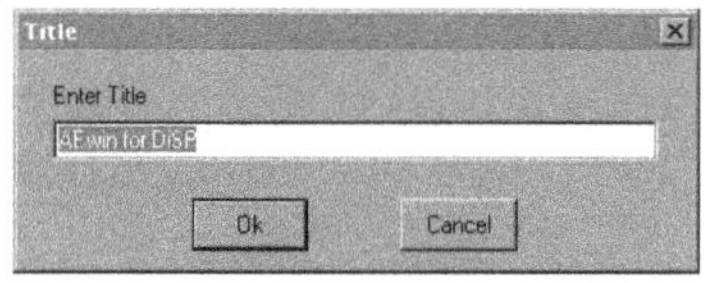

Figure 29. Title

'Title…' is used to set the Test Title. It can be displayed on the statistics bar using the Statistics menu described below.

Statistics…

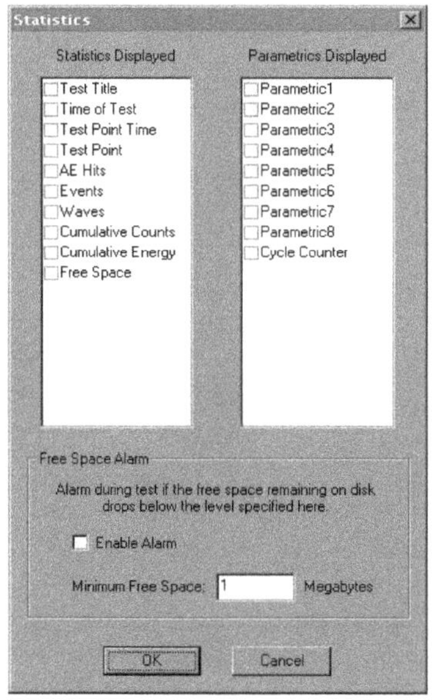

Figure 30. Statistics

'Statistics…' is used to toggle the various test statistics on or off, as well as to customize the Free Se Alarm. Any number of statistics can be on at any one time.

- Test Title: Displays the test title set in the 'Title' dialog box.
- Time of Test: Displays the elapsed time since the start of the test or replay.
- Test Point Time: Displays the elapsed time since the last test point (user-requested time mark).
- AE Hits: Displays the total number of AE hits recorded so far during the current test or replay.
- Events: Displays the total number of located events recorded so far during the current test or replay.
- Waves: Displays the total number of waveforms recorded so far during the current test or replay.
- Test Point: Displays the total number of test points (user-requested time marks) recorded so far during the current test or replay.
- Cumulative Counts: Displays the total number of counts recorded so far during the current test or replay.
- Cumulative Energy: Displays the total number of energy counts recorded so far during the current test or replay.
- Free Se: Displays the current amount of free se left on the disk on which the data file you are replaying or acquiring.
- Parametrics: Displays the value of the specified parametric from the most recent hit or time data message recorded or replayed during the current test or replay.
- Cycle Counter: Displays the value of the cycle counter from the most recent hit or time data message recorded or replayed during the current test or replay.

Free Se Alarm

The 'Free Se Alarm' is designed to warn the user in the event that the amount of free disk se remaining has run low during a test. To enable this alarm, check the 'Enable Alarm' checkbox and set a minimum threshold (in Megabytes), below which the alarm will sound.

Logistics Information…

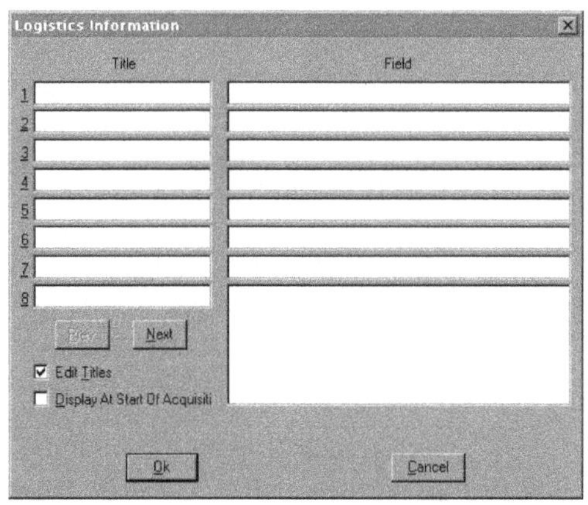

Figure 31. Logistics Information

The 'Logistics Information' dialog box is used to store information about a test. The Title column contains eight editable fields to be used as descriptions of the information that is to be entered by the tester into the field to the right of the title field (Field column). Examples of useful titles would be date of test, location of test, tester name(s), item being tested, identification or serial number, weather condition, etc.

The Field column also contains eight editable fields that may be filled in now, or at test time, with information about the test to be performed. This information can then be preserved in the Layout and Data files.
In addition to these two columns of fields, there are several check boxes and buttons located at the bottom of the screen:

Edit Titles - When this item is checked, the information in the Title column fields can be accessed and edited. When this item is not checked, the information in the Title column fields will be for display purposes only, therefore, all text in the Title column will be grayed out and inaccessible.

Display at Start of Test - Checking this box will cause the logistics information dialog box to automatically be opened after initiating a new test. When this convenient option is checked, you will not have to remember to update the logistics information each time a test is begun, because the program will automatically prompt you for logistics information each time a new test is begun.

Audio Monitor...

The audio monitor board contains the (audible) alarm and audio controls. The circuitry is designed to produce audible tones from high frequency acoustic emission signals. This circuit can be selected to operate in either a homodyne or heterodyne mode. Operation in heterodyne mode allows the user to select the frequency of the AE signal to be monitored. Homodyne operation does not require frequency tuning and allows audio monitoring of AE signals at any frequency.

To activate the dialog box for setup of the Audio Monitor board select 'Acquisition Setup' then 'Audio Monitor...'.

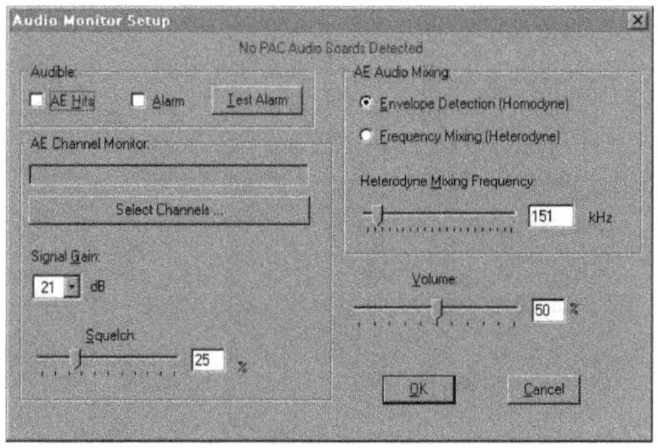

Figure 32. Audio Monitor Setup

In the group titled 'Audible' there are two checkboxes. The first one (AE) enables the board to emit a sound when AE hits are detected. The second one (Alarm) enables the board to produce a tone whenever a front end alarm warning or trip level is exceeded (these alarms are defined in the 'Acquisition Setup', 'Alarms' dialog box). By clicking on the 'Test Alarm' button you can hear how the alarm will sound. Click it a second time turn it back off.

The group titled 'AE Channel Monitor' contains controls for selecting monitored channels as well as setting gain and squelch. Click ‘Select Channels’ to choose which channels are to be used by the audio monitor. The 'Signal Gain' combo box contains a list of valid gain levels for the AE channels. This allows small amplitude AE signals to be monitored. 'Squelch' acts like a threshold circuit to eliminate the sound of background noise. Valid entries are from 0 to 100%. This value should be set to a level so that under conditions of no AE signals there is no audio output. Then as soon as a signal received is above the threshold as set by this control the audio output for that signal will be heard. The scroll bar located under the squelch control can be used as a mouse interface. Clicking on either arrow changes the squelch by 1%, clicking in the area between the thumb and an arrow changes it by 10%.

The group titled 'AE Audio Mixing' contains mutually exclusive checkboxes for selecting the method by which the AE tones are generated. When 'Envelope Detection (Homodyne)' mode is selected each hit produces a click sound. When 'Frequency Mixing (Heterodyne)' mode is selected each hit is mixed with a signal whose frequency is entered in the edit box below the mode selection controls. This limits the audio to hits whose frequency is 10kHz above the selected frequency. For example if the mixing frequency was set to 150kHz then AE signals greater than 150kHz and less than 160kHz will be heard.

The last control is for volume. It controls the volume of both the AE tones and the alarm tone. It ranges from 0 (off) to 100% (full volume).

The changes made to the dialog box will go into effect immediately. Pressing Cancel will restore the audio settings to those in effect before the dialog box was displayed.

Note that the AE detection by the audio board is independent of the threshold levels of each channel so it is possible to set the board up to hear AE which is below the threshold of hits detected during a test.

The top line of the dialog box will display Audio Board Version X.X if a audio board is detected. If one is not, it will show No Audio Boards Detected.

Auto File-Close…

In most tests all data from test-start to test-end is saved to a single data file. In long tests it is sometimes desirable to instead split the data into multiple files . When the Auto File-Close feature is active during acquisition the program can automatically stop the test and start a new test under user-defined conditions. The default result is a numbered set of data files. Optionally it can close the file and begin saving data to another file without stopping the test. The result in this case is a numbered set of 'continued' data files which can be replayed together as if they were one file.

The program will name new files by incrementing the name of the first file. (i.e.: if the name entered for the first file is test0001.dta, then subsequent files will be named test0002.dta, test0003.dta, etc.)

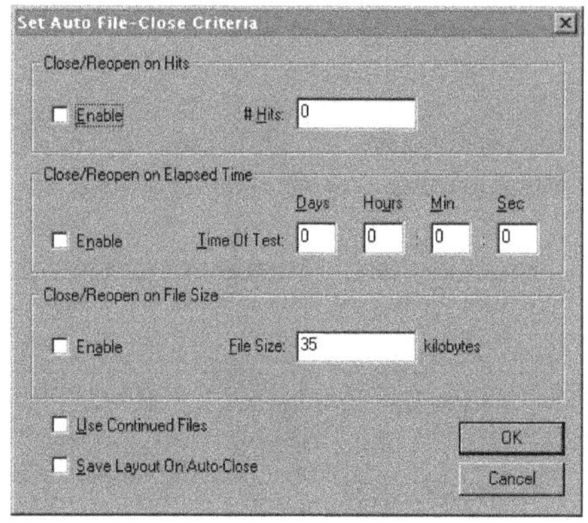

Figure 33. Auto File-Close

If checked the program will also re-save the current layout every time a file is automatically closed. This saves the name of the new file in the layout as the default data file name for future tests.

Enabling and Customizing Auto File-Close

You can enable the Auto File-Close feature and set the conditions under which it will work in the 'Set Auto File-Close Criteria' dialog box seen in the figure above. There are three conditions under which an Auto File-Close operation will take place:

d) #Hits: Whenever the total number of hits in the test equals the #Hits value or a multiple of it, an Auto File-Close operation is performed. (i.e.: If #Hits is 1000 then an Auto File-Close will happen at 1000 hits, 2000 hits, etc.)
e) Time of Test: Whenever the elapsed time equals the 'Time of Test' or a multiple of it, an Auto File-Close operation is performed. (i.e.: If 'Time of Test' is 1hr 10 min then an Auto File-Close will happen every 70 minutes.)
f) File Size: Whenever the current file size equals 'File Size' then an Auto File-Close operation is performed. (i.e.: If 'file size' is 64kb then you will end up with a string of files, each one approximately 64,000 bytes in size.). Note that file sizes will vary especially at high data rates.

You can enable any or all of these conditions.

Using Continued Data Files

If the 'Use Continued Files' box was checked the program will close the file in 'continued' form and open a new file, without stopping the test. The resulting set of files are referred to as a continued file set. They differ from normal data files as follows:

- Every file in the file set (except the last one) has an internal link to the next one. This means that when you replay one of these files it will look for the next file(s) and replay them also. (It can only find them if they are in the same folder.)

Replaying Continued Data Files

d) To replay the entire file set in order: Replay the 'first' file (the one with the name you gave it when you started the test) in the file set. The entire file set (if present in the same folder) will replay as if it were one file.
e) To replay the file set from a certain file onward: Replay the file from which you want to begin replay. The file set from that point on will replay as if it were one file.
f) To replay a single file only: Since the files are linked the program will always try to find and replay all files that come after the selected one in order. To replay only one file you must first copy it to another folder (or move/delete the others) then replay it as you would any other. Eventually it will reach the end of the file and

will warn you that it can't find the next file. If you want to eliminate this message from future replays of this file you must remove its link using the Time Order/Repair utility (see the next section for details).

Note that if the program encounters a 'gap' in the file set (i.e.: a missing file) it will display the "File Not Found" error described above. It can't replay beyond this point because the link is, in effect, broken. You can pick up where it left off by starting a new replay of the next existing file in the set.

Removing the Link From a Continued Data File

To remove the link from a continued data file (other than the last file which has no link) run the Time Order/Repair utility on it. This will allow you to replay this file without the "File Not Found" warning being displayed. Doing this will prevent any files 'after' this one from being replayed as part of the complete file set, so be sure to make a copy first. Time Order/Repair will ask you if you want to replace the continue message with an abort message. Click 'Yes' in response.

Auto Alarm Reset

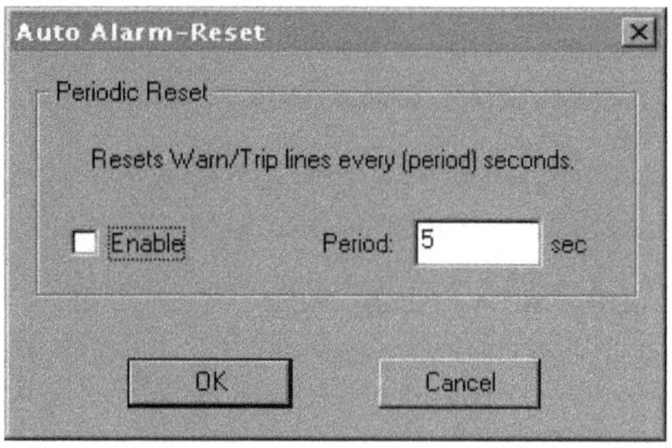

Figure 34. Auto Alarm-Reset

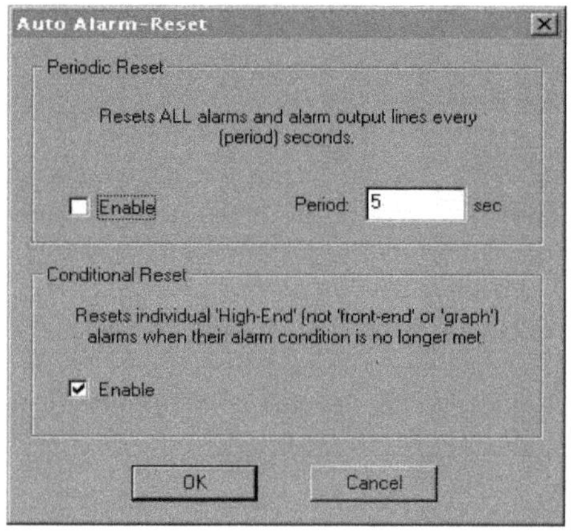

Figure 35. Auto Alarm-Reset

The Auto Alarm Reset dialog box is used to automatically reset alarms based on a user set criteria. This dialog box can be accessed through either the Acquisition Setup menu and the Alarm Status/Test dialog box. There are two types: Periodic and Conditional.

<u>Periodic:</u> If the Periodic Reset feature is enabled it will (during a test) reset ALL active alarm Warn/Trip lines every time a period of time elapses equal to that set in this dialog.

<u>Conditional:</u> The Conditional Reset feature is used in conjunction with 'High End' alarms. High-End alarms include the 'Free Se' alarm (set in the Statistics Setup dialog box), LeakTEC and On-Line Crack alarms (if this option is enabled). If this feature is enabled, a high-end alarm is automatically reset when the condition that caused its triggering no longer exists. i.e. : If the 'Free Se' alarm is triggered because the amount of free se on the disk dropped below 5 Megabytes and the user then deleted enough files to raise the free se remaining above 5 Megabytes the alarm would automatically reset.

Alarm Status/Test

The 'High End Alarm Status/Test' dialog box is used to view and alter the status of various alarms in HMSGHAT2000DDI. A table of alarm types is displayed either on the left or the top of the dialog box. Each cell represents one type of alarm (such as Free Se Low), or condition (such as 'Not in Acquisition'). If a cell is checked it means that that alarm is triggered or that condition is true. You can reset an alarm simply by unchecking it. Note that doing so will not prevent that alarm from triggering again in the future (or automatically clearing itself if Alarm-Reset is enabled).

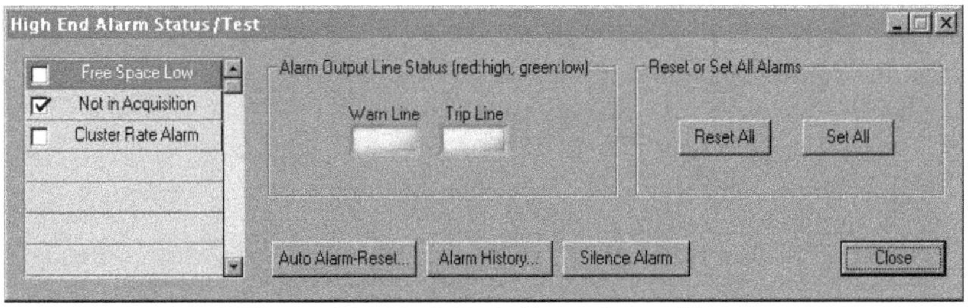

Figure 36. High End Alarm Status/Test

You can reset or set ALL possible alarms by clicking the appropriate button on the right side of the dialog box. In the center of the dialog box are colored indicators that display the status of the Alarm Output Lines (warn and trip). They are colored green if they are reset (low) or red if they are set (high). You can silence the audible 'beeping' sound without resetting the alarm(s) by clicking 'Silence Alarm'.

ASL Alarm...

ASL Alarms are used in conjunction with Internet Monitoring. Unlike Front-End or Graphical Alarms, ASL Alarms do not sound an audible alert or display a dialog box when triggered. Instead they save their alarm data in the STA files.

You can set alarm criteria on a channel by channel basis. Any channels not checked are not evaluated for alarms. The Low Level and High Level you set define the acceptable (inclusive) non-alarm range.

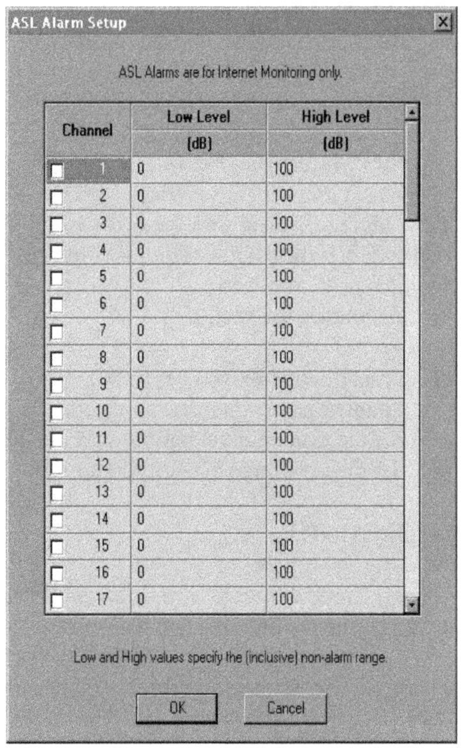

Figure 37. ASL Alarm Setup

Voltage Time Gate Setup...

Voltage Time gate is used for parametric based AE data filtering. It is controlled by parametric 1 and allows the user to eliminate the collection of all AE data that occurs when the parametric falls outside a pre-selected range of values and other criteria.

AE data collection is either Enabled or Disabled. It specifies what to do when the voltage-time gate is generated.

Each threshold is specified with a level in Volts and >= or <=, which indicates whether the signal has to be above or below the level in order to satisfy the threshold. Activation of each threshold is independent of the other; the user can activate either, both or no threshold at all.

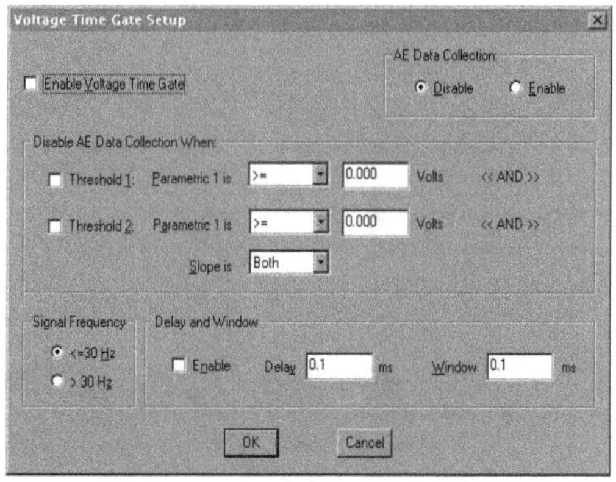

Figure 38. Voltage Time Gate Setup

The slope control can be POSITIVE, NEGATIVE or BOTH. It is satisfied if the parametric is rising, falling or at any time.

The slope and all active thresholds must be satisfied in order to generate a voltage-time gate. This allows the user to restrict the gate to crossings of thresholds with the parametric specifically changing a certain way (rising or falling).

Delay and Window specify an optional time window (the 'time' part of voltage-time gate). When enabled they cause the gate to be a time window of controlled duration, starting with some delay after the moment that the threshold and slope criteria are met.

Signal Frequency specifies a low-pass or high-pass filter that is applied before the parametric is fed into the voltage-time gate processing circuit. This provides the ability to respond to signals that change slowly (most cases) or quickly (fatigue cycles, etc). There are two settings, one for parametric signals below 30 Hz and one for above 30 Hz.

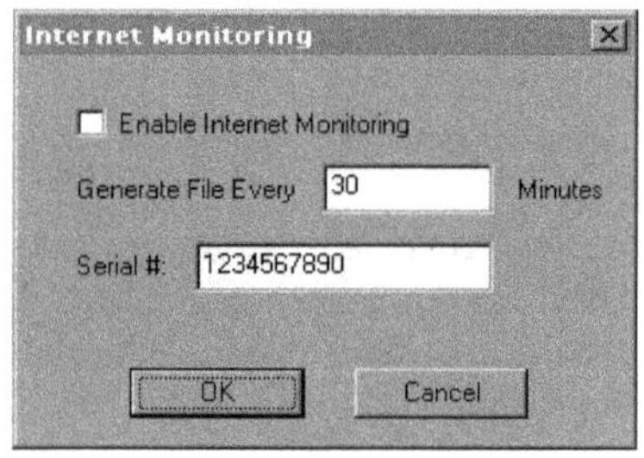

Figure 39. Internet Monitoring

Internet Monitoring Setup...

This dialog box allows you to setup the Internet Monitoring feature of HMSGHAT2000DDI. This is used in conjunction with other 'Remote Monitoring' applications purchased separately. Please refer to their documentation for details.

From this dialog box you can:

- Enable/disable internet monitoring.
- Set the time interval between file generations.
- Set the serial number for your unit.

Display Mode...

Display mode allows you to select the form of display for the AE data, as well as the frequency of graph updates, via the Acquisition/Replay Modes dialog box.

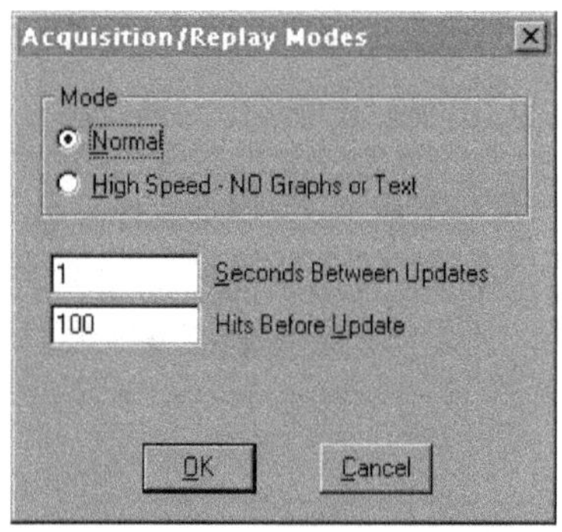

Figure 40. Display Mode

There are two Acquisition/Replay modes available in HMSGHAT2000DDI:

- Normal: This is the standard mode for acquisition and replay. Statistics, graphs, tables and the line display are all active in this mode.
- High Speed: This mode is used to achieve maximum data acquisition speed in tests where a very high emission rate is expected. Graphs, tables

and the line display are inactive in this mode. Only selected statistics will be displayed and updated; the priority task is transferring data to disk for analysis later.

Seconds Between Updates, Hits Before Updates

During data acquisition and replay, graphs and tables are redrawn (updated) at intervals determined by these settings. They are updated whenever the specified time has elapsed or the specified number of hits has occurred since the last update, whichever comes first. Depending on the anticipated data rate and test duration, you will want to select an update rate that is fast enough to keep you current without having the screen constantly redrawing. Type in the desired values in the text boxes provided. Recommended values for initial use are 3 seconds, 100 hits. Since screen updates affect overall data acquisition rate, in high data rate situations, the user may want to use larger intervals.

Line Display…

The Line Display shows a data listing in numeric format. This listing shows all the AE parameters, the parametric and other test messages as numbers in a row. There is one row for each message, plus a title row every time it is needed. The title row shows the user which parameters are being displayed and where they are being displayed. This depends on which active parameters were selected for display in the Line Display Setup dialog box.

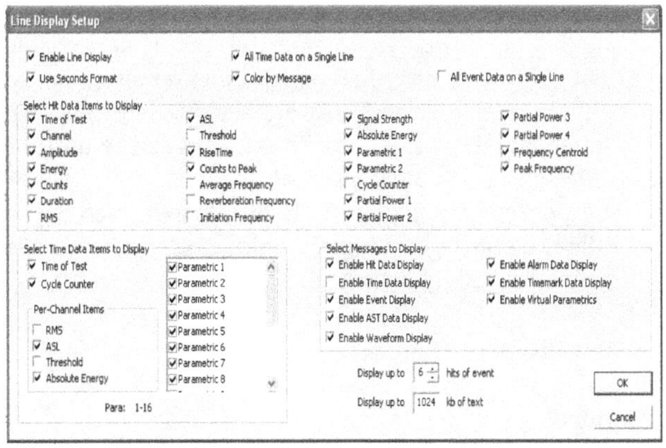

Figure 41. Line Display Setup

Here is a typical illustration:

ID DDD HH:MM:SS.mmmuuun PARA1 PARA2 CH RISE COUN ENER DURATION AMP

- The ID is the message ID. This will be 1 for a data message, 2 for a time driven message, 172 for a waveform message and 128, 129 or 130 for a test control message. A test control message is a RESUME, PAUSE, or STOP message.
- The TIME column is the time of hit read from the time of test counter. This time is displayed in days, hours, minutes and seconds from the start of data acquisition, with resolution to 250 nanoseconds.
- The PARA1 column is for parametric number one. The listed reading is scaled according to the values entered for Multiplier and Offset in the Parametric Scaling dialog box.
- The PARA2 column is for parametric number 2. Again the reading is scaled according to the values entered in the Parametric Scaling dialog box.
- The CH column is for the channel number if the message was of type 1 or an AE data set.
- The RISE column is for RISE TIME. The range of this variable is from 0 to 65535 μseconds.
- The COUN column is for threshold crossing COUNTS. The range of this variable is from 0 to 65535 counts.
- The ENER column is for ENERGY. The range of this variable is from 0 to 65535 energy counts.
- The DURA column is for signal DURATION. The range of this variable is from 0 to 16,777,762 μseconds.
- The AMP column is for AE AMPLITUDE. The range of this variable is from 0 to 100 decibels.

In the Line Display Setup dialog box you can:

- <u>Enable Line Display:</u> Note that the display can be visible (see section on View menu) but not enabled. Turning the display off can boost performance on very high data rate tests.
- <u>Use Seconds Format:</u> With this option you can display TIME in seconds (total) rather than HH:MM:SS.mmmuuun format.
- <u>All Time Data on a Single Line:</u> With this option Time Data messages (ID 2) are displayed using only a single line (each) rather than several shorter lines.

- All Event Data on a Single Line: With this option Event Data Messages (ID 128) are displayed using only a single line (each) rather than several shorter lines.
- Color By Message: With this option you can have the various messages colored by message ID. This can make the display easier to read, though it is not recommended for very high data rate tests.

The Line Listing Display can be made visible by pressing F7 or through the View menu.

Preferences...

The Preferences dialog box is used to modify settings that are not specific to any layout. These settings are remembered from session to session so you generally only need to set them once. They are:

- Only Show Installed Channels: With this item enabled, Hardware Setup will only display entries for channels that represent hardware installed on the unit. If it is disabled then Hardware Setup will display entries for up to 80 channels, including those that do not exist. This can be useful if you want to create layout files for use on other machines.
- Pre-enable Autodump: With this item enabled the 'Auto-Dump' checkbox in the Data Storage (Acquire) dialog box will be checked by default, otherwise it is unchecked by default.

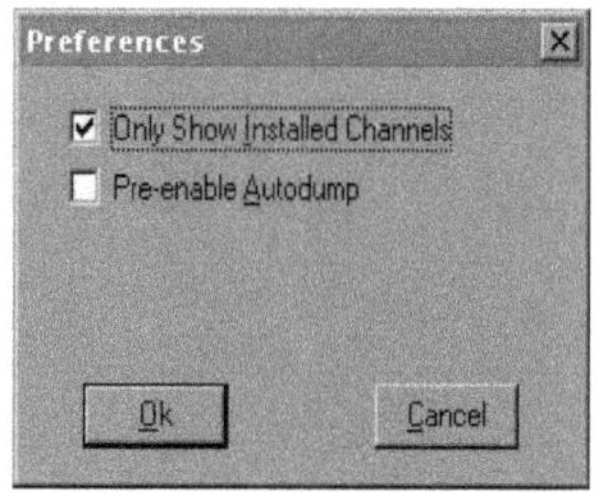

Figure 42. Preferences

Acquire/Replay Menu

Acquire...

An AE test is begun via the Acquire... menu item or by pressing F9. This will display a dialog box prompting you for information before the test begins.

This dialog box is very similar to the Save Layout dialog box discussed earlier, in which you are prompted for a filename, in this case to store AE data. The exception here are the checkboxes in the upper left corner:

- Auto-Dump: If this box is checked, it means that Autodump is ON and the test data will be recorded to the filename you specify in this dialog box. If you want a permanent record of your data, make sure that Auto-Dump is ON. If it is OFF, no data will be written to disk and your record will be limited to the real-time graphic displays.
- Enable Event Linking and Cluster/Area History: If this box is checked then support for these features will be enabled in acquisition. If it is not checked then event linking and cluster/area history will not function for the duration of the test, but the system will exhibit better performance at high data rates. If you expect high data rates then be sure to uncheck this and save the data to a file (Auto-Dump on). You can then use Event Linking and Cluster/Area History during a replay if needed.

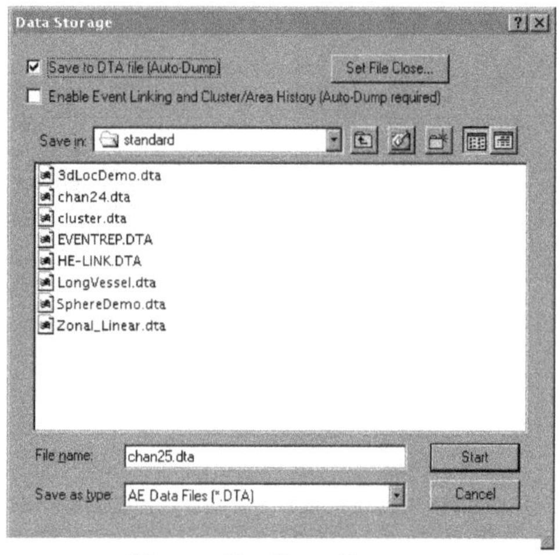

Figure 43. Data Storage

If Auto-Dump is ON, the program will save data to the filename specified. The program will then automatically increment the last letter or digit of the filename specified in the Test Storage dialog box each time you enter acquisition with AUTODUMP ON without specifying a different data filename.

The data file is automatically closed and reopened every second. This assures that the worst case data loss is one seconds worth if power fails. A manual update to disk can be accomplished by pressing pausing/resuming the test.

Replay…

This box is essentially the same as the Acquire dialog box discussed in the last section. Because Autodump is not needed here, it is not an option in this dialog box.

Once the filename has been selected, the controls are essentially the same as for Data Acquisition. During the replay, the system reads data from the disk, updating the graphs/tables at the rate specified in the Display Mode dialog box. The replay automatically ends when it finishes replaying the requested data file. If the file specified data file is a 'Continued file' (described in sections 3.3.2.8.2 and 3.3.2.8.3) then it replays the next file in its continued file set.

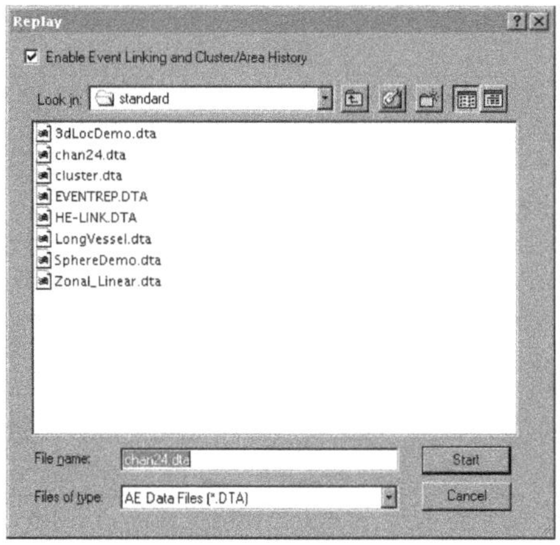

Figure 44. Replay

Resume/Pause/Stop/Abort

These four commands are used to control the test or replay.

- Pause: Click Pause to temporarily halt a test or replay.
- Resume: Click Resume to continue a paused test or replay.
- Stop: Click Stop to terminate a paused test or replay. This is the proper way to end a test. Any buffered data is saved (if auto-dump is on).
- Abort: Click Abort to terminate a non-paused test. This is a quick way to exit a test and can be used even if the hardware is not responding. It is recommended for use only in emergencies or when data isn't being saved to disk as any buffered data is lost.

Time Mark

Click this command to add a Time Mark (a.k.a. 'Test Point') to the data during test. This message represents a point in time of significance to the tester (i.e.: beginning of loading cycle, important observation, etc.). It is saved with the data for future replays and is displayed on all time-based graphs (graphs that have 'time' as their x-axis parameter) as vertical bars. It is shown on the Line Listing Display as an ID 211 along with its time of test.

The statistic 'Test Point' displays how many Time Marks (Test Points) have been added to the data.

Turn Waveforms ON/OFF

Click this during data acquisition to toggle waveform transmission on or off. These commands are only available if you have the Waveform option installed.

Display Last Alarm…

Click this to view the Alarm dialog box. This displays information on all alarms detected during the current test (or previous test if no longer in test). Use the drop down list box in the lower left corner to select which alarm to view. You can Clear (all) the alarms by clicking the 'Clear Alarms' button. This resets any alarm output and turns off the audible alarm (if on).

This command is available only if there has been alarm activity during the current test (or the last one if ended).

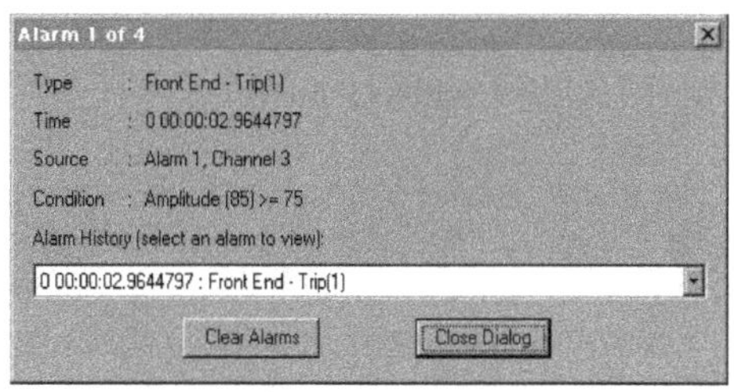

Figure 45. Display Last Alarm

AST

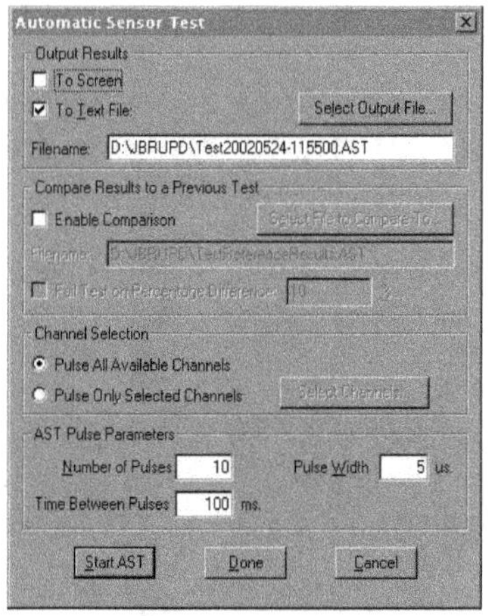

Figure 46. AST

The purpose of the Auto Sensor Test is to verify that nothing has gone awry with the sensors before or after a test. The Auto Sensor test will run a short acquisition test, while pulsing through to each sensor sequentially, using the current settings in the Test and Hardware Setup menus. An AST test should be run before and after an AE test and results compared to make sure there were no sensor sensitivity changes. Also, in order for AST to work, 1220A-AST preamps or Integral Sensors with the AST option (e.g., R15I-AST) are required. The AST Dialog Box is used to setup and begin the AST.

The results of the AST will be displayed on the screen, and/or saved to a file. This is done by selecting the "output Results to Screen" and/or the "Output Results to text file" check boxes. If the "Output Results to text file" box is chosen, you may accept the default filename for the text output file or select a new output filename, either by typing the filename in the Filename text box, or by selecting to overwrite an already existing filename through the Select Output File button.

The results of the AST will be compared to a previous baseline test run if the "Enable Comparison" box is checked. A filename must be supplied either by typing a filename in the Filename text box provided, or by selecting a filename through the Select File button.

The AST pulse is generated automatically by the system following the pulse characteristics shown in the bottom of the dialog box. These characteristics may be changed by typing the appropriate values in the text boxes provided. The characteristics and their legal ranges are:

Number of Pulses: This is the number of times a single sensor will be pulsed. Each active sensor set in the Channel Settings dialog box will be pulsed this number of times. This number must be in the range of 10 to 250.

Pulse Width: This time, specified in μsec, determines the length of the pulse. It must be in the range of 1 to 20.

Time Between Pulses: This time, specified in msec, determines how far apart the pulses are.

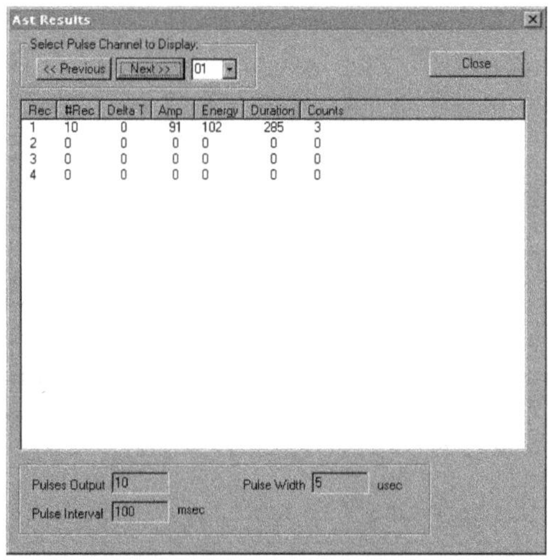

Figure 47. AST Results

Selecting Done will save AST settings, and exit this dialog box without starting the test.

Selecting Start AST will start the Auto Sensor Test. If "Enable Comparison" was chosen, the files supplied in the filename box will be read and used to calculate a percent change for delta-T, amplitude, and energy for all active sensors defined in the Channels Settings of the Hardware Setup Menu.

The results will be displayed, on the screen, written to a file, or both, depending on the options selected in the AST Dialog Box. On screen will be displayed the test setup parameters (Pulses Output, Pulse Width, Pulse Interval) and the actual results one line per receiver channel, one page per pulser channel.

The report file will contain this information plus the date and time of the AST and the baseline report file used for comparison, if one was specified.

Pause Replay On Timemark/Pause

These menu items are used to automatically pause replay whenever a timemark or pause message is replayed. You can toggle these items on or off individually. They allow you to quickly jump to various pre-set points of interest in your data (time marks and pauses set during acquisition). You can resume replay in the normal fashion by pressing F9 or clicking Resume.

Graphing Menu

Graph Setup...

The 'Graph Setup...' dialog box is used to customize an existing graph (created using the 'New Graph' or 'Copy/Paste Graph' commands). You can enter this dialog box by selecting (right clicking) a graph and selecting 'Graph Setup...' from the Graphing menu or by right-clicking the graph and selecting 'Graph Setup...' from the menu that pops up. The appearance of this dialog box will vary depending on the type of graph. The example at right is representative of a simple activity graph.

Section 3.7 details the procedure for creating and customizing graphs. Below is a brief description of the various graph types available and the major pages of the graph setup dialog box.

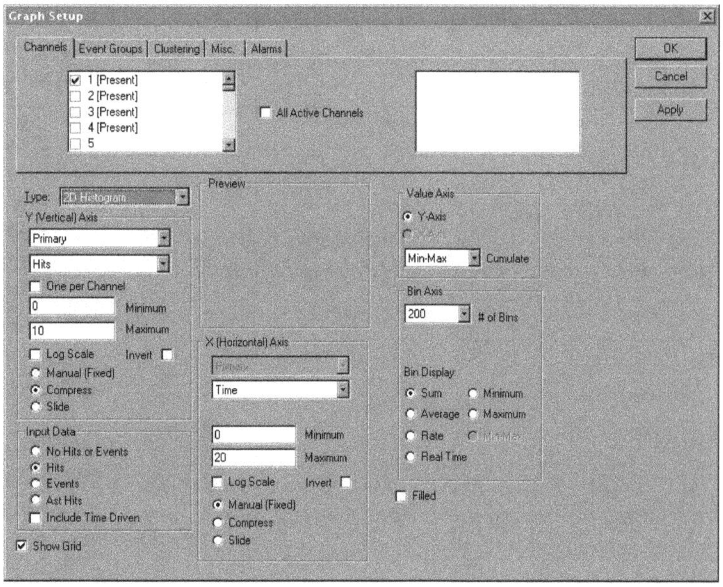

Figure 48. Graph Setup

Graph Types:

- <u>2D Histogram</u> – This is the standard Bar (if 'Filled') or Line (if not 'Filled') graph. You can define multiple plots or use the 'One per Channel' checkbox to automatically create one plot for every enabled channel (max 6). The X-Axis is known as the 'Bin' Axis and is divided into a user-defined number of 'bins'.
- <u>2D Scatter (Point Plot)</u> – This is the basic point plot graph. Multiple plots can be defined.
- <u>2D Binned Scatter</u> – This variant of '2D Scatter' uses a different strategy for data display. It divides the graph area into a grid measuring 200 bins high by 200 bins wide. Thus, every dot represents not a discrete point but actually a small range of values. The advantage of this over '2D Scatter' graphs is that it can display large amounts of data quickly. 2D Scatter graphs can slow down when burdened by more than 100,000 points or so, and will eventually run out of memory around 650,000 points (depending on the setup). This can be a problem during extremely long tests and replays. 2D Binned Scatter graphs do not suffer from this 'data overload'. The disadvantage is a slight loss of resolution. Due to the binned structure there are only 200 possible values across an axis for a total of 40,000 possible distinct points. To calculate the resolution of a graph axis divide its full scale by 200.
- <u>2D Colored Scatter</u> – This variant of '2D Scatter' is nearly identical in operation to '2D Binned Scatter' but uses a single plot which is colored by value. On the right-hand side panel you choose the 'Color Feature' (counts, duration, etc), define the various color level maximums (or choose 'compress') and set the display mode. The color levels define what color a point on the graph is based on its value. For instance assume you set the color feature to be energy and set level 1 (yellow by default) to be 1000 and level 2 (purple) to 2000. If you get a single hit with energy 600 the point displayed will be yellow in color. If you got a single hit with energy 1001 then this point would be purple. Sum, Average, Minimum and Maximum modes display (the color of) the sum, average, minimum or maximum values respectively of all points that occupy the same grid location. 'Count' mode ignores the color feature and simply displays a count of the number of points that occupy the same grid location.

- 3D Histogram – This is the standard 3D bar graph. It operates in much the same way that 2D Histograms do. Both the X and Z axes are 'Bin' axes.
- 3D Scatter (Point Plot) – This is the standard 3D point plot graph. It operates in much the same way that 2D Scatter graphs do except with a depth (Z) axis.
- Waveform – Standard single plot Waveform graph.
- Power Spectrum – Standard single plot Power Spectrum graph.
- Spherical Location – This graph type is used exclusively for the display of spherical events.
- Cylinder Location – This graph type is used exclusively for the display of cylindrical events.

The 'Channels' page (at right) allows you to select which channels and/or channel groups will be active on your graph. Select channels from the left panel and channel groups (if any are defined) from the right panel.

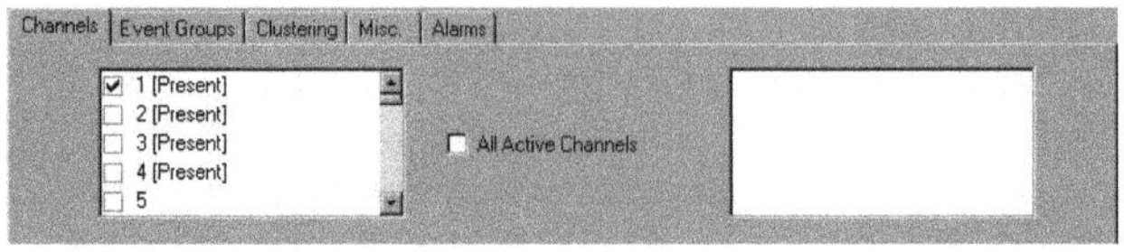

Figure 49. Channels

The 'Event Groups' page (at right) allows you to select which event groups will be active on your graph.

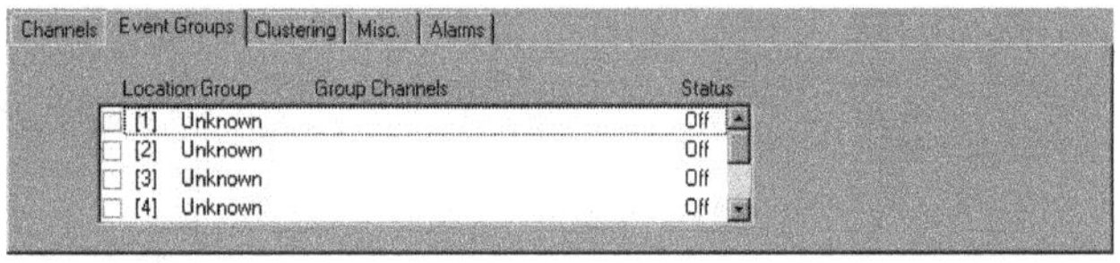

Figure 50. Event Groups

The 'Clustering' page (at right) allows you to customize your graph to display cluster information. See section 3.8 for details.

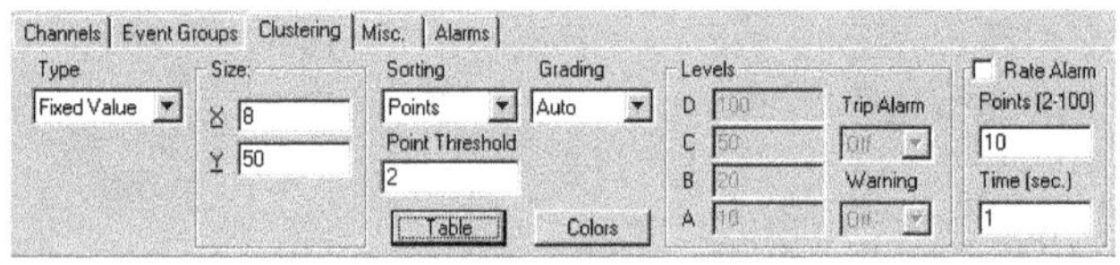

Figure 51. Clustering

The 'Misc' page (at right) allows you to set various graphing options that don't fall into any other category. From here you can turn on the 'color legend'. This displays a small chart on the right of the graph describing what each color means. You can also designate your graph to be a 'history' graph. See section 3.10 for details on how to use the Cluster/Area History feature.

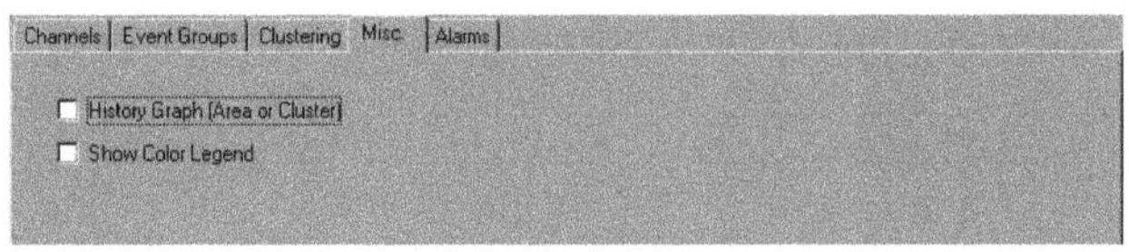

Figure 52. Misc.

The 'Alarms' page (at right) allows you to enable and customize Graph Alarms on your graph. See section 3.11 for details on Graph Alarms.

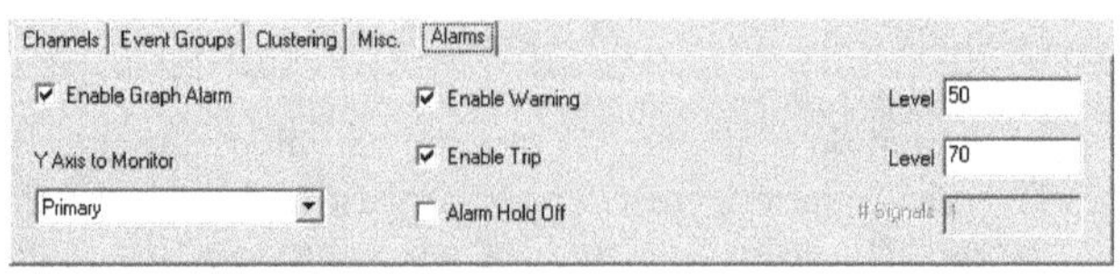

Figure 53. Alarms

Graph Filters…

Graph Filters are used to restrict the data that gets plotted on a particular graph by specifying a graph parameter (such as amplitude, energy, duration, parametric) and a range of values to either plot (accept) or not plot (reject). The parameters can be same as or different from the graph axis parameters of the graph to which they are applied. They only affect the data being plotted, they do not affect the data going to the disk. The Graph Filter dialog box is launched by either right-clicking on a graph and selecting 'Graph Filters' or by selecting Graph Filters from the Graphing pulldown of the main menu.

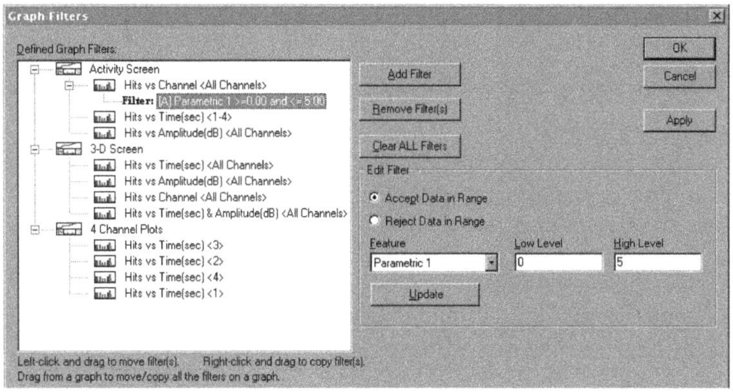

Figure 54. Graph Filters

The left side of the dialog box contains a tree control showing the 'Defined Graph Filters'. This control displays information in three levels. At the highest level are the currently defined pages of graphs. The next level shows all the graphs that belong to a page. The final level shows all the filters that are defined for the graph. If a given level has a '+' sign to its left then it contains items which are hidden. Left-click with the mouse on the '+' to expand the level. Expanded levels have a '-' to the left of them. Left-click on the '-' to compress the level. The text for the screen is the same text that is used on the 'Page Bar' at the bottom of the page. The text for the graph is the same as the caption above the graph. When the dialog box is first displayed it highlights the graph that is currently selected. So if the dialog box is displayed by right-clicking a graph and then selecting 'Graph Filters' the graph that was clicked on will be the graph that is highlighted. Filters are displayed with the format [A/R] Feature >=Low Level <=High Level. [A/R] will be [A] for accept filters and [R] for reject filters. The buttons to the right of the tree control are enabled or disabled depending on which item is highlighted in the control.

Add Filter: Will be enabled when a graph or filter is highlighted. Selecting it creates an empty filter under the highlighted item. The user must then use the controls in the 'Edit Filter' group to define the filters. The 'Update' must be pressed to change the filter.

Remove Filter(s): This will be enabled as long as a filter is defined in any Screen or Graph that is highlighted or if a filter is highlighted. Pressing it will remove either the single filter that is highlighted or all the filters that belong to a highlighted graph or all the filters in all the graphs of a highlighted screen. When the button is pressed another dialog box will appear that will prompt for verification. The filter removal operation can be canceled at this point.

Clear ALL Filters: This will be enabled as long as a filter is defined. Selecting it will remove all filters. When the button is pressed another dialog box will appear the will prompt for verification. The filter removal operation can be canceled at this point. The filter is defined by using the controls in the Edit Filter group. These controls will only be active if a filter is highlighted in the 'Defined Graph Filters' control.

Accept/Reject: Determines if the data will be accepted (plotted) or rejected if it passes the filter test (i.e. falls in the range of the low and high levels).

Feature: Selects the feature used as a filter. The choices are the same as the graph parameters. Note that all graph parameters are listed here. The user must make sure that the selected parameter makes sense. Only hit data can be filtered (all time driven data will plot). If the graph is an event graph, then only the features from the first hit sensor will be used in the test and the event will plot based on the result of the test on the first hit data.

Low/High Level: Determines the range of the filter. The range is inclusive of the limits (as denoted by the <= and >= in the graphical representation of the filter in the 'Defined Graph Filters' control). Any floating point value can be entered and the range is not validated.

Update: This button must be pressed to change the values in the 'Defined Graph Filters' control. The Apply button takes the filters as they are defined in the 'Defined Graph Filters' control and applies them to the system. Pressing the Cancel button after pressing Apply will not restore the filters to the state they were in before the dialog box was displayed.

Drag and Drop:

The text below the 'Defined Graph Filters' control explains how to move or copy a filter or group of filters.

If a filter is dragged with the left mouse button it can be moved to another graph by dropping (release the mouse button) on a target graph.

If a filter is dragged with the right mouse button it can be copied to another graph by dropping (release the mouse button) on a target graph.

If a graph is dragged then all the filters in the graph will be moved/copied to the target graph.

If the target graph is off the display, hold the dragged item very close to the top or bottom of the control to initiate auto-scrolling.

If the target graph is in a collapsed page or graph ('-' to the left of the target), hold the dragged item over the collapsed page or graph and it will automatically expand.

Tips:

- Reject is used over accept. If one filter on a graph is defined as [R] Amplitude >=0 <=50 and another filter on the same graph is [A] Amplitude >=45 <=50 then no hit will be plotted. The [A] filter winds up filtering out all hits with Amplitudes > 50 and <45. The [R] filter removes all the hits in the [A] filter range because it includes the entire range of the [A] filter. Think carefully about what is to be accomplished and test the graph first whenever possible.
- No more that 20 filters per graph can be defined.
- The graphs indicate if they have any filters defined by placing 'Flt' on the caption at the top of the graph. To quickly make sure that there are no filters, bring up dialog and check the Clear Button status. It will be disabled (grayed out) if there are no filters on any graphs.
- Don't forget to press the Update button whenever a filter is added or modified.

Colors

The 'Colors' submenu is used to set the various colors used in HMSGHAT2000DDI.

- Cluster Colors: Set the colors of the various cluster grades used on cluster graphs.
- Colored Scatter Graph: Set the colors of the various levels used by Colored Scatter Graphs.
- Plot Colors: Set the colors used by the plots of all other graph types. 'Plot 1' is used by all graph types other than Colored Scatter Graphs. 'Plot 2' through 'Plot 6' are used only by graphs that can use multiple plots (2D Histogram, 2D Scatter, 2D Binned Scatter).
- Sensor Color: Set the color of sensor icons used on event-based graphs.
- Structure Color: Set the color of the structure used on event-based graphs.
- Legs Color: Set the color of the legs used on spherical location graphs.
- Graph Background Color: Set the background color of the graph frames.

New/Move/Close Graph or Table

- New Graph: Use this command to create a new graph. The new graph will be a copy of the currently selected graph. If no graph exists then the new graph will be a simple 2D Histogram of hits vs. time. Use graph setup to customize it.
- Move Graph or Table: Use this command to move a graph or table using the mouse.
- Close Graph or Table: Use this command to close the selected graph or table.

Cut/Copy/Paste Graph or Table

- Cut Graph or Table: Use this command to cut a graph or table, presumably for later pasting on another page.
- Copy Graph or Table: Use this command to copy a graph or table, presumably for later pasting on the same or other page.

- Paste Graph or Table: Use this command to paste a cut or copied graph or table onto the current page.

Export Data to File...

This option on the right-click menu allows you to export the numerical values used in plotting a 2D Histogram onto a text file. These values are outputted to the corresponding x-y values found on the graph and are properly tabbed. This allows the user to import the values onto an Excel graph for instance for proper analysis. When the option has been selected on the right-click menu, the following dialog box will appear:

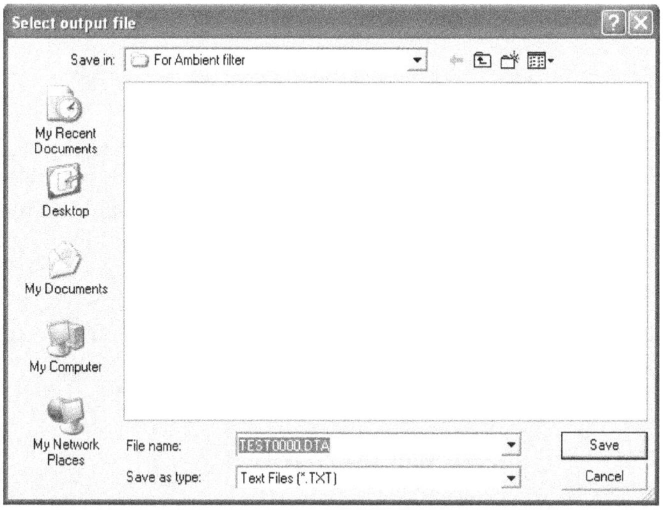

*Figure 55. Select **Output** File*

Here, the user can save the text file containing the numerical data onto any location of their choosing.

Note – This option is only enabled for 2D Histogram graphs. If the user attempts to right-click on any other type of graph, this option will be disabled.

Turn Waveform Threshold Lines On/Off

Waveform graphs can be displayed with horizontal lines denoting the threshold and a vertical line denoting the time of first threshold crossing. Use these commands to turn the display of these threshold lines on or off on all waveform graphs. You can also turn threshold lines on or off on individual graphs by right clicking the graph and checking (or unchecking) 'waveform thresholds'.

Tables Menu

With this menu you can create specialized tables. A table is a window that displays mainly text-based information instead of graphical information. Like graphs, tables can be moved, resized and closed from commands on the 'Graph' menu.

View Menu

With this menu you can select which toolbars and information bars you would like to display on the screen. There are fixed positions for each of these toolbars defined on the HMSGHAT2000DDI screen, but some (Toolbar, Acq. Controls, Line Listing Display) can be moved according to the user's preference by grabbing (left clicking the mouse) the handle (vertical bar) at the left side of the toolbar and moving it to the desired location. The settings on this menu are saved internally (i.e.: not in the layout file).

Utilities Menu

This menu contains a number of useful 'utility' programs for post-test analysis of datafiles. See the Appendix at the end of this chapter for details.

Page Menu

With this menu you can create insert, delete and rename screen pages. This is an HMSGHAT2000DDI screen page management menu. You can also perform these functions with the icon keys and a right mouse click on the screen page tab.

Window Menu

This is a typical "Window" menu that you would find on any common WINDOWS program and therefore the menu contents should be familiar. It allows you to arrange your graphs in cascade or tile patterns as well as edit their window settings. Select 'Maximize Graph' to enlarge the active graph , and 'Restore Graph' to return it to its normal size.

Help Menu

With this menu you can get some on-line help to the operation of certain features of the software. In addition, the "Options" selection informs the user which options have been installed in his/her HMSGHAT2000DDI software and the "About HMSGHAT2000DDI" informs the user to the release version and date of HMSGHAT2000DDI.

Acquiring AE Data with HMSGHAT2000DDI for the First Time

This section is provided to help you understand the sequence of getting into and running an AE Test for the first time using the provided Layout (.LAY) files and also gives you insight into a more detailed sequence of carrying out an AE data acquisition test. Before starting in acquisition, you should have your AE system already set up as per the "Getting Started" section of your AE system manual. This usually entails, connecting sensors, preamplifiers and cables to the AE system, connecting up your AE system, computer keyboard, mouse and monitor and booting up your system into HMSGHAT2000DDI as previously instructed. If you are not ready to connect up the AE system to conduct some type of AE test or at least any AE experimental test on a laboratory test, then skip this section and proceed to the next major section on Replaying an AE test with HMSGHAT2000DDI.

The key to carrying out an AE test is in setting up the hardware set-ups, graphics, location setups and display modes to your particular requirements prior to entering the test. Although there are many set-up items and they might seem confusing for first time users, supplies good default set-ups for most typical tests and they are an excellent starting point. As you get more familiar with the system you can modify these set-ups a little bit at a time and save your new layout (.LAY) setup files for your next AE test. This small step-by-step "Trial and Error" approach builds up your knowledge base until you are totally familiar with all the set-up options and what they mean. Until you get this knowledge please depend on the prepared .LAY files.

The following are the steps that should be taken when carrying out your first AE test using the supplied layout files. First we will outline the steps and next we will carry them out.

1. Load the appropriate system Layout (.LAY) file into the system in order to set-up the system to your desired operation and display conditions. We will be using the standard **Layout4.LAY** (or **Layout2.LAY** on Spartan and PCI-2 systems) layout file in this example which sets up 4 (or 2) channels for acquisition.
2. Enter the Acquire menu.
3. Select the datafile name which you want to save to disk.
4. Start Acquisition.
5. Look at various screens, graphs and data listings during acquisition.
6. Exit data acquisition.

The following is the step by step detail on starting and running your first AE test. We assume that you have sensors set up and will be exciting them in order to generate signals for AE generation or have them attached to a structure, bar or plate with a pulser or some type of simulated AE source.

Loading a Layout File into HMSGHAT2000DDI

From the Menu bar, use your mouse to select **Files** then select **Open Layout...** A menu window called **Open** comes up. With your mouse, select the **Layout4.LAY** (or **Layout2.LAY**) file from the list of layout files so that this file name appears highlighted in the file box next to **File Name:**. This selects it as the layout file that will set up the AE hardware for a 4 (or 2) channel operation with a good, complete default set-up. To finish the sequence, move the mouse to the **Open** box area on the right side of the **Open** window box and click on it (left mouse key) or press <ENTER>. Upon executing this menu, the selected Layout file is read into the program. If you had made any changes to the current layout file, a message box will pop up asking you if you want to save the those changes. If yes, select yes, if no, select no. HMSGHAT2000DDI will then load and configure itself to the selected layout's setup. This completes the reading of the LAY (HMSGHAT2000DDI setup) file.

Entering Acquisition

Entering acquisition involves steps 2, 3 and 4 of the above sequence. or the Mouse operated icon. To begin, left-click the mouse on the **Acquire/Replay** main menu item then click on the **Acquire** selection. Alternately, you can enter the Acquire function by pressing the **F9** function key (on the keyboard), or you can select the Acquire icon in the Acquire/Replay toolbar. The Acquire icon is shaped like a traffic light with the green light showing indicating that HMSGHAT2000DDI is ready to go into acquisition. This toolbar is usually to the right of the main icon toolbar or just below it. If you do not see this small toolbar (like that shown below), it may not be turned on from the **View** menu. To activate it, simply select the Tools menu and make sure the "Acq. Controls" entry is checked. If not, click on it and it will be activated and will show up with the other toolbar icons.

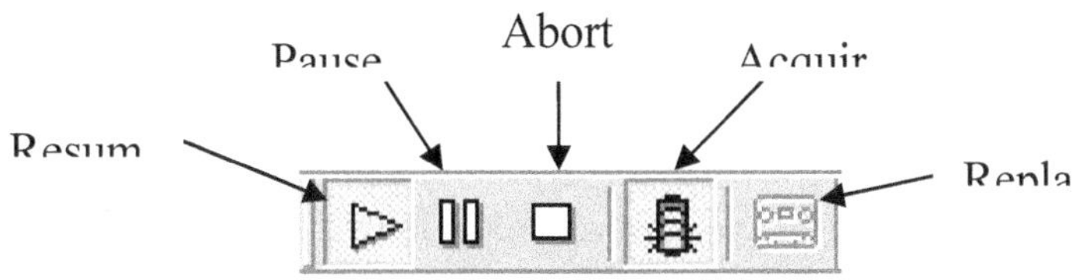

Figure 56. Acquire/Replay Icon Toolbar

At this point the Acquire function, a **Data Storage** files menu box will appear to signal you to enter a data file name and whether you want to save the file to disk. (An example of the Data Storage file menu box is shown in figure 60.)

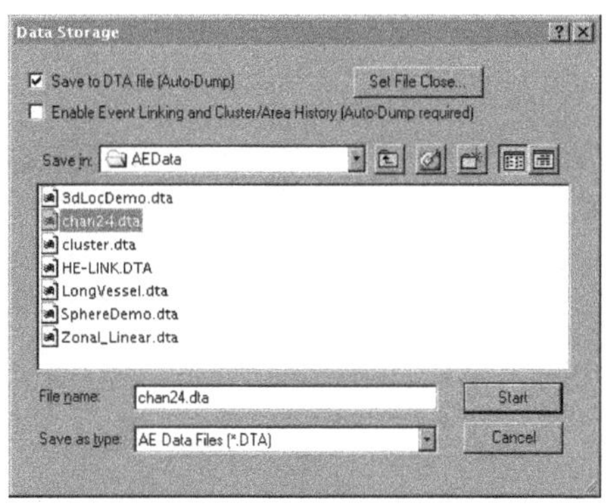

Figure 57. Data Storage

First, make sure the "Save to DTA file **(Auto-Dump)**" box is checked if you want to save the data you will be collecting in this exercise. In HMSGHAT2000DDI, this is always automatically checked and you usually have to deselect (uncheck) it to prevent the generation of an AE data file.

Next, check the name shown in the **File Name:** box. If that name is acceptable then you are ready to start the test by clicking on the **OK** button. If you want a different name then enter the new name by typing it in when the **File Name** box is highlighted. Once you enter the name (you do not have to enter the .DTA extension) you can hit <ENTER> on the keyboard or click the **Start** button (located to the right of the Filename) with the left mouse button to begin acquisition. At this point HMSGHAT2000DDI enters the acquisition screen and waits for you to press the <ENTER> key to start the test. Press <ENTER> to start Acquisition.

The system is now in data acquisition and is collecting and displaying AE information and displaying it on the graphs. Tap the sensor against the table your instrument is set up on or scratch it with your fingernail.

At first you will see no change in your Data Acquisition Graph but once the amount of time specified in **Test Setup → Display Mode** menu parameter "Seconds between Update" time has elapsed (it is usually set for 1 – 2 seconds), you will begin to see a visual representation of your data. Continue tapping the sensors to see the changes to the graphs to verify that you are getting waveforms and graph updates. An example plot showing the first AE graph screen page (selecting the tab, Activity Screen), is shown in figure 61. You should be seeing data visualization like that shown, where graph 1 shows a group of 4 point plots including Counts, Energy, Amplitude and Rise Time (on the Y-axis) versus Duration (on the X-axis), graph #2 (to the right) is showing #AE hits and #waveforms collected versus time,

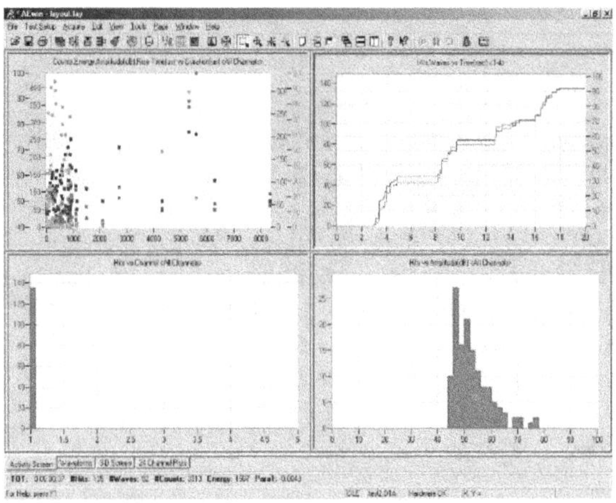

Figure 58. *Example Activity Screen*

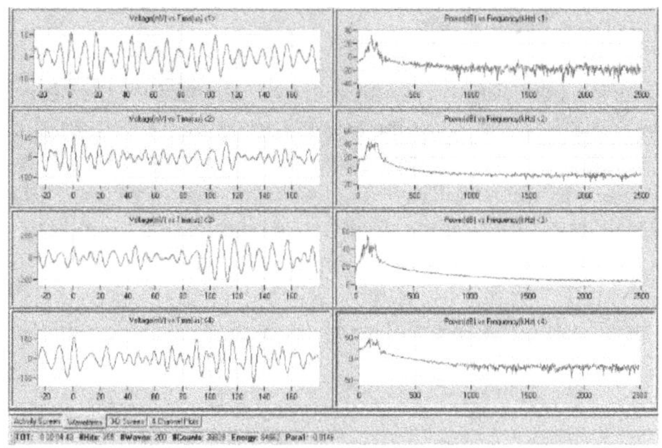

Figure 59. *Example Waveforms Screen*

Graph #3 (below graph 1) is showing Hits versus channel and Graph #4 (lower right) is showing the Hits versus Amplitude (or Amplitude Distribution).

Actions within Data Acquisition

There are several actions you can take in acquisition to view the graphs and other modes in the instrument. Try these items to see what happens.

Selecting and Viewing the Different Screen Pages

You can scroll between screens to view the activity between the different AE graph screen pages. This can be done simply by left clicking the mouse on any of the AE graph screen page Tabs located just below the graphs on the left. In the example Layout file, we have defined four different graph screen pages. They are the Activity Screen (shown in figure 58), Waveforms (figure 59), 3-D Screen (figure 60) and the 4 channel plots (figure 61).

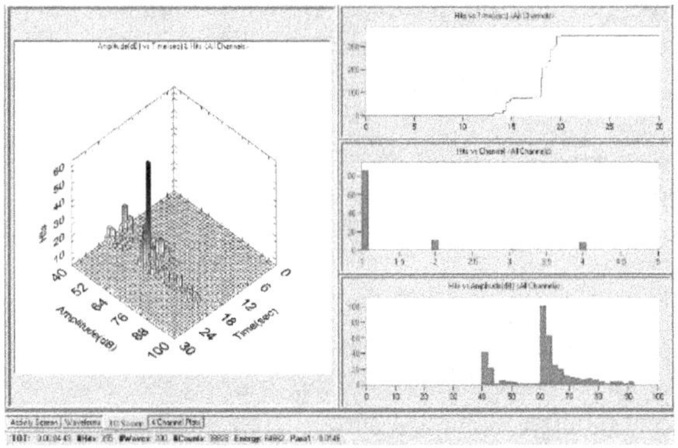

Figure 60. Example 3-D Screen

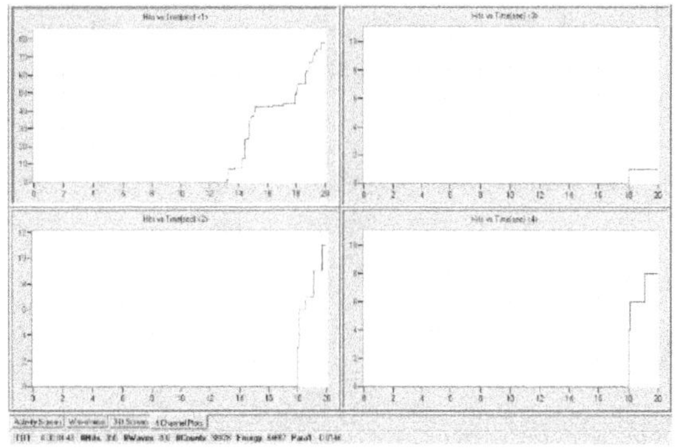

Figure 61. Example 4 channel screen

These graph screen pages give a good idea to the capabilities and the graph types that you can create with HMSGHAT2000DDI. On the Activity Screen page, you can see multiple point plots on one graph, multiple line graphs (AE Hits and Waves versus time) as well as bar charts. On the Waveforms Screen page (figure 59), you can see waveforms on each of 4 channels with the associated FFT located alongside the waveform for each of these channels. On the 3-D Graphs screen page (figure 60), you can see a nice 3-D plot with 3 associated graphs alongside. This graph shows that you can setup 3-D graphs (which can be rotated) with HMSGHAT2000DDI. It also shows that you can size your graphs differently. On the 4 channel plots (figure 61) you can set up graphs with multiple plots on them, one for each channel so you can link AE with other test parameters. This can all be done in acquisition. As you observe and change graph screen pages, the test continues.

There are many other things that can be done within acquisition. This includes maximizing a graph to full screen, zooming in on certain areas of a graph, viewing cursors and cursor readouts, printing graphs, saving graphs to a data file or the clipboard, just to name a few.

Zooming and Panning on a Graph Region

To zoom a graph, please note the graph associated toolbar icons for zoom, pan, and cursor select. This set of toolbars is near the center of the icon toolbar. This toolbar section is shown and defined in figure 62.

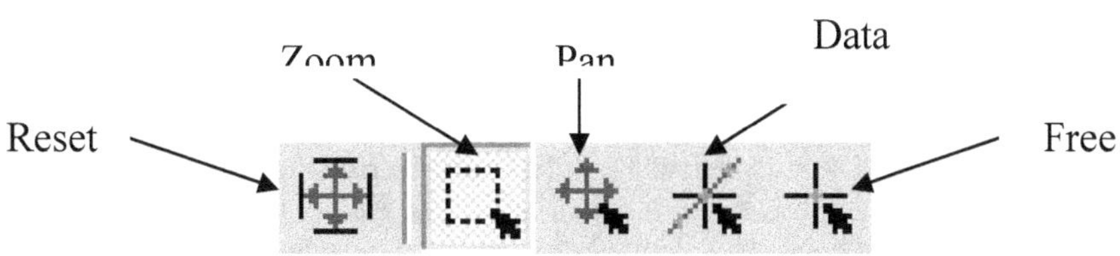

Figure 62. Zoom, Pan, Cursor Control Toolbar Icons

To use, select a graph by clicking on it. A good graph to demonstrate these functions on is the Activity Screen Point plots (see upper left graph of figure 58). Select the graph by clicking on it. Next, click on the zoom icon. This is probably already selected. Next, with the mouse cursor, define a zoom rectangle by pointing to the starting point on the screen where you want one corner of the zoom rectangle to be. Next, press and hold the left mouse button and drag the mouse to the point where you want the zoom box to end. You will see the dotted box as you drag the mouse. As soon as you let go, the graph will zoom and only the area that you selected in your zoom rectangle will be displayed in that graph. You will see that the axis scales have changed as well as the number of points in the point plot. You can now pan the plot (move around the zoomed area) by selecting the Pan icon button, moving your mouse pointer to the center of the graph screen that you have zoomed, click the left mouse button and move the mouse around. You will see the graph pan as you move the mouse. You can continue panning in one direction by letting go of the left mouse button, repositioning the mouse and pressing down on the left mouse button again, and moving the mouse. You can even pan un-zoomed graphs (but there is not a lot of sense in doing that). Zooming and panning can be done on any 2 dimensional graph.

Once done and you want to return that graph back to the unzoomed state, simply press the "Reset Axes" icon. This is the leftmost icon shown in figure 62. This will bring the graph and its axes back to the normal setting.

Zooming, Panning and Rotation of 3-D Graphs

Zooming, Panning and Rotation of 3-D graphs can be accomplished but not in the same way as that for 2-D graphs. For 3-D graphs, this requires a combination of the mouse and keyboard to accomplish. Before starting this exercise, click the Graph Screen Page tab "3-D Screen" and follow along to get familiar with the controls. You should see a graph like that shown in figure 60.

Rotating the graph is the easiest function. All you need to do is place your mouse pointer somewhere near the middle of the 3-D graph, press the left mouse button and while holding the left mouse button down, move the mouse slightly around in a circle. You will see the rotation of the graph change. You can position it anywhere you want.

To zoom, simply place the mouse pointer somewhere near the center of the 3-D graph and when ready, press the **ALT** button on the keyboard, and while holding that down, press and hold the left mouse button down. Then slightly move the mouse down (while holding the ALT button and the left mouse button depressed). As you move the mouse down the 3-D graph will zoom. As you move the mouse up, the graph will shrink.

To pan the 3-D graph, place the mouse pointer somewhere near the center of the 3-D graph, depress and hold the **SHIFT** button on the keyboard down and at the same time depress and hold down the mouse left button. Then slightly move the mouse around to see the graph pan.

Once done and you want to return that graph back to the un-zoomed state, simply press the "Reset Axes" icon. This is the leftmost icon shown in figure 62. This will bring the graph and its axes back to the normal setting just like in the case of the 2-D graphs.

Using the Position Readout Cursor

Two cursor control icons are shown in figure 62. The cursor readout controls are active in both acquisition and replay just as zoom and pan controls are. Each cursor provides a slightly different function. The "Data Cursor" identified above in figure 62, provides a discrete cursor movement only to legal data points on the screen. For example, for a waveform or a point plot, the Data cursor would only be able to identify actual existing data points. This therefore follows actual points on the screen. The other cursor, called the "Free Cursor", identified as the rightmost cursor in figure 62, will be able to be moved to any location on the screen and provide a continuous position that is graph based and not data based.

The Cursor readout position is located on the status bar at the lower right side of the HMSGHAT2000DDI screen. The position readout is identified by the **X,Y =** readout box. The position readout will be in actual graph value readings such as Hits, Amplitude, etc. If the graph is a multiple plot graph, the data point read out will be for the exact graph point, so it will be correct regardless of how many plots are on a graph. If the graph is a Binned Scatter or Colored Scatter graph then the readout will display the range of values represented by the data point.

The cursor readouts can be used on any graph screen. Simply select a graph screen by clicking on it. For this example, let's use the Point plot again on the Activity screen. First click on it. Next, select one of the cursors. Once a cursor is selected, it usually is drawn with the center at the origin of the graph (the lower left corner). Note the small hollow circle which identifies the center. Next you need to position the mouse pointer over the center of the cursor and grab the cursor by pressing and holding the left mouse button down. Once done you can move the cursor around the graph. If you have selected the Data cursor, you might have a little trouble moving it around the first few times but you will get used to moving from point to point. The "Free Cursor" is a little easier to operate. Anyway, as you are moving the cursor around (with the left mouse button depressed), you can see the graph coordinates changing as you move. This is a very handy feature to have.

Maximizing a Graph to Full Screen

Sometimes you have many graphs on a screen and it is a little difficult to see the axis values clearly. You can expand any graph to full screen by right-clicking the mouse in the desired graph area, and selecting the menu item "Maximize Graph" from the graph menu. You can select the same function from the Main Menu bar, **Window → Maximize Graph** selection. To restore the graph to its normal size and place on the screen, simply, right-click the mouse in the graph area and select "Restore" or **Window → Restore Graph**.

Printing any or all Graphs

You can print out any graph, any graph screen or all graph screens, using the various Print commands located in the File menu. To print a given screen of multiple graphs or to print out all screens, simply select the "Print Page(s)" from the files menu. To print out a single graph first click on the graph, then select the "Print Graph" menu item from the files menu. You can also export a graph or screen to a JPEG file by selecting the "Export to JPG" menu item on the files menu or you can "Copy Screen to Clipboard".

Pause, Resume, Abort The Test

The Pause, Resume and Abort functions are provided on the Acquisition/Replay Icon toolbar as shown in figure 56. The icons are fashioned after similar visual controls used in the Audio tape recorder business. To pause your test, simply press the Acquisition/Replay icon labeled **Pause**. In pause you can see the test status box change to "Test Paused". You can also see the Time of Test clock stop. At this point, the system will stop collecting AE data, will flush all buffers to disk, but the test will not be exited and the AE data file will remain open. Someone might want to Pause during a test to ignore some event that might add noise to the test (such as pressure or loading ramp-up's, etc.). To resume, just press the **Resume** icon button located next to the Pause button. This restarts acquisition and the Time of Test clock starts back up again showing the new and current time, taking into account the pause time. To Abort the test, simply click the button labeled **Abort** (see figure 56). Selecting this function stops the test and saves only the data to the data file that already exists on disk. The only difference between the Abort and the standard exiting of a an AE test is that the buffer contents are ignored in the case of an abort while ALL data is saved and written to the AE data file in a normal test exit. This function was mainly implemented for SPARTAN products as it could take the ICC's many seconds (when AE data rates were high) to empty their buffers. The abort function was convenient in those cases where the user wanted to quickly stop a trial test without waiting for data in the buffers to clear. The abort function is not needed in newer products as the transfer rates are very fast.

Exiting Data Acquisition

To exit data acquisition, first Pause the test by pressing the **Pause** icon button (see figure 56). Then stop the test by pressing the **Abort/Stop** key (the Abort/Stop key becomes a STOP function key, whenever the system is in Pause). Upon receiving the Pause, all buffers are flushed to disk and upon receiving the Stop, the AE data file is closed and HMSGHAT2000DDI exits acquisition. The Status box (lower center of screen) indicates "IDLE" meaning that HMSGHAT2000DDI is ready to perform another test or a replay.

If you have made any setup changes before, during or after acquisition and you want to update the .LAY (layout) file (which you should), you may do so by going to the file menu and choosing **Save Layout...** or **Save Layout as...** menu item. "**Save Layout**" will overwrite your existing layout file with the new information, while "**Save Layout as...**" will allow you to rename the layout file as a new file. Then you can follow the guidelines to save the new .LAY file. Enter a name such as TEST1.LAY. It is not recommended to overwrite the existing Layout.LAY file since you may always want to keep this default layout file in case of problems in the future. As computer experience dictates, you should always save your work before carrying out a new task such as acquisition in case for any reason the power or computer fails or that you simply forget after the test to save the LAY file. This is always a good practice.

Replaying AE Data

This section helps you understand the sequence of preparing for, recalling and replaying your previously recorded AE test using the provided system Layout (.LAY) files and AE data (.DTA) files. It also gives you insight into a more detailed sequence of data analysis and manipulation of data during replay.

Replaying an AE test is almost identical to that of acquiring data. As a matter of fact many of the same steps that have been carried out to perform data acquisition are repeated. The following are the sequential steps that should be carried out when replaying an AE test using the supplied Layout and AE data files. First we will outline the steps and next we will carry them out. For more details on the actual operation of software and menu items please refer to the proceeding sections of this manual.

8. Read the appropriate system Layout (.LAY) file into the system in order to load the system to your desired set-up and display conditions. We will be using the **Layout4.LAY** (or **Layout2.LAY** on Spartan and PCI-2 systems) layout file in this example.
9. Enter the Replay function.
10. Select the data file (with .DTA extension) name which you want to replay. We will be calling the datafile "Chan24.DTA" for our replay.
11. Start Replay.
12. Move between the various screens, graphs and data listings during and after replay.
13. Carry out some data manipulation functions during replay.
14. Exit Replay.

The following is the step by step detail of replaying an AE data file using the provided **Chan24.DTA** data file. For a replay test, you do not need to have any sensors connected to the instrument since you will be playing previously recorded AE tests.

Loading a Layout File into HMSGHAT2000DDI

This is identical to section 3.4.1 but is repeated here. From the Menu bar, use your mouse to select **Files** then select **Open Layout...** A menu window called **Open** comes up. With your mouse, select the **Layout4.LAY** (or **Layout2.LAY**) file from the list of layout files so that this file name appears highlighted in the file box next to **File Name:**. This selects it as the layout file that will set up the AE hardware for a 4 (or 2) channel operation with a good, complete default set-up. To finish the sequence, move the mouse to the **Open** box area on the right side of the **Open** window box and click on it (left mouse key) or press <ENTER>. Upon executing this menu, the Layout.LAY layout file is read into the program. If you had made any changes to the current layout file, a message box will pop up asking you if you want to save those changes. If yes, select yes, if no, select no. HMSGHAT2000DDI will then load and configure itself to the Layout.LAY setup. This completes the reading of the LAY file.

Entering Replay

Entering replay involves steps 2, 3 and 4 of the above sequence. To begin, click the mouse on the **Test Setup** main menu item. On the submenu, click on **Display Mode**. Make sure the **Normal** mode is selected. Also verify that the **Seconds Between Updates** is set for 1 or 2 seconds (this selects the graph update rate) and that the **Hits Before Update is 1000**. If not, set these values by clicking the mouse onto the particular box whose value you want to change and edit those numbers by typing new ones and deleting old one's with the delete or backse keys on the keyboard. If you want, set in your desired values. When done, click the **OK** window on the bottom of the menu to execute the changes.

Now you are ready to enter the begin replay. This can be started by clicking that icon (that looks like a cassette tape, see figure 56), or pressing the <F10> Function key on the keyboard or by selecting **Replay** from the **Acquisition/Replay** menu. Once selected, a **Replay** window comes up to signal you to select a data file name for replay. Check the name highlighted in the **File Name:** box. If that name is acceptable then you are ready to replay the test by clicking on the "Start" button. If you want a different file name to replay, then use the mouse to select it in the listed data file names. In this case you want to select the name **Chan24.DTA**. Click the mouse on that entry. Then, to begin replay you can hit <ENTER> on the keyboard or click the "Start" button on the menu. At this point HMSGHAT2000DDI begins replay and is displaying AE information as it appeared during data collection on the test.

Actions within Replay

All the actions carried out during acquisition can be carried out during replay. Please refer to section 3.4.3 for details relating to replay actions. There are several actions you can take in replay to view the graphs and other modes in the instrument. Try these items to see what happens.

Exiting Replay

There is no need to exit replay, when the file is finished replaying, HMSGHAT2000DDI returns to the Idle status waiting for the user to continue manipulating the graphs in replay, enter replay to analyze another file, make changes to the setup before continuing or enter acquisition.

Using Line Listing Mode

Line (Dump) listing mode is a very familiar feature and function of 's AE software. has provided a "line by line" listing capability with all its AE products for many years. This has also been referred to as "Line dump" in the past but now we refer to it as "Line listing mode". The line listing mode allows the user to monitor any and all types of messages being processed by the AE instrumentation. The information is in a "line by line" text format listing offering an additional monitoring capability over the graphing of AE data. With Line listing mode, the user can watch messages related to AE Hits, Events, Time driven data (parametrics), comments, Auto Sensor Test (AST), waveforms and administrative messages. Watching the line listing scroll by, shows how the data coming in is being processed, giving the user an idea as to the AE activity and severity. In HMSGHAT2000DDI, the Line listing has been improved with the following features over the capabilities found in our other legacy software programs:

- the ability to scroll the line listing up or down (used to only scroll down),
- the ability to select which data can be shown on the line listing
- flexible display of the line listing with the AE graphs, the line listing can be adjusted to take up a small part of the screen, operate within a floating textbox or take up the entire screen.

These features do add some extra utility to the line dump listing that makes it even more important and useful in AE testing. The rest of this section explains how to use these capabilities in HMSGHAT2000DDI.

The line listing mode can be used in acquisition or replay, and offer the same flexibility in both, however, it is recommended to use this mode sparingly in acquisition as its use might slow down the data acquisition speed by a significant amount. It is therefore recommended to use this when data acquisition rates are low, before a test as a prescreening method or after a test in replay.

The line listing mode color-codes the messages in order to make them easier to spot as they are scrolling by. Please note the following message types and colors as identified below.

Line Listing Message Type	Line listing Color
AE hit data	Blue
AE Event data	Violet
AE waveforms	Blue
Time Driven Data (e.g. Parametrics, ASL, RMS, etc.)	Green
Administrative messages (Pause, Resume, Alarm, etc.)	Red

Enabling Line Dump Listing for Viewing

In normal use, HMSGHAT2000DDI is kept in graph mode and the line listing is not displayed (as shown in figure 56). In order to view the line listing, two steps need to be performed. First, one has to "Enable" the Line listing mode, and then the line listing mode needs to be selected for viewing. This is done as follows (whether in acquisition or replay).

To Enable the line listing mode, simply click the mouse as follows;

Test Setup → Line Display → Select "Enable Line Display" Checkbox

To Display the "Line Display", simply press the **F7** function key or click the mouse as follows;

View → Line Listing

As mentioned above, the function key **F7** provides a shortcut to toggling the line listing on and off, once the Line Listing mode has been "Enabled" in the Test setup menu.

Note that whenever the line listing is displayed, it displaces the existing graphs based on its size and location. To restore the graphs simply press one of the "Tile" functions (Tile horizontally or vertically).

The line listing shown above shows various messages. The top line shows a message ID 2 or Time Driven message, while the message ID 1 is an AE hit message and ID 173 is a waveform message. As can be seen, all messages start with a time of test stamp and all data is in time order. In addition, the Time driven message shows some time driven based parametric readings while the AE hit message shows various AE hit based features.

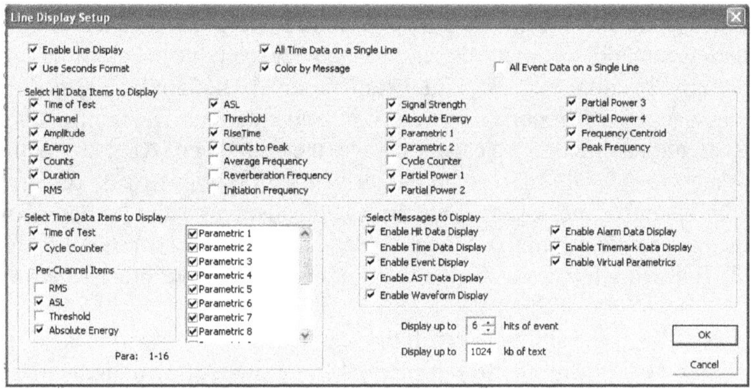

Figure 63. Line Display Setup

Modifying The Line Display

Many times the user wants to view a limited set of messages in the line listing screen. For example, maybe it is only important to look at AE hit messages or even just a few features in an AE hit message, even though it is desired to collect much more comprehensive data then that to be viewed. The Line Display Setup menu (click the mouse on Test Setup menu, then select the Line Display menu) allows the user to select exactly which data is to be viewed in the line listing display. The figure at right shows the Line Display Setup menu. There are 4 sections to this menu. The Enable Line display at the top of the menu was discussed in the previous section, make sure the checkbox is clicked if you want to enable the line display mode. The next section is called the "Hit Data Items". These are all the AE features that could be displayed on the line listing, if all the features were enabled in the **Test Setup → Hardware → Data Sets/Parametrics** menu. Simply check those AE hit features you want to view in the line display, even though all the features selected in the Data Sets/Parametric menu will be processed to graphs and the data file.

The "Time Data Items" (shown on the bottom left of the Line Display Setup menu), allows the user to select which Time driven data will be displayed with the Time driven data message. The final "Messages" area (to the bottom right of the Line Display Setup menu), allows the user to select which messages will be displayed. In all cases, just click those checkboxes to enable their display in the Line Listing display.

This menu is accessible in acquisition or replay and can be changed at any time.

Sizing and Moving the Line listing Display

The line listing display, when selected to be viewed, pops up as a full width, anchored text box and anchor's itself just below the HMSGHAT2000DDI toolbar. However, the Line listing display can be moved, floated or re-sized, simply by using standard WINDOWS functions. To float, just "double click" on the border of the window. Once floated, the Line Listing Display window can be sized using typical WINDOWS sizing capabilities of moving the mouse to the edge of the window border until the double sided arrow appears. Then the user can left click the mouse and drag the window side to be larger or smaller as desired. The user can also move the floating Line Listing Display window anywhere on the screen simply by left clicking the mouse when the mouse cursor is in the title bard (the blue title bar at the top of the Window) and can also anchor the Line Listing Display anywhere on the screen top, bottom, side, by moving one of the Line Listing Display Window edges to intersect with the side where it is to be anchored.

Once the Line Listing Display has been moved, this position will be remembered.

Clustering Setup and Operation

The purpose of clustering is to automatically identify and associate groups of hits or events which meet user controlled "sing" criteria. These clusters are drawn as rectangles around the data on a point plot screen to visually identify active region(s). Cluster data can also be shown in a tabular form, allowing a quantitative assessment of the activity. Clustering offers the user computer-aided processing to identify regions of AE activity.

Cluster analysis can be activated on any 2D point plot (2D Scatter, Binned Scatter or Colored Scatter), regardless of whether it is a Hit or Event based graph. Clustering is also available on the Spherical, Conical & Cylinder Location graphs types & 2D Histogram graphs allow 1-dimensional clustering along the X Axis. Clustering on a location graph identifies active areas that require follow-up for defect growth, while clustering on a Hit based graph usually identifies hits & events with similar AE feature pattern's which can lead to hit classification of that data.

Clustering Settings

Clustering is controlled from the Clustering tab at the top of the Graph Setup dialog box shown below. There are a number of parameters related to clustering that can be adjusted including the type, cluster size, point threshold, sorting & grading options. In addition there are options for generating alarms based on clusters based on either the cluster grading (level alarms) or cluster growth (rate alarm). Each of these settings are described below.

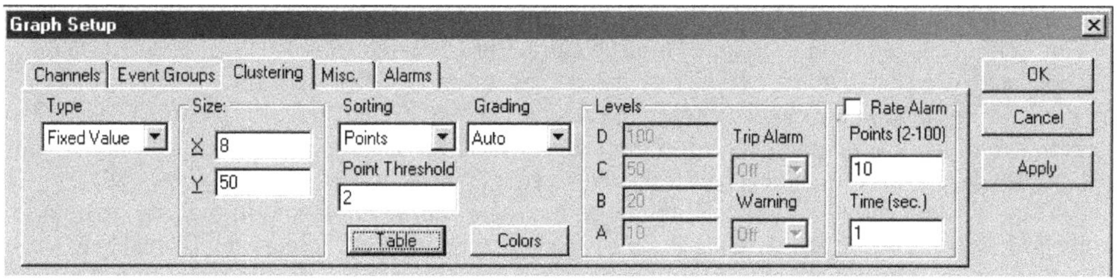

Figure 64. Clustering setup menu (at top of Graph Setup Menu)

Type & Size: Type controls how we want the Size parameter interpreted. Setting Type to None turns clustering off. Using Fixed Value allows you to enter the cluster dimensions you want directly into the Size X & Y field in graph axis units. In this case, this is the distance that 2 points must be within to form a cluster. Once the cluster is formed, the distance from the center of cluster to any new points must be less than half this value. Axis % allows you to enter the cluster size you want as a percentage of the graph axis range. For example, a Size X entry of 10 means the cluster can be as large as 10% of the X axis. For example, if Size X were 10 & the X Axis Maximum & Minimum are set to 10 & 50, then to form a cluster 2 points must be within 10% of 40 from one another or within a distance of 4. For graphs using autoscaling & Axis %, the internal size used does not change as the graph scale changes. Instead it is fixed at the value corresponding to the percentage of the initial axis limits.

Point Threshold: Point threshold is the minimum number of data points that must be in the cluster before it is reported. Internally the software maintains a list of up to 100 clusters on a graph, but it only report clusters that are above the threshold up to a limit of 26 reported clusters. If your Input Data is set to Hits, the number of points in the cluster is the number of hits. If the Input Data is set to Events, then it corresponds to Events. The minimum number of points that can be in a cluster is 2, but clusters containing just 2 points are not usually of much interest. You can increase the Point Threshold beyond 2 (up to 100) to decrease the number of clusters reported in the Cluster Table.

Sorting: Sorting controls how clusters are reported in the cluster table. The available choices are Points, Energy, Counts, Counts to Peak. The cluster table displays clusters sorted from highest to lowest by one of these quantities. In order for Counts To Peak to be useful, you must have actually collected Counts To Peak as part of your data set. Otherwise the software simply treats it the same as sorting on Counts. When a cluster table is displayed in the current graph window, you can use Ctrl-F6 to control the sort. Pressing Ctrl-F6 toggles the sorting through the list of available options.

Table: The Table button brings up the Table Setup dialog box shown at the right. The Display checkbox controls whether or not a cluster table is shown in the graph window. The F6 key is a keyboard shortcut for turning the cluster table on & off. The Placement setting controls where the table appears relative to the graph. The Shift-F6 key shifts the table through each of the 4 positions. The Window % setting controls how much of the graph window allocated to the cluster table. If the text in the table grows too large for its window, scrollbars will appear to allow you to see all the text.

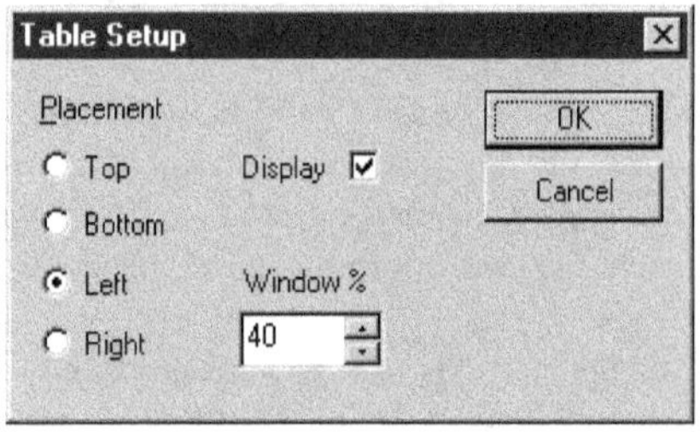

Figure 65. Table Setup

Colors: The Colors button brings up the Cluster Colors dialog box shown at the right. You may also access this dialog directly using the Graphing pulldown Colors submenu. Look for the Cluster Colors menu option. This dialog allows you to configure what colors are used in the cluster table & when drawing the cluster rectangles on the graph. These colors are used by all graphs doing clustering. Double clicking one of the entries will cause a standard Windows color dialog to appear allowing you to select a new color for the selected grade.

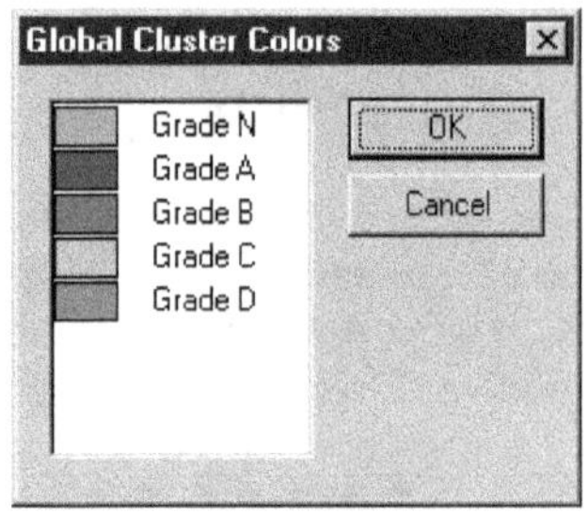

Figure 66.

Global Cluster Colors

Grading: The Grading selection controls how clusters are assigned a color. After the clusters are sorted, a grade level is assigned to each cluster based on the sorted quantity. Choosing None causes all clusters to be classified as N (not significant) & they are drawn & reported in the Grade N color. Choosing Auto grades clusters by their percentage of the maximum observed in the quantity used for sorting the clusters. It just divides the maximum observed value into 5 equal ranges (<20%, <40%, <60%, <80%,<=100%). For example, if the clusters are sorted on Points & the largest clusters had 250 points in it, then the grading levels used would be 50, 100, 150 & 200. If the Grading is set to Manual, then the Levels section will enable & you can enter your own grading levels. That can be convenient when you want to compare the cluster results from 2 different graphs as it allows you to enforce the same grading levels on both graphs.

Levels: The Levels section is used when the Grading parameter has been set to Manual. It disables when Grading is set to Auto or None. Clusters are drawn in up to 5 colors corresponding to 5 different grades. This section controls the boundary levels between the manual grades. If you are sorting on Points, have Grading set to Manual & entered levels of 10, 20, 50 & 100, clusters with less than 10 points are graded N & by default colored green (see Colors to change). Clusters with 10-19 points would be graded A, 20-49 points are B, 50-99 are grade C and any with 100 or more points are considered grade D.

Warning & Trip Alarm: The Warning & Trip Alarm settings allow the user to generate alarm messages in the line display & set the Alarm Warning and/or Alarm Trip output lines high when a cluster reaches a user determined grading level. The Warning & Trip Alarm settings require that Grading be set to Manual before they will enable in the dialog box & they only have an effect during acquisition. Both Warning & Trip Alarm can be set to Level A, B, C or D or to Off. The Warning setting must always be lower than the Trip Alarm setting if the Trip Alarm setting is not set to Off. During acquisition when a cluster level alarm is generated, an alarm message will be saved to the DTA file. If the line display is enabled, a text message containing the details will be displayed & the alarm dialog box appears with the details. The Alarm Warning & Alarm Trip lines will be set high as appropriate. The output voltage of the Alarm Warning & Alarm Trip lines are never altered in replay mode,. When a DTA file is replayed, the same cluster level alarm messages will reappear that were observed when the file was recorded. These messages are not controlled by the current Warning & Trip Alarm settings during a replay.. The appearance of these alarms is only controlled by the settings that were in use when the DTA file was recorded.

Rate Alarm: The Rate Alarm can be used to monitor a graph for concentrations of activity that occur at high rates. This type of cluster alarm works identically during acquisition & data replay & does not have to be enabled during data collection. The checkbox allows for simple on/off control. The Rate Alarm has 2 parameters, Points & Time. The Points setting specifies how many points must be observed within the Time period to set off the cluster rate alarm. The number in the Points field can range from 2-100, while the Time setting is specified in seconds & can be any floating point number. For proper functioning, a Time value greater than 0 should be entered or else the alarm will trigger anytime data is added to a cluster. Likewise a large entry like 3.4E+38, the floating point maximum, causes problems as well as it would be so long that the alarm would never trigger. If the software receives enough points in one cluster within the alarm time, a cluster rate alarm will be generated. For the settings above, that would be 10 points within 1 second of one another.

The cluster rate alarm is implemented to not generate extraneous messages. If the user says to check for 10 events in 1 second, the cluster has to collect at least 10 points before the rate is ever checked. Once the 10th point comes in, the software adds 1 sec. to the time of the first point & checks to see if the 10th occurred before or after that time. If it occurred before that time, then the rate is higher than the user setting & an alarm message is generated. When the cluster gets an 11th point, its time will be compared with that of the 2nd data point. If the new data maintains the rate above the user cluster rate & the cluster has already generated an alarm message, the cluster stays in the alarm state and does NOT generate a new alarm message. If a cluster is in an alarm state & a new data point comes in after the alarm time, then the cluster alarm state will be reset to no alarm. If the cluster later on gets more data that is again above the user specified cluster rate, it will generate another cluster alarm message. Basically, if the system is flooded with hits, one message will be generated & nothing else until the hit rate drops below the user threshold to reset the cluster & then the hit rate comes back up. When a cluster rate alarm is generated, the software will generate a status message in the line display reporting the cluster ID & current cluster information.

Clustering Example

With so many settings, it may be useful to see an example of how to set up clustering. The software includes example layout and data files, **Cluster.LAY** and **Cluster.DTA**. The steps in setting up and running a clustering application are described below.

Step 1. If you wish, start by using New Layout under the File pulldown menu. Reconfigure the default Hts vs. Time 2D Histogram to meet your requirements. For example, you could set up a Y-Position versus an X-Position location graph set up as a point plot, or set up an AE feature versus AE feature (e.g. Amplitude versus Duration, etc) plot. When we plot an AE feature versus a feature, we usually call it a "Correlation" graph, referring to the way we are trying to "correlate" one AE feature against another to see how and if they are related. Our example layout **Cluster.LAY** and data file **Cluster.DTA,** shows an example of a Duration versus Amplitude point plot. If you were starting with a new layout file & graph, you would set the graph type to 2D Scatter & set the X & Y Parameters to Amplitude & Duration. The example layout uses Manual scaling on both axes, with the Y Axis limits set to 0 & 600, while the X Axis limits are set to 0 & 120. The Input Data is set to Hits. Alternatively you can just recall the example layout file & use the graph settings in it.

Step 2. Next we must configure the Clustering function. We do this in the Graph Setup menu by selecting the tab labeled "Clustering" at the top of the Graph Setup menu (see figure 64). Start by picking Fixed Value from the Type dropdown listbox.

Step 3. Next set the Size X setting to 8 & the Size Y setting to 50.

Step 4. Leave the Sorting set to Points, but change the Point Threshold to 3 to eliminate the smallest clusters.

Step 5. Select Auto from the Grading dropdown list. Leave the Rate Alarm disabled & ignore the Table & Colors buttons for the moment. Clicking on the Graph Setup OK button will save the settings. If this is a new layout for you, you may wish to use Save Layout under the File pulldown menu before starting a replay.

Step 6. Now we are ready to run a test and view clustered data. Use the F10 key or select Replay under the Acquire/Replay pulldown menu to open a DTA file for replaying. Select the Cluster.dta data file and click on OK or press Enter to replay the data. If you use the setup as provided in the Cluster.lay layout file with the Cluster.dta data file, you will see results similar to those shown below.

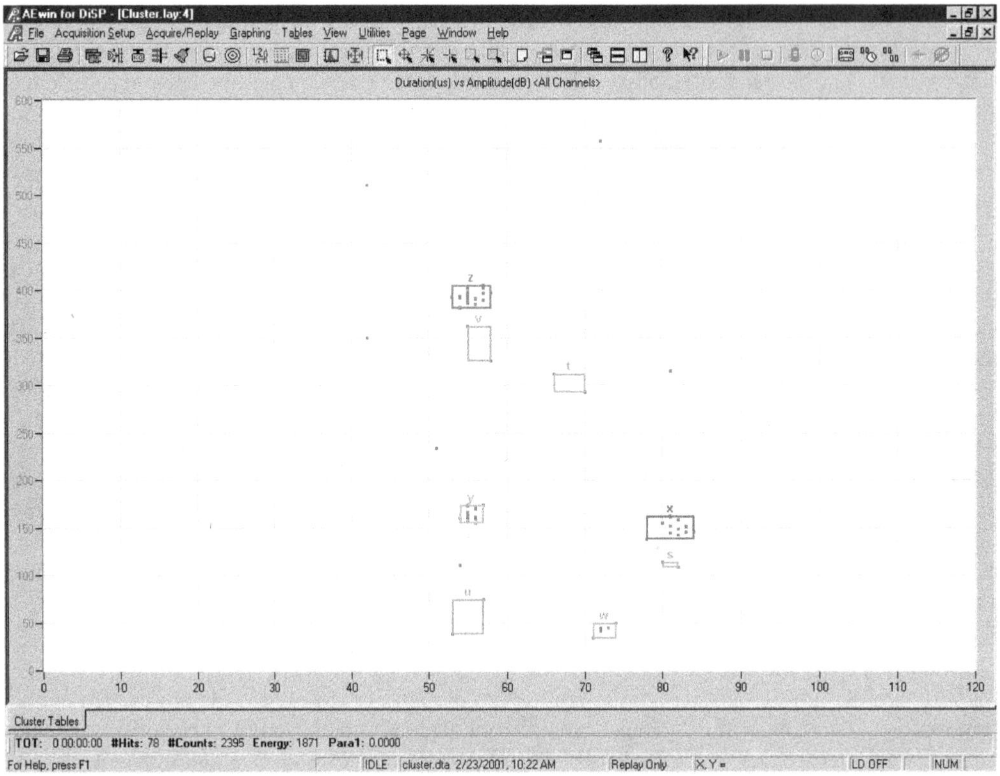

Figure 67. Cluster Graph

From the graph, we can see that we have identified 8 clusters based on the cluster configuration that we have set up. We can also see that there are cluster letters associated with each cluster and that each cluster is colored. Clusters are labeled with a letter starting with "z" as being the most intense and "a" being the least intense. Up to 26 clusters can be reported with the clustering capability built into the software.

Step 7. You can adjust the colors of the clusters by using the Colors submenu of the Graphing pulldown menu. Look for the Cluster Colors menu option. When you select it, the Global Cluster Colors dialog box will appear as shown in Figure 66. You can also get to this dialog box using the Colors button on the Clustering tab of the Graph Setup dialog box. The Global Cluster Colors dialog box controls the color of each grade level. If you double click one of the entries a standard Windows Color dialog box will appear to allow you to select a new color for that particular grade level.

Step 8. This graphical representation of the clustered data is nice, but we may also want to see a more quantitative tabular listing. Press the F6 function key from the keyboard. By pressing the F6 key, you toggle the cluster table on/off. This cluster table shares the window with the cluster graph. When we use F6 to turn the cluster table display, we get the table as shown in figure 68.

The Cluster table has the following columns. The leftmost shows the cluster letter, starting with z as the highest. Next is the cluster ID, this is the identification number assigned when the cluster was formed. As new points are added to the graph, the a-z cluster letter can change, but the cluster ID always stays the same. The next column is the Points. This indicates how many points (hits or events) are in a given cluster. The Total Energy is the accumulated energy in the cluster from all the hits or events. The Total Counts is also the accumulated number of AE Counts for the cluster. The Xrange & Yrange min & max values identify the boundary points of the cluster while the Avg. X and Y values identify the center of the cluster. Each line of the cluster table is colored by the grade level of the cluster so it matches the color of the cluster box on the Cluster graph.

Cluster	ID	Points	Total Energy	Total Counts	X Avg.	Y Avg.	min Xrange	max Xrange	min Yrange	max Yrange
z	2	28	309	1539	55.43	392.36	53.00	58.00	381.00	405.00
y	1	18	68	415	55.50	164.33	54.00	57.00	155.00	174.00
x	3	13	1206	294	81.62	150.23	78.00	84.00	138.00	162.00
w	5	5	64	33	72.40	42.60	71.00	74.00	34.00	50.00
v	4	2	14	55	56.50	343.50	55.00	58.00	325.00	362.00
u	6	2	5	3	55.00	56.50	53.00	57.00	38.00	75.00
t	7	2	13	3	68.00	303.50	66.00	70.00	293.00	312.00
s	8	2	109	33	81.00	111.00	80.00	82.00	108.00	114.00

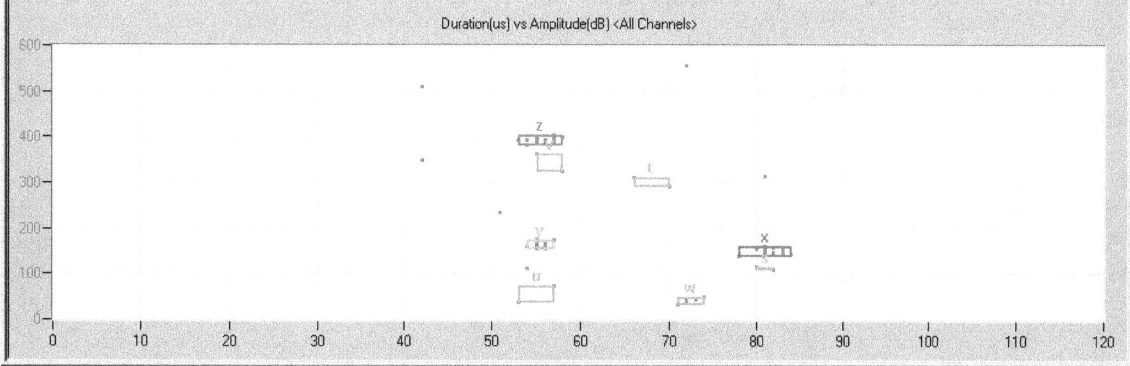

Figure 68. Cluster Table

Step 8. Try pressing shift-F6 while the table is displayed & the cluster graph is the active graph. Shift-F6 causes the cluster table to shift its position around the graph from top to bottom & then from left to right.

Step 9. Try pressing Ctrl-F6 to control the sort order in the cluster table. Each time you press Ctrl-F6, the table will switch between sorting based on Points, Energy & Counts & immediately update.

Step 10. The final step in the process of setting up and running the clustering function is adjustment of the setting for optimal results. This is accomplished by changing some of the clustering configuration settings & replaying the test until you feel that you have the most logical results. If you have too many small clusters, make your Size X & Y settings bigger. Alternatively, you can adjust your Point Threshold higher to only display the largest clusters & ignore the smallest clusters. You might wish to try setting Grading to Manual & modifying the Level A thru D values to adjust the grading results.

In our setup, we set the cluster threshold at 2 which is a bit low. We can also see by figure 68 that there are 4 of the 8 clusters with only 2 hits within the cluster. Let's get rid of these by going back to the graph setup menu and changing the Point Threshold to be 3 instead of 2. After we click the OK button, we'll find the cluster table has been updated & now only hold 4 of the original 8 clusters. Note also that those points that are not part of the cluster seem to be randomly sed, which tells us that it does not seem that these points or hits should have been identified as a cluster. Experiment further to get familiar with the clustering function.

Graphing in HMSGHAT2000DDI

The graphing capability in HMSGHAT2000DDI is very impressive, as it should be since data visualization is very important in AE testing. HMSGHAT2000DDI is able to provide multiple graphs, multiple plots within a graph, multiple graphs on a screen (or "Page" using Windows terminology), and multiple screens (Pages) full of graphs. The graph types can multi-dimensional (2D and 3D) graphs, multi-colored graphs and multiple graph types including: Point plot, Histograms, Waveforms and Spectrum graphs. Graph axes can be set up to be fixed scaling (always stays at a fixed minimum and maximum value), automatic scaling (compression or sliding scale) logarithmic or linear scaling. The graphs can be set up to plot anything versus anything, limited only by the operators imagination.

Before getting into setting up Graphs in HMSGHAT2000DDI, it is important to define the terminology used in graphing on HMSGHAT2000DDI. This will be used throughout the manual.

Plot: A plot is defined as a single graph within a graph (or graph window). Since a graph can be set up with multiple graphs inside, we need to distinguish the difference between the word graph as a graph window and the fact that there can be multiple graphs in a single graph window. We therefore call this a plot. So now we can say there are multiple plots within a graph or graph window.

Graph: A graph is a single Window on a Screen or Page. A graph is controlled by the Graph setup menu. The graph can have multiple plots of the same type within it, but these are all set up by a single specific graph setup menu.

Screen or Page: A screen or Page is the graph(s) that show up in the display area of the HMSGHAT2000DDI screen. This the area between the top and bottom tool bars (see figure 4). A screen or page has a page tab located at the bottom of the screen to select between the screen pages.

Now that we have definitions we can provide an example. Figures 58 –61 show examples of different screens. Four are shown, Figure 58 shows what we have labeled as an activity screen. The page tab at the bottom of the screen has been labeled "Activity Screen" to let the user know what types of graphs are set up on this screen page. Figure 59 shows a waveform and FFT screen page. Its tab also identifies it as "Waveforms". Figure 60 shows an example of a 3D screen page with the appropriate Screen Page tab labeling. Figure 61 shows an example of some channel by channel line plots with an appropriate screen page tab.

In regards to graphs, figures 58 – 61 show multiple graphs on each screen. Figures 61, 63 and 64 shows a 4 graph screen while figure 59 shows an 8 graph screen. There is no limit to the number of graphs that can be placed on a screen, the only limitation is a physical one, pertaining to the readability of the graphs. As far as multiple plots on a graph, the top two graphs of figure 58 shows graphs with multiple plots. On the top left plot there are 4 different point plots on one graph, while in the top right graph there are two line plots.

The figures 58 – 61 which are provided by the example **Layout4.LAY** and the data file **Chan24.DTA** give a good idea as to the functionality of HMSGHAT2000DDI as far as graphing is concerned. Since all the major examples for graphs and graph setups can be found in this **Layout4.LAY** file, it provides an excellent reference for how to set up the different types of graphs. Please use this example as such. The following subsections will provide information on how to generally set up a graph and screen page.

Graph Setup Procedure

The following is the step-by-step process in setting up graphs in HMSGHAT2000DDI. This assumes that you are starting from a very simple single screen, single graph arrangement, into a multiple graph, multiple screen setup.

8. Before starting, give a little thought to the number of screen pages your application will need, what the theme will be for each screen page (for tab labeling considerations) and how many and what type of graphs you would like on each screen. The better you plan this out, the easier and faster the graphing setup process will be. Use as an example the theme provided (and theme tabs) in the **Layout4.LAY** (or **Layout2.LAY** on Spartan or PCI-2 systems) layout file whose graphical results are shown in figures 58 – 61.
9. Find an existing Layout file that gets you partially to your goal. You can then edit that layout file to meet your exact needs. It is always easier to start with an existing layout. You can easily copy and paste graphs that have already been setup, saving a lot of time. If this is not possible, select **File → New Layout** which will start with 1 screen page (labeled Page 1) and one graph (a Hits versus Time graph).
10. Create the number of screen pages that you need by adding and deleting screen pages, then edit the Page tab titles to the desired name or description of what that page will contain.
11. Create the number of graphs you need in each screen page by adding or deleting graphs.
12. Size your graphs as desired and tile them as desired so that they are laid out in the order that you wish.
13. Individually setup each graph in each graph screen to get the desired result. Enter the graph setup menu for each graph by right clicking the mouse when it is inside the graph, then select the first pop-up menu called "Graph setup" to enter the graph mode.

14. Once you are done setting up your graph(s), test them out in acquisition or replay to make sure they act as expected. There will always be the need for some adjustment, you will rarely get things perfect the first time.

Screen Page and Tab Management

As with all Windows operations, there are various ways to add and edit Screen pages. The most convenient is by Right-clicking your mouse on an existing page Tab, adjacent to where you want to add the page. A Page/Graph setup menu will appear. To add a New Page after the selected page, simply select **the New Page** menu item. To insert a new page before the selected page, select the **Insert Page** menu item. In either case, a new page and page tab will appear. The page tab will be labeled as "Page X".

There are other ways to add a page including selecting the **New Page Icon** from the icon toolbar (looks like a blank page), which will always place a new page at the end of the screen page tab list. You can also go to the main menu bar and select the **Page** Menu to make the **New Page** and **Insert Page** selections.

Once you have created a new page, it will be empty (no graphs). You will add the desired graphs and will label the page tab. You may add as many graphs as you wish to the screen page.

To label the page tab simply right-click on that tab and select the Rename Page menu item. This will allow you to enter a page name. There are no major restrictions to naming a page tab. However do not make the page tab title too long as it might take up too much se.

There is one other screen page editing function, which allows you to remove or delete a screen page. Simply right click on the screen page tab and select the **Delete page** function. The screen page and tab will immediately disappear. If you had graphs in the screen page that you deleted, they will be deleted as well.

Adding, Placing and Sizing Graphs on a Screen Page

Once you have a Screen Page set up, you need to begin adding and placing graphs on it. If you know how many graphs you want and what types they will be you can quickly add and edit the graphs. If all the graphs are to be the same style (e.g. waveform graphs, one for each channel), you should start by adding a graph, defining it and then copying it as many times as you need. Then you can size the graphs and create the layout (position of the graphs) as desired and then you can edit the graph setup to finalize the graph function. If each graph is to be different, you can either copy similar graphs and paste them into the new screen page or you can simply add as many graphs as you want on the screen and edit the graph setup for each graph added.

To add a graph to the graph screen, simply right-click you mouse anywhere in the screen page and select the **New Graph** menu item. Alternatively you can left-click the new graph icon from the icon toolbar menu. This icon shows 3 staggered graphs. Additionally you can add a new graph by selecting the **Graphing → New Graph** main menu selection. Whichever way you select a new graph, a single graph will appear. To add multiple graphs on the screen, simply perform this function multiple times.

Once you have selected the number of graphs you want on the screen, you need to tile them. Tiling allows you to place them next to each other so that each part of every graph can be seen. There are two tiling options. You can use **Tile Horizontal** or **Tile Vertical**. The most convenient way to perform this function is to select one of these functions from the icon tool bar. The Tile Horizontal icon is shown as two graphs stacked on top of one another while the **Tile Vertical** icon shows two graphs side by side. Alternatively you can select these from the **Window** main menu. Tile Horizontal will favor a length-wise positioning of the graphs and will usually make the graphs longer than taller. Tile Vertical will give preference to a vertical direction. Choose one of these to place the graphs in a nice ordered manner.

Once the graphs are placed on the screen, we can alter their sizes. We do this using the standard Windows conventions of moving the mouse to the edge of a graph until we see a double arrow and then left-click the mouse and move the edge in or out to make the graph larger or smaller. We can also move graphs by right-clicking on the graph and selecting the **Move Graph** menu item. An example of sizing graphs can be seen in figure 63 where we have sized the 3D graph to be larger than the associated graphs.

The final function that we need to perform is that of editing or setting up the graphs to perform the desired display function.

Graph Setup Function

This section discusses the graph setup menu and how to set up a graph. First, we select the graph that we want to modify. We do this by right clicking the mouse when it is inside the graph region of the graph that we want to work on. Once we do that a menu appears. This menu gives us many choices but we will select the top one called "Graph Setup". Upon selecting this a graph setup menu appears like that shown to the right.

This shows us how we have setup the Graph setup menu for a Hits versus Time graph. It gives us a lot of insight into how the graph setup menu works and how to set it up. Lets note first the layout of the menu. We see at the top of the menu, a selection area for Channels, Event Groups, Clustering, Misc and Alarms. These settings allow us to configure those items. When in the Channels Tab, we can select which channels and/or channel groups will be displayed on the graph that we are setting up. In this case we are selecting all active channels for viewing in the graph. You can see that we have no groups set up in Event Groups if you select that tab. Clustering allows you to set up parameters that link closely sed hits/events together for point plot graphs. Misc allows you to set various options like turning on/off the legend control. Alarms allow you to create and customize graph alarms on this graph.

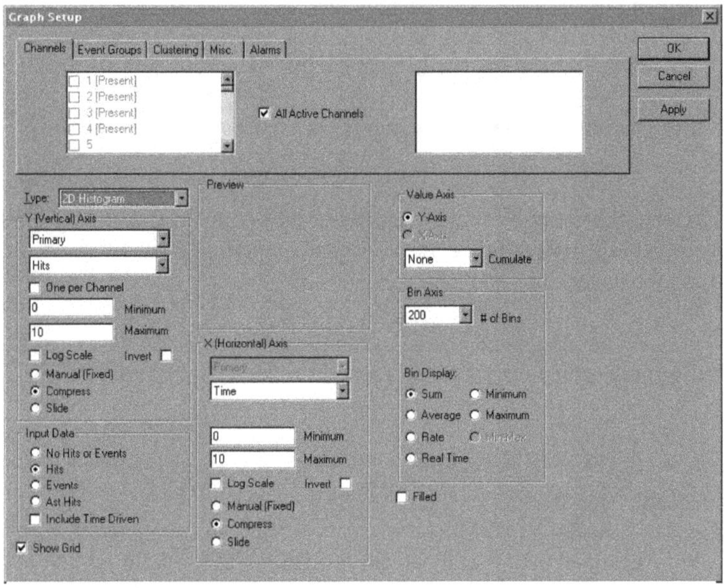

Figure 69. Graph Setup Menu for Hits versus Time Graph

The next menu area is the Graph Type selection box. This is located just below the Channel, Event Group and Clustering selection area. Various graph types can be selected included 2D Histograms, 2D Scatter (Point Plots), 3D Histograms, waveforms, Power Spectrum plots, etc. With the selection of the graph type the graph setup menu below this may change to offer the special configurations needed by that graph type. For the hits versus time application we will use the 2D Histogram.

Below the Graph Type selection, is the Y-Axis setup menu area. Going down one item at a time in this menu is the Plot # selection within the graph. This selection allows you to select more than one plot to be graphed. The one being selected for setup is the one shown. We have only selected one plot in this graph called the "Primary".

Next is the Y- Axis parameter to be plotted on the graph. By clicking this, you can see many selections. However, we have selected Hits. This will display the number of hits (on all channels) detected, on the Y-Axis.

The checkbox "One per Channel" allows us to select a unique per channel plotting color that will give us a multiple color by channel plot. In our case for hits and all active channels, if we selected this checkbox, we will get a multiple plot graph, each plot a different color based on the hits from that one channel, up to 6 plots on one graph. If not checked, we will only get one color plot.

The next two edit boxes allow us to set the starting Y-axis settings (minimum and maximum).

The next checkbox "Log Scale" allows us to set a log or linear graph scale.

The next 3 radio buttons are related, allowing you to select one type of graph scaling. "Manual (Fixed)" will force the graph limits as set above to remain fixed. "Compress" will compress the graph so that the minimum graph value remains but the maximum keeps changing as the graph limit is exceeded. And the "sliding" selection will allow both the minimum and maximum limits to change or slide as data is received that is beyond the current limits.

The next box allows us to select what type of Input data we are plotting. In our case since we are plotting hit data, we want to choose the Hit data for plotting.

Below this is a checkbox for displaying the a grid on the graph.

The next column to the right, shows the graph preview at the top and then the X-Axis settings.

The X-axis settings have similar setup boxes as were described for the Y-Axis and will not be described here. Just note how they have been set up to display the Time parameter and that we have set up the minimum and maximum graph values and a compression scaling so that we can see all data from the start of the test.

The last column on the right changes according to the graph type being set up. If a 3-D graph were being set up, this would be the column for the Z-axis setting. If a Point plot were being set up, there would be no extra column as the points do not need to be conditioned for display.

Since we have selected a graph type of 2-D Histogram, a "Value and Bin Axis" setting menu appears. The Value axis sets which axis the histogram or bars will grow. For our example we are selecting the Y-Axis as the Hits will grow in the vertical direction. We set a cumulate mode if we want the data in the histogram to build upon the previous bin or bars.

The bin axis settings allow us to set the resolution and how the data that makes up a bin get counted in the histogram bar (or what we call a bin). In general, the larger the number of bins, the finer the bars. If your bin axis is time you usually want to have a very high number. If your bin axis is channels, you would like to match the number of bins and channels. In terms of the bin display, you can choose whether you want to sum up all the data in the bin or if you want to average the data in the bin or if you want a rate to be displayed (e.g. Hits/second). To understand, this assume that we have a test for 100 seconds and that we have 100 bins. Each bin will contain one second's worth of data. We can decide to "add it up" and display it as a bin sum, average the amount of data received in that time period or display a rate. In this case since the bin size is 1 second, all the numbers would be the same for this setting, but for any other case, they would be different.

Learn more about these settings by making changes and seeing the effect.

At the bottom of the third column is a histogram setting that allows you to fill the histogram bars or keep them hollow.

Please note and try to make sense of all the selected menu settings. Although it might look complex at the start, if you have gone through these settings, one at a time you will see that all is quite simple. Exit the graph menu by selecting the OK button at the top right of the Graph setup menu.

By looking at the top of the graph we see its graph title of "Hits vs. Time (sec) <all channels>". This summarizes what is being displayed. First is the graph type "Hits vs. Time (sec)". Next we are displaying data from all channels.

Now that we are finished, we see that we have set up a Hits versus time graph and are ready to test it. Test it and make any corrections, until you get the desired result. If we have other graphs, we set them up in a similar way, one at a time until all the graphs are set up.

Now we are ready to proceed with a test or a replay. As a reminder, always save your Layout file before trying it out. It is not fun to have all your work disappear because power went down or something happened to lose the layout.

Hit/Event Linking and the Data Lookup Display Dialog Box

Hit/Event Linking is a way to retrieve hit/waveform or event information on specified points on all types of scatter graphs ("2D Scatter", "2D Binned Scatter", "2D Color Scatter", "3D Scatter", "Spherical Location", etc.). Once a point or area is selected a dialog box is displayed showing a mini-linedisplay containing the hit or event messages linked to that point (or all points in that area). The associated waveform(s) or power spectrum(s) (if any) are displayed above it. Furthermore, a simple location plot is available with most location modes that displays the location of all selected event points. The user can hide/display and customize this plot as desired.

During hit/event linking small cursors (crosshairs) are placed on all other valid 2D scatter/binned/color graphs highlighting the selected hit or the first hit of the selected event.

Hit/Event linking can be performed at any time during or after a test that has autodump ON or a replay. A data file MUST be present for this feature to be available. Changing location settings will disable hit/event linking until a new test (with autodump on) or a replay is performed.

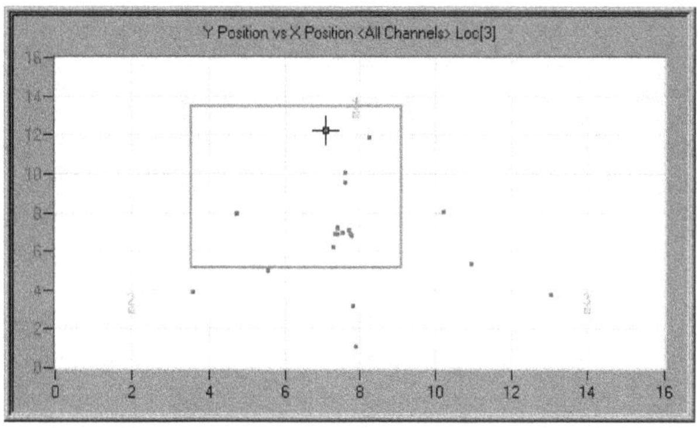

Figure 70: Area-Based Hit Event Linking

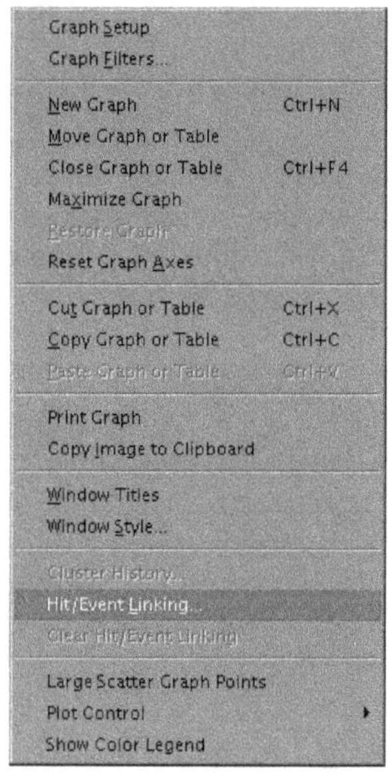

Figure 71: Point-Based Hit Event Linking

Procedure

6) Start a test with autodump ON or begin a replay.
7) Select a graph to work with by left-clicking on it. Its border will darken.
8) Select the point(s) to work with. There are two ways to do this:
    c) Point-based: You can select an individual point by using the data cursor. Activate the 'Data Cursor' from the toolbar. Position the cursor over the desired point. Then right-click the graph and select 'Hit/Event Linking' from the menu that pops up.
    d) Area-based: You can designate a rectangular area on a 2D graph. Activate 'Area Hit/Event Linking Mode' from the toolbar. If you are working with a 2D Scatter or 2D Binned Scatter graph that has multiple Y axes defined it will prompt you to select which Y Axis to work with. Then use the mouse to select an area on the graph (as if you were zooming).

    This begins the Hit or Event linking process. Depending on the graph and the size of the datafile this may take some time. Once finished, the Data Lookup Display dialog box will be displayed and the cursors will be positioned on the appropriate graphs.
9) The cursors can be cleared by right clicking any graph and selecting 'Clear Lookup Cursors', or by exiting the dialog using the 'Exit and Clear Cursors' toolbar button. Exiting the dialog using the 'Exit' toolbar button will leave the crosshairs on the graphs.
10) You do not have to exit the dialog to perform another hit/event link operation. Repeat steps 2 and 3 at any time to refresh the dialog with new data.

Controls

The Data Lookup Display dialog box has a number of controls, most of which are located on a toolbar at the top of the dialog. These are described below. As you move the mouse cursor over each toolbar button a 'tooltip' will popup

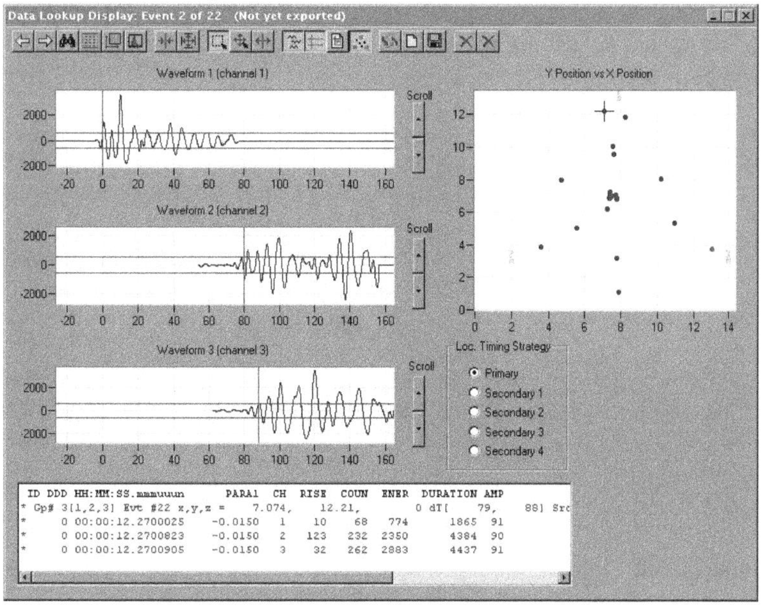

Figure 72: Data Lookup Display Dialog Box

to display the button's name. Some of these controls are only available in event linking. Other controls are only available if the Waveform Option is present.

Prev/Next Buttons: If there are multiple hits/events associated with the selected point(s) then you will be able to view all of them. Use the prev/next buttons to step through them sequentially. The cursors (crosshairs) on any 2D point plot graphs will move appropriately.

Select Button: Instead of using the Prev/Next buttons to step through a large number of hits/events, the 'Select' button can be used to jump to the desired hit/event by simply entering its hit/event number.

View All Hits/Events Button: You can view all selected hits or events together by clicking the 'View All' button. This will popup a dialog box with a large line display showing all hits or events associated with the selected point or area. If you are working with a Hit-based graph (as opposed to an Event-based one) then you can export these hits to a new data file by clicking "Export to DTA". This cannot be done with events here. You can export events one at a time from the main dialog box. See section 3.9.3 below.

Set Power Spectrum Width Button: Clicking this button will open a dialog box allowing you to set the power spectrum width. You can set a specified width or accept the default.

Location Graph Setup Button: Clicking this button will open the 'Location Graph Setup' dialog box. From here you can customize the location graph, by selecting axis parameters, limits and autoscaling.

Reset Unexported User Changes Button: Clicking this button will reset any unexported changes you have made by use of the time-of-arrival cursor. Once you 'Export' the current hit/wave/event the change becomes permanent in the exported file and cannot be reset.

Reset Axes Button: Clicking this button will allow you to reset the axes of one or more graphs. A dialog box will be displayed for you to specify which graphs. You can also reset a waveform graph by right clicking it and selecting 'Reset this graph's axes'.

Select Zoom Mode Button: Clicking this button will select the 'Zoom' mode for all graphs and clear the previously selected mode. In Zoom mode you can use the mouse to select an area of a graph to view in greater detail. Use the 'Reset Axes' button to view the original graph again.

Select Pan Mode Button: Clicking this button will select the 'Pan' mode for all graphs and clear the previously selected mode. In Zoom mode you can use the mouse to drag the viewable window of graph to view data that is currently beyond the graph axes limits. Use the 'Reset Axes' button to view the original graph again.

Select Cursor Mode Button: Clicking this button will select the 'Select Cursor' mode for all graphs and clear the previously selected mode. In 'Select Cursor' mode you can use the mouse to drag the red 'Time of Arrival' cursors on waveform graphs forward or backward to effectively change its time of arrival. While in this mode the mini line-display is replaced with a table showing the time of arrival for each graph. If you do this and then click the 'Force Recalculation of Event' button the event location will change and the new (modified) location will be displayed on the location graph as a black X. Use the 'Reset Unexported User Changes' button to undo these changes (but only BEFORE you export them).

Synchronize Waveform Display Button: When this button is checked the waveforms will be displayed in the same time reference. i.e. If the second waveform in an event has an arrival time 20 us after the first waveform then the entire waveform will be positioned 20 us to the right so its time of arrival (red line) will be positioned at 20 us. If the button is not checked, then all waveforms are left justified on the graphs (their time or arrivals will be positioned at 0 us). This does not modify the waveforms in any way. It is simply a display preference.

Display Threshold Lines Button: When this button is checked, red horizontal lines appear on the waveform graphs showing the threshold. When this button is unchecked these lines are hidden.

Display Power Spectrum(s) Button: The power spectrums can be viewed by clicking the 'Display Power Spectrum(s)' toolbar button. This will replace the waveform graphs with corresponding power spectrum graphs. Click it again to return to viewing waveforms.

Show Location Graph Button: The location graph can be displayed by clicking the 'Show Location Graph' toolbar button. Clicking it again will hide the location graph. Note that this graph is only usable in event linking and only when the event being viewed has more than one hit. You can customize this graph by clicking the 'Location Graph Setup' toolbar button.

Force Recalculation of Event Button: After changing the time of arrival of one or more waveforms in an event, click this button to recalculate the event location. If successful, the new event will be shown in the mini line display and the location graph will be modified to show an X at the location of the modified event. If a new event location could not be created using the new time of arrivals then the mini line display will show a message indicating the problem. Use the 'Reset Unexported User Changes' button to undo these changes (but only BEFORE you export them). Note that certain actions can automatically force a recalculation of the event. These actions include: Changing the location timing strategy and exporting the event.

Export As (New File) Button: Clicking the 'Export As (New File)' button allows you to create a new datafile and export the currently selected hit/wave or event to that datafile. See section 3.9.3 for details.

Export Button: Clicking the 'Export' button allows you to export the currently selected hit/wave or event to a datafile. If you have already specified a datafile for export then it will use that datafile. If not then you will be prompted to specify a new datafile name as if you had clicked the 'Export As (New File)' button. See section 3.9.3 for details.

Exit Button: Clicking this button will close the Data Lookup Display dialog box without clearing the cursors (crosshairs) placed on other graphs to denote the current hit (or first hit of the current event).

Exit and Clear Cursors Button: Clicking this button will close the Data Lookup Display dialog box and clear any cursors (crosshairs) placed on other graphs to denote the current hit (or first hit of the current event).

<u>Loc Timing Strategy Radio Buttons</u>: The bandpass filters and timing strategies defined in the 'Timing' page of the 'Location Setup' dialog box can be applied to the waveforms/events. Use the radio buttons in the 'Loc. Timing Strategy' section to select a strategy. The waveform/power spectrum graph(s) and the mini-line display will be refreshed to show it after filtering. The default 'Primary' strategy does not apply a filter. Note that the selected strategy must have an enabled filter for a filter to be applied. See section 3.9.7 for information regarding the 'Custom Arrival' timing feature, which is particularly useful in this dialog box.

<u>Scroll Buttons</u>: The 'Scroll' buttons (next to each graph) are used to change which waveform is displayed in a particular graph. They allow you to 'scroll' through the list of waveforms up or down as desired. This is particularly useful when the selected event has more than 3 waveforms.

Exporting Hits, Waves and Events

One feature available in the Data Lookup Display dialog box is the ability to 'export' the current hit/wave or event to a new datafile. This can be used as a powerful and flexible way of filtering data, as you can decide on a case by case basis what to export to the new file. If you have made any changes to the time of arrival(s), then these changes are exported as well (see section 3.9.4). To export the currently selected hit/wave or event simply click the 'Export As (New File)' button and enter a filename for the new datafile when prompted. After you do this the title of the dialog box will indicate that the currently selected hit/wave or event has been exported. Subsequent exports to the same file can be accomplished by selecting a new hit/wave or event (prev/next/select buttons) and clicking the 'Export' button.

For Systems that have the Moment Tensor Option: When you perform an export on a system that has the Moment Tensor Option you will be asked if you want to export to PT files as well. This is used for moment tensor (Sigma3D). If you aren't using this then click 'no'. See the file 'Waveform Event Synchronization.doc' for details.

Changing the Time of Arrival of a Waveform and Recalculating an Event

Another powerful feature of the Data Lookup Display dialog box is the ability to 'Modify' a hit/waveform's time of arrival, and (if event-linking) recalculate the event location. To do this, simply use the 'Select Cursor' mode and use the mouse to drag the (red) time of arrival cursor on a waveform graph to the desired point in time. As you do this you will see the new time of arrival updated on a table in the mini line display. If you are in event linking then you can recalculate the event location by clicking the 'Force Recalculation of Event' button. This will recalculate the event using the new time of arrivals. If your event has more than 3 hit/waves then you might have to use the scroll controls to access some of the waveforms to modify.

If you export a hit/wave or event that has been modified in this manner then the changes will be saved to the new datafile and accessible using the 'Custom Arrival' timing feature in location (see section 3.9.7).

Other Features

3) All settings in the Data Lookup Display dialog box are saved when the layout file is saved. This includes the various on/off preferences ('display threshold lines', 'show location graph', etc), location graph setup and custom power spectrum width.
4) If the hit/event linking dialog box is open when a 'cluster/area history' operation is performed then the dialog box will be updated with its data.

Notes

6) Try loading the 'HE-LINK.LAY' layout file and replaying 'HE-LINK.DTA'. Hit/Event linking works well on this example dataset.
7) "2D Binned Scatter" and "2D Colored Scatter" graph types store their data in a 'binned' grid structure much like a Histogram (though in 2 axes). This means that each 'point' on these graphs represents an area rather than an absolute point. The cursor display (near the bottom right corner of the screen) shows the extent of the area represented by the point. Hit/Event linking will locate all hit/events that fall into that area (lower bound inclusive, upper bound exclusive).

8) Selected hit/events are not highlighted with crosshairs on "3D Scatter" and "Spherical Location" graph types, though hit/event linking can be initiated from them.
9) It is possible to perform hit linking on TRA datafiles even though they lack hits. In this case only the waveforms will be displayed.
10) Event Linking functionality can be disabled by unchecking the "Enable Event Linking" checkbox in the Acquire/Replay menu (this setting is NOT saved in the layout). Doing this may result in a modest performance improvement in high data rate tests with large numbers of events. Once disabled it cannot be re-enabled until a new test or replay is performed. Disabling Event Linking will not disable Hit-Linking functionality.

Custom Arrival Timing Feature

When a user-modified waveform is exported it is saved in its unmodified state to the new datafile. The user-modified TOA (time of arrival) is also saved for future use. This can be accessed through the use of the 'Custom Arrival' timing feature.

Before starting the replay, enter location setup. Set one of the timing strategies in the timing page of the location setup dialog box to the 'Custom Arrival' feature. If you want location plots to use the modified TOAs then select this strategy in your location group's 'Timing Strategies'. Or, you can simply set the Primary strategy to 'Custom Arrival' so all location groups will use it. Since 'Custom Arrival' only works in replay, this will default to 'FTC' (First Threshold Crossing) during acquisition.

<u>If you want to see the modified TOAs in hit/event linking:</u>
When the Data Lookup Display dialog box is opened it will plot the locations and waveforms based on the currently selected 'Loc Timing Strategy'. Note that the dialog will remember your previous setting and use it by default the next time it is loaded. If this strategy uses a 'Custom Arrival' timing feature then all the locations will be plotted using the modified TOAs. If this strategy doesn't use a 'Custom Arrival' timing feature then all locations will be plotted using the original TOAs.

Once the Data Lookup Display dialog box is loaded you can switch strategies through the use of the 'Loc Timing Strategy' radio buttons. Doing so will NOT re-plot all locations but it will re-plot the current one AND reposition the TOA cursors on all current waveforms. You can switch back and forth between a timing strategy that uses 'Custom Arrival' and one that doesn't to view the difference between the two.

Cluster/Area History

Cluster History and Area History provide the ability to view detailed graphical information about AE activity in a specified area on a 2D Scatter, Binned Scatter or Colored Scatter graph (called the 'source' graph). This area can be defined by selecting a cluster (Cluster History) or by designating an area manually (Area History). This information is displayed on any number of graphs defined and designated by the user as 'History Graphs'. These graphs remain inactive until the user starts a Cluster or Area History process. When the user does so the software 'replays' the data contained in the selected cluster or area onto all history graphs. The history graphs then remain active for the remainder of the test or replay, accepting/displaying any incoming data that would fall within the original bounds of the cluster or area at the time the process was started. This process can be repeated any number of times during a test or replay.

Setup

4) To use Cluster History you need to create at least one cluster graph before starting the process, usually before starting the test or replay. Area History does not require this.

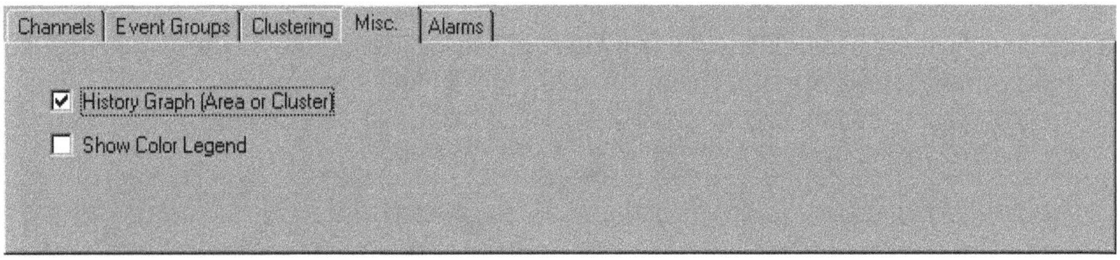

Figure 73. History Graph Checkbox

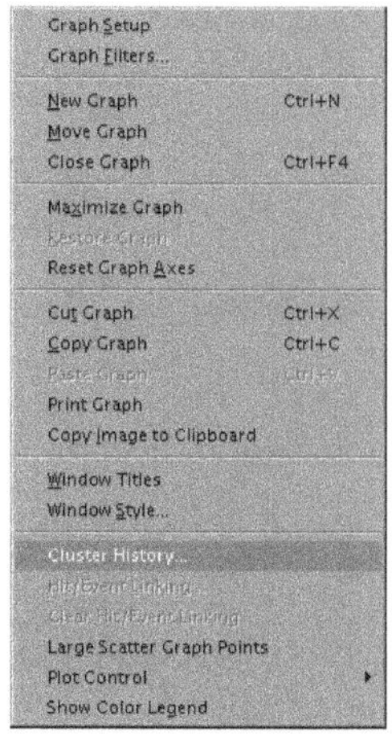

Figure 74. Context Menu

5) Create as many history graphs as desired. This can be done at any time before you start the history process. Any graph can be designated as a 'History Graph' by checking the 'History Graph (Area or Cluster)' checkbox on the 'Misc' page of graph setup. The graph's title will then include the '(Hst)' tag.
6) Start an acquisition test (with autodump on) or a replay.

Starting a Cluster History Process

Cluster History is started by right clicking on a cluster graph and selecting 'cluster history' from the menu that pops up. If this option is grayed out then it means one of the following:

5) There are no clusters on the specified graph.
6) You disabled 'Event Linking and Cluster/Area History ' from the Acquire/Replay dialog when you started the current test or replay.
7) You started a test without autodump on.
8) You started a history process or Hit/Event Linking process that hasn't completed yet.

If there are more than one cluster available on the selected cluster graph then the program will prompt the user to select one from a list.

Starting an Area History Process

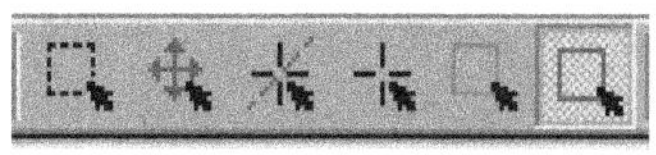

Figure 75. Area History

Area History involves manually designating a rectangular area on the graph. Activate 'Area History Mode' (blue square) from the toolbar.

If this option is grayed out then it means one of the following:
5) The active graph is of an invalid type (i.e.: histogram, waveform, 3D graph, etc).
6) You disabled 'Event Linking and Cluster/Area History' from the Acquire/Replay dialog when you started the current test or replay.
7) You started a test without autodump on.
8) You started a history process or Hit/Event Linking process that hasn't completed yet.

If you are working with a 2D Scatter or 2D Binned Scatter graph that has multiple Y axes defined it will prompt you to select which Y Axis to work with. Then use the mouse to select an area on the graph (as if you were zooming).

Execution

The Cluster/Area History process is detailed below:
6) All history graphs are re-initialized. Any data on them (from a previous history process) is cleared.
7) The setup of all history graphs are modified to reflect the setup of the source graph.

- Channel sets, location group sets, and the data source (hits/events) are made identical to the source graph.
- The min/max limits of any graph axes which share the same graph quantity as on the source graph are changed to reflect the limits of the cluster or area.

8) Special graph filters are set on all history graphs to reject any future incoming data that does not fall within the (original) cluster or area from being displayed. These filters persist until changed by another history process, the user turns off the 'History Graph' checkbox on the graph or until the end of the test/replay.
9) All history graphs are enabled. This allows them accept and display historical and future incoming data.
10) Historical data (the data within the cluster or area) is displayed on all history graphs.

Notes

4) A Cluster/Area History process can be started at any time during or after a test that has autodump ON or during/after a replay. A data file MUST be present for this feature to be available.
5) It is important to note that the bounds of a history graph reflect the bounds of the cluster or area at the time when the process was started. These bounds do not change if the cluster itself grows.
6) Cluster/Area History on event based source graphs is disabled if the 'Enable Event-Linking and Cluster/Area History' checkbox on the Acquire/Replay dialog is disabled before the current test or replay begins.

Graph Alarms

'Graph Alarms' is a feature that enables certain graph types to trigger an alarm whenever data is plotted on them above one or two pre-set thresholds ('warning' and 'trip'). They are available on 2D Histograms, 2D Scatter graphs, Binned Scatter graphs and Colored Scatter graphs during data acquisition.

How to setup a graph alarm:

An 'Alarms' tab is available in graph setup (for graph types that can use them).

6) Check the 'Enable Graph Alarm' checkbox to turn on graph alarm functionality for this graph.
7) Check the 'Enable Warning' checkbox to enable the alarm 'warning' level. Check the 'Enable Trip' checkbox to enable the alarm 'trip' level. Either or both may be used at the same time.
8) Set the vertical axis thresholds for the alarm levels you have enabled. These thresholds are inclusive, so a threshold of 80 will be triggered by a signal of 80 or above.
9) Select which vertical axis controls the graph alarm (Primary, Secondary-1, etc).
10) If desired, you may enable the 'Alarm Hold-Off' and specify its '#Signals'. See the section on 'Alarm Hold Off' below for information on this feature.

How an alarm is detected/triggered:

Graphs that have an enabled graph alarm will display horizontal lines at the levels of the thresholds set in graph setup (Green for warning, Red for trip).

A graph alarm 'Warning' will trigger when all of the following occurs.

- The 'warning' level is enabled for that graph.
- New data is displayed on this graph at or above the green threshold line.
  - This needs to occur a number of times consecutively equal to the specified '#Signals' if used.
- No graph alarm (Warning or Trip) is currently active (on this graph).

An alarm 'Trip' will trigger when all of the following occurs.

- The 'trip' level is enabled for that graph.
- New data is displayed on this graph at or above the red threshold line.
  - This needs to occur a number of times consecutively equal to the specified '#Signals' if used.
- No graph alarm trip is currently active (on this graph).

'New' data is defined as any data that is plotted after the alarm was last reset. Note that the alarm is automatically reset at the start of the test.

What happens when an alarm is triggered:

When a graph alarm is triggered several things happen:

- The audio monitor (if installed) sounds an audible beeping sound. This continues until the alarm is cleared or the test ends.
- The alarm output (WARNING or TRIP as appropriate for the level) is set high. This stays high until the alarm is cleared or the test ends.
- An alarm message (16) appears in the Line Display in red, and is saved to the datafile if autodump is on.

A small dialog box appears, displaying the conditions of the alarm. The alarm can be cleared by clicking the 'Clear Alarm' button. You can exit this dialog without clearing the alarm if you wish. It can be accessed later by selecting 'Display Last Alarm…' from the Acquire/Replay menu.

Alarm Hold Off

Normally (when the "Alarm Hold Off" is disabled) an alarm is triggered when a <u>single</u> data point ("signal") is plotted above threshold.

By enabling 'Alarm Hold-off' you can delay alarm triggering until a specified number of <u>consecutive</u> data points ("signals") are plotted above threshold. This is useful when you wish to avoid having data spikes trigger alarms. It is important to remember that the string of data points must be consecutive for an alarm to be triggered. Any data points below threshold will 'reset' the internal counters.

For example, if you set '# Signals' to 5 then an alarm would only be generated until 5 consecutive data points are plotted above threshold. Let's say your threshold was 70. The following sequence:

- 70, 80, 75, 80, 70 would trigger an alarm.
- 70, 80, 60, 80, 70, 80 would NOT trigger an alarm, because even though 5 signals passed threshold, they were not consecutive.

Alarm 'Warning' and 'Trip' both maintain their own separate internal counters (though both use the same '#Signals'), so a particular sequence might trigger an alarm 'Warning' without triggering an alarm 'Trip'.

Tips:

7) Graph alarms are checked when the graphs are refreshed. If you want to increase the rate at which they are checked then you should increase the rate at which the graphs are updated. See the settings in 'Display Mode…' under the 'Test Setup' menu.
8) Note that data does not have to be in the visible portion of the graph to trigger an alarm. Panning/Zooming of the graph will not hinder graph alarm operation.
9) Setting #Signals to 1 is identical to disabling 'Alarm Hold Off'. In fact, if you do so, the system will automatically disable 'Alarm Hold Off' for you.
10) Caution is recommended when using Compression on the horizontal axis of histograms, particularly when using "Sum" Bin Display mode. In this scenario the bins will become taller as the graph compresses horizontally, an effect of the bins being combined as part of the compression process. Be aware that data is only checked against alarm criteria only when it is first plotted, not when the graph is refreshed during compression. A bin that suddenly grows above threshold due to compression will not trigger an alarm.
11) Caution is recommended when using 'Alarm Hold Off' on graphs that have all the following settings selected:
    - Graph is plotting RMS, ASL, Threshold or Absolute Energy.
    - The 'Include Time Driven' checkbox is checked in the 'Input Data' section of Graph Setup.
    - Graph is plotting more than one channel.

    In this case you are plotting channel-based 'time driven data', which can plot multiple points from a single time driven data message. 'Alarm Hold Off' will advance its internal counter at most once per message, so this can be tricky if your graph is plotting more than one channel. If you want to use 'Alarm Hold Off' with Graph Alarms on the type of graph listed above it is advised that you create multiple single channel graphs instead.

12) It is recommended that 2D Scatter (not Binned/Colored) graphs that have a graph alarm enabled use only Manual scaling on the Y (vertical) axis. Using compression or sliding scale won't hinder graph alarm operation but it may slide the threshold line off the visible portion of the screen.

List of Shortcut Keys

The following is a list and short description of all the shortcut keys available in the software. Many of the shortcut keys follow in the tradition of 's DAQ and LOC software programs and should be familiar. Others follow standard WINDOWS conventions. Menu items usually do list the shortcut key, but the full list is provided below for convenience.

F1	Help Menu
Shift + F1	Context Sensitive Help
F2	Hardware Setup
F3	Graph Setup
F4	
Ctrl + F4	Close (Delete) Graph
Shift + F4	Delete Page (Entire Graph Screen)
F5	
F6	View Cluster Table (Toggle)
Ctrl + F6	Controls Cluster Table Sorting
Shift + F6	Shifts Cluster Table Position
F7	View Line Dump (Toggle)
F8	Location Setup
F9	Acquisition
Shift + F9	Acquire to Existing File
F10	Replay
F11	
F12	
Ctrl + C	Copy Graph
Ctrl + D	File New
Ctrl + H	Window Tile Horizontal
Ctrl + N	Add New Graph
Ctrl + O	Open Layout File
Ctrl + P	Print Page
Ctrl + S	Save Layout
Ctrl + T	Window Tile Vertical
Ctrl + V	Paste Graph
Ctrl + X	Cut Graph
Ctrl + W	Waveform Collection (Toggle)
Ctrl + Shift + C	Copy to Clipboard
Ctrl + Insert	Copy Graph (same as Ctrl + C)
Shift + Insert	Paste Graph (same as Ctrl + V)
Ctrl + Alt + O	Options Dialog

LOCATION

The purpose of this section is to familiarize you with setting up and using the software location modes. Location is the process of collecting incoming hits into events & analyzing the arrival times of the hits in an event to produce a source location. Individual location modes have several key differences which are summarized in the table below.

Location Mode	Hits Used	Description	Analysis Type	Required Option
Zonal	1	Trivial solution to location. Identifies the sensor closest to the source using the relative arrival times of the hits. Returns coordinates of the first hit sensor as source location.	Single	Standard
Linear	2	Designed for 1 dimensional location on a line. Performs linear interpolation between 2 sensors' coordinates based on differences in the arrival times of the first 2 hits in the event.	Single	Standard
LinearXY	2	Linear without the assumption of sensors in a straight line & events allowed between any sensor pair. For example, sensors could be in a square with sources locating on the sides or diagonals.	Single	Standard
Tank Bottom	3	Designed for analyzing evenly sed 2D circular arrays of sensors.	Single	Tank Bottom
2D Planar	3-8	Designed for an arbitrary array of sensors on a 2D plane.	Regression	Full
2D Planar (xy) 2D Planar (yz) 2D Planar (xz)	3-8	2D Planar designed to complement 3D Location. Locates using just 2 of the 3D coordinates & ignores the 3rd. Useful for some events that fail normal 3D Location.	Regression	3D
2D Anisotropic	3-8	2D Planar allowing for differing velocities in the x & y directions.	Regression	Full
Cylinder	3-8	2D Planar analysis wrapped around a 3D cylindrical shell. The ends of the cylinder may be open or have flat, spherical or elliptical endcaps. Assumes sound travel through cylinder shell rather than its interior.	Regression	Full
Anisotropic Cylinder	3-8	Cylinder allowing for differing velocities in the x & y directions. Only supports location on cylinder body, not on the endcaps.	Regression	Full
Conical	3-8	2D Planar wrapped around a 3D conical shell.	Regression	Full
Spherical	3-8	2D Planar wrapped around a 3D spherical shell again assuming sound travel via the shell only.	Regression	Spherical
3D Location	4-8	2D Planar extended to 3D coordinates assuming sound travel directly through se rather than via an external shell.	Regression	3D Location
3D T0 Location	5-8	3D Location with time zero input. Assumes all events are signaled via one channel that indicates the absolute start time of an event.	Regression	3D T0 Location

The most important difference in the location modes is the geometry they are designed for. The geometry in turn affects the number of hits used per event to calculate the location. The type of analysis also affects the number of hits used in an event. The single analysis modes use just enough hits to create one set of simultaneous equations which are then solved to extract the source location coordinates. A more complex approach uses more hits & multiple regression analysis to produce the source location that best fits all the available data. By using data from extra sensors, the regression modes yield a more accurate source location with improved immunity to errors in the individual arrival times. For more details, see the section 4.1, Theoretical Background.

Two location modes, zonal & linear, are standard features. Others are location enhancements that require additional licensing options. If you are not sure what locations options are enabled in your software, look for the Options dialog available on the Help pulldown menu. It will show a checkmark next to any purchased options. If you wish to purchase additional license options for other location modes, contact , Physical Acoustics Corp. at 1-609-716-4000.

Theoretical Background

The following section provides a brief overview of the math behind location. It is definitely not required that the average user to comprehend this material. Its purpose is to educate the inquisitive & allow the user additional insight into what location is doing behind the scenes.

The fundamental basis for the location calculation is just the simple time-distance relationship implied by the velocity of the sound wave. The absolute arrival time, t, of a hit in an event can combine with the velocity, v, of the sound wave to yield the distance, d, from the sensor to the source.

$$d = v * t \tag{1}$$

The distance between 2 points depends on the geometry of the problem. The majority of the location modes are a variation of 2 dimensional source location in a plane, although in many cases the 2D plane will wrap around a 3 dimensional object. For 2 points in a flat plane, the distance equation is just the Pythagorean theorem expressed in Cartesian coordinates:

$$d = \sqrt{(x_2 - x_1)^2 + (y_2 - y_1)^2} \tag{2}$$

This calculation is complicated by the lack of knowledge of the exact time the event originated. To get around that problem, all the times are considered relative to the first hit in the event. Each arrival time difference implies a difference in distance to the sensor relative to the distance to the first hit sensor. For the second hit sensor relative to the first hit sensor, a difference equation can be written as:

$$t_2 - t_1 = (d_2 - d_1)/v \tag{3}$$

The distance equation (2) can be combined with the difference equation (3) to yield:

$$t_2 - t_1 = [\sqrt{(x_2 - x_S)^2 + (y_2 - y_S)^2} - \sqrt{(x_1 - x_S)^2 + (y_1 - y_S)^2}]/v \tag{4}$$

where x_S & y_S are the unknown coordinates of the source. This equation contains 2 unknowns & cannot be solved by itself. To get a second equation with the same 2 unknowns, a 3rd hit is added to the event producing analogous equation:

$$t_3 - t_1 = [\sqrt{(x_3 - x_S)^2 + (y_3 - y_S)^2} - \sqrt{(x_1 - x_S)^2 + (y_1 - y_S)^2}]/v \tag{5}$$

These simultaneous equations can then be solved for x_S & y_S, although the algebra is fairly complex. Tank Bottom location uses this particular approach to produce a source location from 3 hits as did the original version of 2D Planar location before it was extended to include multiple regression analysis.

If there are additional hits added to the event, there is a question of how to use the extra information. In the case of Tank Bottom location, the algorithm simply takes the extra hits & tries them 3 at a time to produce extra source locations. It iterates through all the 3 hit combinations of the hits in the event until it exhausts them or runs into a user specified maximum. The problem with this approach is that it produces more than one source location per event & if there is any error in the timing values, the source location can be wildly incorrect. A better approach would be to average the data somehow to produce a single location.

Multiple regression analysis does just that, although it does not actually average the results of multiple 3 hit calculations directly. Rather it searches for the location that best fits all the all the available data. Each additional hit basically adds an extra equation to the set of simultaneous equations given above as (4) & (5). This can be generalized as:

$$t_i - t_1 = [\sqrt{(x_i - x_S)^2 + (y_i - y_S)^2} - \sqrt{(x_1 - x_S)^2 + (y_1 - y_S)^2}]/v \tag{6}$$

If one defines

$$\Delta t_i = t_i - t_1 \tag{7}$$

one can write:

$$\Delta t_i = [\sqrt{(x_i - x_S)^2 + (y_i - y_S)^2} - \sqrt{(x_1 - x_S)^2 + (y_1 - y_S)^2}]/v \tag{8}$$

Equations (7) & (8) give 2 ways of calculating the Δt of the ith sensor. Using the known arrival times of the hits, equation (7) calculates the observed time difference, $\Delta t_{i,obs}$,. For a given set of source location coordinates, equation (8) defines the calculated time difference, $\Delta t_{i,calc}$. Multiple regression analysis is a general purpose algorithm that minimizes the difference between 2 quantities which in this case are the observed & calculated Δt values. To do that a quantity called χ^2 (chi square) is calculated. In our case, it is defined as a sum across all sensors in the event:

$$\chi^2 = \Sigma(\Delta t_{i,obs} - \Delta t_{i,calc})^2 \tag{9}$$

This quantity is also called the fit value and is dependent on the x,y coordinates of the source location . The sum is recalculated for each potential source location. Assuming no error in the data, χ^2 will have a value of 0 at the source location. The location code searches for the values of x_S & y_S that minimize the value of χ^2. The process is an iterative search, because it is not possible to directly write down simple equations for the values of x_S & y_S that minimize χ^2.

The search is performed using one of 2 algorithms, either a Simplex search (3D or 2D Planar modes) or Powell's Method (Spherical, Cylinder & Conical Location). Both search methods are based in general on code & algorithms described in Numerical Recipes in C, 2nd edition (NRC). The code has obviously been heavily adapted for the problem of acoustic emission source location, so there are significant variations from the NRC originals. The interested user may, however, find more information about Simplex & Powell fitting & searching in NRC in chapter 10 & more information on regression analysis in chapter 15.

The math gets more complicated when extended to 3 dimensions or involves wrapping around a 3D object, but the approach is still basically the same. An important point is that the regression approach is no longer over determined in the case of the event containing the minimum number of hits, i.e. 3 hits for 2D Planar. In that case, however, regression analysis turns out to be mathematically equivalent to solving the complicated algebra of equations (4) & (5) & yields the same location coordinates.

Location Setup Dialog

When setting up location groups, you use the Location Setup dialog box (F8 shortcut key). It has 2 tabs labeled General and Timing. Source locations are produced by analyzing events. Events are just collections of hits & their associated time of arrival. The settings on the General page control how hits are combined into events & how events are analyzed by the location modes to produce source locations. The settings on the Timing page control how the time of arrival of each individual hit is determined and even whether or not a hit is considered significant or excluded from the location analysis.

There are several features to make setting up location groups easier. First you should be aware of the F5 function key. It allows the user to quickly set all rows in a column of the grid to the value of the current cell. It works on both the General & the Timing tabs. In the example setup shown below, pressing F5 would set all rows of the Overcal column to the current cell's value of 200. Alternatively, you can right click a cell & select Set Column from the context menu that appears. The context menu for the General tab, also has Copy Group & Paste Group options. Right click & select Copy Group on source row, then right click the destination row & select Paste Group. All of the group settings & the associated sensor & structure settings will be copied to the new row. By default, Paste Group will use the first group as the source, if you did not use the Copy Group option to specify a source row.

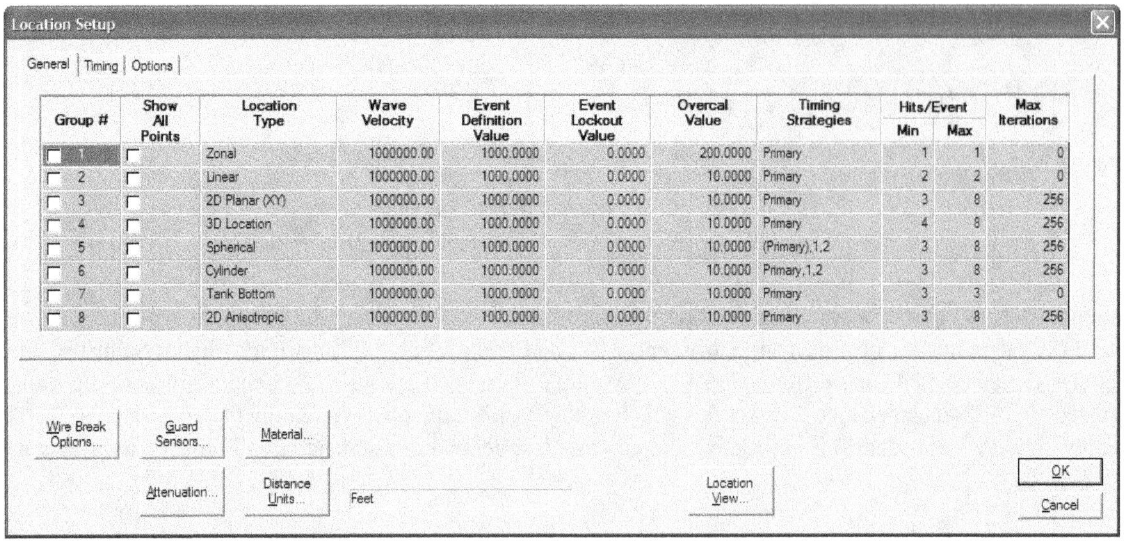

*Figure 76. **Location** Setup – General Page*

General Tab Settings:

The layout for the General tab is shown above. The grid on the General page of Location Setup contains a list of location groups & their type. If you click an entry in the Location Type column, an arrow appears allowing you to change the type. As mentioned above, many location types (3D, Spherical, Cylinder) appear only if you have purchased the appropriate option.

Group #: This column acts as an On/Off setting. Unchecked groups are ignored when doing a replay.

Show All Points: This column allows the user to show all source results for events that this group can successfully locate. Independent of which groups have Show All Points checked, the software always allows all enabled groups to analyze a given event to get the best results for each group and then compares the results from all successful groups to come up with a best result. For example, linear location is always presumed to give a better result that zonal location, if there are enough hits(2) to produce a location. If no groups have Show All Points checked, the software will report one source location per event & it is the best result that is shown on graphs & in the line display. There are times when a side-by-side comparison of the results of 2 locations groups are useful. In that case you can mark the groups as Show All Points & then whatever source locations they can calculate will be reported. The software will appear to sometimes produce more than one source location per event & if you look at the line display, you will see extra event listings with the same first hit times. The extra location results will also show up as extra points on your graphs too, but they only count as 1 event in the Statistics bar. In all cases you will see the software's best result as one of the points. There will be times when an event cannot be located by more than 1 group, so having Show All Points does not guarantee a result from a given location group. It is only a guarantee that you will see it, if it can produce a useful result.

Location Type: The Location Type column lets you specify which location mode to use with this particular location group. If you click on a cell in this column, a down arrow appears on the right hand side of the cell. If you click the down arrow, a drop down list of the available location choices will appear. The available choices do depend on the software licensing options as mentioned in the introduction, so you may not see all the location modes that are discussed in this manual.

Wave Velocity: The Wave Velocity is the expected rate of propagation of the acoustic emission signal you wish to monitor. It is entered in the user's Distance Units per sec. For example, if the Distance Units are in Inches, then the Wave Velocity takes units of Inches/sec. If the user's Distance Units are time based, this column is ignored by the location algorithm. The Materials button will allow you to access a database of known velocities for various materials.

See the section on the Material Properties dialog for more information. It is always recommended that you determine the appropriate value by measuring the velocity on your test structure using lead breaks, for example.

Event Definition Value: The Event Definition Value (EDV) is the length of an event in the user's Distance units. If the user's Distance Units are time based, then the EDV is directly the maximum time allowed between first & last hits in an event. If the EDV is distance based, then it is the physical length between the first and the last hit sensor of the event. The goal is to select an Event Definition value that will give enough hits to perform source location with the desired accuracy. For example, in the figure below, the 2D location requires a minimum of 3 hit sensors to locate, however more sensors can provide a more accurate source location, therefore, the inner circle radius defines the Event Definition Value and in the case shown, it will detect channels 9, 10, 15, 16, 17, 21, and 22, so up to 7 channels (hits) will be able to form the event if the EDV value as shown is entered.

Since the location calculations use time information, any distance units will be later converted to time units using the velocity entered in the Wave Velocity column. The velocity units are the user's Distance Units per second. For example, if the Distance Units is set to meters, then the velocity units are meters/sec.

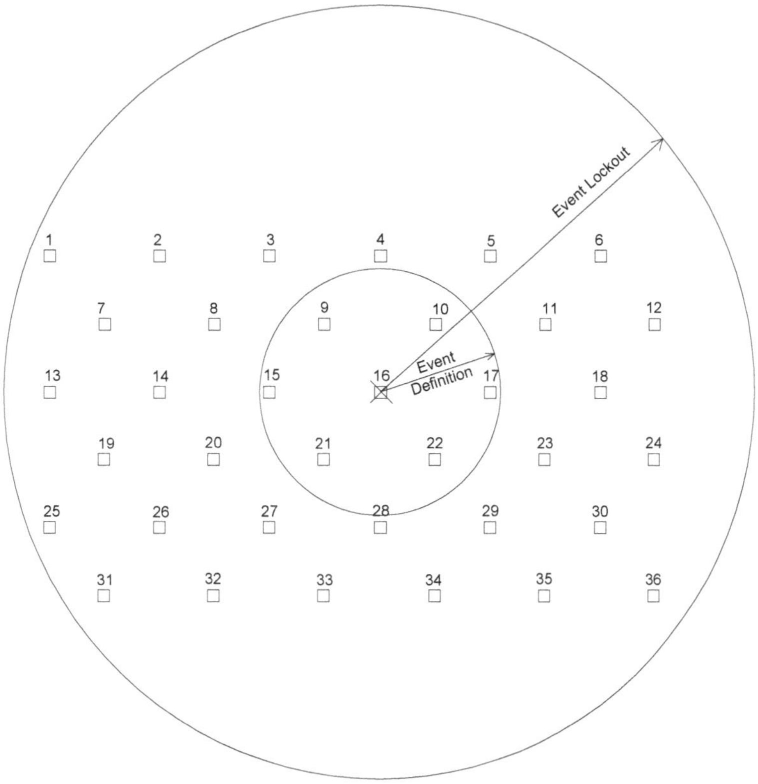

Figure 77. Relationship between Event Definition and Event Lockout.

Event Lockout Value: The Event Lockout Value is also a length of time or distance entered in user units, but it controls the interval between consecutive events. The Event Lockout of the last located event blocks hits in all location groups until it has expired. The Event Definition Value & the Event Lockout Value are both specified relative to the start of the event (time of test of the first hit). An Event Lockout Value of 0 allows overlapping events in 2 *different* location groups. Overlapping events are not permitted in the same location group no matter what the Event Lockout Value is. An Event Lockout Value larger than the Event Definition Value allows hits to be ignored after an event, for example, if reflections could generate extraneous acoustic signals. In the example shown in the figure above, all AE hits between the Event Definition Value and the Event Lockout will be ignored from the point of view of being part of the event in progress or the next event.

Overcal Value: The Overcal Value or more fully the over-calibration value is also a length that is specified in the user's Distance Units. Before data acquisition or a replay begins, the software calculates a table of time differences between sensors using their known coordinates & the user entered wave velocity. The Overcal Value is also converted to a time value if needed. After the first hit in an event, these calibration times are compared to the time differences of later hits relative to the first hit sensor. In order for a hit to be included (or accepted) as part of the event, its time difference must be less than this calibration time, but the Overcal value provides an allowance for a margin of error. The time difference can be higher than the calibration time by the Overcal Value & still be included in the event.

Timing Strategies: The Timing Strategies column is used to specify which strategies listed on the Timing page will be applied to a particular location group. You set up the strategy on the Timing page, but you control whether it is used with a particular location group using the Timing Strategies column. See the section on the Timing page for information on the settings for strategies & use of the Timing Strategies column.

Hits/Event Min & Max: The Hits/Event (Hits per Event) columns are related to the regression modes. In theory, the multiple regression can sum across any number of sensors (hits) in a given event when calculating a fit, but there are limits in the code. First the absolute minimum number is determined by the mathematical requirements specific to the location mode. For Spherical, Cylinder Location and all 2D Planar regression modes, the event must contain at least 3 hits to produce a location. For 3D Location a minimum of 4 hits are required to analyze an event & produce a location. All regression modes can make use of up to 8 hits in an event.

The Hits/Event Min & Max settings allow the user control over the analysis between these limits. For example, in 2D Planar, a minimum of 3 hits are required to produce a location result, but then this is not an over determined result. If there is an error in the timing data of any one of the 3 hits, the location calculated will be in error. With 4 or more hits, the regression analysis produces a best fit location that averages out errors in the data. Thus, the user might want to adjust the Hits/Event Min setting to 4 for 2D Planar to eliminate analysis of 3-hit events. This will reduce the number of events observed & generally reduce scatter in the results. In some cases, the user might want to only look at events with a specific number of hits. Setting the Hits/Event Min & Max settings to 4 will limit the analysis to events with 4 or more hits using only the first 4 hits to form the event.

Max Iterations: The Max Iterations column is also related to the regression modes. In this case, the setting controls the maximum number of locations evaluated in the search for the best fit location. It is not strictly speaking the same as the number of times through a search loop for 2 reasons. First some searches can test a variable number of locations during one pass though the loop. Second other searches use loops within loops. Due to the variable nature of the search algorithms, the actual number of locations tested can be slightly higher (0-3) than the Max Iterations setting. A lower Max Iterations setting will terminate long location analyses sooner, but the higher speed results in lower accuracy in the best fit location produced.

When the Location Type is not set to an regression mode, the Max Iteration column & the Hits/Event Min & Max columns will be disabled. In that case the Max Iterations column displays 0 as iterations are not used for the other modes. The Hits/Event Min & Max settings will be updated with the number of hits required for the location mode. For example, the Linear location mode requires 2 hits to produce a location and if Location Type is set to Linear, Hits/Event Min & Max will both display 2 & the user will not be able to adjust the setting.

Timing Tab Settings:

Timing strategies processes hits and their accompanying waveform to produce a time of arrival. A list of arrival times a series of hits is grouped together to form an event. The location modes analyze the arrival time differences (delta t's) in the event to calculate the source location. For an enabled location mode, the software always analyzes the event at least once. The first analysis is performed using only the arrival times from the Primary Strategy. If you use additional strategies with one of the regression modes, the software performs a second event analysis to refine the source location by adding the arrival times produced by the secondary strategies (1-4) to the event.

The settings for the strategy are on the Timing page, but you control which location groups use them with the Timing Strategies column on the General page. When you click on a cell in the Timing Strategies column, the Select Strategies dialog appears as shown below.

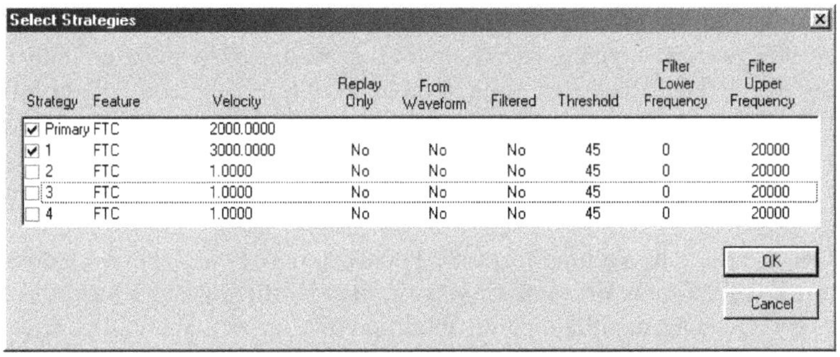

*Figure 78. Select **Strategies** Dialog*

Simply check or uncheck the strategies you want to use. The dialog summarizes the current strategy settings from the Timing page. After you click on OK, the cell in the Timing Strategies column will update to list the enabled strategies. A notable exception is the Primary Strategy which always appears in the list as either Primary or (Primary).

With additional strategies enabled, unchecking the Primary Strategy causes the arrival times from the Primary Strategy to be omitted from the second event analysis, but they are still used to perform the first analysis. If the location mode cannot produce a useful source location after the first analysis, the software will not attempt to refine the location with the secondary data. In that respect the Primary Strategy acts like a filter for the second analysis even when the Primary results are not being used in the second analysis. To reflect this implied filtering, the Timing Strategies column displays (Primary) whenever the Primary Strategy is not to be included in the secondary analysis. In the figure showing the General settings page above, Location group 5 has Strategies 1 & 2 turned on with the Primary turned off and displays "(Primary),1,2". Unchecking all the strategies is equivalent to checking only the Primary Strategy. In both cases the event is analyzed just once with the Primary Strategy and the Timing Strategies column just displays "Primary"

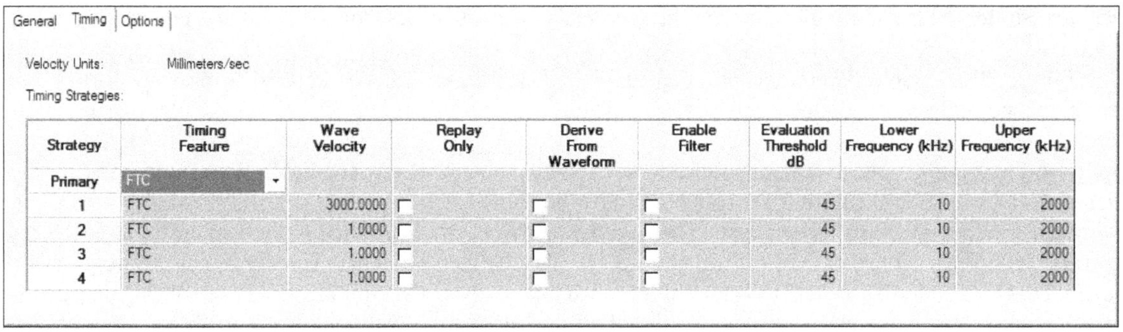

*Figure 79. **Location** Setup – Timing Page*

The current layout for the Timing page of the Location Setup dialog is shown above. For each strategy there are a number of options available. Those to the right of Derive From Waveform are all related to waveform processing.

For the Primary Strategy the only available option is the Timing Feature option. The other columns are not used because they are unneeded or not allowed for the Primary Strategy. For example, the Primary Strategy always uses the velocity associated with the location that is listed in the Wave Velocity column on the General page. The Primary Strategy is always applied during both acquisition & replay. The Primary Strategy relies on the results reported by your hardware in a single hit message, so waveform processing options are not available. The software later searches & pairs each hit message with an accompanying waveform message making the waveform processing options available for the secondary strategies.

Timing Feature: The Timing Feature column controls how the arrival times for a hit are determined. There are 2 basic choices: FTC (First Threshold Crossing) and Peak Timing (PT). All the arrival times are based on the time of test associated with the hit and a variable offset. During data acquisition, the hardware actually sets the time of test of a hit based on a threshold crossing where the Threshold level is controlled by the user in the Hardware Setup dialog. The FTC setting actually just defaults the arrival time to the time of test for the hit with no offset. The Peak Timing parameter adds the risetime (RT) as determined by the hardware to the hit time of test. Risetime is defined as the delay between the first threshold crossing and the highest amplitude peak during the Peak Definition Time (PDT). The risetime feature must be enabled in the Data Sets tab of the Hardware Setup dialog for this option to work.

In addition to FTC and PT there is a third timing feature called Custom Arrival. This is used when replaying a datafile that has had the time-of-arrival of some or all of its waveforms modified through hit/event linking. If this timing feature is used when replaying such a datafile then the modified time-of-arrival's will be used instead of the original ones. For any other timing feature, the original time-of-arrivals will be used. Custom Arrival is designed for use in replay. If it is used during acquisition it will default to FTC behavior.

Wave Velocity: The Wave Velocity column contains the velocity used during analysis by the location modes. The regression modes use a velocity to convert arrival time differences into differences in distances to the sensors. There is a velocity associated with each hit arrival time. For the Primary Strategy that velocity always comes from the location group Wave Velocity setting. During the second analysis, the velocity from the secondary strategies are used when the arrival times are added to the list of hits in the event.

Replay Only: The Replay Only checkbox controls whether the strategy is applied during both data acquisition and replay or only in replay mode. Since secondary strategies cause events to be analyzed twice & the arrival times may be extracted from a digitally filtered waveform, secondary analysis can slow data acquisition rates. In cases where high data acquisition rates are important, you may prefer to postpone the secondary analysis until after data acquisition is over. The Replay Only flag just makes it easier to disable the strategy during acquisition without having to continually access the Location Setup menu to toggle the strategy on or off.

Derive from Waveform: The Derive From Waveform checkbox controls whether the strategy attempts to extract the arrival time from the waveform. If it is disabled (unchecked), the arrival time can still vary from that produced by the Primary Strategy if the Timing Feature for the secondary strategy varies from the Primary Timing Feature setting.

Enable Filter: The Enable Filter checkbox controls whether a digital filter is applied to the waveform before the arrival time is determined.

Evaluation Threshold: The Evaluation Threshold is similar to the Hardware Setup Threshold used during data acquisition. If the Timing Feature is set to FTC (First Threshold Crossing), the Threshold directly imts the arrival time or even whether an arrival is detected. The Evaluation Threshold is applied to the waveform after the optional digital filter is applied. Note that the Hardware Setup Threshold affects what data is recorded during acquisition. The user can set the Evaluation Threshold to values higher than the Hardware Setup Threshold used during data acquisition & simulate the results that would have been recorded during acquisition with a higher Threshold. Beware, though, setting it to lower values cannot reveal data that was not recorded in the first place.

Upper & Lower Frequency: Lower Frequency & Upper Frequency columns control the bandpass cutoff frequencies. Turning on the digital filter & adjusting the filter cutoff frequencies can help isolate individual acoustic emission modes with differing wave velocities. Note that you should be aware of the sampling rate used during acquisition and avoid setting the Lower Frequency cutoff above ½ the lowest sampling rate on any channel. You should also avoid targeting frequency ranges that your sensors do not respond to or that have been removed by your hardware's analog filters. Otherwise you could be analyzing meaningless waveforms.

Options Tab Settings:

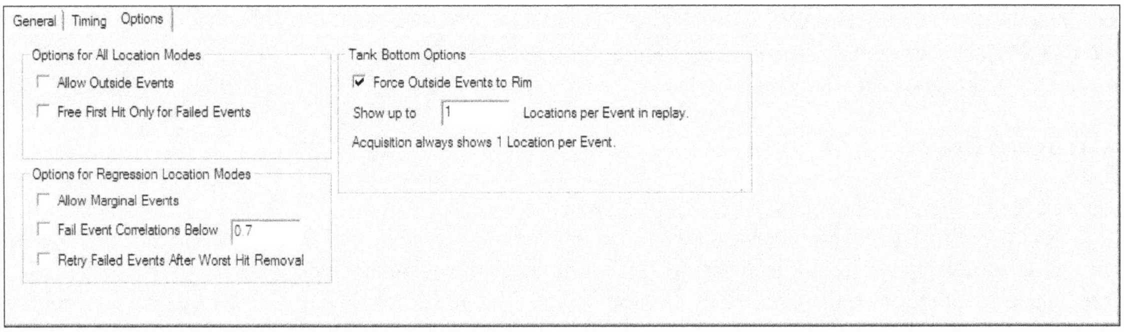

Figure 80. Location Setup – Options Page

There are a number of options available on the Options Page of Location Setup that control how location processing occurs. There are 3 different groups of options. Some apply to all location modes, some to only the regression location modes & some are specific to Tank Bottom Location. Here is a description of each group of options.

Options for All Location Modes

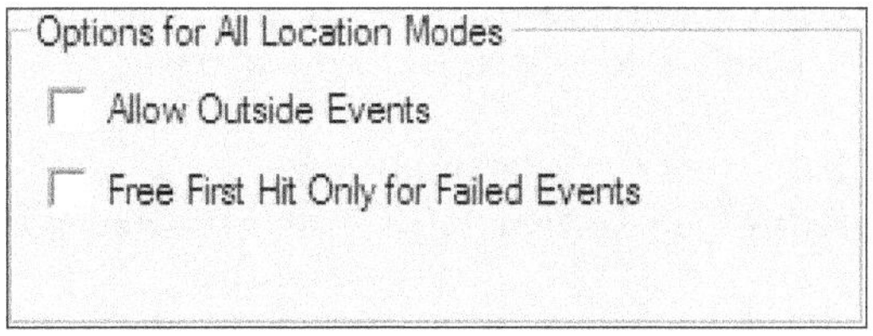

Figure 81. General Location Options

Allow Outside Events

The Allow Outside Events checkbox controls what happens to events that produce source locations outside the structure. Usually this is due to timing errors from an odd combination of hits like the 2nd hit sensor being further away from the 1st hit sensor than the 3rd hit sensor. The default behavior is to reject them as failed events & not display the calculated source locations, but the user can optionally display them. This option does not apply to the Zonal or Linear location modes, but it does apply to Tank Bottom location & all the regression location modes. For example, if 2D Planar location is paired with the Plate structure, events that locate outside the Plate limits will be rejected. The option has no effect for 2D Planar or 3D Location when they are set to use the Free structure. The software does allow a margin of error at the structure limits, so that events that locate very close to the limit still show. In general that margin of error is an extra 2% of the structure dimension. For the Plate structure, that is 2% of the length and 2% of the height. For the Tank Bottom structure, it is 2% of the radius. When you have Allow Outside Events enabled, you may wish to use manual axis limits on any 2D location graphs as outside events can sometimes have coordinates with very large values.

Free First Hit Only for Failed Events

The software assembles groups of hits into event candidates which are then analyzed by whatever location groups the user has set up. When these potential events failed to produce a valid location for whatever reason, the software has always freed up all hits in the candidate event in the past. This meant it would not reuse those hits in other candidate events. A single stray hit due to noise can easily obscure a true event that occurs right after the noise hit. Now there is an option to only free the first hit of a failed event. The other hits will be allowed to continue to be considered for use in other candidate events. Checking *Free First Hit Only for Failed Events* will allow the software to find more potential events at the expense of more processing time slowing down the replay of the user's data. This option is ignored during acquisition. During acquisition all hits in a failed event are always freed to minimize the imt of location processing on the data acquisition speed.

Options for Regression Location Modes

There are 3 options specific to the regression based location modes. These are 2D Planar (& XY, YZ & XZ variants), 3D Location, Spherical Location, Cylinder Location & Conical Location, 2D Anisotropic, Anisotropic Cylinder.

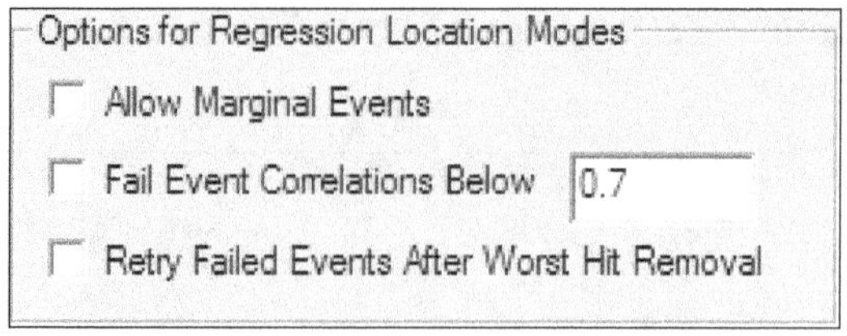

Figure 82. Regression Location Options

Allow Marginal Events

The Allow Marginal Events checkbox has the largest imt on 2D Planar & 3D Location. By default this checkbox is not enabled. Marginal events are events that the software expects to have low accuracy. These are linear events for 2D Planar & planar events for 3D Location. See the Notes sections on those location modes for a more detailed discussion of what linear & planar events are & why they are problematic.

This option is useful for comparing with the results of 's older DOS kages that did not automatically filter out these kinds of events. If the user wishes, he can enable this checkbox & see the results of these events, but be aware that it is not recommended as the calculated results have a 50% or greater chance of being incorrect. The incorrect locations can sometimes be useful, though. For example, if a 2D linear event show a result that is +5 from the line of sensors, then -5 is an equally valid answer according to the mathematics. This is also true for distance from the plane for 3D events. In both cases the software has no way of deciding between the 2 locations, but you could always do a visual inspection at both locations. Note that marginal events often produce locations outside the structure, so you may want to enable Allow Outside Events, if you are using Allow Marginal Events.

Fail Event Correlations Below
Retry Failed Events After Worst Hit Removal

All of the regression location modes now calculate a correlation coefficient. These values are reported as "q" (= quality) in the Line Display:

```
* Gp# 1[2,3,1,4,5,6] x = 11.45, y = 20, q = 0.9902 dT[ 68, 71, 302, 316, 323]
```

Correlation is also available for use on graph axes. For events with 4 or more hits (5 or more for 3D Location) the reported correlation values range from 0 to 1 with 1 being a very good event & 0 being a bad event. For events with 3 hits (4 for 3D Location) the correlation coefficient should be very close to 1.0. Note that for these events, a value of 1 does not guarantee a high quality event as timing errors cannot be detected with only 3 hits. Other location modes like Linear & Zonal will always report correlation coefficients of 0 as they do not calculate them.

There are 3 location settings relevant to correlation coefficients. There is a Fail Events Correlations Below checkbox along with an edit field to set the failure level. The regression location modes can optionally reject events with low correlation coefficients which are less likely to produce useful location information. The failure level can be adjusted based on the user's personal preference.

The third new setting is a checkbox labeled Retry Failed Events After Worst Hit Removal. This allows failed events to be reanalyzed after the worst hit of the event has been removed. Worst hit removal & reanalysis can be applied to events that fail due to low correlation values & can significantly improve the correlation coefficient which implies improved location accuracy as well. Here are a few examples:

Worst Hit Removal & Reanalysis on Events with Low Correlations

Time of Test						
* 0 00:00:23.3043055						
– Gp# 1[5,6,4,7,3,8,2,9,1,11]	x =	7.897	y =	-4.841	q =	0.483
+ Gp# 1[5,6,4,7,3,8,2,9,1]	x =	4.4,	y =	-3.246	q =	0.9967
* 0 00:02:11.5468993						
– Gp# 1[2,1,3,12,4,11,5,6,10,9,7,8]	x =	4.129	y =	7.312	q =	0. 4556
+ Gp# 1[2,1,3,12,4,11,5,6,10,9,7]	x =	2.835	y =	5.095	q =	0.9822
* 0 00:02:29.9029040						
– Gp# 1[5,6,4,7,3,8,2,9,1,11]	x =	9.131	y =	-6.009	q =	0.413
+ Gp# 1[5,6,4,7,3,8,2,9,1]	x =	4.521	y =	-3.38	q =	0.9968
* 0 00:03:22.0759543						
– Gp# 1[8,9,7,10,6,11,5,12,4,1,3,2]	x =	-6.047	y =	-7.491	q =	0.3598
+ Gp# 1[8,9,7,10,6,11,5,12,4,1,3]	x =	-3.915	y =	-4.642	q =	0.987

– = No Correlation Filtering, **+** = Correlation Filtering + Worst Hit Removal + Reanalysis

Worst hit removal & reanalysis can also be applied to failed outside events when Allow Outside Events is not checked. Reanalysis after worst hit removal can shift the results of the event analysis from outside the structure to inside the structure sometimes with the removal of just 1 hit. Here are some results for events that originally occurred way outside the user's structure:

Worst Hit Removal & Reanalysis on Outside Events

Time of Test						
* 0 01:05:04.1100330						
– Gp# 1 [5,6,4,7,3,10,12]	x =	71490	y =	-50270	q =	0.9988
+ Gp# 1 [5,6,4,7,3,12]	x =	15.55	y =	-13.91	q =	0.8813
* 0 01:09:39.3894840						
– Gp# 1 [8,9,10,5,1]	x =	-31600	y =	-31380	q =	0.9899
+ Gp# 1 [8,9,5,1]	x =	-15.99	y =	- 17.58	q =	0.978

– = Outside Events Allowed, **+** = No Outside Events + Worst Hit Removal + Reanalysis

Worst hit removal & reanalysis is never applied to marginal events. Events are considered marginal due to a problem with the geometry of the sensors included in the event. For example, for 2D Planar, an event where all the hit sensors are in a straight line is considered marginal. Removal of any hit, still leaves all the hits in a straight line, so the event is still marginal. For that reason marginal events are never reanalyzed.

Options for Tank Bottom Location

There are 2 options specific to Tank Bottom location. One allows the software to display more than 1 source location for a given event by varying which 3 hits it uses in the calculation. The Force Outside Events to Rim option works in conjunction with the Allow Outside Events to control where outside events are plotted. If Allow Outside Events is disabled, the Force Outside Events to Rim does nothing. If Force to Rim is disabled, the events are plotted at their calculated coordinates. If both Allow Outside Events & Force Outside Events to Rim are enabled, then all the outside events are plotted on the rim of the tank bottom structure. *For examples of how the options affect your Tank Bottom location results, see the notes for the Tank Bottom location mode in section 4.4.9.*

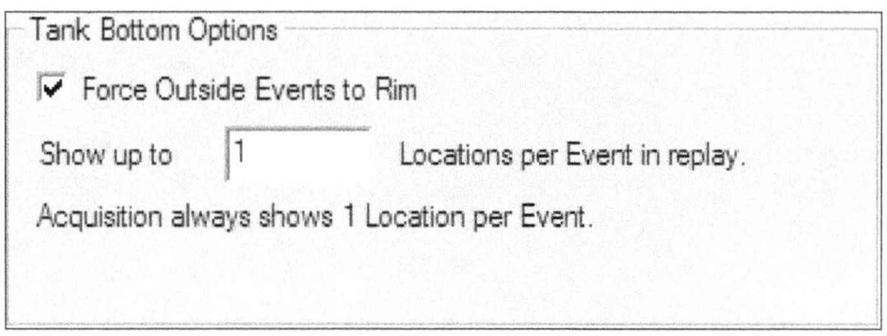

*Figure 83. Tank Bottom **Location** Options*

Distance Units Dialog

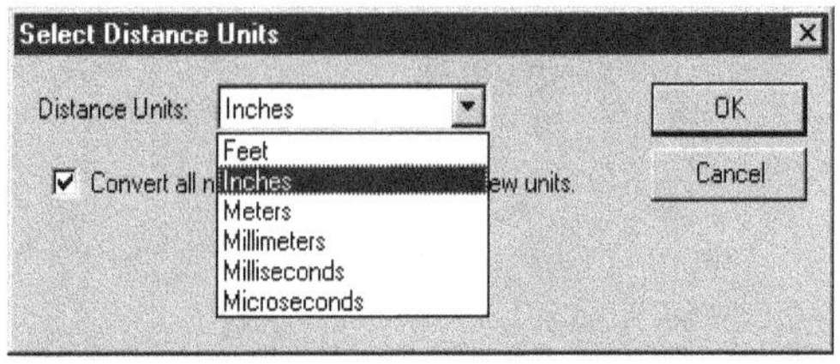

Figure 84. Distance Units

Pressing the Distance Units button opens the Distance Units dialog box. The user can select his desired units from the dropdown list. The user is given an option of converting old numerical values to the new units by simply checking the "Convert all numerical values based on new units." option. This will cause the event definition value, event lockout value, the overcal settings, the wave velocities & sensor positions to be updated as appropriate.

Material Properties Dialog

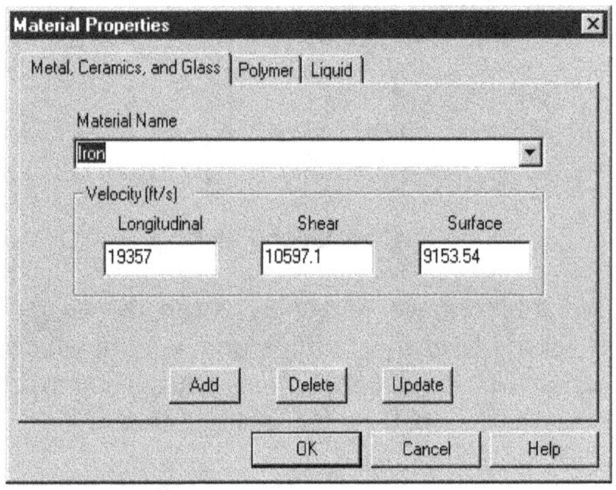

Figure 85. Material Properties

Pressing the Material... button opens the Material Properties dialog box. This dialog box is just a simple database of velocities for acoustic emission in various materials. When you find a useful velocity, simply highlight it & press Ctrl-C, Ctrl-Ins or right click & select Copy. You can then insert it into any field in the Location Setup dialog with Shift-Ins or Ctrl-V.

If you wish to add some of your own numbers, you can simply type a new entry for the Material Name, fill in the velocities & press the Add button. The Delete button removes an entry from the database, while the Update button lets you replace bad values.

Attenuation Dialog

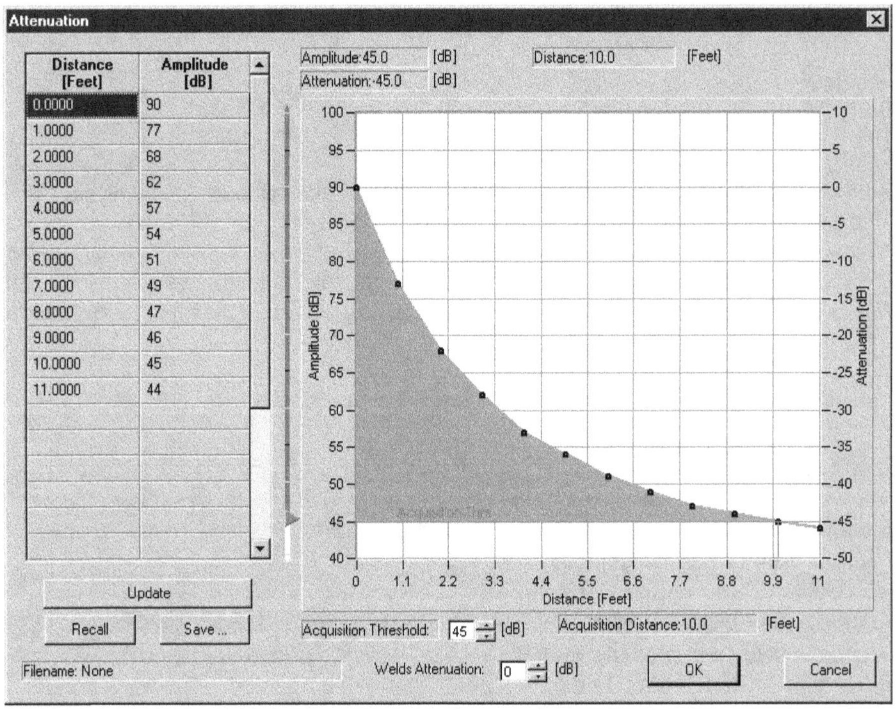

Figure 86. Attenuation Dialog

Pressing the Attenuation... button opens the Attenuation dialog box shown below. The sole purpose of this dialog is to allow the software to account for the decay of the acoustic emmission as a function of distance travelled & to estimate of the initial amplitude at the source. It uses the amplitude observed at the first hit sensor, the distance to the source for the event and the attenuation table entered by the user in this dialog. The attenuation table documents amplitude as a function of distance. The units of distance are controlled by the Location Setup Distance Units setting. The Update button will causes the graph to display the current values in the table. The user may use the Save & Recall buttons to save or recall an attenuation table to/from the disk

The basic procedure uses the table to calculate the difference between an amplitude at the source & at the observed distance from the source. If the source to first hit sensor distance falls within the range of values in the table, an amplitude at that distance is calculated using linear interpolation between the 2 nearest distances in the table. Using the table above as an example, if the source was determined to be 2.5 feet from the first hit sensor, the table has the 2 closest entries as 68 at 2.0 feet & 62 at 3.0. The amplitude predicted at 2.5 feet would be 65 (= 68 – (68-62)*0.5). The correction applied to the amplitude at the first hit sensor would be 25 dB (= 90 - 65). If the first hit amplitude was 60, the line display would report a Source Amplitude of 85 for this example. The amplitudes are not extrapolated beyond the maximum distance. If the distance is beyond the maximum distance in the table, only the maximum correction of 46 (= 90 – 44) is applied. If the user never accesses this dialog to fill in the table, no correction is applied & the source amplitude reported is identical to the amplitude of the first hit sensor.

Guard Sensor Dialog

Pressing this button opens the 'Guard Sensors' dialog box. Guard Sensors can be used to lockout out events on selected location groups whenever a hit is first detected on a guard channel. It is designed for situations in which one wishes to ignore events that originate outside of the location group.

In the example setup shown, sensors 1, 3 & 5 are protecting Location Group 1 from outside events. If there is a hit on sensors 1, 3 or 5, then Location Group 1 will not process any events for the next 1000 us. If the Event Only flag were checked, only the next event on Location Group 1 would be locked out instead of being locked out for the next 1000us. The Lockout Time field would also be disabled as its value would no longer be used.

The Guard Data Filter Setup controls what hits get saved into the DTA file during acquisition, if guard sensors are enabled. The default setting is to just save all hits, but you can restrict the data saved to just those hits that make up events & the guard sensor hits. This can be useful in situations where external noise creates a large number of hits that end up being useless for the purpose of location. Not only do the extra hits take up se in the data file, but they also slow down replays of the data file too.

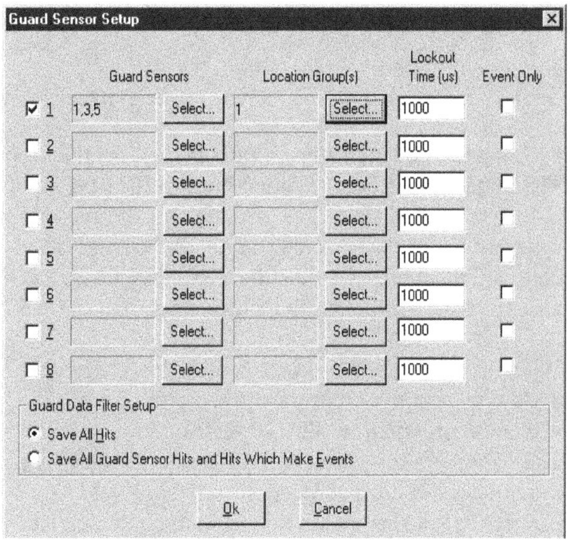

Figure 87. Guard Sensor Setup

Location View & Sensor Placement Dialog

Before discussing the Sensor Placement Dialog, there is one important point about the physical placement of your sensors that you should understand and it applies to all of the regression location modes. *You can place sensors anywhere, but you will get better results with the regression location modes if your sensor layout is made up of an evenly sed triangular array of sensors. The optimal arrangement would be a triangular grid made up of equilateral triangles, but the regression modes do not require exact equality of the triangle legs or even regular sing between the sensors.* ***Avoid creating triangles with one very long or very short side, however, as they do not work well for locating sources.***

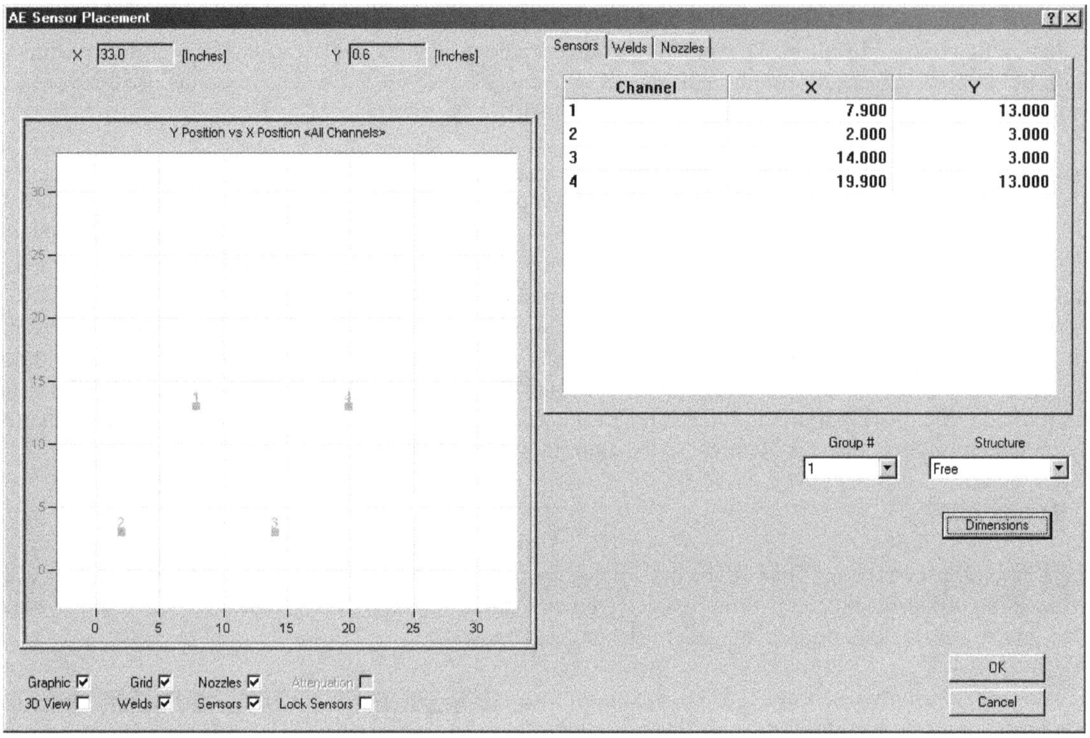

Figure 88. Sensor Placement Dialog

Pressing the Location View button in the Location Setup dialog opens the Sensor Placement dialog box. shown above. The 2 main areas are the actual view of the sensors in the left-hand side of the dialog & a listing of their coordinates in the right-hand side. The grid area for the sensor coordinates also has tabs that will display the coordinates of any welds and nozzles created. You can change the color of the sensors using the Graphing menu to access Sensor Color on the Colors submenu.

When using the Sensor Placement dialog, there are 2 ways to add new sensors. When working in the grid area, pressing the INS key adds a new sensor. The F2 key allows you to edit its coordinates & move it to the correct location. Alternatively you can right click the sensor view area to get a context menu like the one shown to the right. Place Sensor & Next Sensor allow you to put a new sensor at the current position. Next Sensor numbers it automatically using the next available sensor number. Place Sensor will ask for a sensor number. You can also use AutoPlace Sensors to create a grid of sensors with regularly varying coordinates. You work with welds & nozzles in a similar manner. The exact menu you see may differ from that shown here as it depends on the Structure.

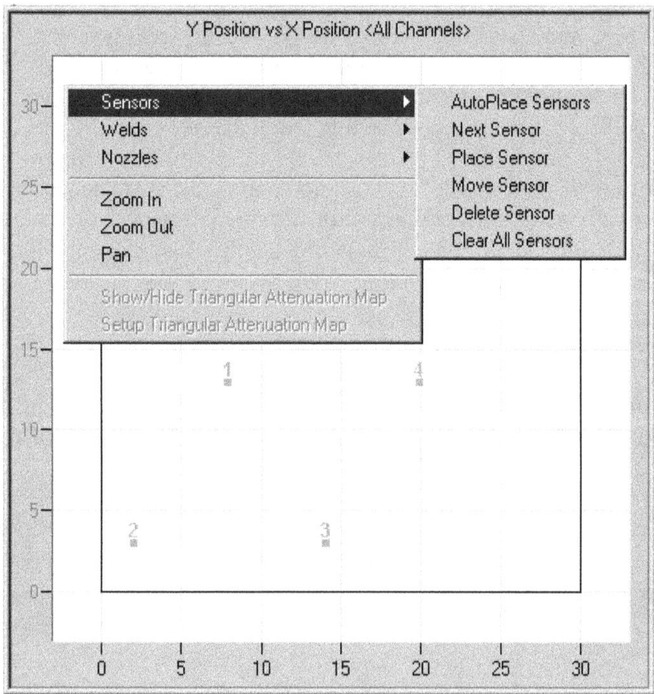

Figure 89. Sensor Menu

The Structure list is just below the grid area. Support for welds, nozzles, manual placement of sensors & autoplacement of sensors varies with the Structure setting. When creating a sensor layout for a new location group, the first thing you should do in the Sensor Placement dialog is to set the Structure type. The Group # setting allows you to quickly switch to editing the sensors for another location group without having to return to the Location Setup dialog. If you made changes to the current settings, you are prompted to save them.

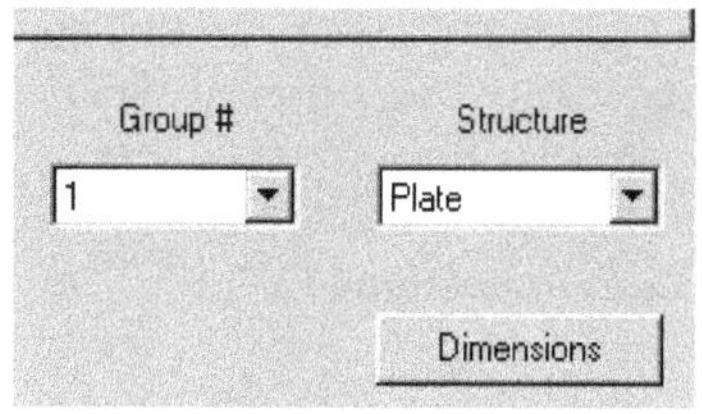

Figure 90. Structure List

Below the Structure setting is the Dimension button. If you click it, one of the dimensions dialogs will appear. They are also specific to the type of Structure. The one shown to the right is for the Plate structure. When working with a new location group, the second thing you should do in Sensor Placement is to fill in the dimensions for the structure you are working with. The dimensions affect the scaling of the sensor view area. If you have incorrect values entered, some sensors may not be visible in the sensor view.

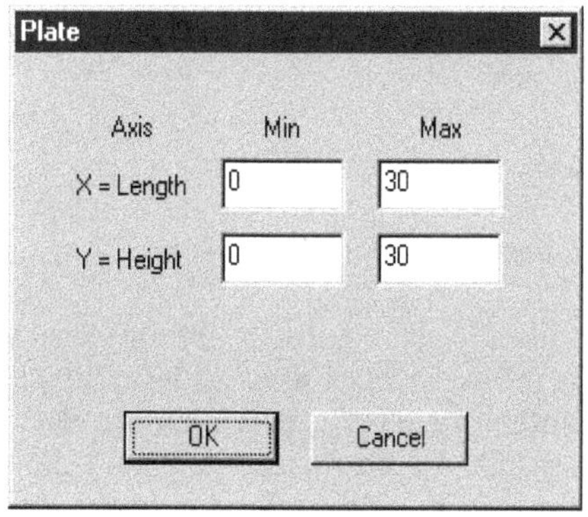

Figure 91. Plate

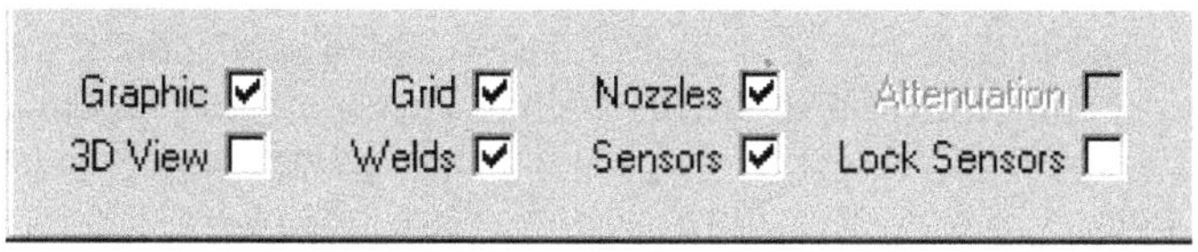

Figure 92. Sensor Placement Checkboxes

Below the sensor view area, there are a number of checkboxes that control the sensor view. The Nozzles, Sensors & Welds checkboxes control whether or not those objects are displayed. Likewise the Grid checkbox controls whether or not the sensor view shows a Cartesian grid to facilitate placing of sensors. The Graphic checkbox controls whether or not a representation of the structure is drawn on the graph. The Lock Sensors checkbox disables moving of sensors with the mouse. The 3D View checkbox will toggle the display between 2D & 3D views of the sensor layout, if that structure type supports both 2D & 3D views. Finally the Attenuation checkbox enables if you have used the Attenuation dialog to enable the calculation of source amplitudes. Checking Attenuation in the Sensor Placement dialog causes the sensor view to display an attenuation map as a series of concentric rings around each sensor as shown to the right.

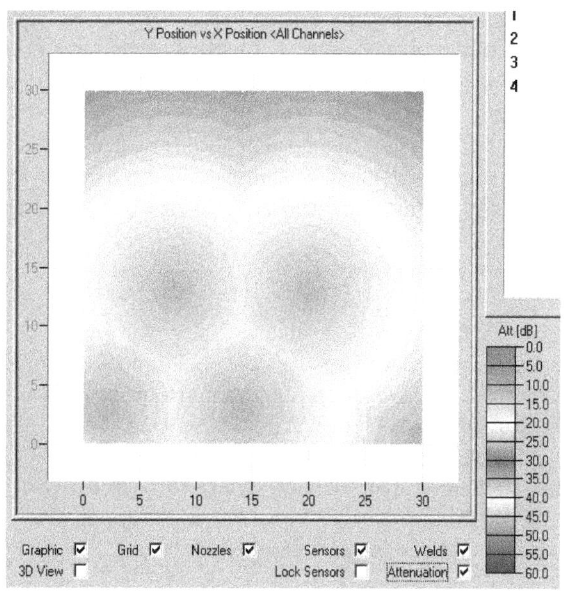

Figure 93. Attenuation

Notes on Individual Location Modes

As mentioned in the section describing the Sensor Placement Dialog, there is a point that all users should understand about the regression location modes & it is important enough to bear repeating. *You can place sensors anywhere, but you will get better results with the regression modes if your sensor layout is made up of an evenly sed triangular array of sensors. The optimal arrangement would be a triangular grid made up of equilateral triangles, but the regression modes do not require exact equality of the triangle legs or even regular sing between the sensors.* ***Avoid creating triangles with one very long or very short side, however, as they do not work well for locating sources.***

In general, you can just read the section on the location mode you are interested in using. All the regression location modes are a variation on 2D Planar, however, so you might also want to read it as well. If you'd like more information on setting up a Zonal, Linear or Tank Bottom location group, you can also see the Examples section.

Zonal Location

Zonal location is the simplest location mode there is. It just returns the coordinates of the first sensor in the list of hits that make up the event. You can place sensors absolutely anywhere & you can get better resolution on source locations by just using more sensors. A better way to improve your source locations would be to switch to a more sophisticated location technique like Linear or 2D Planar. See section 4.5.1 for a detailed example of setting up a Zonal Location group.

Linear Location Modes

Linear Location

Linear location is a step up from Zonal location in that it actually attempts to locate of the source from the timing data in the event rather than just restricting it to a region of se around the first hit sensor. Linear location uses 2 hits & the difference in their arrival times to calculate the source location using linear interpolation of the coordinates for the 2 sensors. Its drawback is that in a 2-dimensional area it restricts the source to points on the line segments that connect the sensors when real sources do not face that restriction. Still that can be good enough when looking for a source along a long series of sensors on, for example, a pipeline.

Linear location requires the sensors to be laid out in a line, although the line does not have to be perfectly straight. You could lay them out in an arc. It is not possible, however, to have the line double back on itself, so laying out the sensors in a square or circular array would not work. Linear location relies on a list of the sensors sorted by the x coordinate when deciding which 2 hits to use in an event. You can also use a vertical line. If the x coordinate of the first & last sensor in a Linear location group are equal, the list of sensors gets sorted by the y coordinates. Linear location only forms events between adjacent sensors, so there are only 2 possible 2nd hit sensors for any 1st hit sensor. When there are hits on both, it decides which hit to use based on the ratio of the Δt value of the hit divided by the maximum Δt calculated from the sensor sing & the user entered velocity. The hit with the lowest relative Δt value becomes the 2nd hit of the event used for calculating the source location. The logic is that the 2 sensors on either side of the source will have the shortest relative Δt value, while the Δt value between the 1st hit sensor & other sensors will always be higher. See section 4.5.1 for a detailed example of setting up a Linear Location group.

LinearXY Location

LinearXY location is related to Linear location in that the source location is still calculated by linear interpolation between the arrival times of the 2 hits in the events. The difference between LinearXY & Linear location is that LinearXY removes the assumption that the sensors will be laid out in a line & the sorting based on coordinates. With LinearXY location you can put the sensors literally anywhere in 2-dimensional se, so it is perfectly acceptable to have your sensors laid out in a square, a circle or a rectangular array. The advantage & at the same time the drawback to this method is that it allows events to form between any 2 sensors in the array. For a square, not only are events allowed on the sides of the square, but they can also occur on the diagonals. LinearXY does still look for the pair of sensors with the lowest relative Δt values.

Wrapped Linear Location

A common question from users of 's DOS software is how to turn on wrapping in Linear location. The software doesn't have a checkbox to enable wrapping in Linear location like the old DOS software, but there are 2 methods you can use to set it up. The first method is an older long mentioned in the HMSGHAT2000DDI software manual & it works in any version of HMSGHAT2000DDI independent of when it was released or how the user's licensing options are configured. It requires the use of 2 linear location groups. The second method only works in the newest releases of HMSGHAT2000DDI licensed for Full Location. It makes use of a newer Wrapped Linear location mode.

Method #1 — Wrapped Linear Location Using 2 Location Groups

Below are a series of graphs from the sample layout, WrappedLinear.lay, showing how it is done with 2 Linear location groups.

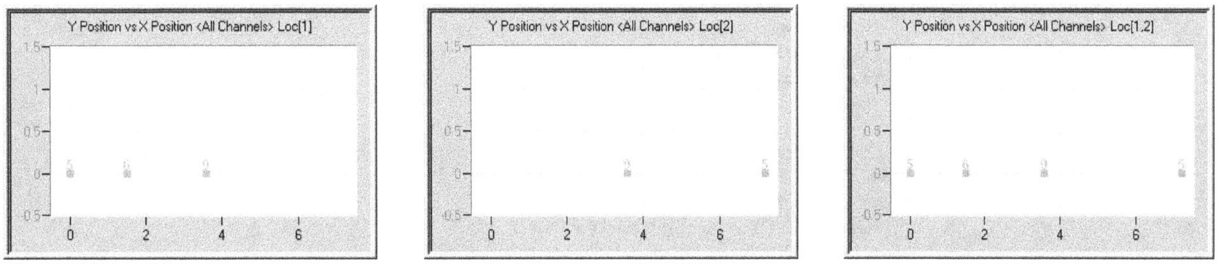

*Figure 94. Wrapped **Linear** Location using 2 Groups*

The graph on the top left is showing location group 1 which consists of sensors 5, 6 & 9. Linear location allows events between sensors pairs 5 & 6 or 6 & 9 in this location group. Turning on wrapping requires allowing events between sensors 5 & 9 and it requires a second set of coordinates for the wrapped location of sensor 5. This is accomplished by creating location group 2 shown in the graph on the top right above. Sensor 9 is common to both location groups at the same coordinates. Sensor 5 in location group 2 is at its wrapped location which is just the position of sensor 5 in group 1 offset by the structure circumference. The graph on the bottom just illustrates that you can superimpose both location groups 1 & 2 on a single graph using Graph Setup. You also need to turn on both groups in Location Setup. Do not enable Show All Points for both groups, though, if you want the software to decide whether events between sensors 6 & 9 for group 1 or events between sensors 5 & 9 for group 2 make the most sense based on relative Δt values. If you were to enable Show All Points, then you could end up with a single set of hits producing an event between sensors 6 & 9 as well as between 9 & the wrapped location of sensor 5.

Method #2 — Wrapped Linear Location Mode

Method #2 is illustrated in the WrappedLinear2.lay example layout. In more recent versions of HMSGHAT2000DDI a Wrapped Linear location mode has been added to the software.

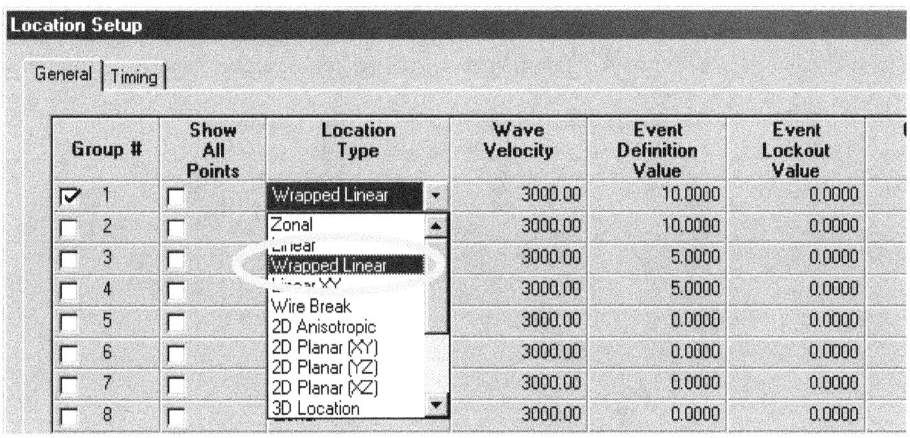

*Figure 95. Wrapped **Linear** in Location Setup*

If you can't see Wrapped Linear in the list of available location modes in the Location Setup dialog, check your licensed options by selecting Options under the Help pulldown menu. The Options dialog will appear: The Wrapped Linear location mode requires the Full Location licensing option.

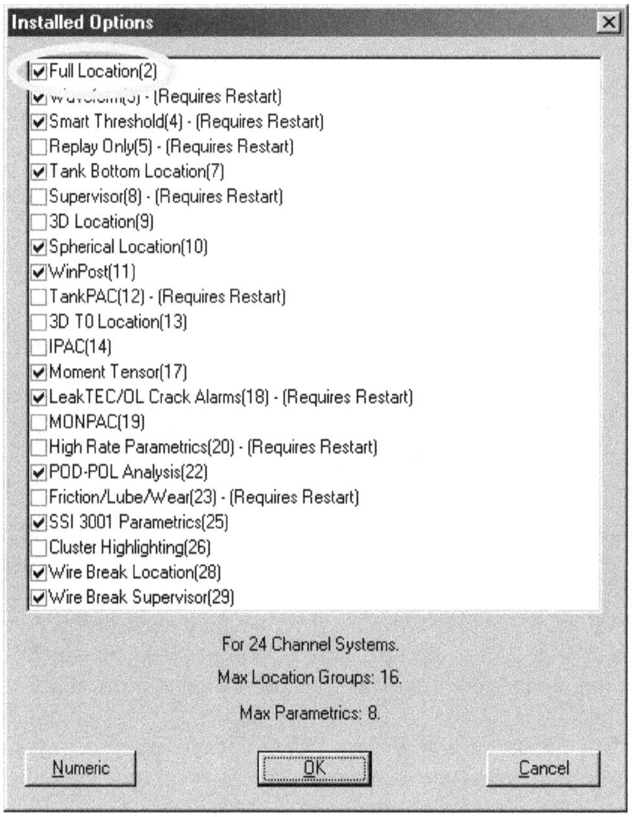

*Figure 96. Full Location Option in **Installed** Options Dialog Box*

If you do have an enabled Full Location license option & still can't see a Wrapped Linear location mode in Location Setup, you will have to either upgrade to the latest version of HMSGHAT2000DDI for your hardware or use Method #1 for doing wrapped linear location. Contact Group Inc, Products & Systems Division, Customer Service (customer.service@group.com) for information any needed upgrades or licensing options.

The Wrapped Linear location mode consolidates the 2 location group method into 1 location group. Here is the location graph from WrappedLinear2.lay which demonstrates the newer Wrapped Linear location mode.

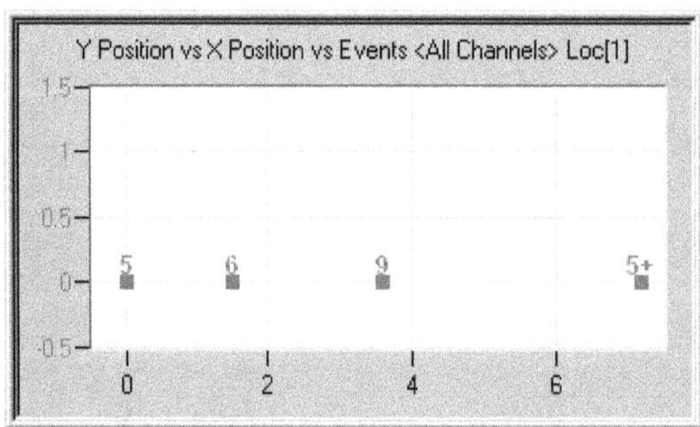

*Figure 97. Graph **Showing** Wrapped Linear Location Group*

Here the wrapped location for sensor 5 is marked on the graph as "5+". Putting the same sensor in 2 locations is a bit of a complication for HMSGHAT2000DDI, because the Sensor Placement dialog is only designed to have 1 set of coordinates for each sensor. The way around this is to invent a new sensor in Sensor Placement that is used for the wrapped sensor location. Here's the Sensor Placement dialog showing the sensors for this location group:

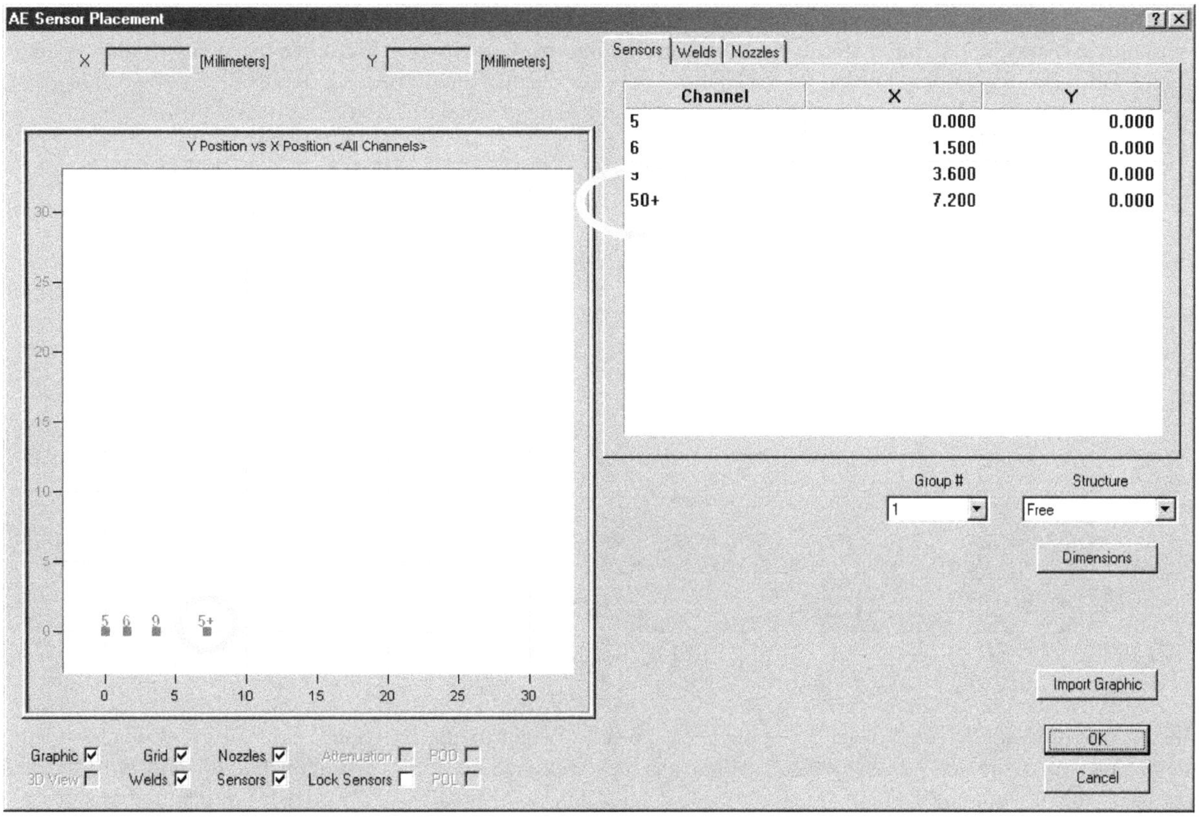

Figure 98. Wrapped Linear Location Group in Sensor Placement

Notice that the table of sensors lists sensor "50+", but the sensor diagram shows sensor "5+" at the location of "50+". These are the same wrapped sensor as indicated by the "+" signs. The only location mode that relabels a sensor on the graph is Wrapped Linear. If this location group is changed to a regular Linear location group, then the "+" signs will disappear in the grid & on the graph & sensor "5+" on the graph would become sensor 50. There is nothing magical about sensor number 50 used in the table. It could have just as easily been labelled sensor 51, 52 or any other unused channel number.

Wrapped Linear location always replaces the label of the last sensor with the number of the first sensor & a plus sign to indicate that it is the wrapped sensor location. It picks the first & last sensors based on their coordinate values, not their sensor channel ID's. Internally it sorts the list of sensors based on the X coordinate to find the sensors with the lowest & highest X coordinates & then labels the highest X, rightmost sensor on the graph with the ID of the lowest X, leftmost sensor & the "+" sign. It also adds the "+" sign in the table to let you know which sensor it picked as the last sensor. If you input a vertical array of sensors where all the sensors have the same X coordinate, Wrapped Linear will sort the sensors based on their Y coordinate to decide which sensors are the first & last sensors to identify the wrapped sensor.

2D Planar & 2D Planar (XY)

There is no fundamental difference between 2D Planar & 2D Planar (XY) location except on the menus. Enabling 3D Location also enables 2D Planar (YZ) & 2D Planar (XZ) location. To distinguish the standard 2D Planar from those 2 modes, it is relabeled 2D Planar (XY) location whenever the 3D Location option is enabled. 2D Planar (YZ) & 2D Planar (XZ) location are discussed in the section on 3D Location.

For 2D Planar location, a particularly difficult problem is locating the source of an acoustic emission event, if all the hits are from sensors in a straight line (linear events). For linear events, even if a distance from the line can be determined, there is no way to distinguish positive & negative directions away from the line. Random error in the data can also make determining the distance from the line highly suspect. By default, the software filters out linear events and does not attempt 2D Planar location on them. You could always enable Allow Marginal Events (See discussion in section 4.2.4), but then you could end up flooding your graphs with lots of low accuracy results. A better solution is to use a layout that minimizes the chances of linear events being observed. An example of a bad layout is shown below. It is bad because of the horizontal sensor sing is much smaller than the vertical sing.

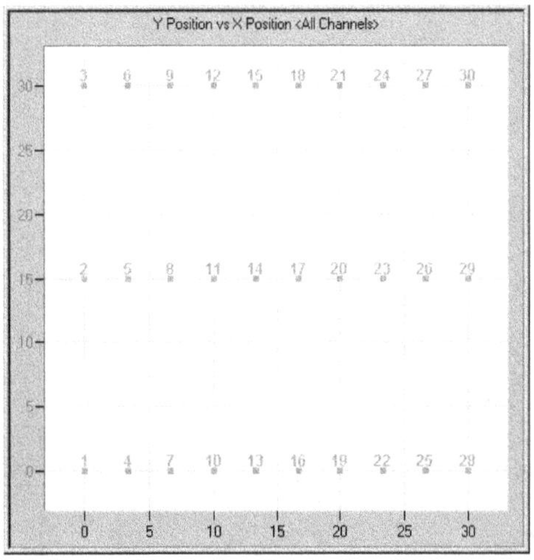

Figure 99. Bad 2D Planar Layout

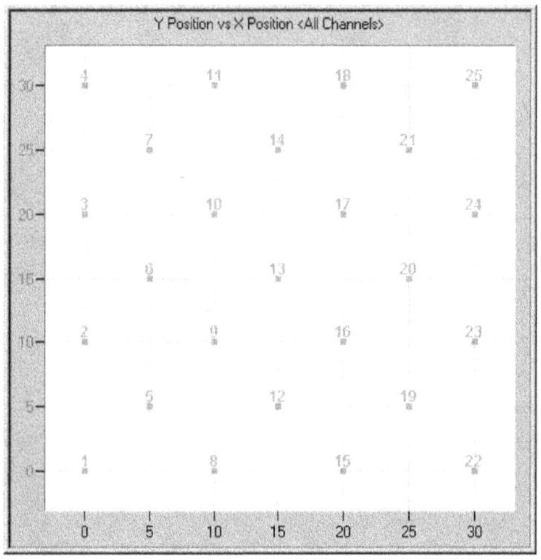

Figure 100. Good 2D Planar Layout

A better layout would have all distances in between the sensors approximately equal. Triangles with one very long side or one very short side can also cause problems as they approach becoming a straight line of sensors. The ideal 2D Planar layout would be made up of equilateral triangles, but exact equality of all sides is not required as shown in the good layout above . As mentioned in the introduction, you should avoid creating triangles with one very long side or one very short side. Events based on triangles with one very long or very short side become linear events when one side goes to infinity or to 0. Short of those limits, though, they can still act similar to the linear events, and won't accurately locate any sources outside of the triangle.

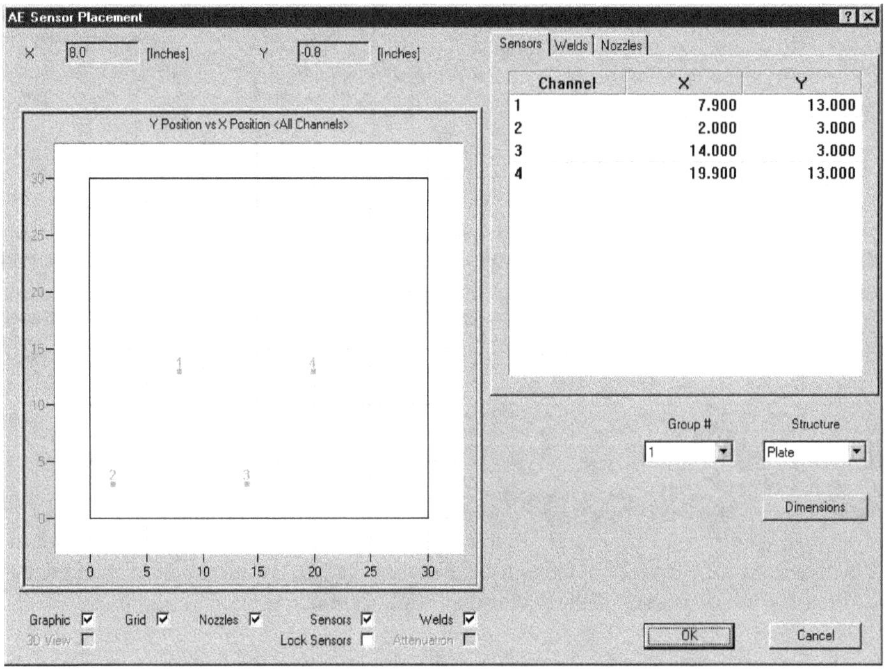

*Figure 101. 2D **Planar** Sensor Placement*

Here is Sensor Placement set up for a 2D Planar mode with the Plate structure. 2D Planar location is designed to work with either the Free or Plate structures. Although there isn't any difference in 2D Planar & 2D Planar (XY) location during analysis, there is one difference in Sensor Placement. A 2D Planar (XY) location group has the option of displaying the sensors in 3D graph, if the user wishes it. For a 2D Planar (XY) location group with the structure type set to Free, the 3D View checkbox enables. The user may use it to toggle between 2D & 3D views of the sensors. The checkbox disables for the Plate structure, though.

The Dimensions dialog for both the Free & the Plate structure contain the max/min limit settings for both the X & Y axes. These settings control the scaling of the graph in the sensor view of Sensor Placement. For the Plate structure, they have added significance. When 2D Planar is analyzing events, it will restrict results location groups using the Plate structure to points on the plate. To allow locations close to the limits to report, the software uses a tolerance of 2% of the axis length when deciding whether it is an acceptable result. For example, if x has a minimum value of 0 and a maximum value of 100, then results with an x coordinate greater than 102 or less than -2 are thrown out as invalid. If you don't want to limit your results, you can either set the Structure type to Free or stick with Plate & enable the Allow Outside Events checkbox in Location Setup. See section 4.2.3 for more details.

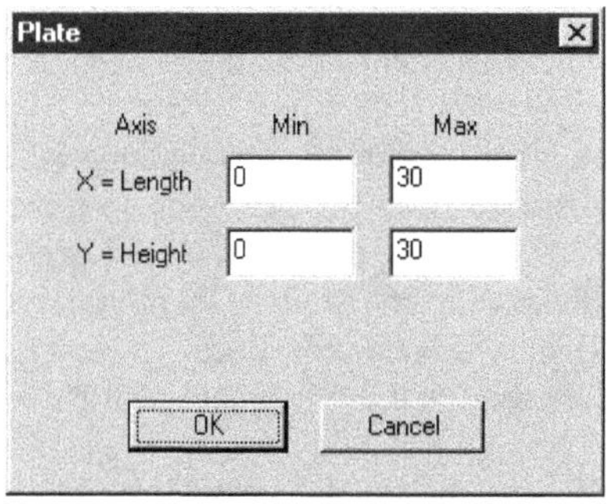

Figure 102. *Plate*

After setting up the sensors & a 2D Planar location groups, you will want to create a location graph. Use the Graphing pulldown menu's New Graph option & then configure the graph using the Graph Setup dialog box (F3 shortcut key). To the right is a 2D X-Y location graph based on the 2D Binned Scatter graph type where the location group is using a Plate structure. If you create a 2D Binned Scatter, 2D Colored Scatter or 2D Scatter graph that uses X & Y Position, the graph will automatically draw the sensors on the plot for any location groups selected on the Event Groups tab. In the Graph Setup dialog, make sure that the Input Data is set to Events and use the Event Groups tab to select which location group(s) to use for source information. Below is the corresponding Graph Setup dialog to illustrate the typical settings.

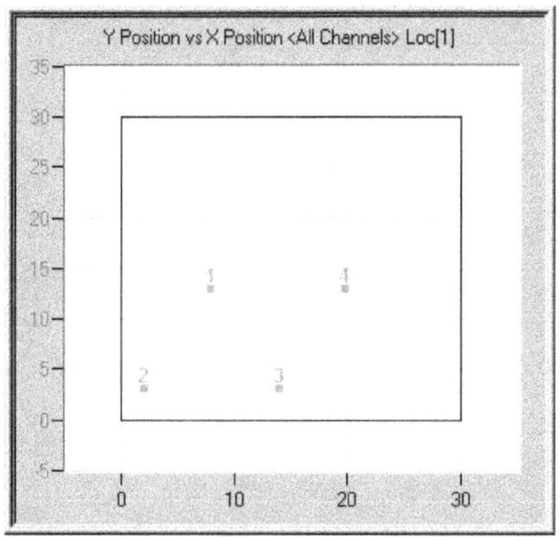

Figure 103. *2D Planar Graph*

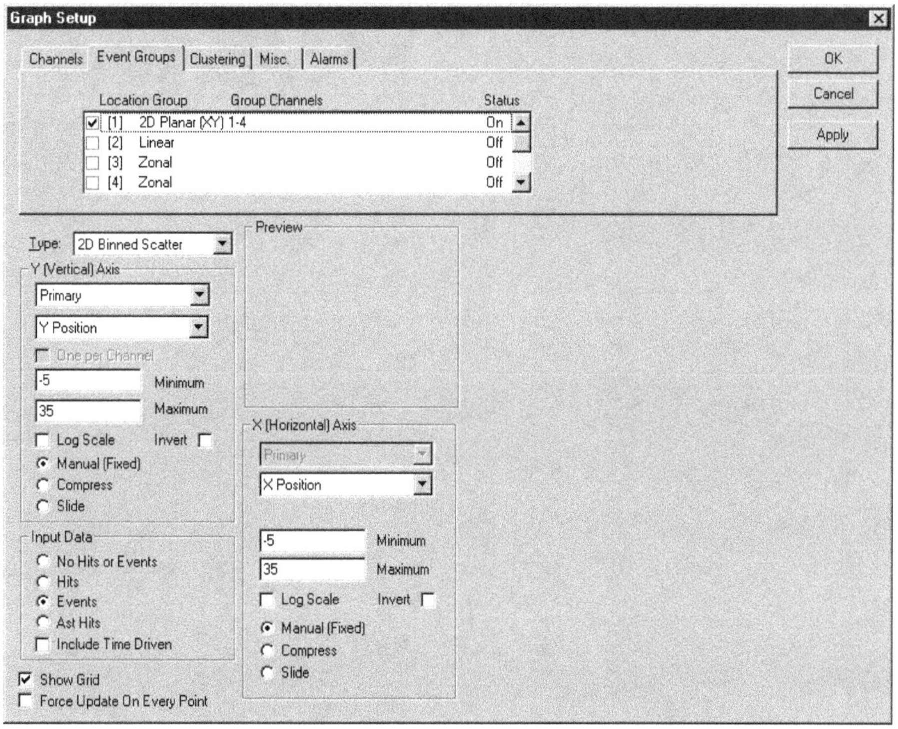

Figure 104. *2D **Planar** Graph Setup*

3D Location, 2D Planar (YZ) & 2D Planar (XZ)

Primarily this section focuses on 3D Location, but 2D Planar (YZ) & 2D Planar (XZ) are related. All 3 location modes are designed to use 3D coordinates. The 2D Planar (YZ) & 2D Planar (XZ) modes are similar to normal 2D Planar & 2D Planar (XY) location with the exception that you can enter 3D coordinates & have the software locate the source while ignoring either the X or Y coordinate. The reason these modes are useful relates to a problem with sensor placement & event handling for 3D Location.

The equivalent of the linear event problem for 2D Planar location is a problem with planar events for 3D Location.

A planar event is one in which all the hits are on sensors in the same plane. For planar events, it is hard to determine distance from the plane not to mention distinguishing sources in the positive direction from those in the negative direction. By default, the 3D Location algorithm filters out planar events and does not attempt to locate them as the distance from the plane is often quite simply useless. You can override this behavior by enabling the Allow Marginal Events checkbox in Location Setup, but it exists primarily for compatibility with older versions of 's DOS software. See section 4.2.4 for more details.

Because of the planar problem, you will get better results if you set up your sensors to minimize the number of sensors that line up across the body of the test structure. Avoid using a rectangular grid layout that lines up sensors on opposing faces of your structure. Instead try to stagger the sensors, so that no 2 sensors are directly across from one another. The bad sample layout above uses a rectangular approach with 3 rows of 3 sensors directly opposite one another. The good layout uses a triangular approach by staggering each row on one side by half the sensor-sensor distance and then it inverts the pattern on the opposite wall. The diagram may not show it well, but sensor 4 is below the midpoint of sensors 1&2, above the midpoint of 7&8 and across from the midpoint of 12&13. Both sides have 3 rows of sensors, but the left-hand side has 3-2-3 pattern of sensors, while the right-hand side has a 2-3-2 pattern of sensors. This staggering minimizes the number of sensor planes through the body of the structure.

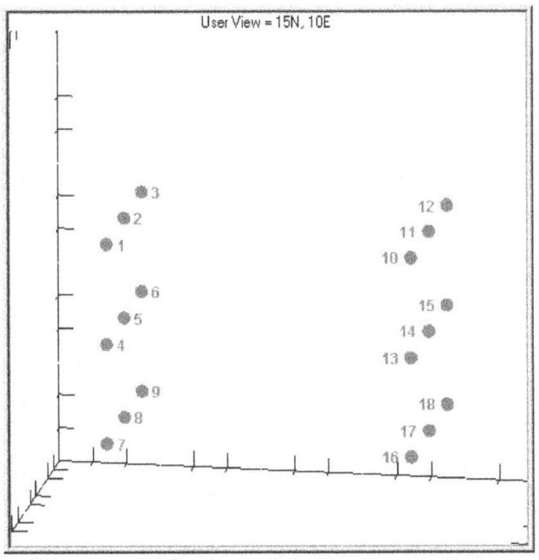

Figure 105. Bad 3D Location Layout

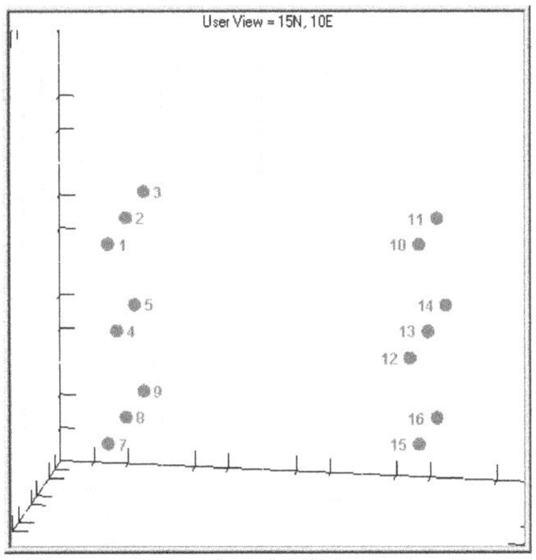

Figure 106. Good 3D Location Layout

One problem with 3D Location filtering out planar events is that pencil lead breaks will not locate with 3D Location unless the lead break is energetic enough to register with a sensor on one of the other faces. If you enable Allow Marginal Events, you may get results far outside the structure. A good alternative in this case are the 2D Planar (XZ) & (YZ) location modes. The distinction between them is what sensor coordinates they use when fitting the data. Both allow you to enter 3D coordinates, but ignore one of the values when locating. 2D Planar (XZ) use the X & Z sensor coordinates & ignores the Y values. 2D Planar (YZ) ignores the X coordinate. They are useful when doing pencil lead breaks by sensors confined to one face of your structure. With the 3D Location option enabled, 2D Planar (XY) will allow you to enter 3D coordinates & only locate using the x & y coordinates. Fundamentally that's no different than the standard 2D Planar location. Both are ignoring the z coordinate, but as a convenience one allows you to enter a non-zero value for Z.

For best performance, any 2D Planar groups using 3D coordinates should be confined to just one face of the test structure. Parallel sensors on an opposing face could appear as duplicate sensors with identical coordinates when the 3rd unused coordinate is ignored & that would lead to errors in the event location calculations. You should also note that during acquisition a result from a 2D location mode does not necessarily imply that the source was confined to the 2D location plane. Just as for planar events in 3D Location, distance from the plane is indeterminate. You can draw a perpendicular to the 2D plane at the best fit location and the source event could be anywhere on the line. Of course for pencil lead breaks, you will know in advance that the source is confined to a given face and be able to check your sensors before starting acquisition.

For 3D Location, the Sensor Placement dialog shown here allows you to directly enter X, Y, Z coordinates for any groups that are considered a 3D Location group with the right structure type. 3D Location is designed to use either the Free & the Box structure types.

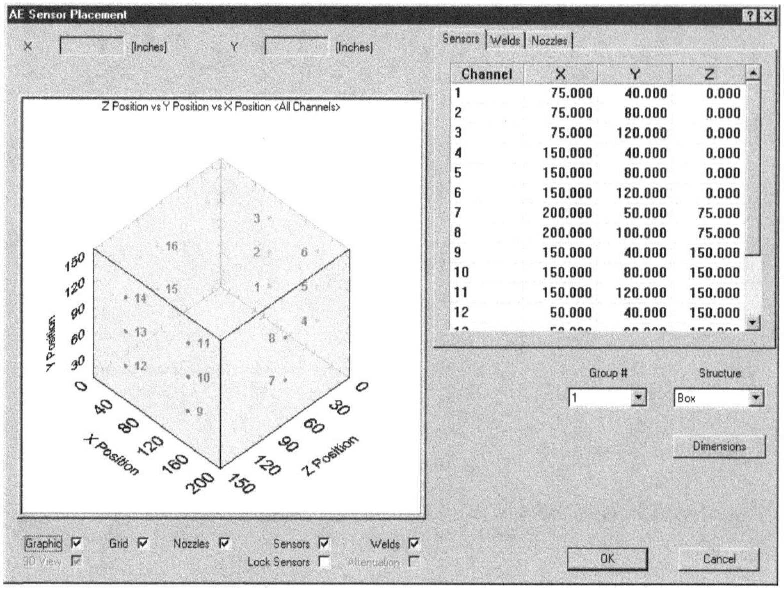

Figure 107. 3D Sensor Placement

The Dimensions dialog for the Box & Free structures is similar to the one for the 2D Free or Plate structures with the addition of max/min settings for the Z axis. Similar to the way the Plate structure limits 2D Planar results, the Box structure acts as search limits for 3D Location. If any location results are more than 2% above or below the limits, the result will be considered bad & is thrown out. For example, if Z Min is 50 & Z Max is 150, then any location results where z is above 152 or below 48 will be filtered out. As with 2D Planar, using the Free structure avoids imposing any limits on the results. Alternatively, you can enable Allow Outside Events in Location Setup. See section 4.2.3 for details.

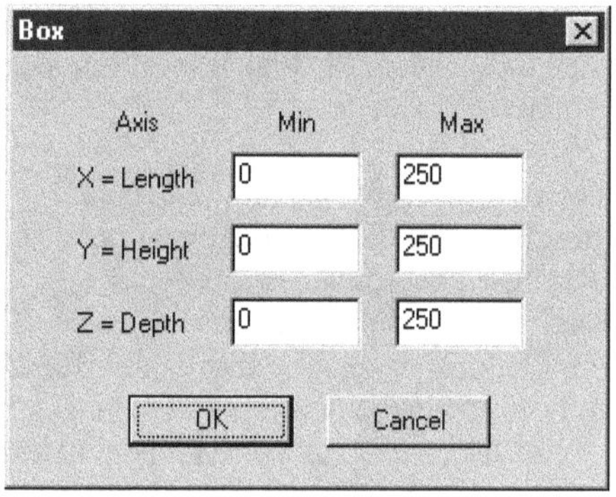

Figure 108. ***Box*** *Structure Dimensions Dialog*

You cannot graphically enter 3D sensor locations in Sensor Placement, but you can graphically see the sensors in the 3D view. If you do not see a grid column for Z or a 3D view of your sensors, go back to the Location Setup dialog & make sure the group type is set to 3D Location, 2D Planar (YZ) or (XZ). Also make sure you have the structure type set to Free or Box.

Use the INS key & F2 to insert a new sensor & edit its coordinates in the grid. As an alternative to creating all the sensors manually, you can also use the AutoPlace Sensors option to create a grid of sensors with varying x,y coordinates & a fixed z coordinate of 0. You would then manually edit the z coordinates of each sensor. Note that you can use AutoPlace Sensors multiple times without erasing the old sensors. That allows you to build up a 3D grid one layer at a time. Right click the graph in Sensor Placement to find AutoPlace Sensors.

To create a location graph for 3D Location, use the 3D Scatter graph & set each axis to X, Y & Z Position which correspond to the x, y & z coordinates of a located source point. The graph to the right is a 3D Scatter graph set up as a location graph. As shown below, Graph Setup has the Input Data is set to Events & the Event Groups tab has the 3D Location group enabled. If you create a graph that uses X, Y & Z Position, the graph will automatically draw the sensors and structures on the plot for enabled location groups.

For 3D Scatter plots, both the sensors & the data points are drawn as small cubes. You may use the Graphing Colors menu to change the colors. The data points use the color assigned to Plot 1 under Plot Colors and are drawn slightly smaller than the sensor points. If you would like larger points, right click your graph & turn on the Large Scatter Graph Points. Both the sensor points & the data points will get larger.

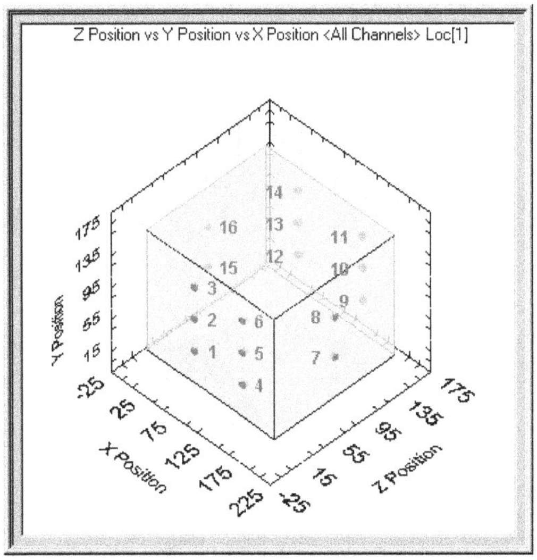

Figure 109. 3D Graph

Graph Setup for the example graph above is shown here. It sets the X (Horizontal) Axis to X Position, Y (Vertical) Axis to Y Position & Z (Depth) Axis to Z Position. You may prefer to remap the axes, if you are used to discussing z as height. For example if you are used to z being vertical, y being horizontal and x being out of the plane of the monitor(paper), set the X (Horizontal) Axis to Y Position, the Y (Vertical) Axis to Z position and the Z (Depth) Axis to X Position. You may also want to change the positive direction of the Z (Depth) Axis. The Invert checkboxes will let you specify whether the positive direction is in or out of the monitor. Similarly you can also invert the horizontal & vertical axes.

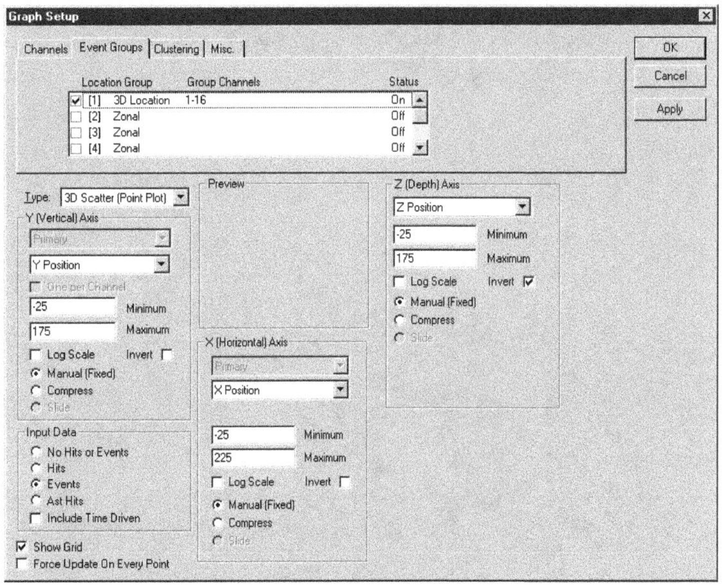

Figure 110. 3D Graph

You can also create 2D location plots using X, Y & Z Position settings on a 2D Binned Scatter graph as done in the section on 2D Planar. If you do, the sensors will also draw on the 2D plots using 2 of the 3 sensor coordinates ignoring the unused dimension. For example, an X & Z Position graph would show the sensors at their x & z coordinates & ignore their y coordinate. This allows you to create front, top and side views of your structure as shown below. These graphs also illustrate another benefit to staggering your sensor locations – legible sensor markers. If your sensors line up on opposing faces, you will end up with overlaid sensor markers on the 2D graphs. That can make them hard to read.

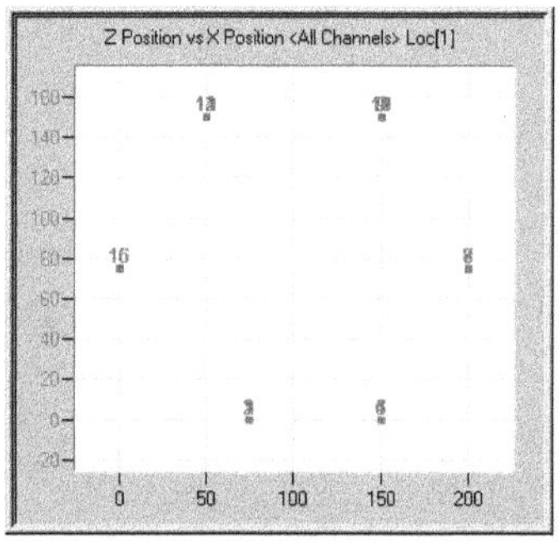

Figure 111. Top View

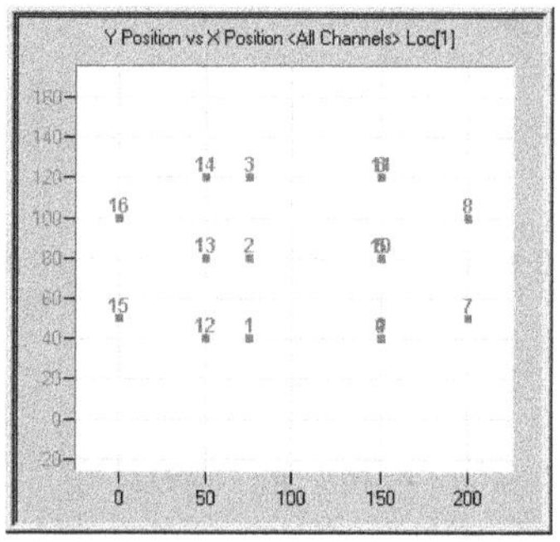

Figure 112. Front View

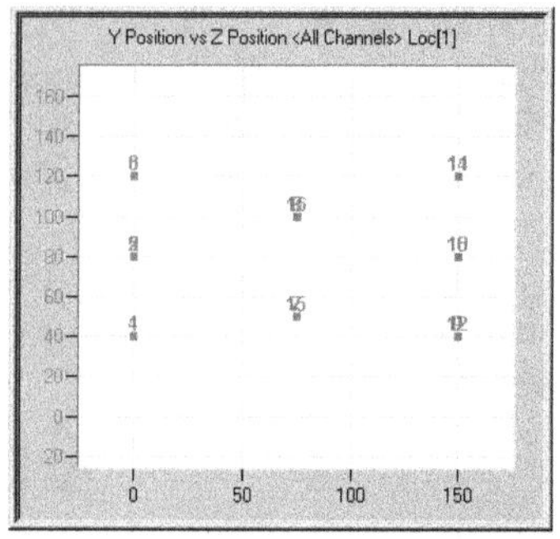

Figure 113. Side View

3D Location on Solid Cylinders

3D Location has been enhanced to allow the use of a cylinder structure to determine the limits of acceptable 3D source locations. The 3dLocCylinderDemo.lay file is a sample layout file included with HMSGHAT2000DDI that demonstrates how to set up & use a cylinder structure with 3D Location. Location Setup is configured as for any other 3D Location group (see 3D Location section above). Sensor Placement controls the user's choice of structure. The user simply selects Horiz. or Vert. Vessel as the Structure setting:

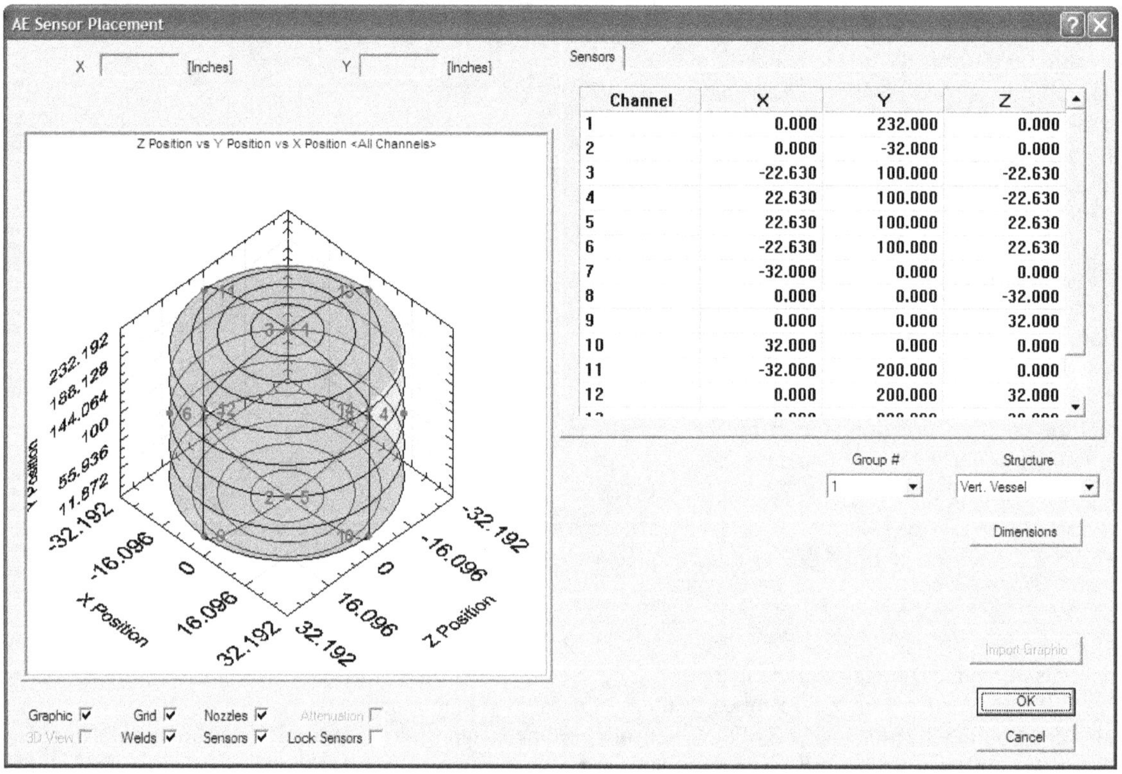

Figure 114. Sensor Placement for 3D Location on a Solid Cylinder

The user then uses the Dimensions dialog to configure the size of cylinder:

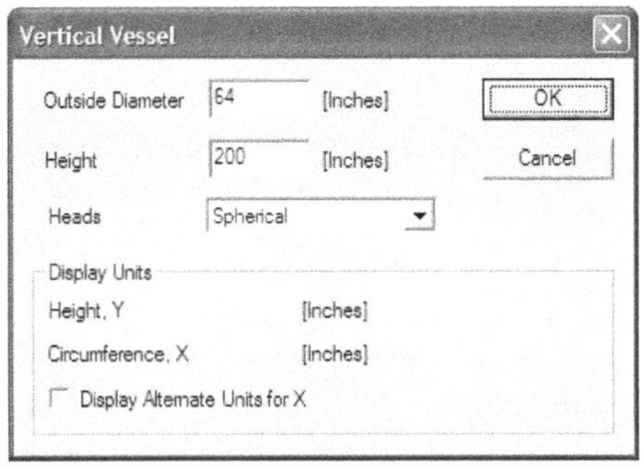

Figure 115. Vertical Dimensions

Sensors are placed on the cylinder using the grid in the Sensor tab of the Sensor Placement dialog box just as for the Box & Free structures. Simply click on the grid & press the Insert key to add a new sensor. Edit the coordinates to values needed. You can highlight a sensor in the grid & press the Delete key to remove an unwanted sensor.

The results of 3D Location on a solid cylinder can be displayed just like any other 3D Location result in a 3D Scatter graph that is set up to display the X, Y & Z Location results on each of the axes:

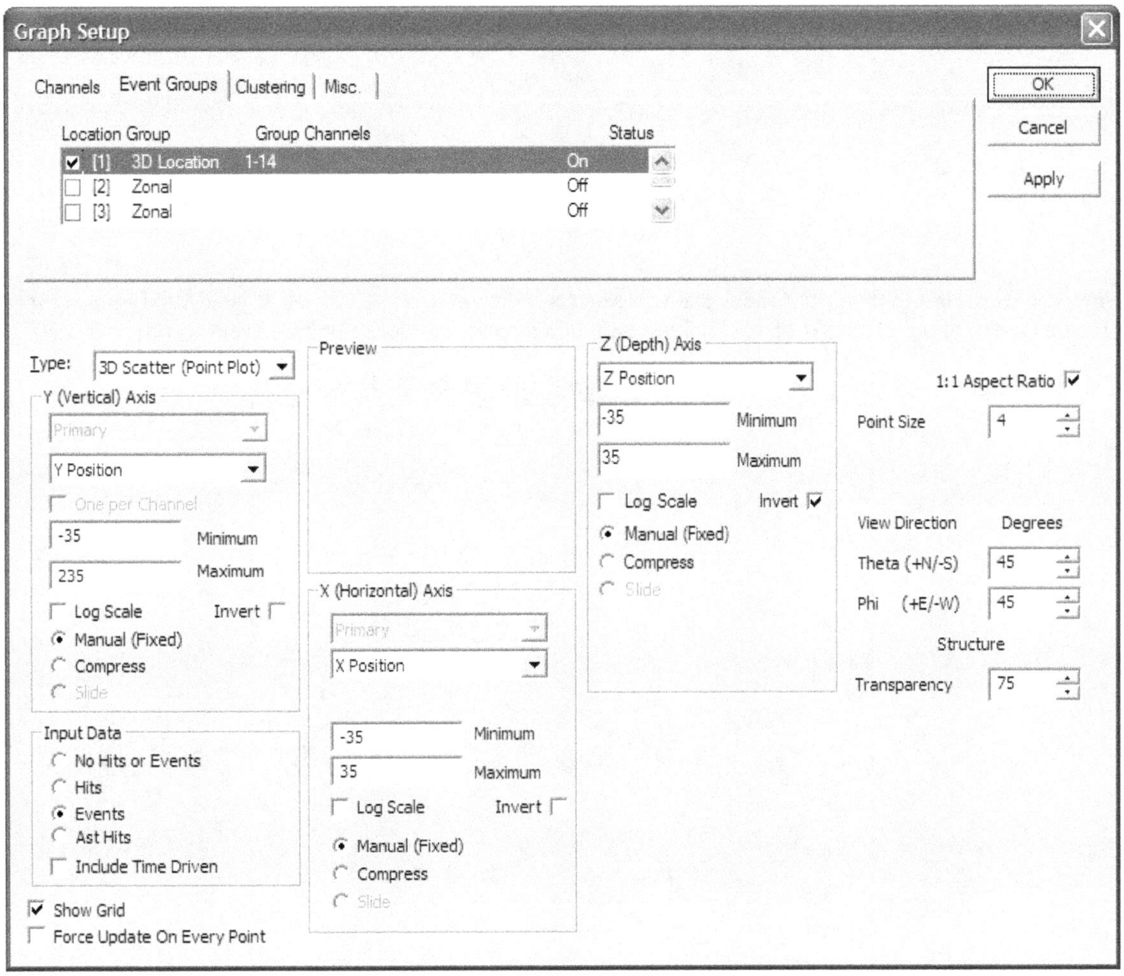

*Figure 116. Sensor **Placement** for 3D Location on a Solid Cylinder*

Note that the user checks the desired location group on the Event Groups tab to determine which location group generates the location results. The user also can control the transparency level of the cylinder and the viewing angle in Graph Setup as shown above.The results display on the 3D Scatter graph after a replay of the data:

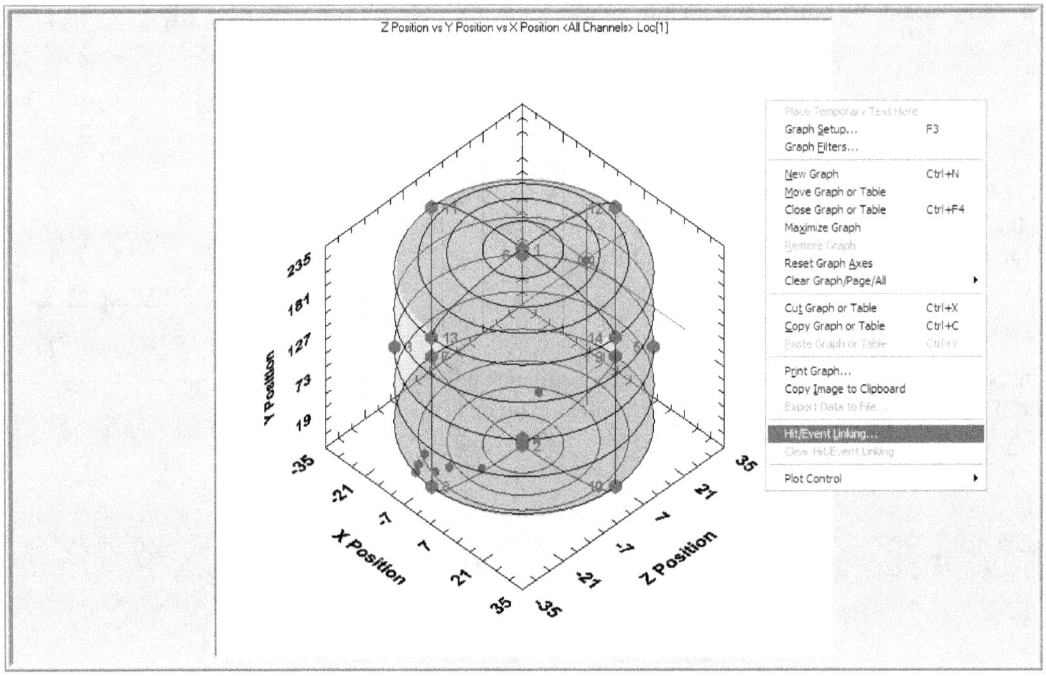

*Figure 117. Sensor **Placement** for 3D Location on a Solid Cylinder*

If desired the user can use the Data Cursor feature to select a data point & launch Hit/Event Linking using the right click context menu of the graph. Hit/Event linking will then display the corresponding event results in its dialog box:

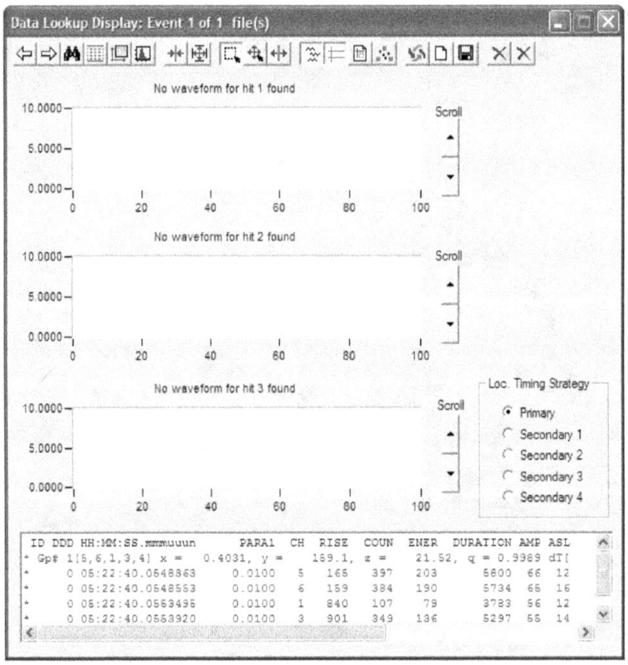

*Figure 118. Hit/Event **Linking** on 3D Solid Cylinder Results*

Spherical Location

Using Spherical location is similar to using the 2D Planar location mode, although there are differences in the coordinate system used & & how a typical location plot is set up. Mathematically, Spherical location behaves like 2D Planar where both axes wrap back around on themselves. There aren't any new pitfalls in sensor placement, but the linear event problem that affects 2D Planar also affects Spherical. If you attempt location using just 4 sensors on the equator of a sphere, Spherical location would always fail as all events would be due to sensors in a straight line. You could override that behavior by enabling Allow Marginal Events in Location Setup at the risk of having many poorly located events. See section 4.2.4 for details. For better results, use at least 6 sensors (north & south poles & 4 evenly sed on the equator). The maximum number of sensors is limited only by the number of available hardware channels. When using more than 6 sensors, it is recommended that you stagger alternate rows of the sensors to yield a triangular grid and produce more locatable events.

The Sensor Placement dialog has been enhanced to allow you to directly enter theta & phi coordinates anytime the structure type is set to Sphere as shown here. Sensor coordinates are entered in degrees. Theta is defined as the north/south angle (latitude) and should be between +90 and –90. +90 is 90 degrees north (top of the sphere), while –90 is 90 degrees south (bottom of the sphere). Phi is the east/west angle (longitude) and ranges between +180 & –180. A phi of +90 is 90 east (right), while –90 is 90 west (left) relative to the origin. +180 (east) and –180 (west) are actually the same location due to the wrapping of the sphere. A white + symbol marks the theta, phi origin of 0,0. You can quickly generate sensor layouts with the AutoPlace Sensors dialog. Just right click the graph in the Sensor Placement dialog.

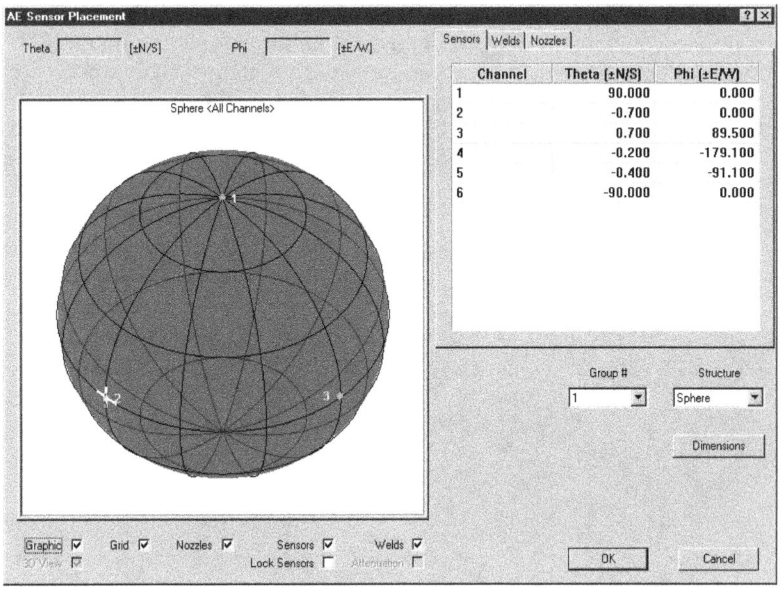

Figure 119. Spherical Sensor Placement

Alternatively you can use the INS key to manually insert a new sensor & then edit the sensor coordinates on the sensor grid.

Spherical location is designed to work only with the Sphere structure type. You will receive a warning if you attempt acquisition or replay with a Spherical location group that attempts to use any other kind of structure. The Dimensions button allows you to set the diameter of the sphere. The drawing of the sphere in Sensor Placement does not change in size with the diameter, but the diameter is a key parameter required when making use of the sensor coordinates in the location calculations. If your diameter is set incorrectly, location will often fail all spherical events.

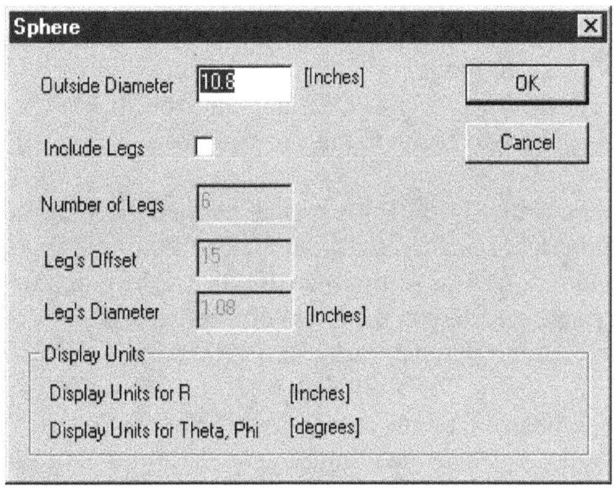

Figure 120. Sphere Dimension Dialog

There is an Include Legs option in the Dimensions dialog that allows the drawing of the sphere to show support columns if desired. Spherical location does not directly support location on the legs, but you could use Cylinder location on the legs, if location on the legs is required. You can specify the number of legs, their diameter and the offset in degrees of the first leg. The rest of the legs are drawn evenly sed around the sphere.

After setting up the sensors & each of the location groups, you will want to create location graphs to display the results. It is possible to use any of the standard 2D graph types and display theta or phi on one of the axes. Just pick Theta Position or Phi Position from the list of available features. The easiest way, however, is to use a Spherical Location graph like the one shown at the right.

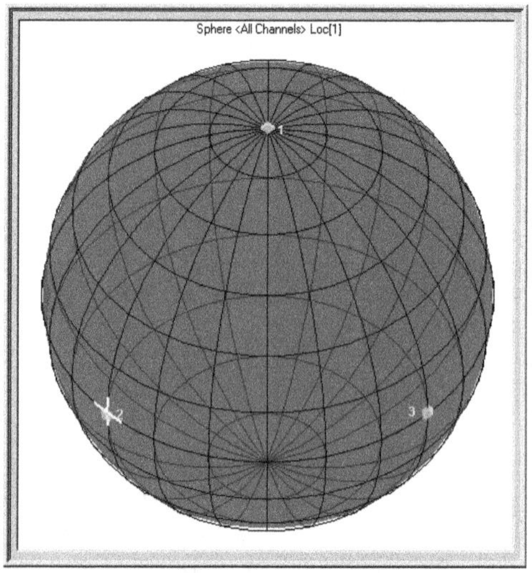

*Figure 121. **Spherical** Location Graph*

In use, a Spherical Location graph can contain a maximum of 4095 data points. During acquisition or replay, the newest data points replace the oldest ones on the graph once the 4095 point limit is reached.

When viewing the Spherical Location plot, you can use the mouse to rotate the sphere by clicking & dragging with the left mouse button. If you use the wheel on a wheel mouse or hold down the Alt key while clicking & dragging with the left mouse button, you can make the sphere larger or smaller. Holding down the Shift key while clicking & dragging pans the sphere up/down/left/right. The Reset Axes button on the toolbar will cause the sphere to revert to the view originally specified by the View Direction settings in Graph Setup.

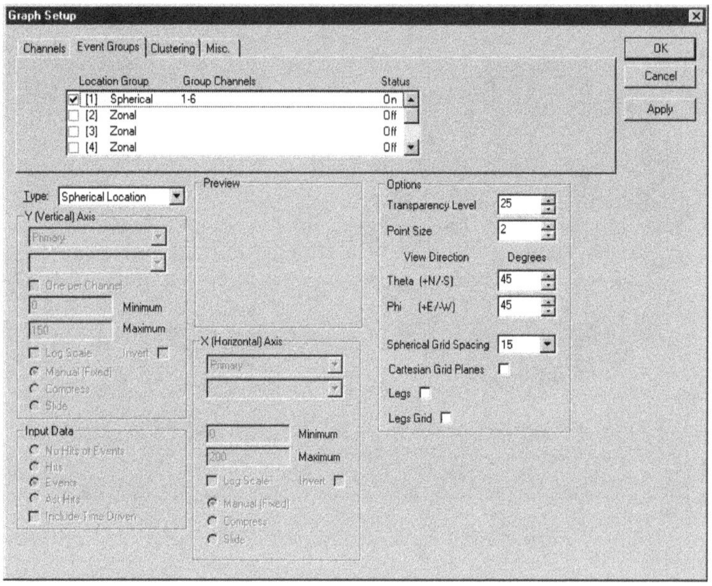

*Figure 122. **Spherical** Graph Setup*

Here is the Graph Setup dialog configuration for the graph shown above. Unlike the normal 2D & 3D graphs, for the Spherical Location plot you do not specify the axis quantities or the data type (always plots events), so the Input Data and X & Y Axis settings are disabled. The graph will automatically draw sensors on the plot for any spherical location groups selected on the Event Groups tab. If you do not see a sphere on the graph after you exit Graph Setup, go back into Graph Setup & make sure at least one spherical location group is selected in Event Groups.

There is an Options section specific to the Spherical Location plot. It contains Transparency Level, Point Size, View Direction, Spherical Grid Sing & Cartesian Grid Planes. Here is brief description of what they do.

<u>Transparency Level:</u> The Transparency Level can vary between 0 and 100%. When set to 0%, the sphere becomes opaque & points that plot on the back side of the sphere are not visible. If you use 25 or 50%, the points on the back side become visible, but are a darker shade than those on the front side.

<u>Point Size:</u> Point Size directly affects the size of the data points & the sensor markers. The Spherical Location plot draws both the sensors & the data points as small diamonds. A bigger Point Size yields bigger data points & bigger

sensor markers. The sensor markers are always slightly larger than the data points. You may also use the Graphing Colors menu to change the colors of the sensors, data points and the sphere (structure color). The data points use the Plot 1 color in Plot Colors.

View Direction: The View Direction setting in Graph Setup controls what angle the sphere is viewed from using the same conventions as the sensor coordinates. This allows you to create top, bottom and side views of your sphere. For example, if you enter Theta = -90 and Phi = 0, the sphere will draw as if you are looking directly at the south pole. Below is a page of sample graphs from the SphereDemo.lay example file.

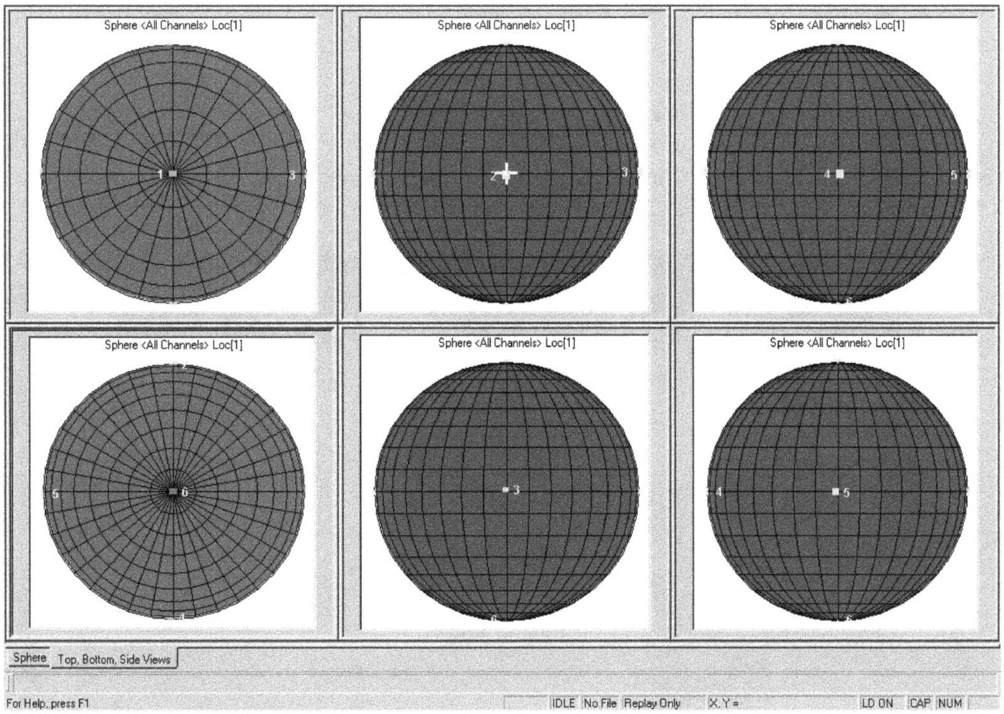

Figure 123. *Spherical Location Graphs From Differing Viewpoints.*

Spherical Grid Sing: The Spherical Grid Sing parameter controls how a grid is overlaid on the sphere. Available choices are 10, 20, 30, 45, 90 degrees and None.

Cartesian Grid Planes: The Cartesian Grid Planes checkbox toggles an X, Y, Z frame of reference on/off.

Legs: The Legs checkbox controls the display on the graph of any support columns defined for the structure.

Legs Grid: This option turns on an additional grid for the legs.

Cylinder Location

Cylinder Location is designed for locating source emissions on a hollow cylinder with either flat, elliptical, or hemispherical endcaps or no endcaps at all. (If you are interested in location on solid cylinders, see section 4.4.5 3D Location on Solid Cylinders.) It is similar to 2D Planar Location with the x, y coordinates being the unwrapped cylinder distances, but one or both of the axes wraps around the body of the cylinder.

When setting up your sensors on your cylinder, there are several guidelines to keep in mind. As with 2D Planar, the software cannot successfully locate any source when all the hits are from sensors in a single straight line. You can force it to do so with the Allow Marginal Events checkbox in Location Setup, but the advice on good vs. bad sensor layouts for 2D Planar applies here too. Avoid rectangular sensor layouts where the distance between rows differs greatly from the distance between columns as shown in the Bad 2D Planar Layout diagram. As stated before, the optimal arrangement would be a triangular grid made up of equilateral triangles. For Cylinder location, also be sure your sensor triangles on the heads are not significantly different in size than the triangles on the body especially if the cylinder has flat heads. If not, you may get spurious event locations clustering along the head-body border.

Sensors that are 180 degrees apart can also cause mathematically invalid events too. Geometrically, it is similar to problem with a straight line of sensors. If an event consists of a sensor on one side of the cylinder & 2 others exactly 180 degrees around the body, there is no way to distinguish source locations in the positive direction from the negative . As with linear events, you can force it to do so with the Allow Marginal Events checkbox in Location Setup. See section 4.2.4 for details. Generally it is recommended that you use 3 sensors around the circumference of the cylinder in a given row. If you must use 2 sensors per ring around the cylinder, definitely stagger the alternate rows or else you will not see any locatable events.

Use the Sensor Placement dialog shown below to set up the cylinder as well as place your sensors. Cylinder location is only designed to be used with the Horizontal or Vertical Vessel structure types. For Cylinder location, the Structure list in Sensor Placement *must be* set to either Horizontal or Vertical Vessel (cylinder) as shown above. A warning displays in acquisition or replay with a Cylinder location group using any other kind of structure.

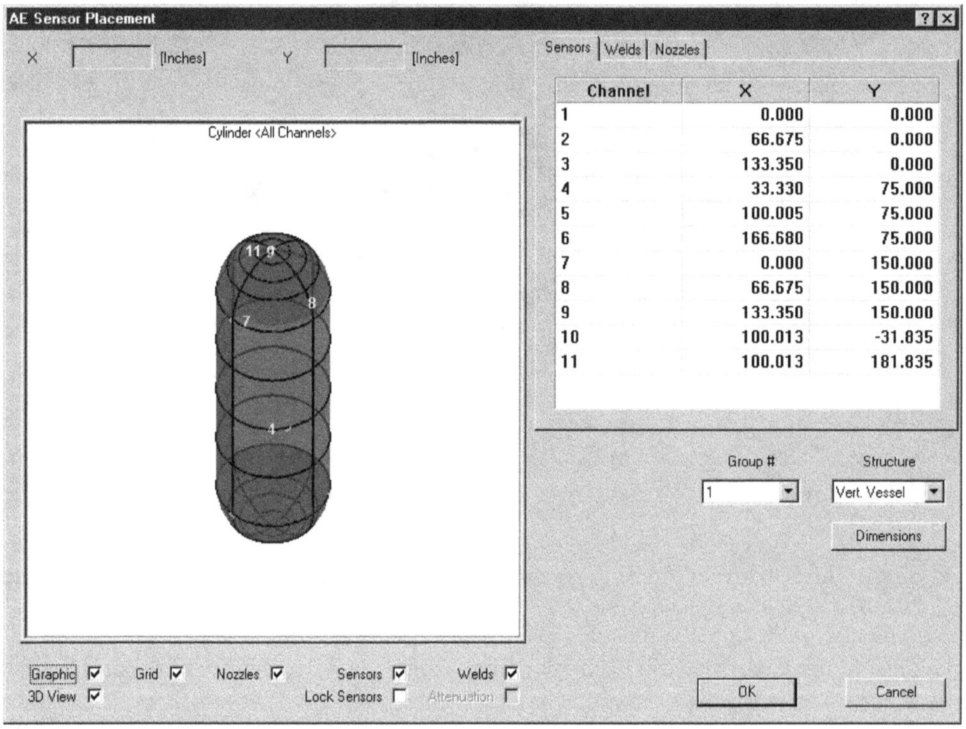

*Figure 124. Cylinder Sensor **Placement** (3D View)*

By default, Sensor Placement shows a 3D cylinder view. If you prefer a 2D view, simply uncheck the 3D View option at the bottom of the dialog & you will be presented with an unwrapped 2D view of the cylinder shown below.

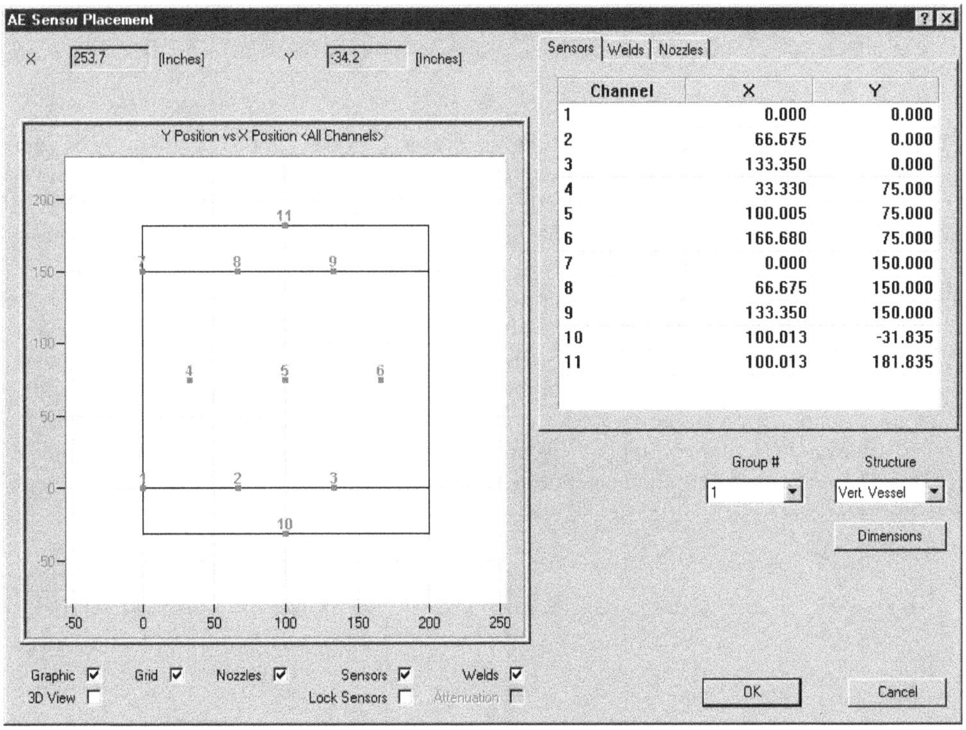

*Figure 125. Cylinder Sensor **Placement***

The Sensor Placement dialog in 2D view & 2D location plots draws a rectangle to represent the body of the cylinder based on the user entered dimensions. You use the Dimensions dialog shown to the right to enter an Outside Diameter and the Height (vertical) or Length (horizontal). For a Horizontal Vessel, the rectangle extends along the x axis from 0 to Length and along the y axis from 0 to the circumference calculated from the user entered Outside Diameter. For a Vertical Vessel, the rectangle extends along the x axis from 0 to the circumference and along the y axis from 0 to Height.

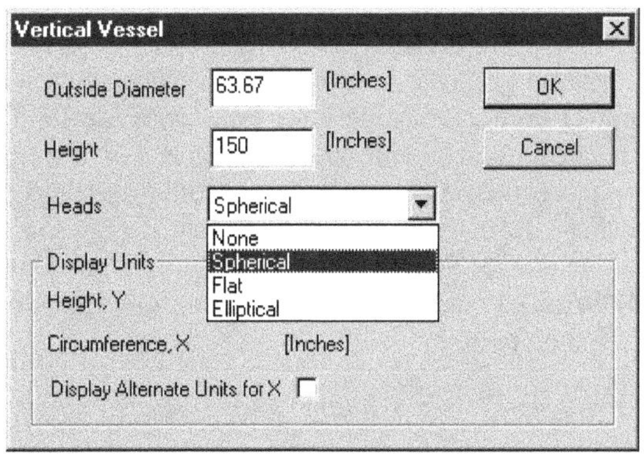

Figure 126. Vertical Vessel

Both dimensions are also vital to accurate source calculations. The Diameter is used to allow for the wrapping around the cylinder circumference. The Height or Length in combination with the Diameter is used in eliminating 2D locations off of the cylinder body caps itself.

There is also a dropdown list to select the type of Head on the cylinder. Available choices are Spherical, Flat, Elliptical & None. If a head type other than None is selected, the body rectangle drawn in Sensor Placement will have 2 additional rectangles on the top & bottom (Vertical Vessel) or left & right ends (Horizontal Vessel). The head rectangles extended the body rectangle by a distance determined by the head type. For Spherical or Flat, it is the length of the cap radius which is also the body radius and half the user entered diameter. In the horizontal case, the rectangle will extend along the x axis from –radius to Length+radius, while in the vertical case the y axis will run from –radius to Height+radius. For Elliptical heads, a 2:1 ratio is assumed between the body radius & the perpendicular head radius, so the distance is ½ the body radius or ¼ the user entered diameter.

You can quickly generate rectangular & triangular sensor layouts with the AutoPlace Sensors dialog available by right clicking on the sensor view. If you are using the 2D view of the cylinder, you can also manually place sensors on the cylinder by right clicking. After using either option, you can edit the coordinates of the sensors if structural features (valves, nozzles, welds) require shifting sensors around. Just enter the x,y coordinates in the user's preferred distance units.

When you enter sensor coordinates & you are not including the caps, express your x, y coordinates with a lower limit of 0 and an upper limit of the circumference or the length/height. For example, when working with a vertical cylinder of height 150 and diameter 63.66 (circumference = 200), express x as a value from 0 to +200 and y as value from 0 to +150. Basically, just make sure your sensor is visible in the body rectangle drawn in the Sensor Placement dialog.

If you are using Spherical or Flat heads, these limits are widened for the y axis by the radius to –31.83 to +181.83. These extensions allow for the entry of sensors on the heads. For this vertical cylinder, a pole sensor on the bottom cap would have a y coordinate of -31.83. The x coordinate for a cap on a vertical cylinder corresponds to a rotation angle, but is entered as a fraction of the body circumference. In this case the limits are 0 to +200 and sensors placed 120 (=360/3) degrees apart are separated by 66.67 (=200/3). For a cap sensor placed directly at the pole of the hemisphere, the rotation angle is unimportant in the source location calculations, but a value of half the circumference will place the sensor halfway up the cap rectangle border. In the case of the vertical cylinder shown above, the pole sensor on the bottom cap is at -31.83, +100 (any x really) and +181.83, +100 for the top cap.

When you are finished setting up your sensors & ready to actually do data acquisition, you will need a location graph to display the results of the source location calculations. You can either use a standard 2D graph as described in the section on 2D Planar location or you can use a Cylinder Location graph. Like the Spherical Location graph, a Cylinder Location graph is limited to 4095 data points with newer points replacing older ones once the limit is reached. A Cylinder Location graph is shown at right and its corresponding Graph Setup dialog settings are below. It has an Options section that is very similar to the one for Spherical Location. See the section on Spherical Location for a description of the options common to both.

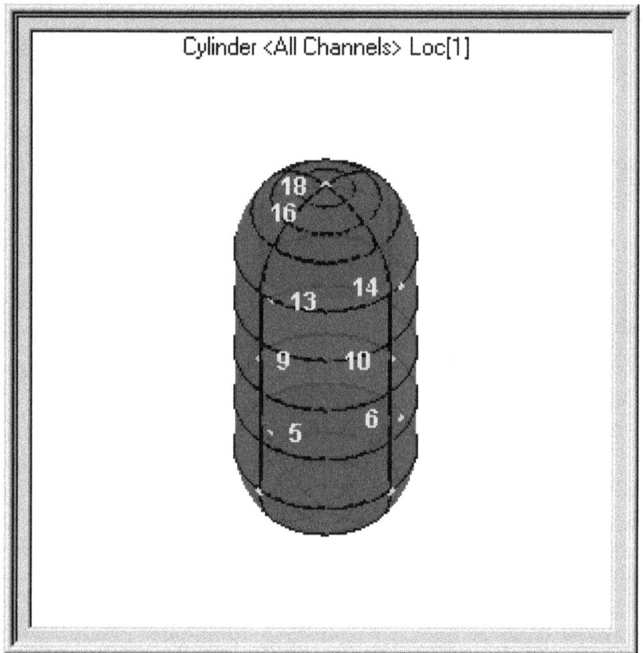

Figure 127. Cylinder Graph

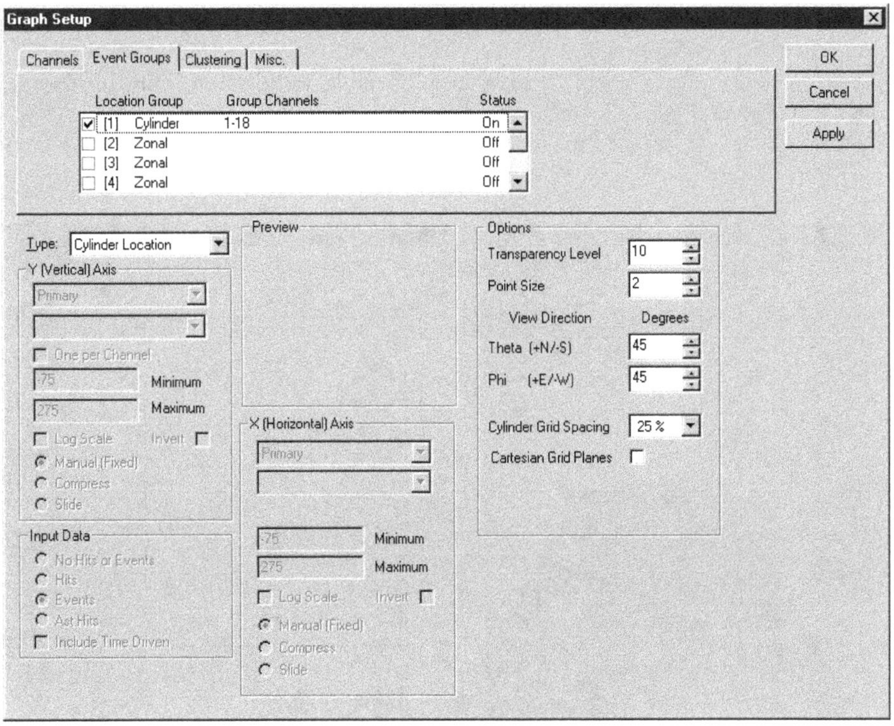

*Figure 128. Cylinder **Graph** Setup*

The Spherical Grid Sing option is replaced by the Cylinder Grid Sing option, although it behaves in a similar fashion. You may set the Cylinder Grid Sing to None, 10%, 20%, 25%, 50% or 100% to control the appearance of a grid overlaid onto the surface of the cylinder. The percentages are applied to the heads & the body separately as the dimensions for the 2 regions can vary greatly. If the user selects a grid sing of 25%, the body will be divided up into 4 equal regions and the head will also be divided up into 4 regions. If the user selects 100%, only the boundary between the heads & the body will be marked. Cylinder Location graphs always use Events as the Input Data, but as for the 2D graphs, make sure you set the Events Groups to the appropriate location group or your graph will not show any data & will not draw the cylinder associated with the group.

Conical Location

Conical location is set up almost identically to Cylinder location mode. In Location Setup, the user selects Conical for the Location Type on the General tab. The other settings in the table are described in the section on the Location Setup dialog. In Sensor Placement, set the Structure list to Vert. Cone or Horiz. Cone as shown on below.

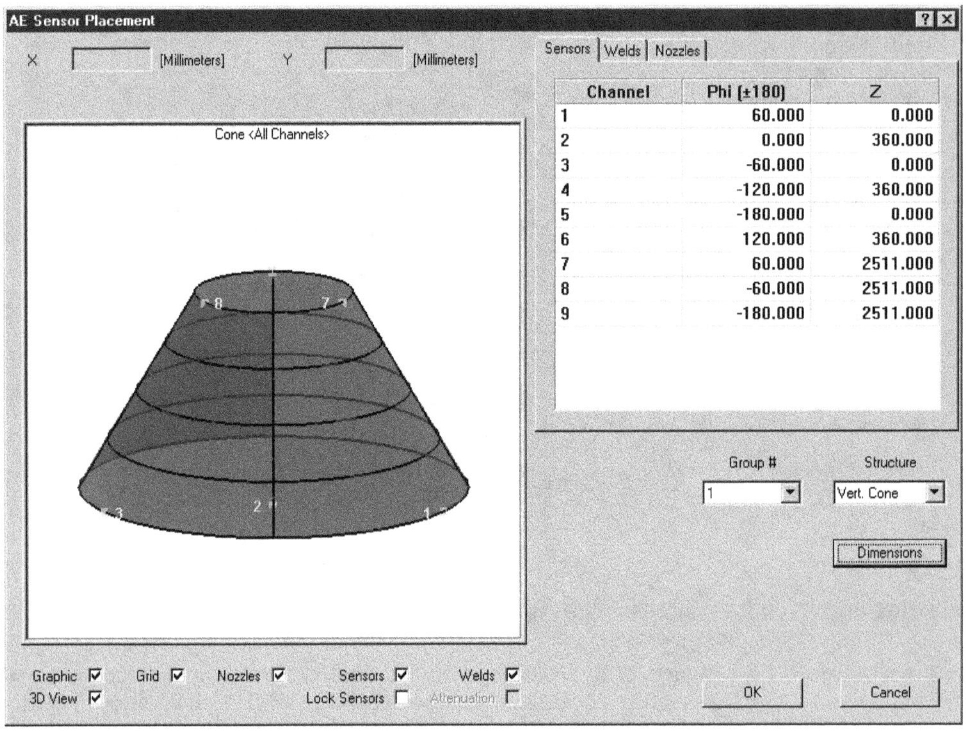

*Figure 129. Conical Sensor **Placement** (3D View)*

The Dimensions button calls up the Cone dimensions dialog allowing the user to enter the upper & lower diameters as well as the height (or the left & right diameters & length, if horizontal). The height or length has added significance when conical location is analyzing data. Conical results are limited to source locations to within +/- 2% of the height. In other words, points located wildly off the cone body are thrown out as bad results.

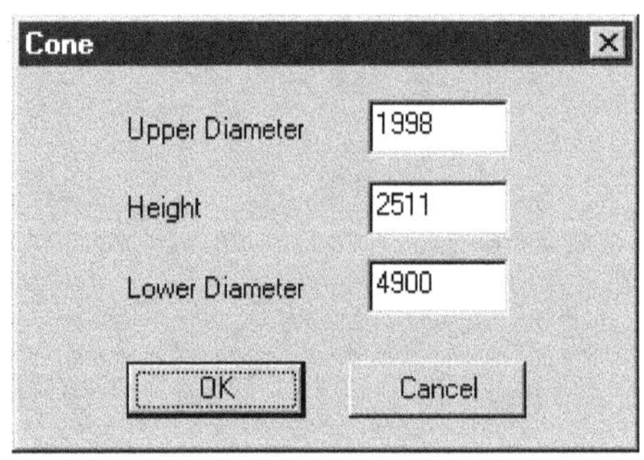

Figure 130. Cone

Sensor coordinates are entered as Z & Phi. The Z axis is the cone axis independent of horizontal vs. vertical configuration. Z starts at 0 at the bottom of the cone & goes up to the height of the cone (or 0 at the left & up to the length). Phi should be entered between -180 to +180 degrees to keep the sensors within the 2D drawing of the unrolled cone, although location will work with values from 0-360 as well.

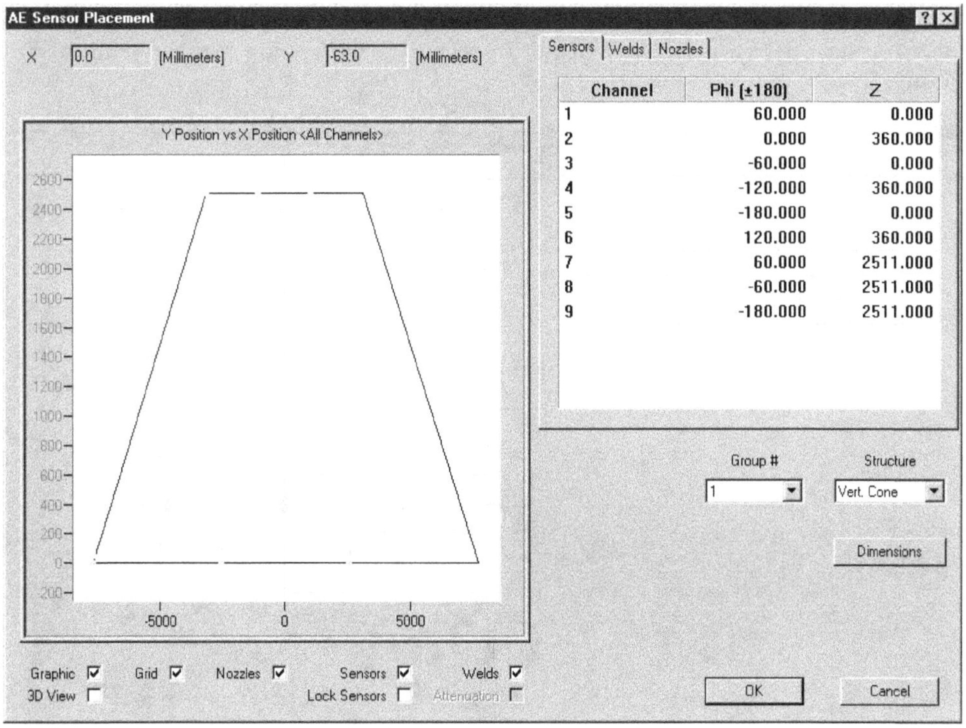

Figure 131. Conical Sensor Placement (2D View)

You can toggle between 2D & 3D graphs in Sensor Placement with the 3D View checkbox. The 2D view translates the phi value into a fraction of the circumference. The trapezoid is drawn from +circumference/2 to -circumference/2 where the diameter depends upon the z value in the cone.

There is also a 3D Conical Location graph similar to the Cylinder Location graph as another graph type in Graph Setup. It has options similar to those for the Cylinder Location graph & the Spherical Location graph. See their sections for a description of these options.

In use the 3D Conical Location graph looks like the 3D graph in Sensor Placement. The sample layout, ConicalDemo.lay, gives an example on the 3D Cone page. A Conical Location graph can contain a maximum of 4095 data points. During acquisition or replay, the newest data points replace the oldest ones on the graph once the 4095 point limit is reached.

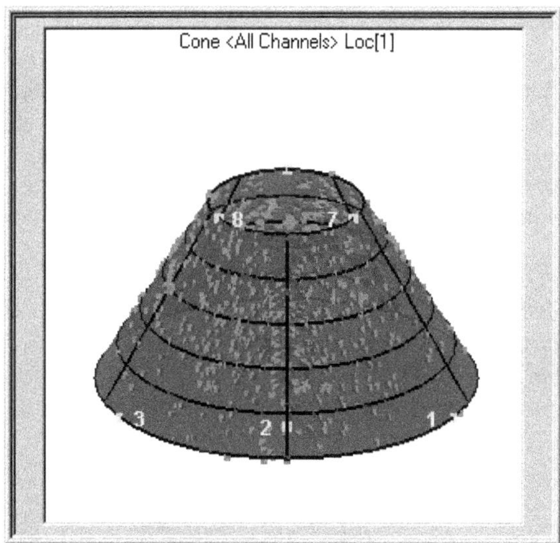

Figure 132. Conical Graph

Anisotropic Location Modes — 2D Anisotropic & Anistropic Cylinder Location

These location modes allow for differing velocities in the x & the y directions. They are almost identical to their corresponding isotropic modes, 2D Planar & Cylinder Location. The velocity in the Wave Velocity column of the

General tab corresponds to the velocity along the X axis for these modes. When one of these location modes is selected in Location Setup, another field will appear in Sensor Placement as show below that allows the user to enter the Y axis wave velocity, Vy.

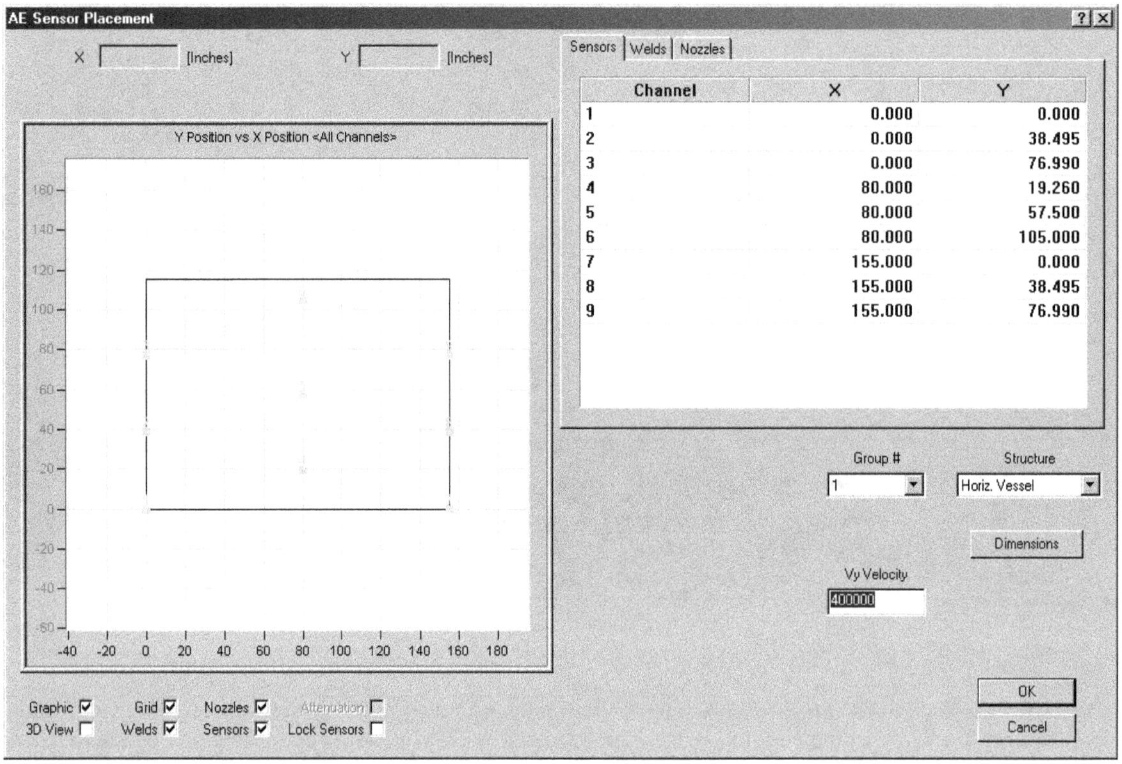

Figure 133. Sensor Placement for Anisotropic Location Modes.

These location modes do not support the use of Timing Strategies other than Primary & the software will not let you select other strategies if you try. The reason is that the alternate strategies only have one velocity associated with them, but the anisotropic modes require 2. The anisotropic location modes rely on a coordinate transformation to enable the standard location modes, 2D Planar or Cylinder location, to analyze the data. That coordinate transformation is based on the ratio of the X & Y axis velocities being constant, but allowing alternate timing strategies with varying velocities would violate this assumption. Anisotropic cylinder location also does not support location of the cylinder endcaps for much the same reason. One might reasonably expect that if the velocities along the X & Y axis differ substantially, then the velocities on the endcaps might also differ. For example, flat endcaps might use the velocity down the length of the cylinder, while spherical endcaps might need the velocity observed around the circumference of the body of the cylinder. The coordinate transformation cannot allow for those kinds of variations, so location on the endcaps is not supported. Except for those limitations, the anisotropic location modes, 2D Anisotropic & Anisotropic Cylinder, function just like the their isotropic counterparts, 2D Planar & Cylinder.

Tank Bottom Location

Tank Bottom location is designed for use with evenly sed 2-dimensional circular arrays of sensors. You must use it with the Tank Bottom structure. A direct consequence of requiring evenly sed arrays of sensors is that the user is not allowed to manually place sensors on the Tank Bottom structure. Instead one uses the Dimensions button in Sensor Placement to bring up the Tank Bottom dimensions dialog shown to the right. Unlike other dimensions dialog boxes, this one adds a section for automatically generating the evenly sed circular array of sensors. By default the sequence starts with sensor 1 placed at 0 degrees or 12:00 on the circle, but the user can start with other channel numbers and specify the rotational offset to the first sensor in degrees. You set the diameter of the tank bottom circle in this dialog box as well. After you click on OK, you will be asked if you want to delete the previously placed sensors. You must answer Yes to have the changes take effect. If you answer no, the changes will be cancelled. Once you have things configured correctly, Sensor Placement will reappear with the display updated based on your settings as shown below.

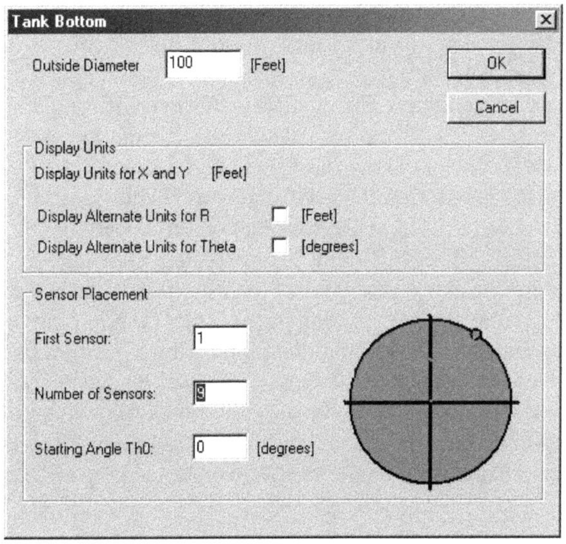

*Figure 134. Tank Bottom **Dimensions** Dialog*

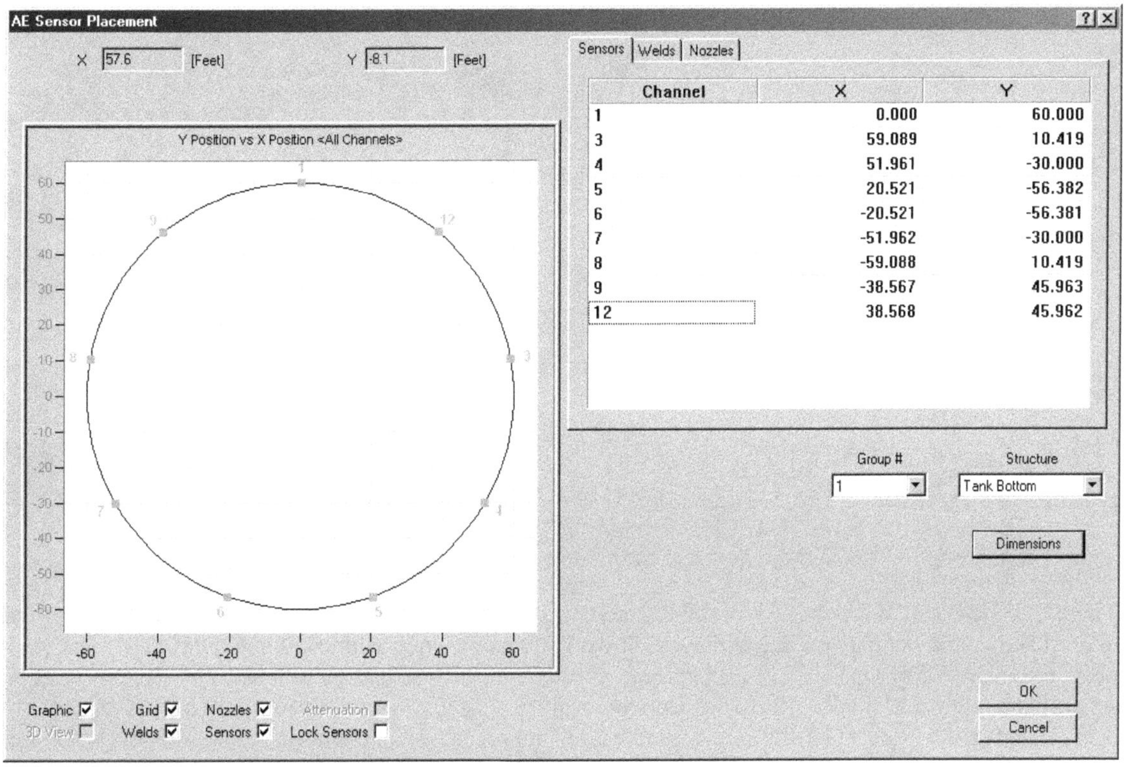

Figure 135. Sensor Placement for Tank Bottom Location.

Note that you cannot edit the X & Y coordinates of the sensors in the sensor grid array when using a Tank Bottom structure. You can edit the channel numbers, though, to allow non-sequential ordering of the sensors on the tank. In the example above sensor 2 was renamed 12, even though, the X & Y coordinates have been altered.

Tank Bottom location determines the source location by directly solving the time-distance equations generated from an event with 3 hits. It cannot directly make use of more than 3 hits when calculating a source location. When there are more than 3 hits available, however, it can vary which 3 hits are used in the calculation. The user can tell the software to do this by using the Tank Bottom Options dialog available thru the Tank Bottom Options button in Location Setup. The top field in the dialog controls whether or not variations of the 3 hits are calculated & the results displayed on location plots. By default, the software will only show 1 location per event & variations are never shown in acquisition mode. When limited to 1 location per event & there are more than 3 hits, the algorithm uses the first 3 available hits on the assumption that those sensors were the closest to the source. This setting has no effect on events that only have 3 hits available. Below are 2 graphs showing how the results vary for a data file with about 600 events. The graph on the left is limited to 1 location per event, while the one on the right allows up to 10 locations per events. Note that the software counts these variations as extra events, so the total events for the graph on the right is reported to be about 2500 events.

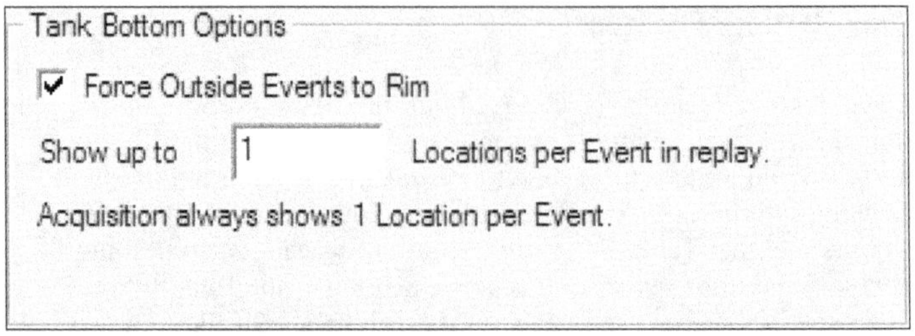

Figure 136. Tank Bottom Options

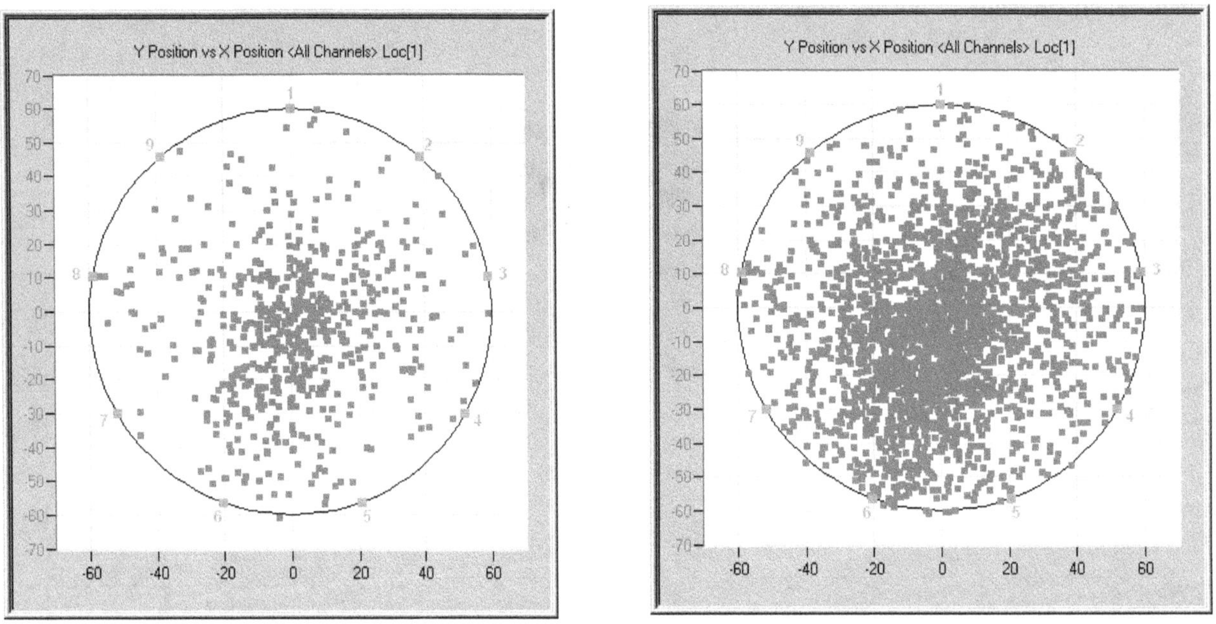

*Figure 137. Tank Bottom Location **Results** with 1 (left) vs. 10 (right) Locations per Event.*

The Force Outside Event to Rim checkbox at the bottom of the Tank Bottom Options dialog controls how events that locate outside the tank bottom area are displayed. It works in conjunction with the Allow Outside Events checkbox in Location Setup. Allow Outside Events applies to all the regression modes as well as Tank Bottom location, but Force Outside Events to Rim only affect Tank Bottom location. Allow Marginal Events does not affect Tank Bottom location. The effects of Allow Outside Events combined with Force Outside Events to Rim are shown in the 3 graphs below which are restricted to showing just 1 location per event.

As discussed in the section on the Location Setup dialog box, "Allow Outside Events" controls whether or not the software reports events that locate outside the structure. If it is off, "outside events" are considered failing events and in general are not reported. Some events may occur very close to or just outside the tank rim & still be considered inside events, because the algorithm uses a tolerance of ±2% beyond the radius when deciding whether an event is inside or outside the tank. That 2% tolerance allows for rounding errors that can push events at the rim just outside the tank bottom. The results of Tank Bottom location with Allow Outside Events disabled are shown below in the graph on the right. If Allow Outside Events is disabled, the Force Outside Events to the Rim setting has no effect on the results. If Allow Outside Events is enabled & Force Outside Events to Rim is disabled, all events plot at the calculated source locations as shown below in the middle graph. If both Allow Outside Events & Force Outside Events to Rim are enabled, then events that would plot outside the tank bottom area are redirected to the rim of the tank as shown below in the graph on the right. The software draws a line from the center of the tank to the calculated source location, finds the point this line segment crosses the tank rim & reports it as the source location. Although this location is not an accurate source location, it does allow the user to see that an event occurred. A drawback to enabling Force Outside Events to Rim is that it can obscure events that actually do locate at the edge of the tank. It is provided primarily for backwards compatibility when comparing with results observed in 's older DOS applications.

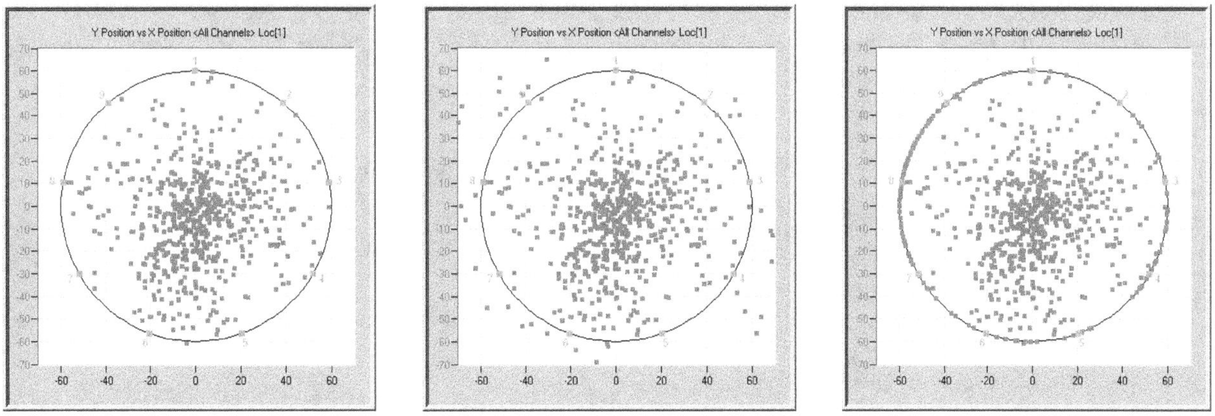

*Figure 138. Tank Bottom Location Results with Allow **Outside** Events disabled (left) & enabled with Force Outside Events to Rim disabled (middle) & enabled (right).*

Location Examples

It is not feasible to provide detailed examples of all location modes in this manual. There are, however, example layout files & sample data files for each location mode. When you are starting out with a new location mode, you can simple recall the sample layout & inspect the location related menus (Location Setup, Sensor Placement & Graph Setup dialogs). After you learn how the example is set up, you can make changes to better reflect your test setup & save the modified layout under a new name. Below is a table summarizing the example layouts & data files for the various location modes.

Location Mode	Layout File	Data File
Zonal	Zonal_Linear.lay	Zonal_Linear.dta
Linear	Zonal_Linear.lay	Zonal_Linear.dta
(Wrapped) Linear	WrappedLinear.lay	WrappedLinear.dta
Tank Bottom	TankBot.lay	TankBot.dta
2D Planar	2D-Planar.lay	HE-Link.dta
Cylinder	OpenCylinderDemo.lay	LongVessel.dta
Conical	ConicalDemo.lay	ConicalDemo.dta
Spherical	SphereDemo.lay	SphereDemo.dta
3D Location	3dLocDemo.lay	3dLocDemo.dta

In addition to the example files, there are detailed examples given here to illustrate how to use Zonal, Linear, 2D Planar & Tank Bottom location.

The steps in setting up and running a location test are pretty similar no matter what location mode you are using. Here is an overview of those steps:

11. Create a layout file with the hardware set up appropriately (e.g. the correct number of sensors turned on, detection threshold set properly).
12. Bring up the Location Setup dialog (=F8 key) and start by setting the distance units.
13. Next fill in the location group information. Enable the group & select the Location Type. Enter the Wave Velocity, Event Definition Value & Overcal Value. If this is your first time working with location, leave the Event Lockout Value, Timing Strategies, Max Iterations, Min & Max Hits/Event at the default values. That applies to all the settings on the Timing page as well. These parameters are more fully described in section on the Location Setup dialog under the headings of General Tab Settings & Timing Tab Settings.
14. Next define the sensor locations on the structure using the Sensor Placement dialog. Use the Location View button in Location Setup to display it.
15. In Sensor Placement, select the Structure type, then click on Dimensions & enter the structure information.
16. Now create your sensors by right clicking the sensor view area & or using INS & F2 in the sensor grid.
17. When finished, click OK in both the Sensor Placement dialog & in the Location Setup dialog.
18. Next create a suitable location graph using the Graphing menu (New Graph) & configure it using Graph Setup (=F3 key). See the notes on individual location modes for information on the appropriate location graphs.
19. Now we are ready to run a trial AE test using pencil lead breaks to verify that everything is set up properly. If everything works, you're ready for acquisition.
20. If you end up staring at an empty graph, here is a list of things to double-check. Some are obvious, but personal experience shows them to be easy mistakes.

g. Is the Input Data set to Events in Graph Setup?
h. Is this location group checked on in the Event Groups tab of Graph Setup?
i. Is this location group checked on in the Group # column of Location Setup?
j. Are there any events from the group in the Line Display?
k. If there are events, are the axis limits in Graph Setup wide enough to see the positions?
l. If no events, are the Location Setup settings reasonable? Did you look up a velocity like 10,000 ft/sec & enter 10,000 when it should be 120,000 because the Distance Units are Inches? Verify the Event Definition & Lockout Value are reasonable given your sensor sing & structure type. They all use the same units, so the values should be related to sensor-sensor sing.

Set Up and Testing of a Zonal or Linear Location Group

As an easy to set up layout, we will be using two sensors located 30 inches apart. This sensor layout will be used to illustrate Zonal location in this section & Linear location in the next section. We will be using example layout and data files called Zonal_Linear.lay and Zonal_Linear.dta which are installed in the AEData directory under your software program directory, for example "C:\Program Files\Physical Acoustics\HMSGHAT2000DDI for DiSP\AEData". If you do not have these sample files, we will be showing the setup screens so that you can duplicate the setup and conduct your own acquisition test.

In this section we will be setting up a Zonal Location Test following the steps identified above. For the example, we are assuming that we have two sensors in Channels 1 & 2 of the AE system.

<u>Step 1.</u> Start by using the File pulldown menu to access the Open Layout menu option or press Ctrl-O. Select the layout file named, "Zonal_Linear.lay" & recall it.

<u>Step 2.</u> Enter the location setup menu by either pressing the F8 function key, clicking on the Toolbar icon that looks like a "bullseye" target called "Location Setup" or select Location on the Acquisition Setup pulldown menu.

The Location Setup dialog will be displayed as shown in this section. This is the first menu that needs to be set up in our step by step procedure. By default, the General page displays. Notice the General tab near the top left of the dialog box. The General page of Location Setup allows you to control the setting of the individual location groups. Up to 8 location groups can be set up, although an additional licensing option provides for up to 32 groups.

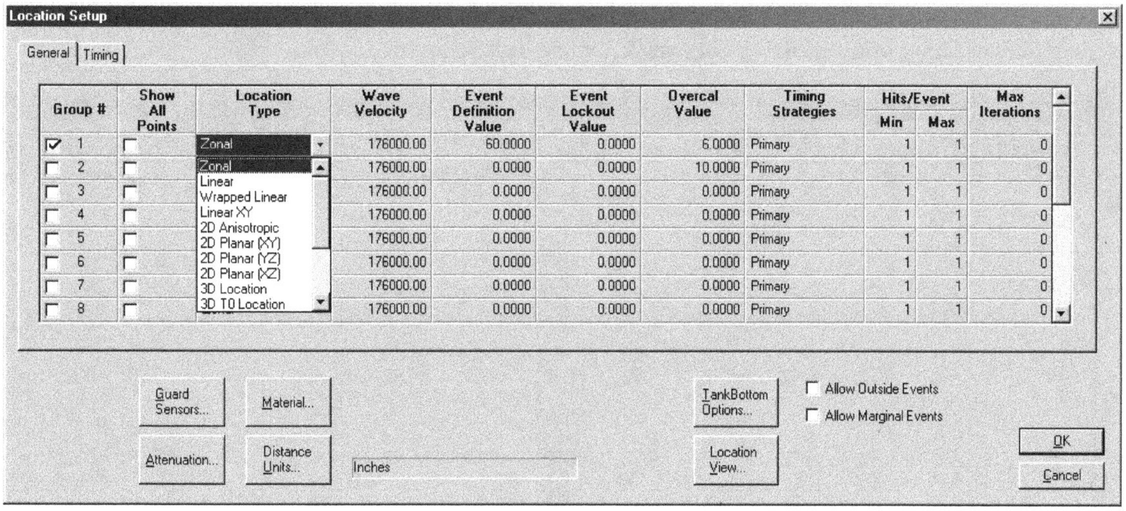

Figure 139. Setting up Location Group Information

In addition, near the bottom center of the dialog, the user can set up the location distance units. Various selections including Metric, English and time units can be selected for distance units. Selecting a real distance value is most appropriate since the values read out can be directly measured when trying to locate a source on a structure. Use the Distance Units button to select inches as in this example.

Step 3. In this example, we will select a Zonal location group as group #1. Follow the Group #1 line and enter the information as shown. First select the zonal group. This is done by left-clicking the mouse on the down arrow on the right side of the Group #1, "Type" box. A dropdown list will appear to allow the user to select from the various alternatives (Zonal, Linear, etc.). Select Zonal.

For the Wave Velocity, the velocity entered will be related to your Distance Units setting. In this case, we've selected inches as our units, so the velocity is in inches/second. You can look up a velocity using the Material Properties dialog by clicking on the Material button. The database contains values for Longitudinal, Shear & Surface velocities. It is often better to directly measure the velocity using the AE system. Simply use the Line Display mode to determine the actual transit time (Δt) to traverse between two sensors. You can then divide the distance between the two sensors by the Δt to arrive at the estimated velocity. In order to determine this correctly you really need to simulate the type of defect you are trying to detect. Enter the desired velocity in the edit box. In our example, we used an aluminum bar and a pencil lead break as a source. Although the Longitudinal velocity is 249,000 inches/second and the Shear velocity is 123,000 inches/second, we experimentally determined a test velocity of 176,000 inches/second.

For the Event Definition Value, we want to select a generous Event Definition using the same units as the Distance Units selected. This should be set to make sure that from any point in the location array, a minimum desired number of sensors will be hit, forming an event. In our example, two sensors 30 inches apart, the Event definition should be at least 30 inches to make sure that the two sensors get hit within a single event. We have entered 60 in the example above.

For Event lockout Value, we enter a value (in the selected distance units) to assure that the there is enough time (or distance) for all additional hits and reflections related to the event have had a chance to die down. The amount of time or distance selected will be the "dead" time as we wait for the event to die down. In this example, we will leave it at the default of 0 as reflections are not expected to be a problem for this setup.

The Overcal value is also set in the selected distance units and is usually set at an amount of approximately 10% to assure that minor variations in wavepath velocity are not ignored by the location algorithm. When you are just learning to use location, it is best to leave the settings on the Timing page & the Timing Strategies column at the defaults. The Max Iterations, Min & Max Hits/Event cannot be changed for Zonal as they are not used by it. They only apply to the regression location modes. All of these parameters are more fully described in section on the Location Setup dialog under the headings of General Tab Settings & Timing Tab Settings.

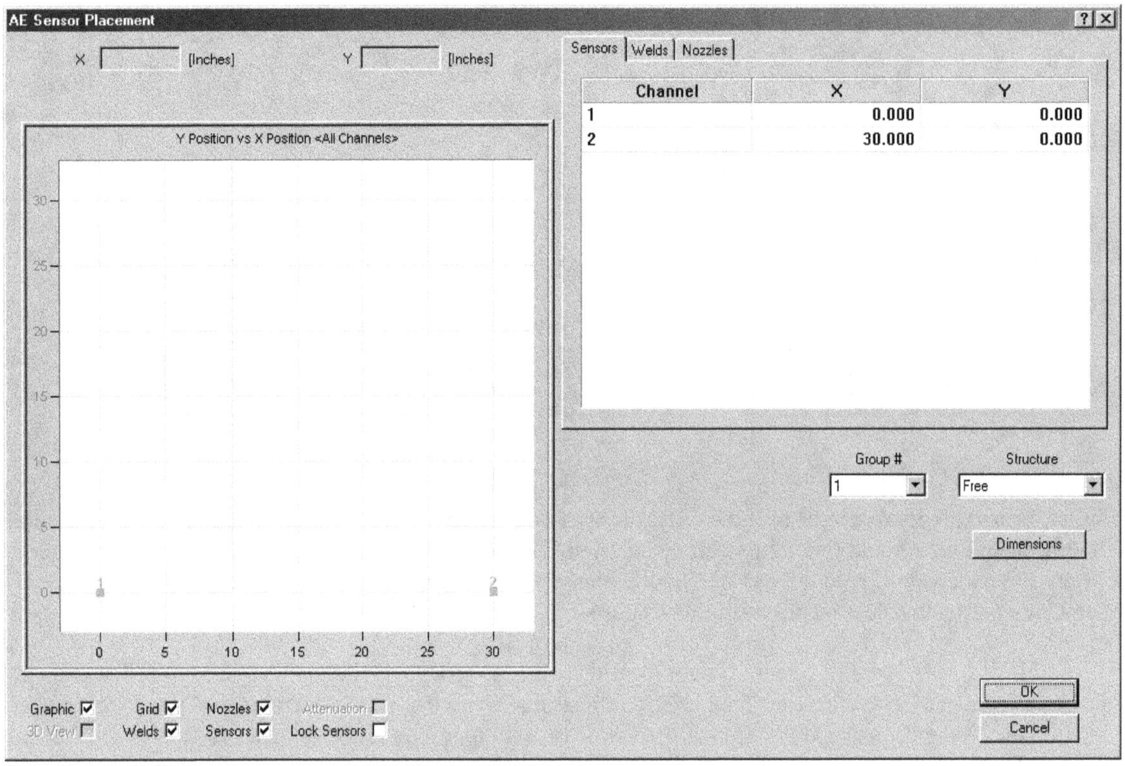

Figure 140. Sensor Placement Dialog

Step 4. Once the location group settings have been entered, we can define the sensor setup using the Sensor Placement dialog. Click on the Location View button on the bottom right side of Location Setup to display it. The figure above shows the Sensor Placement dialog box. There are 2 main area to the dialog box. The sensor layout view on the left-hand side and the sensor grid with the coordinates of individual sensors on the right hand side.

When we click Location View, the sensor layout for that particular location group is displayed. If we had group 1 highlighted, but wanted to display the sensors for group 5, we could change the group displayed with the Group dropdown listbox just below the sensor grid. Select Group 1, if it is not already selected.

Step 5. Next, we need to select the "Structure". The Structure setting is also just below the sensor grid & just to the right of the Group setting. This is currently set to Free. Free allows sensors to be placed anywhere without worrying about the specific structure geometry such as a plate, box or sphere.

Next, we set the Structure dimensions. For the Free structure, this isn't required for location, but the dimensions control the axis limits of the graph in the sensor view area. To change them we click the "Dimensions" button and the Free dimensions dialog shown at the right appears. Here we've set the X and Y minimum and maximum values to 0 & 30.

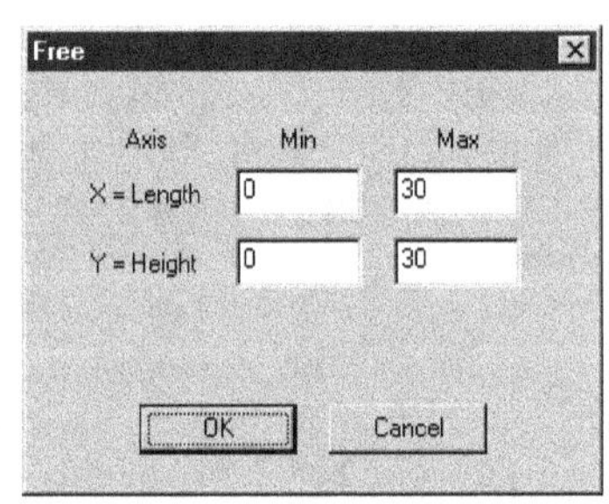

Figure 141. Free Dialog

Step 6. We are now ready to create our sensor layout. In Sensor Placement example above, you can see that the sensors have already been placed at the bottom of the graph area, at locations of (0, 0) and (30, 0) using an (X, Y) arrangement. The software makes it easy to place and move sensors on the screen and to get them in an exact position.

To place a new sensor, we right click inside the graph region. The context menu to the right appears. The menu options we want to concentrate on are on the Sensors submenu allowing you to perform various functions related to sensor placement.

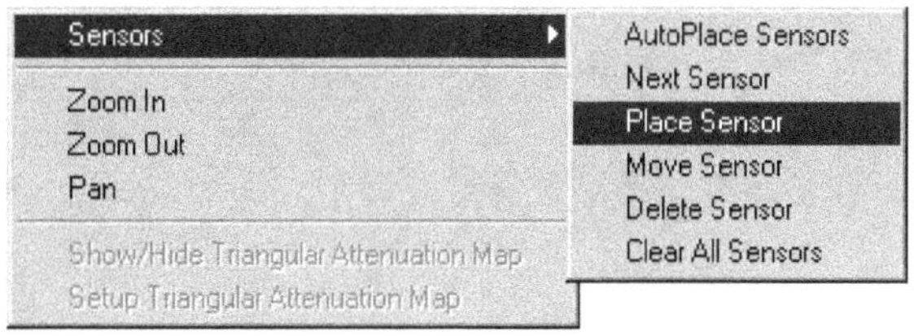

Figure 142. Place Sensor

Since we want to place a sensor, simply click on the "Place Sensor" selection. Upon doing so a small menu box comes up asking you to verify the number of the sensor to be placed or to change it. When you're done, click OK and the mouse cursor will change to a red sensor icon with a crosshair cursor. Each time you click you will generate another sensor in your layout. Notice the x,y coordinates of the mouse cursor are shown above the sensor view area. When you are finished placing sensors, right clicking will return the mouse to its normal arrow shape. Simply follow the right click with a left click off the menu, if you don't need any of the context menu options. Whatever you do, don't press the ESC key. That is equivalent to pressing the Cancel button for the dialog & you'll lose all your changes to the sensor layout.

There are 2 ways to move sensors around to correct bad placement. First, the right click Sensors menu has a Move Sensor option. When you select it, the mouse cursor changes to a red sensor icon with a 4-headed crosshair cursor. To use it, click on the sensor you want to move. The sensor will turn red. Now click a second time at the spot where you want to move the sensor & it will shift to the new location. Right clicking ends the Move Sensor mode.

If after graphically placing or moving a sensor, its placement is not accurate enough, you can alter the position to an exact value by editing the X or Y value in the sensor grid. Left-click the mouse to highlight the desired box to be changed and enter the exact value, followed by the Enter key. Alternatively, press the F2 key to edit the current value.

Note the other sensor placement and edit capabilities provided by the submenu. You can automatically place the "Next" sensor by number by pressing the "Next Sensor" button. In this case the sensor channel number selection box does not come up. It simply enters the Place Sensor mode using the next highest number sensor. Please note and experiment with these sensor placement management selections.

By now you may have added several sensors to your setup that you want to delete. Do this by right clicking over the sensor you want to delete and then select "Delete Sensor". The mouse cursor with change to a red sensor with a crosshair labeled "del". When you click on a sensor, a "Delete this sensor?" message box asks for confirmation. Click "Yes", to delete the sensor from the graph and from the table. Again right clicking puts the cursor back to normal. Return to the original sensor layout as shown above when you are finished exploring the Sensors menu options.

<u>Step 7.</u> In the previous steps we have fully defined the location setup. The next step in the process is to set up graphs to see our location in acquisition or replay. We do this by first saving our sensor placement & location setup. Click the OK button in Sensor Placement & then in Location Setup. DO NOT CLICK Cancel in either one as you will lose any changes you made while in these dialog boxes. This is a very good time to save your layout on disk as well using the File pulldown Save Layout As menu option. When you don't need to rename the layout, you can just use Ctrl-S to update the current layout on disk.

<u>Step 8.</u> Next we need to create a suitable location graph. If you have been following along with our example layout file called Zonal_Linear.lay, you will be seeing a screen like that shown below. In this screen you can see two graphs. The first or top graph is set up already as a location graph. Since this is being set up for zonal location, we will only see the activity in a histogram or bar chart type of arrangement.

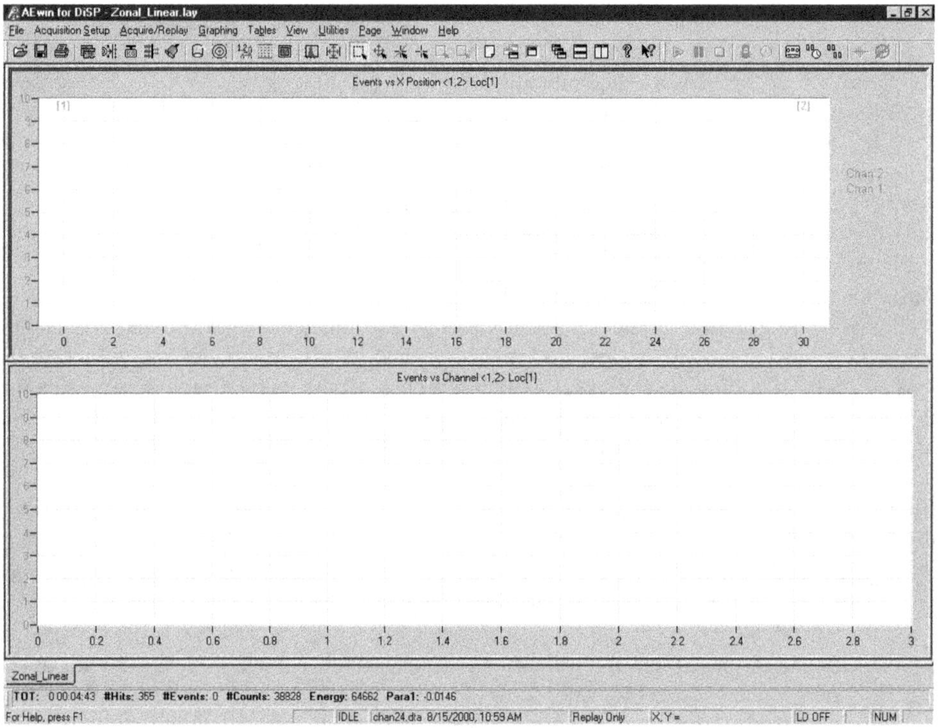

*Figure 143. **Zonal** – Linear Location Graph*

Assuming that this graph was not already set up in the desired location mode, let's go through the steps of setting up the graph for a Linear or Zonal location. First, we select the graph that we want to change to the location graph. We do this by right clicking the mouse when it is inside the graph region of the graph that we want to work on. Once we do that a menu appears. This menu gives us many choices but we will select the top one called "Graph Setup". Upon selecting this a graph setup menu appears like that shown below.

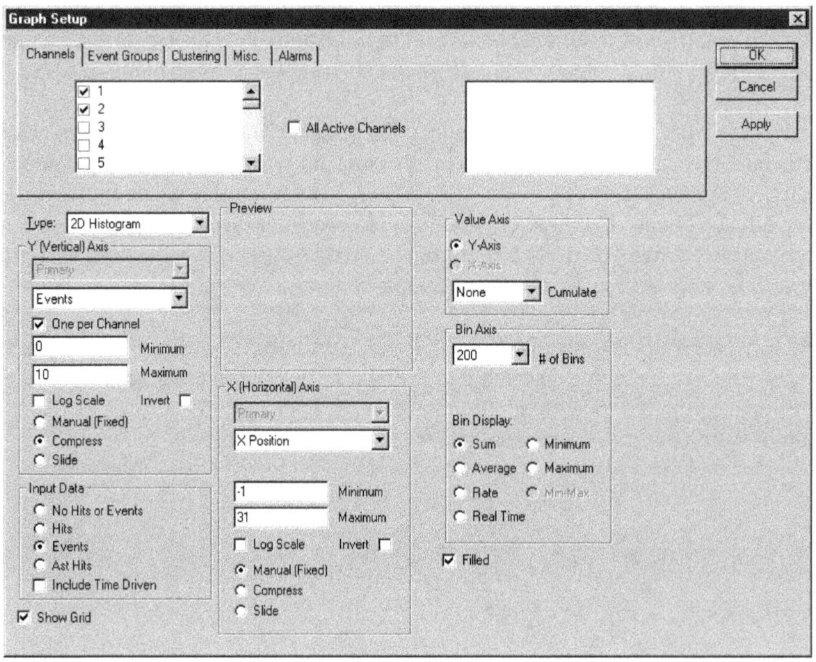

*Figure 144. **Graph** Setup Dialog*

This screen capture shows us how we have set up the Graph setup menu for a linear location. This gives us a lot of insight into how the graph setup menu works and how to set it up. Lets note first the layout of the menu. We see at the top of the menu, a selection area for Channels, Event Groups and Clustering. These settings allow us to select exactly those items. With the Channels Tab, we can select which channels will be displayed on the graph that we are setting up. In this case we are only selecting channels 1 and 2 for viewing in the graph. You can see that we have set group #1 in Event Groups if you select that tab. Clustering allows you to set up parameters that link closely sed hits/events together.

The next menu area is the Graph Type selection box. This is located on the left just below the Channel, Event Group and Clustering selection area. Various graph types can be selected, but for this application we will use a 2D Histogram.

Below the Graph Type selection, is the Y (Vertical) Axis setup menu area. Going down one item at a time in this menu is the plot # selection within the graph. This selection normally allows you to display more than one plot on a graph, but it is disabled for this 2D Histogram type, because we have checked the One per Channel checkbox (see below). Basically the selection (Primary, Secondary-1, 2 etc.) controls which plot the Y Axis settings apply to.

Next is the Y-Axis parameter to be plotted on the graph. By clicking this, you can see many selections. However, we have selected Events. This will display the number of Events detected on the Y-Axis.

The checkbox One per Channel allows us to select a unique per channel plotting color that will give us a multiple color by channel plot. In our case for events and 2 channels, we will get a different color for the channel that was hit first if this is checked. If not we will only get one color plot.

The next two edit boxes allow us to set the starting Y-axis Minimum and Maximum settings.

The next checkbox Log Scale allows us to set a log or linear graph scale. Beside it is an Invert checkbox which allows the user to reverse the positive to negative direction on the axis.

The next 3 radio buttons are related, allowing you to select one type of graph scaling. Manual(Fixed) will force the graph limits as set above to remain fixed. Compress will compress the graph so that the minimum graph value remains but the maximum keeps changing as the graph limit is exceeded. And the Slide selection will allow both the minimum and maximum limits to change or slide as data is received that is beyond the current limits.

The next box allows us to select what type of Input Data we are plotting. In our case since we are plotting a location graph, we want to choose the Event data for plotting.

Below this is a Show Grid checkbox for displaying a grid on the graph background.

The next column to the right, shows the X (Horizontal) Axis settings. The X-axis settings have similar setup boxes as were described for the Y-Axis and will not be described here. Just note how they have been set up to display the X-Position for location and that we have set up the minimum and maximum graph values to include the sensor positions of 0 and 30 inches. Also note that we have selected fixed scaling on the X-axis because we do not expect the X axis position to change. If this were a time axis we would be using a "compression", or sliding graph scaling.

The last column on the right changes according to the graph type being set up. If a 3-D graph were being set up, this would be the column for the Z-axis setting. However, since we have selected a graph type of 2-D Histogram, settings for the Value Axis and the Bin Axis appears. Again, for now just note the settings and you can get familiar with what they do by making a change and seeing how it affects the graph later.

At the bottom of the third column is a checkbox that allows you to decided whether the histogram shows unfilled, outlined bars or filled solid bars.

Please note and try to make sense of all the selected menu settings. Although it might look complex at the start, if you have gone through these settings, one at a time you will see that all is quite simple. Exit the graph menu by selecting the OK button at the top right of the Graph setup menu.

By looking at the top of the graph we see its graph title of "Events vs. X Position <1,2> Loc(1)". This summarizes what is being displayed. First is the graph type "Events vs. X-Position". Next we are displaying data from channels 1 and 2. Finally, we are displaying results from location group #1.

Now that we are finished, we see that we have set up a Linear location graph and are ready to test it. However, before we do, we have a second graph located below the first one. Let's take a look at it. By the title, we see that it is an Events vs. Channel graph, plotting channel 1 and 2 data and also working with data from location group #1. We can look further to see how this graph is set up by entering the graph set up menu (right clicking within the graph area and selecting the "graph menu" selection at the top of the submenu).

Take a look to see how this has been set up. The main difference is in the X-Axis graph parameter and the Bin Axis setting # of bins. This should help give a little insight into how you can set up different graphs.

Now we are ready to proceed with a test or a replay. As a reminder, as a precaution always save your Layout file before trying it out. It is not fun to have all your work disappear because power went down or something happened to lose the layout.

Step 9. The next step after setting up the graphs is to test the layout & verify that it functions properly during acquisition. We can do this by carrying out a trial acquisition and location test if we have set up our sensors or we can do this by replaying our test file called **Zonal_Linear.DTA** . To do this, refer to the sections on going into acquisition or replay (section 3.4 and 3.5). In our example we will replay the DTA file by going to the Replay function, selecting the file "Zonal_Linear.DTA and clicking Start. The data file eplays and the results should look like the screen below.

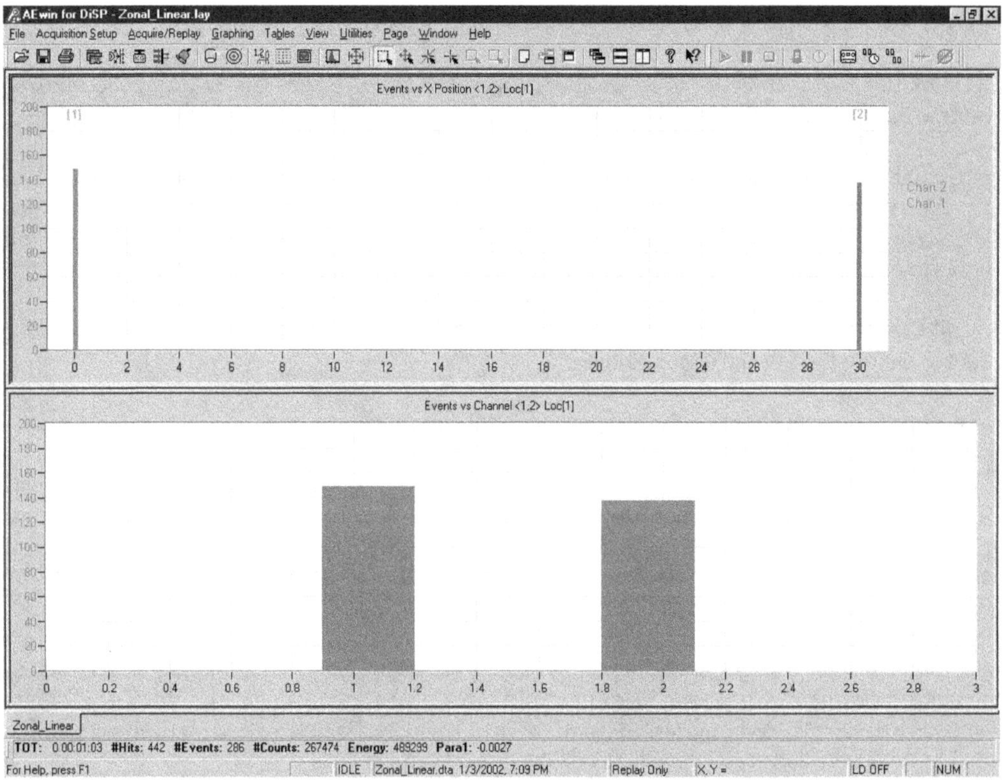

*Figure 145. **Zonal** Location Graph*

The top Events versus position graph shows that all the locations occurred at each of the sensor positions, even though in reality the events were distributed along the length between the two sensors. You can also see that there are fewer events on the right side, or from sensor #2. This is confirmed in the Events versus channel graph below the location graph. The reason for the events showing up at the channel of a Zonal location group is that again the zonal group provides information on which sensor was hit first.

From the above display it seems that our setup is good and we could proceed on to test with a setup like this. However, you can modify some of the settings in the graph setup menu & repeat the replay of the data to see how the graphs are affected by changes and to get familiar with the graphing capabilities.

To illustrate the ease with which changes can be made, we'll now modify the location setup & transform the Zonal location group into a Linear location group. Simple use F8 to bring up the Location Setup dialog & change the Location Type of group #1 from Zonal to Linear, Next use F10 to replay Zonal_Linear.DTA again. Note how the results on the location graph below compare with the results for the zonal location shown above using the same Zonal_Linear.dta data file. There are now locations along the length of the top graph indicating that some events have located in between the 2 sensors..

This example shows one of the very nice capabilities built into the software, the ability of changing a setup and visualizing a different result upon replay. Remember that simply by changing the Zonal setup to a Linear location setup we got two different graphical results. This is very useful when analyzing data.

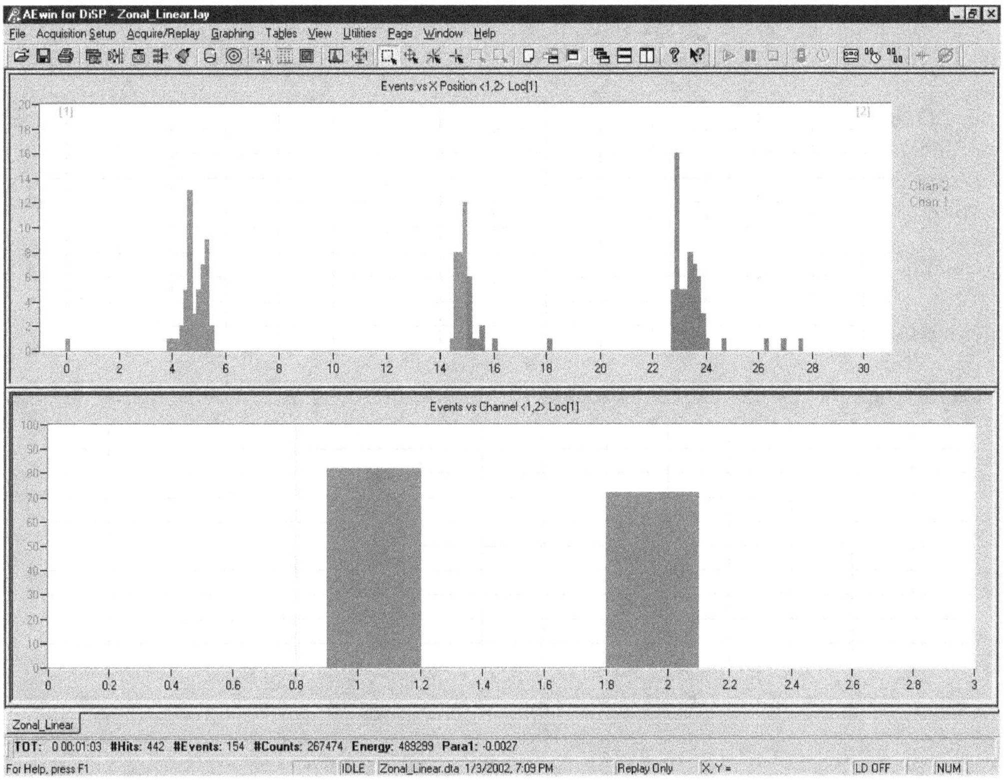

Figure 146. Linear ***Location*** *Graph*

Setup and Testing of a 2D Planar Location Group

As a detailed example, we will be setting up a 2D Planar location group with 4 sensors. We will be using the 2D-Planar.lay and chan24.dta example layout and data file and which are in the AEData directory by default. If you do not have these examples, we will be showing the setup screens so that you can duplicate the setup and conduct your own acquisition test.

As explained at the beginning of this chapter, 2D Planar location is part of the full location option. You can check for the full location option using the Options dialog available on the Help pulldown menu. Look for a checkmark next to Full Location. Alternatively you can just look for 2D Planar in the list of available location modes under the Location Type column in Location Setup.

2D Planar location lets you place multiple sensors anywhere on your test structure without restricting you to a fixed grid arrangement. This is convenient as many structures have pipes or valves that may make a fixed triangular or rectangular arrays of sensors impossible to set up. The position of the source event is calculated from the differences in arrival times between the first 3 valid hits, $\geq 2\ \Delta t$ values, & can use as many as 8 hits in the calculation. The positions of the sensors for first 3 hits always determines a triangle. The ideal arrangement for 2D Planar would be a layout made up of equilateral triangles, but exact equality of the triangles & sides of the triangles is not an absolute requirement. You will get better results if you avoid creating sensor triangles where one side of the triangle is much larger or smaller than the other 2.

Setting up for 2D Planar location is very easy in the software and is similar to setting up Zonal or Linear location. The steps are identical to those outlined above, only some of the menu selections are changed for the 2D Planar location. Let's provide a quick run-through of setting up 2D Planar location using the same steps outlined above for the other location setups. We will be using the example layout file, 2D-Planar.lay.

Step 1. Use the File pulldown menu's Open Layout option (Ctrl-O) to recall 2D-Planar.lay.

Step 2. Enter the Location Setup dialog (F8 key) & use the Distance Units button to select Inches.

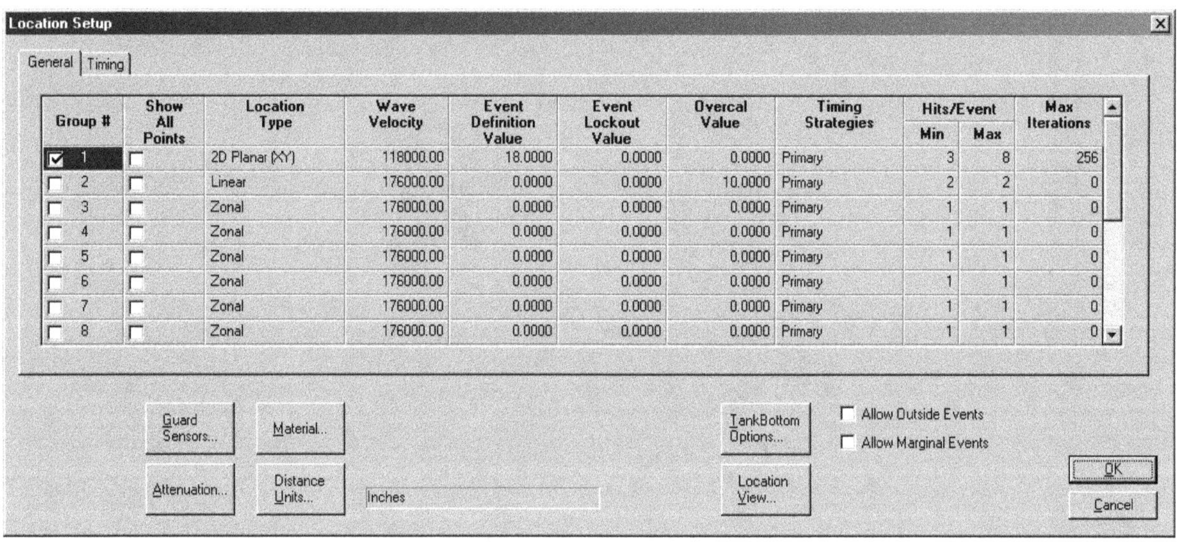

Figure 147. 2D Planar Location Setup

Step 3. Now configure the settings for location Group #1 as shown above. Enable the group by putting a check next to #1. Set the Location Type to 2D Planar. Enter 118,000 for Wave Velocity. Set the Event Definition Value to 18, Event Lockout to 0 & Overcal to 0. Leave the Timing Strategies, Min & Max Hits/Event, Max Iterations & all settings on the Timing page at the defaults.

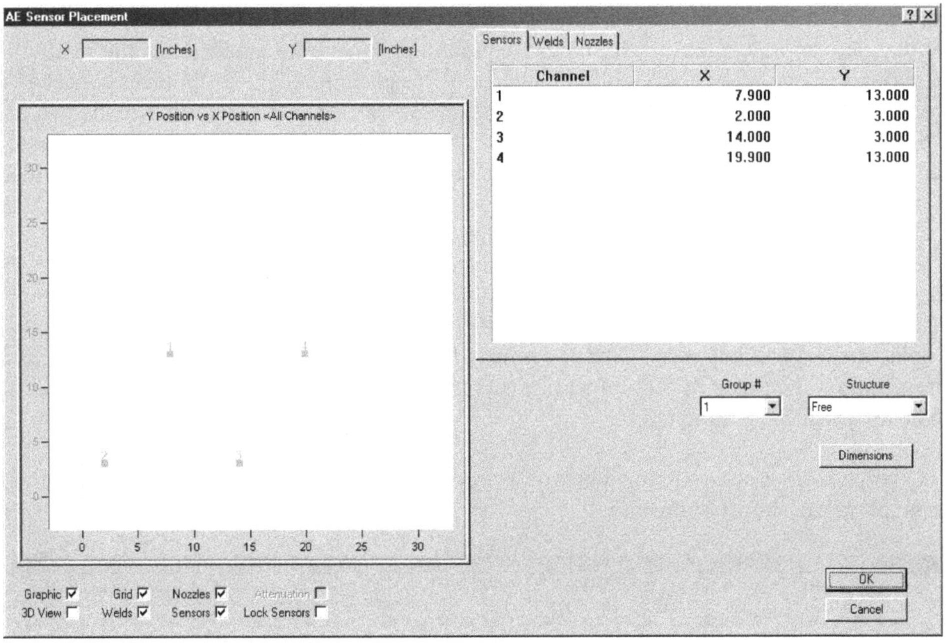

*Figure 148. Sensor Placement for 2D-**Planar**.lay Location Setup*

Steps 4-5. Click on Location View to bring up the Sensor Placement dialog. Set the Structure type is to Free & use Dimensions to set 0 & 30 for the Min & Max values for both x & y.

Steps 6-7. Set up the sensor locations on the screen using the same procedure described in the previous section for zonal location. The only difference now is that you will set up 4 sensors in a 2 dimensional layout. An example is shown above along with the coordinates of the 4 sensors. Even though we are showing a triangular arrangement, 2D Planar is not restricted to a specific sensor pattern. When you are done placing sensors, click OK in Sensor Placement & Location Setup to return to the main menu.

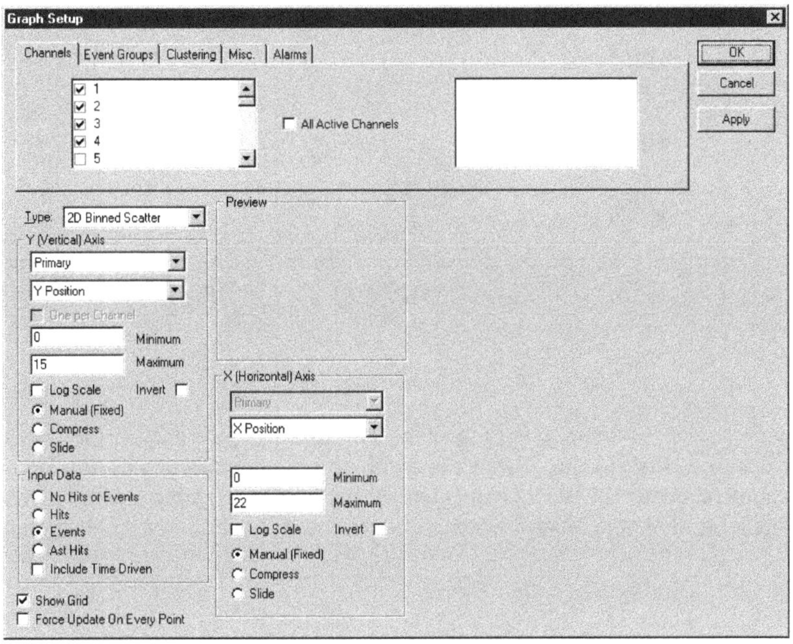

*Figure 149. Graph Setup for 2D-**Planar** Location Plot*

Step 8. Next, define a 2D X-Y location plot using the graphing menu following the procedure described in the section for zonal location . The screen capture above shows the graph setup menu for the location plot in 2D-Planar.lay while the one below shows the graph (top graph) for that layout. You can compare the differences between the setup above and the graph setup for the zonal and linear location graph shown earlier. First, you can see that we have enabled all 4 sensors for display that have been setup in the location view menu. Next, you can see that we have selected a 2D Binned Scatter (point plot) for the graph type. Next, you can see that we have changed the Y-Axis to plot "Y-Position" instead of Events. You can also see that we have changed the axis coordinates in both the X and Y axes. Finally, you can see that there is no "bin or value axis settings" to the right of the X axis setup as there was for the 2D Histogram graph setup. 2D Binned Scatter graphs don't require any user choices about the binning strategy. All 2D Binned Scatter graphs use a fixed 200x200 binned grid when plotting the data. The point of using a 2D Binned Scatter graph rather than a 2D Scatter graph is speed. Its binning strategy decreases the number of data points actually plotted, so that multiple point at one location graph as just one point. The 2D Colored Scatter graph is similar to the 2D Binned Scatter, but it will allow you to plot the points in different colors that reflect how many original data points are superimposed at any particular location.

The result of the graph setup is shown in top graph of the screen below. In this graph you can see the sensor locations, although your graph will be empty at this point.

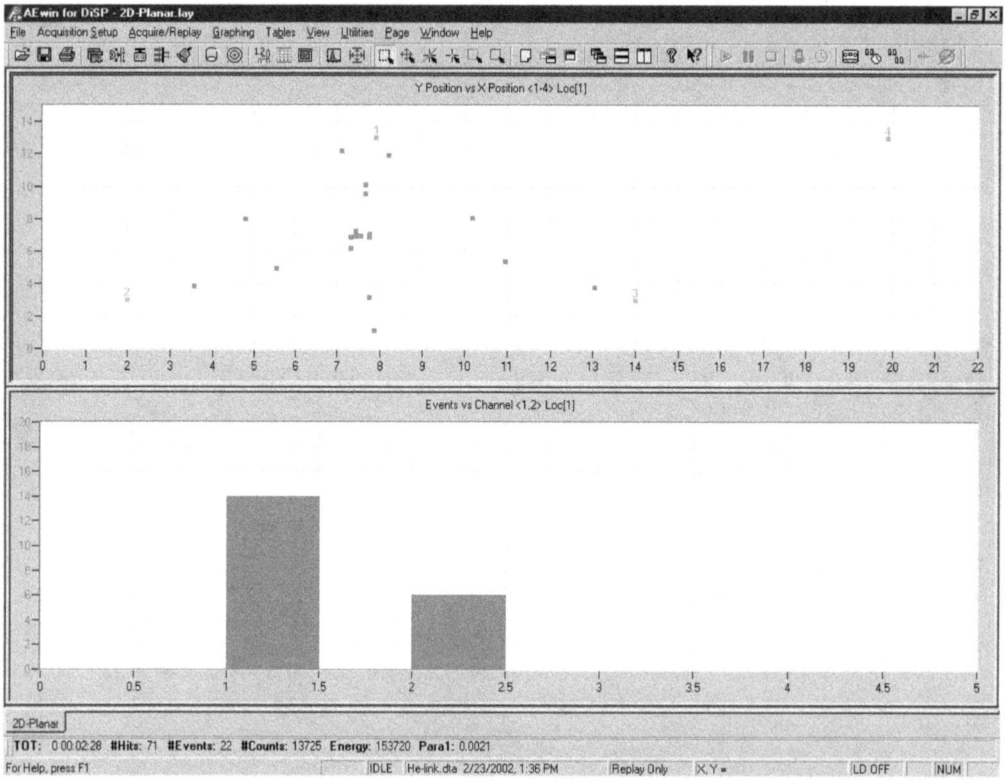

Figure 150. 2D-Planar Location Point Plot

Step 9. Now perform a test of the setup using the same procedure as described in the previous section for zonal location. Replay the data file HE-Link.dta. Results in a point plot will be plotted as dots on the screen where each dot represents the location of a detected event. Now your graphs should resemble those shown above. You can right click on the graph & look for the Large Scatter Graph Points for better visibility of the data points, if you prefer.

Setup and Testing of a Tank Bottom Location Group

Tank Bottom location is a specialized AE source location mode, developed specifically for monitoring the floors of large cylindrical vessels. The software assumes a tank with a circular bottom with sensors equally sed around the circumference of the tank. Tank Bottom location should not be confused with TANK. The difference between them is that Tank Bottom software performs a location calculation and projects the result of that location determination onto the tank floor graphic. In contrast, TANK© is a knowledge & experienced based procedure for assessing and grading the integrity of tank floors.

If you have purchased the Tank Bottom location software option, this section will help you to setup for your first Tank Bottom test. Please first verify that you have the Tank Bottom software option by selecting the Options dialog from the Help pulldown menu. If you have this option, it will be checked. If the option is not present and you desire it, please contact the Customer Service Department at customerservice@ndt.com.

Tank Bottom location follows along the same location setup, menus and setup procedure as do the other location modes. Setting up Tank-Bottom location is a simple 3 step process including;

5. Accessing the Location Setup menu (F8), from the "Acquisition Setup" main toolbar menu, and setting up a Tank Bottom location group.
6. Accessing the Location View menu to set up the Tank bottom structure, dimensions and sensor placement.
7. Setting up a Tank Bottom location graph in the Graph setup menu.
8. Acquiring and/or Replaying data.

Assuming one is familiar with the software, the following is a step by step procedure for setting up and running the Tank Bottom Location software.

Configuration of Location Setup

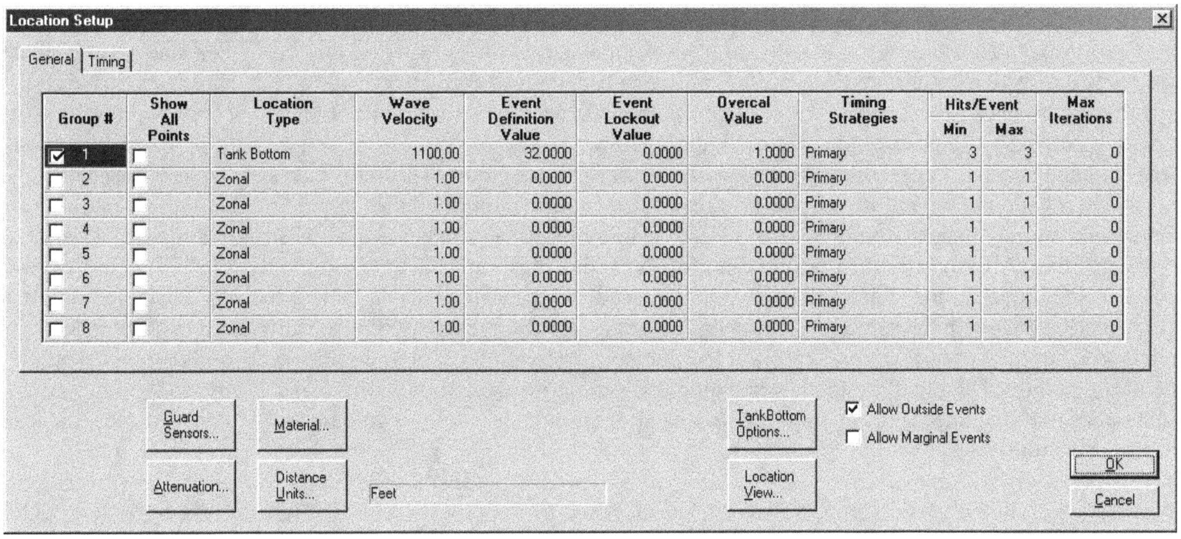

Group #	Show All Points	Location Type	Wave Velocity	Event Definition Value	Event Lockout Value	Overcal Value	Timing Strategies	Hits/Event Min	Hits/Event Max	Max Iterations
1		Tank Bottom	1100.00	32.0000	0.0000	1.0000	Primary	3	3	0
2		Zonal	1.00	0.0000	0.0000	0.0000	Primary	1	1	0
3		Zonal	1.00	0.0000	0.0000	0.0000	Primary	1	1	0
4		Zonal	1.00	0.0000	0.0000	0.0000	Primary	1	1	0
5		Zonal	1.00	0.0000	0.0000	0.0000	Primary	1	1	0
6		Zonal	1.00	0.0000	0.0000	0.0000	Primary	1	1	0
7		Zonal	1.00	0.0000	0.0000	0.0000	Primary	1	1	0
8		Zonal	1.00	0.0000	0.0000	0.0000	Primary	1	1	0

Figure 151. Location Setup Menu

The first step in setting up for a Tank Bottom location application is to access the Location setup menu. This is done by selecting Location from the Acquisition Setup pulldown menu (or by pressing the F8 key). The Location Setup menu appears as shown above. As an example we have shown a Tank Bottom location group as group number 1. We give an overview of setting up this location group below, but you may also wish to read section 4.2 on the Location Setup dialog box for more detailed information on the individual settings.

The first step in setting up a location group is to first select the distance units that you want to use for your work. You will know what distance units you are working in by viewing the textbox at the bottom center of the Location setup menu. In the example above, the current measurements are in feet. To change from the default distance units, left-click the "Distance Units" selection box at the bottom of the Location setup menu. Upon clicking this button, you will see the dialog shown to the right appear. Select whichever units you desire to work with and click "OK". Upon clicking OK, you will see the units textbox at the bottom of the location setup menu show these new values.

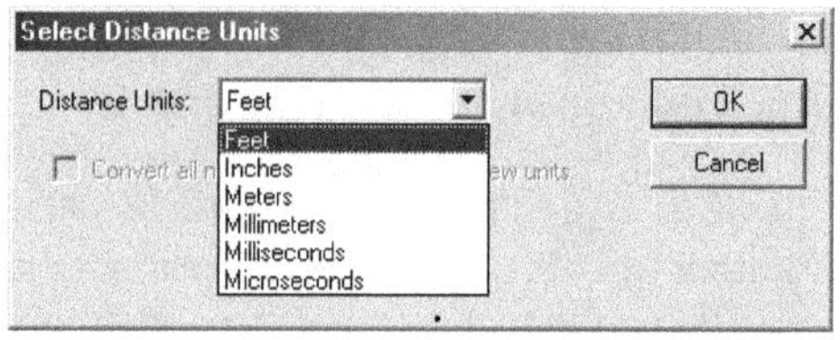

Figure 152. Distance Units Dialog

It is important to setup your distance units as early as possible into the location setup because many the location setup entries including "Wave Velocity", "Event Definition Value", "Event Lockout value", and "Overcal value" all use the units selected or in the case of velocity it will be units/second. By dealing in direct units, the user can think in terms of distance values when setting up these entries.

Returning to setting up the Group #1 (as shown by the first row of the Location setup menu), we will address each column entry below.

<u>Group#:</u> The first column shows that we have checkmarked the Group #1 Location group to be an active location group. Putting a check in this box activates that location group and tells the software to perform a location calculation on based on the setup data in this menu, the Location setup "Timing "menu (see the upper tabs) and the "Location View" and "Tank Bottom Options".

<u>Show All Points:</u> This allows the user to show all source results for events that this group can successfully locate. Independent of which groups have Show All Points checked, the software always allows all enabled groups to analyze a given event to get the best results for each group and then compares the results from all successful groups to come up with a best result. For example, linear location is always presumed to give a better result that zonal location, if there are enough hits(2) to produce a location. If no groups have Show All Points checked, the software will report one source location per event & it is the best result that is shown on graphs & in the line display. There are times when a side-by-side comparison of the results of 2 or more location groups are useful. In that case you can mark the groups as Show All Points & then whatever source locations they can calculate will be reported. The software will appear to sometimes produce more than one source location per event & if you look at the line display, you will see extra event listings with the same first hit times. The extra location results will also show up as extra points on your graphs too, but they only count as 1 event in the Statistics bar. In all cases you will see the software's best result as one of the points. There will be times when an event cannot be located by more than 1 group, so having Show All Points does not guarantee a result from a given location group. It is only a guarantee that you will see it, if it can produce a useful result.

<u>Location Type:</u> In the next column, we have selected "Tank Bottom location" as the location type. By left clicking your mouse in the Location type area on this first row, a selection box will appear with all the available location option choices that are enabled within the software. "Tank Bottom" will be one of the choices, select this.

<u>Wave Velocity:</u> For Wave Velocity, we suggest that you use the velocity of the fluid that is in the tank as a starting point. It is recommended you perform some tests to try to empirically determine what that actual velocity is once you get on the tank. This can be done by creating an imt signal near one sensor, that travels from one side of the tank to the other and monitoring the time it takes for the signal to travel the liquid path to the sensor on the opposite side. You can then measure the time it has taken to arrive and use that with the knowledge of the diameter to arrive at a velocity. However, you must be careful not to mistake the faster metal path arrival, traveling around the tank.

If you do not know the velocity of the fluid in the tank & cannot easily measure it, you can access a built in "Material Properties" database within the software. To access the "Material Properties" database, simply click the "Material" button at the bottom of the Location Setup menu. The Material Properties dialog box will appear as shown below. Along the top of the menu, you can select what type of materials group that you want. In our case, the "liquid" database has been selected. By clicking on the down-arrow in the Material Name text entry box, a list of materials (liquids) in our case will be shown. Simply scroll down and select the desired material (or liquid). In our case, we have selected Motor oil. You will note that the velocity will automatically read out in the distance units that you have selected in the Location setup menu. If the material selection you want is not in the database, you will be able to enter this into the material database by using the "Add" "Delete" and "Update" keys near the bottom of the Materials database menu. You will note that for the example of the Motor oil in the figure the velocity is 5708. We can make use of this velocity by entering it into the "Wave Velocity" column of the location group setup row. The direct way is to simply left-click the cell in Location Setup, type the appropriate value into it and hit the "Enter" key. Remember to enter the value in the same units as that listed at the bottom of the Location setup menu. As an alternative, you can Copy & Paste the value by highlighting it in the Material Properties dialog box & pressing Ctrl-C. In Location Setup, highlight the cell, press F2 to edit its contents & then press Shift-Ins

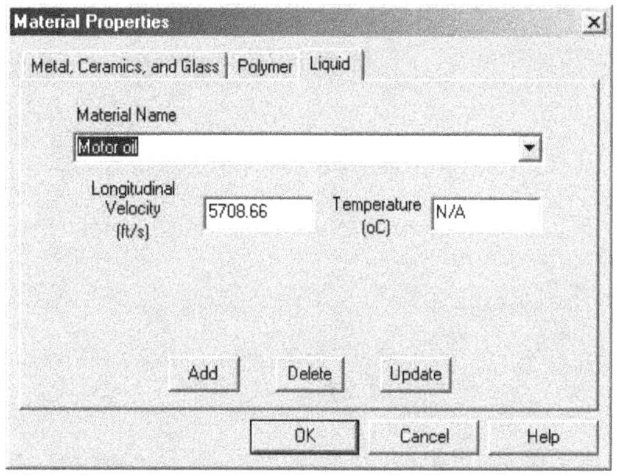

Figure 153. Material Properties Dialog

Event Definition Value: The Event Definition Value is the length of an event in the user's distance units. The Event Definition Value should be set to make sure that an AE source, from any point in the location array, will intersect a minimum desired number of sensors (3 is the minimum), forming an event. In our example of a 30 foot diameter tank, we have selected an Event Definition value of 32 which assures that all the sensors in the tank will have the capability of detecting an event regardless where it originates.

Event Lockout Value: The Event Lockout Value is also entered in the user's distance units. The Event Lockout Value controls the interval between consecutive events. Enter an Event lockout Value starting from the beginning of the Event and long enough to assure that the there is enough time (or distance) for all additional hits and reflections related to that event, to have had a chance to die down. In order for the Event Lockout Value to be effective, it needs to be set longer than the Event Definition Value, since they both start at the same time. There are no rules of thumb for setting Event Lockout Value, since the event related activity is dependent on the structure and sensitivity settings of the system. It is best to experiment to find the best selection.

Overcal Value: The Overcal value is also set in the selected distance units and is usually set equal to approximately 10% of the Event Definition Value. This is to assure that the location algorithm does not ignore minor variations in the wavepath velocity.

Timing Strategies: The user can select from some different timing strategies as set up in the "Timing" tab located at the top of the Location Setup menu. Please note that the Tank Bottom location algorithm does not support multiple timing strategies like the advanced regression location modes can.

Hits/Event (min/max): This entry is used with the regression location modes which are able to use more than 3 hits when calculating the source location for an event. Tank Bottom location is not, however, a regression based location mode. It only uses 3 arrivals to determine location, so these fields are fixed at 3 & are not adjustable for Tank Bottom location.

Max Iterations: Again this setting only applies to regression location modes not Tank Bottom location, so you cannot modify this setting for Tank Bottom.

Once you fill out all this information, you should proceed to the "Location View" menu where you can setup up the Tank Bottom "Structure" and dimensions and set sensors around the tank. An important point with all these settings is that they do not affect the data that is stored in the DTA file during acquisition, because the location results are not stored in the DTA file. They are recalculated each time you do a replay of the data, so you can change any of the above entries at any time & reanalyze the data to try to improve the location accuracy.

Configuration of Sensor Placement

The Sensor Placement dialog is entered from the Location Setup by clicking on the "Location View" button. Upon selecting this option, the dialog box shown below will appear. This menu is the base starting point for setting up the

Tank bottom and other location based structures. The Tank Bottom structure is one of the easiest location setups to create since the sensors can be placed and automatically. Please perform the following steps to set up all the aspects required in the Sensor Placement dialog shown below.

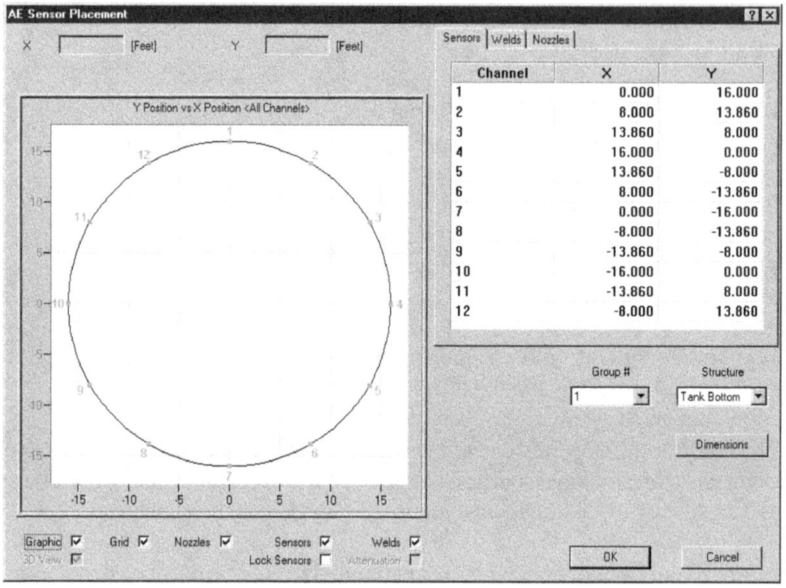

Figure 154. Tank Bottom Sensor Placement

5. First, make sure that you have selected the location group that you wish to work with. In our case, we had set up location group #1 in the Location Setup menu, so make sure Group #1 is selected in the Group selection text entry box.
6. Next, we go to the "Structure" menu and select the "Tank Bottom" structure as the desired structure. This gives us a chance to talk about our "structure" based location capability. Since the user is setting up his application on a particular type of structure (such as a plate, a pipe, a vessel, a sphere, a tank bottom, etc., etc., it is important for the location to think in terms of structures as well. This is what has been implemented in the software and is an extremely advanced concept that only is doing.
7. Once you have selected the Tank Bottom as your Structure type, you should then click the "Dimensions" button. The dialog box shown to the right will appear. This menu is a customized user configuration menu for the type of structure you have selected and allows the customer to set up certain aspects of that structure. In the case of a Tank Bottom, the menu allows the user to enter the Tank diameter (which we have set to 30 feet in our example, allows you to set alternate units and also allows you to enter information about the automatic sensor placement. First it lets you select the starting sensor number as well as the total number of sensors that will be equally sed around the tank and the starting phase, assuming that 0 degrees is that the top of the tank bottom (or North). In our example, we have set up for our sensors to start at sensor #1, starting at the top or 0 degrees and the placement of 12 sensors. When accepted (by clicking the OK button), the sensors will be automatically placed in a clockwise fashion at equal angular sing, on a circular tank bottom in the Location View area. This allows the user to view the sing. Please note the table of sensor positions that is

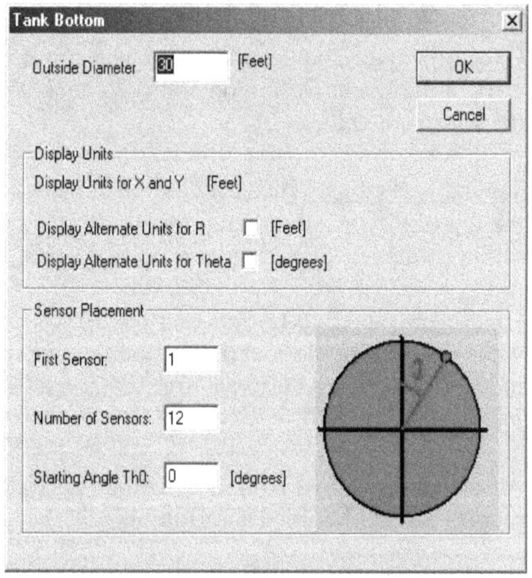

Figure 155. Tank Bottom Dimensions

automatically generated. This table can be exported to the clipboard (by right clicking on the grid and selecting 'copy to clipboard') for insertion into a report file.

8. After the sensor placement step, the user should then go and set up the sensors on the tank in the locations indicated by the Location view menu. Once done and the sensors have been connected to the appropriate AE channel number, you are ready to proceed to the setting up the Location graph and begin acquisition.

Setting Up a Tank Bottom Location Graph

The next step in the process of getting ready for carrying out a Tank Bottom location test is to set up the Tank Bottom Location screen. This is a straightforward process which is carried out in conjunction with the Graph Setup menu. The following are the steps in setting up a Tank Bottom Location Graph.

12. First, exit the Sensor Placement dialog and the Location Setup menu by clicking the OK button. Please note that if you click "Cancel" that you will lose all the setup changes that you have entered. Make sure you click the OK button before proceeding. Upon getting out of the Location setup menu, you will be back to the main screen showing some graph setups.

13. First, let's add a graph page by either right-clicking on one of the graph tabs at the bottom of the screen or by clicking on the "New Page" icon on the icon toolbar (the icon looks like a sheet of paper). This action will add a new tab to the bottom of the graph selection tabs and will create a blank screen. Next, lets name the new Graph Tab, "Tank Bottom" by right-clicking on the tab and selecting the "Rename Page" selection.

14. Next, lets add a graph to the page by either right-clicking in the new "Tank Bottom" page or selecting the "New Graph" icon on the icon toolbar menu at the top of the screen (its next to the "new page" icon). Once you do this, a graph window will pop up. Maximize the graph to be full screen by clicking on the "Tile Horizontal" or "Tile Vertically" icons on the top icon toolbar. You are now ready to create your Tank Bottom Location graph.

15. To setup for a Tank Bottom Location graph, first you need to enter the Graph Setup menu by right-clicking in the Graph area of the graph you want to setup. This is the topmost selection in the context menu that comes up when you right-click. Please note the Graph setup dialog box to the right.

16. Next, select the "Event Groups" Tab at the top of the Graph setup menu and Check the location group #1 which is identified as a Tank Bottom location group (if you have properly followed all previous steps).

17. Next, select the type of Graph. In the case of a location graph, you want to select a 2D Binned Scatter Plot. This text entry box is just above the Y-vertical axis settings.

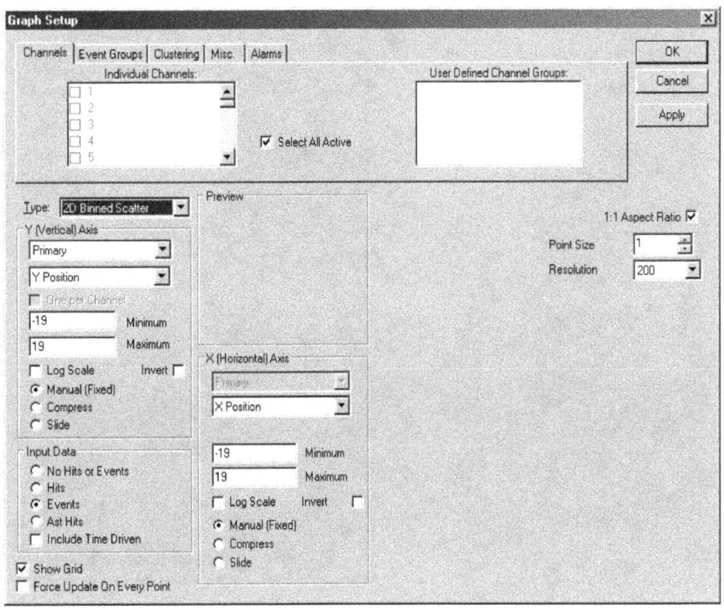

Figure 156. *Tank Bottom Graph Setup*

18. Next, setup the Y-(vertical axis) and X-(horizontal axis) for display. First set the Y vertical axis to be a Primary graph (this means that you will only be plotting one graph on this screen plot. The next scrolling textbox allows you to select the feature or value that you want to plot. Since this is a location plot you want to plot the Y position versus the X position, so select these for your Y and X settings.

19. Next, you need to select the graph scale. Tank Bottom location assumes that the center of the tank is the 0,0 position, so the axis limits should be a little larger than the tank radius,. In our case the radius is 15, so we'll use ±19 as the limits for both X & Y. Since we want our Tank Bottom outline to look like a circle rather than an ellipse, we've checked the 1:1 Aspect Ratio. This checkbox controls whether the software is allowed to stretched the graph to fill the available se or whether it is required to keep the X & Y axis the same length.

20. The last important setting is to select Events as your Input data to the graph. Events will show the result of the location determination on the incoming AE data.

21. With all these settings done, we can exit the graph setup menu by clicking the OK button. If you did everything correctly, you will get a nice looking tank bottom screen like that shown above. You can have other graphs on this same screen, you are not limited to just the one graph, if you want to just show a portion of the tank in different regions.

22. Note that the structure and the sensors were automatically plotted on the screen. This is because the software was instructed that the structure of the Location group #1 was a Tank Bottom location and was automatically able to draw this structure as soon as it saw that the graph which was being set up was an X Position versus Y Position location graph. If you do not see the Tank Bottom outline in the graph, go back to step 5 & make sure Event Groups has location group #1 checked & that it says Tank Bottom next to it.

You have now totally completed the setup of a Tank Bottom location test and are ready to acquire or replay data.

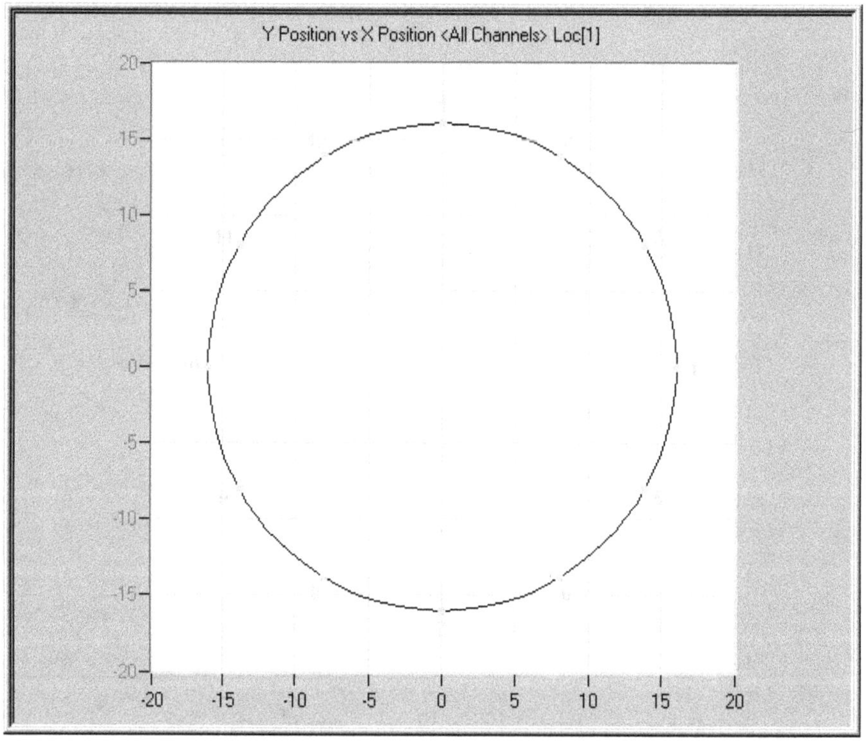

Figure 157. Tank Bottom Graph

Acquiring and Replaying and Displaying Data

Now that the entire setup is complete, you are ready to acquire or replay data from your Tank bottom test. To acquire data, simply enter data acquisition and begin recording data. You should always perform a test to determine if the location algorithm is operating properly before actually performing the true acquisition test. Verify that the location is performing properly and the AE hardware and software settings are all correct to minimize noise and maximize the types of signals you want to record. Once satisfied, start acquiring test data.

Once acquisition is completed you can start replaying and analyzing data. There is a Tank Bottom data file named TankBot.dta in your AEdata directory for you to experiment with. You can use it to test Tank Bottom location. If you do a replay of TankBot.dta using the TankBot.lay layout file, you should see results like that shown here.

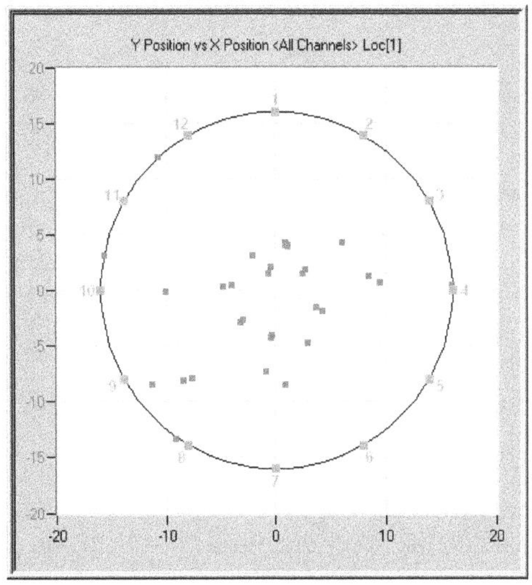

Figure 158. Tank Bottom Results

All standard analysis and filtering tools can be used on the data to eliminate noise and assure relevant data.

There is another option that has been built into the Tank Bottom Software that allows multiple location calculations and plots to be displayed on the Tank Bottom location screen. To access this processing feature, look for the Tank Bottom Options button in Location Setup (F8 or select Location on the Acquistion Setup pulldown menu). This button is just above the "Location View" button as shown on figure 139. The resulting Tank Bottom Options dialog box is shown here.

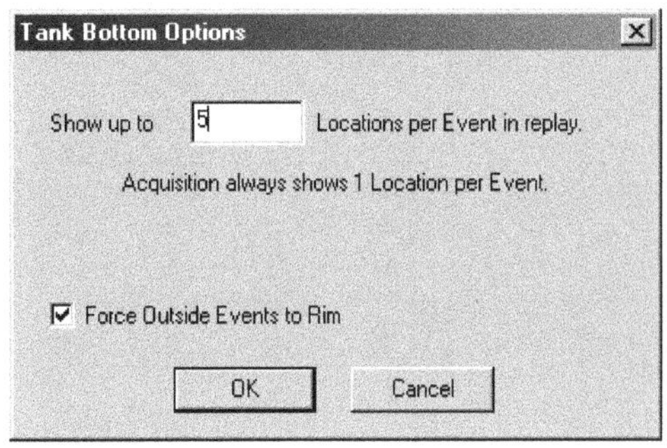

Figure 159. Tank Bottom Options Dialog

This option allows you to choose to display more than 1 location per event. Normally, the number is set to 1 as a default but we have set it to 5 to allow up to 5 location calculations per event to process and display on the screen. This uses other hits in the event in order to calculate location in case the first hits of the events have some problems. Note that this option only affects replay mode to keep acquisition at its highest speeds. Once this is selected and the file is replayed, you can see the difference by comparing the graph at the right with the one above. As can be seen, there are more events displayed on the graph to the right, as there should be. What we learn by this is that the location is pretty good and matches pretty closely telling us that the original location data in the previous graph is valid.

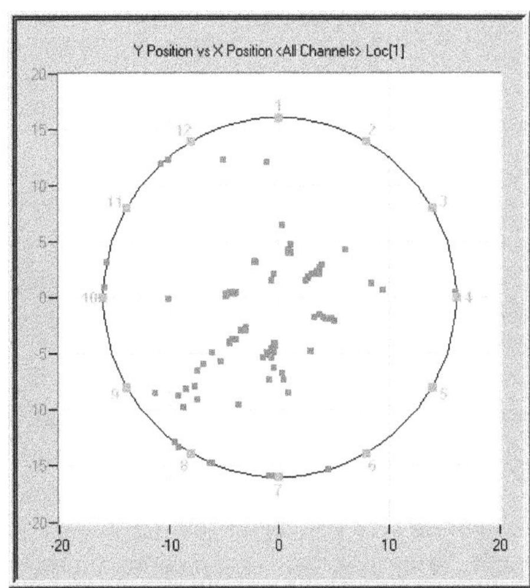

Figure 160. Tank Bottom Results

For a discussion of the Force Outside Events to Rim checkbox , its interaction with Allow Outside Events in Location Setup & how that can affect your results, see the Tank Bottom Location notes in section 4.4.9.

VIRTUAL PARAMETRICS

Virtual parametric data is data that the software collects from hardware usually manufactured by another company & treats it similar to the parametric data normally collected with the hardware manufactured by Physical Acoustics from the SAMOS & PCI2 boards & the Sensor Highway Smart Monitor system. The DiSP board does not support use of virtual parametrics. This readme file documents how virtual parametrics are set up & used in HMSGHAT2000DDI in general along with specific details on using the individual instruments. The instruments supported included the Vaisala Weather Station, the Corr Instruments' NanoCorr, Measurement Computing's PCI-DAS6033 & USB-2533 64 channel boards & the Group 9380-2114 Strain Gauge Module. Note that the Strain Gauge Module is only supported by the Sensor Highway Smart Monitor system. It is not available for the SAMOS or PCI2 systems.

Configure Virtual Parametrics Dialog Box

To make use of virtual parametrics in HMSGHAT2000DDI it is necessary to tell the software what kind of virtual parametric hardware you will be using before acquisition starts. This is done with the Configure Virtual Parametrics menu option under the Acquisition Setup dropdown menu.

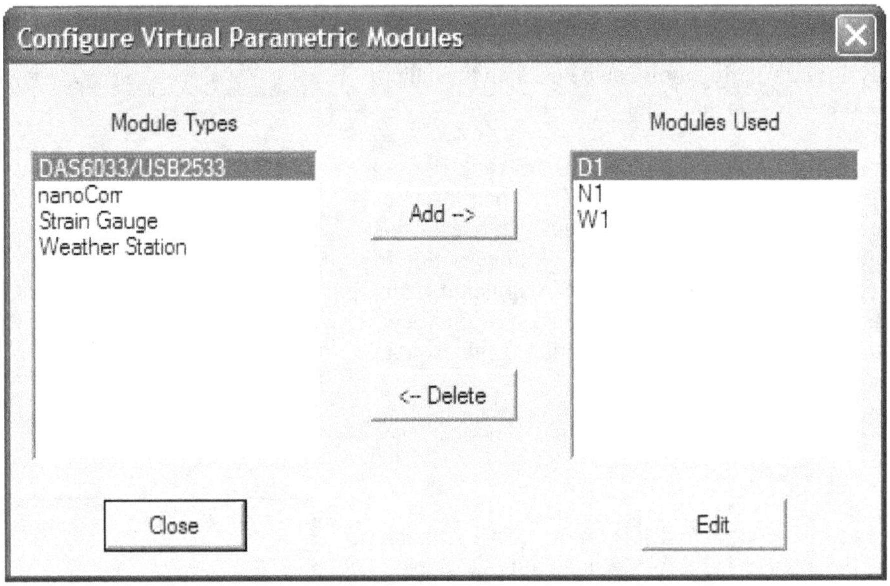

Figure 161. Configure Virtual Parametrics Dialog Box.

This dialog box can be accessed several ways. First it is available directly on the Acquisition Setup dropdown menu. You can also use the hotkey combination Ctrl+F2 to bring it up when you are at the main menu. It can also be accessed thru a button on the Virtual Parametrics tab in the Hardware Setup dialog box. More on the Virtual Parametrics tab in Hardware Setup below.

The column on the lefthand side of this dialog box displays the Module Types list. This is just a list of the types of virtual parametric hardware that is currently supported by HMSGHAT2000DDI. Simply highlight the module you want to use & click on the "Add –>" button. Alternatively, you can doubleclick the module type to add it as well. Each module type will lead to a different Virtual Parametric Module Setup dialog box specific to that hardware. More on each dialog box below. Once the module has been configured & the user clicks on OK, the module name will appear in the Modules Used column. If you no longer wish to use a module, you may highlight it in the Modules Used column & click on on the "<– Delete" button.

Modules listed in the Modules Used column can be doubleclicked to edit their settings. Alternatively, you can highlight the module name & click Edit to launch the appropriate Virtual Parametric Module Setup dialog box. For some virtual parametric types, you may wish to work with more than one instrument. For example, you might want to make use of 2 NanoCorr instruments for collecting data. If you wish, you may copy the setup for an existing module by highlighting it in the Modules Used column & clicking on the Copy button. A new setup dialog box will open with an exact duplicate of the settings.

Before you can successful save the new modules settings, you must change the Module Name & for many modules you need to change the Module ID. See the Virtual Parametric Module Setup dialog box for the DAS6033/USB2533 for an example. The Module ID is often used as an address or board identifier when communicating with the device. For example, Weather Station instruments can communicate via a shared usb port & come configured by default to address 0. If you wished to use more than one, the second Weather Station would have to be reconfigured to appear at another address like 1 or 2. The Module ID is how the user tells the software which instrument to access. It is not a problem if a NanoCorr module & a Weather Station module both use an ID of 0, but if you were using 2 Weather Stations, they cannot both have ID values of 0.

You can have up to 4 modules of a given type like a NanoCorr instrument. There is also a maximum limit of 256 total channels of virtual parametric data that are allowed. Weather Station modules have 22 channels of data, NanoCorrs have 46, Measurement Computing DAS6033 & USB2533 boards have 64 channels & Strain Gauge modules have 4 channels. You may mix & match modules as needed, but you are limited to no more than 4 modules of a given type & no more than 256 channels total from all modules in use.

There are many settings that are common to all the Virtual Parametric Setup dialog boxes. Even if you are not using a Measurement Computing board to collect virtual parametric data, it is suggested that you read that section as well as the sections appropriate for the hardware you are using, because the common settings are described in this first section.

Measurement Computing's PCI-DAS-6033 & USB-2533 Boards

The Measurement Computing boards are designed for 64 channels of analog input. The DAS-6033 uses a PCI slot in your computer, while the USB-2533 is an external board that connects to one of your computer's USB ports. They can monitor any instrument or sensor that outputs a voltage between +/- 10v making them very versatile for virtual parametric input. Here is a description of the available options in the setup dialog box shown below.

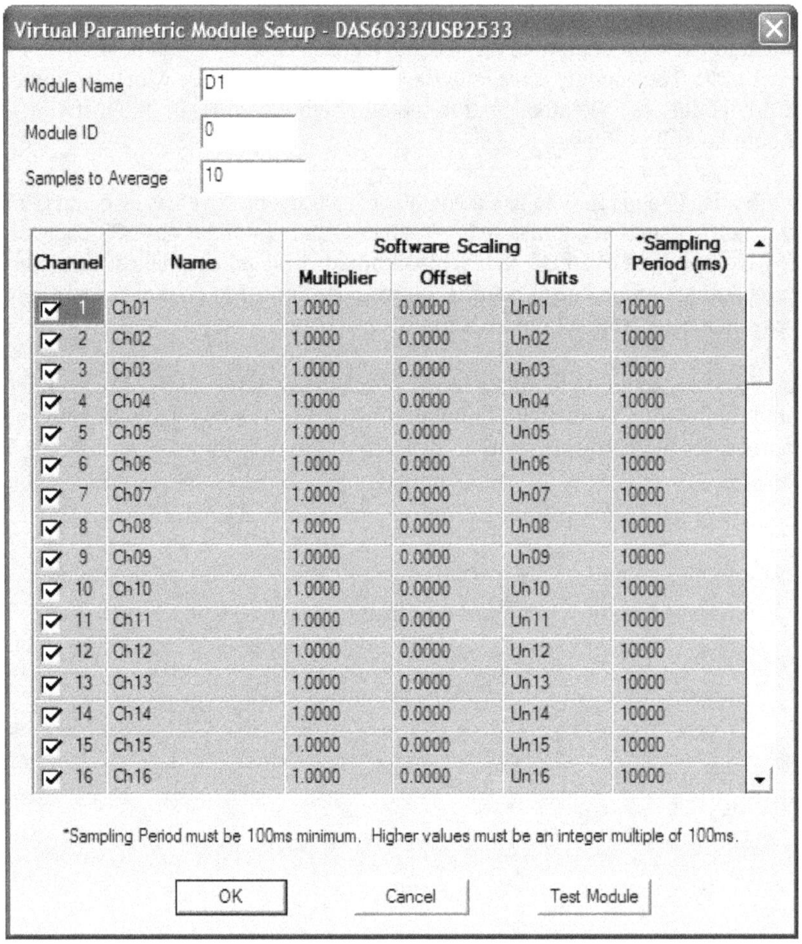

Figure 162. Configure Virtual Parametrics Dialog Box.

Common to All Modules

Module Name – Module Name is just a string used to identify the source of virtual parametric data in the Line Display. Because of the potential for confusion, you can not use the same name for 2 modules. It is also recommended that you keep the Module Name short as it does affect how much data can be displayed on one line in the Line Display.

Module ID – Module ID is an integer used to uniquely identify the module in the software. You may not have 2 modules of the same type with the same ID. In the case of the PCI-DAS6033 & the USB-2533 boards, this number corresponds to the board number used in Measurement Computing's InstaCal program. This defaults to 0 for the first board. If you have multiple boards, the second board number is usually 1. Enter the same number here that is used in InstaCal to access the hardware.

Channel On/Off Checkbox – This is just a simple On/Off toggle that specifies whether or not to collect data from this channel. If you wish to turn all the channels on or off, simply check the box for the first channel & then press the F5 key. F5 sets all values in the column to the value of the selected cell.

Channel Name – This is a string that can help you identify what data a channel is collecting. For example, channel 9 could be a voltage reading that represents the current temperature reading. In that case you might prefer to replace the default label of Ch09 with Temp.

Multiplier, Offset & Units – These 3 settings are identical to those for regular parametrics in HMSGHAT2000DDI. The Multiplier & Offset allow the user to rescale the reading from the hardware to units more to the user's liking. Multiplier & Offset are floating point numbers. The raw reading is multiplied by the Multiplier & then the Offset value is added to it. The Units setting holds a string to represent the new units. For example, if the original reading was in $^{\circ}$C & the user want the reading in $^{\circ}$F, then the Multiplier would be 1.8 (= 9/5) & the Offset would be set to +32. Units could then be set to F. For the DAS6033 & USB2533 all raw reading are actually expressed in volts, but the same principle applies if you are convertings volts to loading pressure or a rotation speed.

Sampling Period – All modules have a Sampling Period setting, but some modules have only one setting for all channels. The DAS6033 & USB2533 boards can, however, vary the sampling rate for each channel independent of the others, so there is a column in their setup dialog box to allow a value for each channel. The values are entered in milliseconds & the minimum value is 100 ms. Higher values must be an integer multiple of 100 ms. If you wish to set all channels to the same sampling rate, type the value desired into one cell in the column & then press F5.

Test Module – This is not an option per se, but this button is useful for testing whether or not HMSGHAT2000DDI can successfully communicate with your Measurement Computing board. It will either tell you "Could not detect the board" or "Board successfully detected". If it was not detected, verify that the Module ID setting is set to the same value as the board number assigned in the Measurement Computing's InstaCal program. If InstaCal can't see the hardware, HMSGHAT2000DDI won't be able to either. If it's a USB-2533, you can doublecheck that the USB cable hasn't come loose. If it's a DAS6033, make sure the board is fully seated in its PCI socket.

Specific to PCI-DAS6033 & USB-2533 Boards

Samples to Average – The other modules do not have a Samples to Average setting. It is specific to the DAS6033 & USB2533 boards. You can enter a number from 1 to 256. This is the number of samples that the board acquires when making a reading. The samples are collected as fast as possible for the DAS6033 & USB2533 (about 640ms/reading) & then averaged to give the reading reported for that sampling period.

Vaisala Weather Station

The Vaisala Weather Station is designed to collect information on current weather conditions such as temperature, rainfall amounts & windspeed.

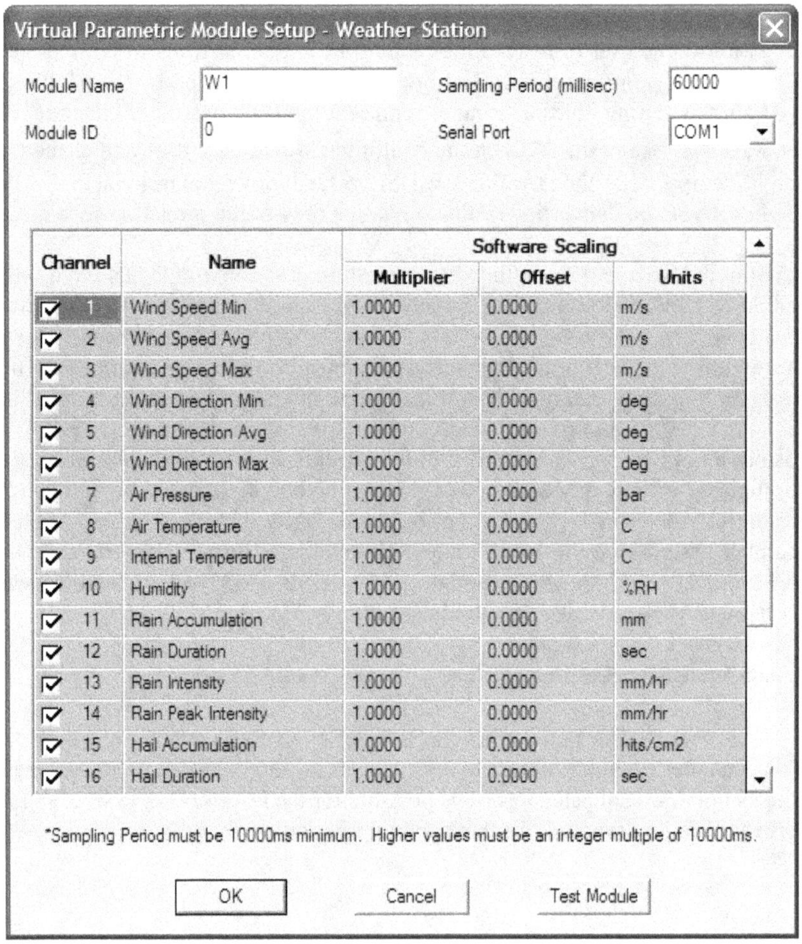

Figure 163...Weather Station Setup Dialog Box.

Common to All Modules

Module Name, Channel On/Off Toggle, Channel Name, Multiplier & Offset – These are described in the section on the Measurement Computing boards. Please refer back to that section for detailed descriptions.

Module ID – Module ID is an integer used to uniquely identify the module in the software. You may not have 2 modules of the same type with the same ID. In the case of the Vaisala Weather Station, this number corresponds to the device address on the shared serial connection. This defaults to 0 for the Weather Station. If you were using multiple Weather Stations, the second one would have to be configured to use another address like 1. You can use the Vaisala Configuration Tool to change the device address using its Device Settings dialog box. Enter the same number here that is used in the Vaisala Configuration Tool to access the hardware.

Units – The Units string is common to all virtual parametric modules, but a discussion of the Weather Station is not complete without a few comments on the units. The default units listed in the Weather Station setup dialog are the units the data is transmitted in from the instrument over a serial port connection. HMSGHAT2000DDI specifies what units the Weather Station should use when it initializes the instrument at the start of data acquisition. The user can use the Vaisala Configuration Tool to modify the instruments default units, BUT HMSGHAT2000DDI will override any changes made in the configuration program. Do not try to change the units this way. For example, if you desired wind speed expressed in km/hr instead of m/s & you enter km/hr here & try to use the Vaisala Configuration Tool to do switch to readings in km/hr, you will get a reading in m/s & it would be misleadingly labelled with km/hr. Instead set the Multiplier & Offset settings here to 3.6 (3600s/hr * 1km/1000m) & the Offset to 0 to scale the incoming m/s readings to km/hr. Don't forget to update the Units string to km/hr as well.

Sampling Period – For the Weather Station, the Sampling Period must be at least 10,000 ms. You can set a higher value, but not a lower one. Higher values should be integer multiples of 10s (10000ms), so valid numbers are 10000, 20000, 30000, but not 15000, 25000.

Specific to Vaisala Weather Station

Serial Port – The Weather Station is designed to communicate over a serial connection. It can be configured to communicate over an RS-232, RS-422, RS-485 or via USB connection. HMSGHAT2000DDI expects the instrument to communicate at the defaults of 19200 baud, 8 data bits, no parity, 1 stop bit. The only user configurable parameter in HMSGHAT2000DDI is what serial port is connected to the Weather Station.

Test Module – You may use the Test Module button to verify that HMSGHAT2000DDI can properly communicate with the Weather Station. You may receive the cryptic Windows error message "The system cannot find the file specified.". Windows treats serial ports like a file for reading & writing information which is why it says it cannot find the file. In this case the file is the serial port & what Windows is saying is that it cannot open that port. Most likely you've used a serial port that does not exist on your computer. If you receive a "Weather Station communication failed." error message, the port exists, but the Weather Station did not respond to the test message.

Troubleshooting Problems with an RS-485 Connection

If you happen to be using an RS-485 connection, there is one more instrument setting that may affect your ability to reliably communicate with the Weather Station. If you start up the Vaisala Configuration Tool & it cannot maintain a connection with your Weather Station, HMSGHAT2000DDI will not be able to maintain a connection with it either. In that case, you may need to adjust the RS-485 line delay parameter. It cannot be adjusted by HMSGHAT2000DDI, but it can be adjusted with the Vaisala Configuration Tool, if the connection is somewhat reliable. Look for it in the Device Settings dialog box. If the Configuration Tool cannot maintain a connection, you will need to use Windows HyperTerminal to adjust the parameter.

Configure HyperTerminal to use the same serial port that the Vaisala Configuration Tool is using. In HyperTerminal this is done using the File Properties menu option. On the Connect To tab, set the serial port to use 19200 baud, 8 data bits, no parity, 1 stop bit. Next on the Settings tab, click the ASCII Setup button. Check "Send line ends with line feeds", "Echo typed characters locally" & "Append line feeds to incoming line ends." Click on OK to save your changes & then again to exit the properties dialog box.

When you exit the properties dialog box, HyperTerminal will attempt to open the port. Assuming the port opened successfully, you should be able to type a command to the Weather Station instrument & have it respond. Simply type the device address, 0 by default & press Enter. The instrument should respond with the same value. If you don't know the device address, type ? and then press Enter. The response should be the device address. All commands to the Weather Station start with its device address & commands are case sensitive. Assuming this instrument is using the default address of 0, you can check the line delay parameter by typing:

```
0XU<enter>
```

The instrument should respond with something like:

```
0XU,A=0,M=P,T=0,C=2,I=0,B=19200,D=8,P=N,S=1,L=0,N=WXT520,V=1.00<enter>
```

Look for the L=# entry. This is the line delay parameter in millisecs. You can update the value to 100 ms by typing:

```
0XU,L=100<enter>
```

which the instrument should respond to by echoing the command back to you:

```
0XU,L=100
```

You can doublecheck that the setting was stored by using the 0XU command again. Once you have adjusted the line delay, disconnect from the port in HyperTerminal & try the Vaisala Configuration Tool. If it still cannot maintain the connection with the instrument, close the Configuration Tool, return to HyperTerminal & readjust the line delay. Valid line delay values range from 0 to 10000 ms. Start low & slowly make it larger by 50 or 100 ms at a time. Otherwise you may jump over the range of values that work. Most likely you will only need a delay between 0 and 250 ms, but it does depend on the length of your RS-485 cable so the exact value is hard to predict.

Corr Instruments' NanoCorr

Corr Instruments' NanoCorr is designed to measure corrosion rates of metal probes. It is unique in that it is the only instrument that the software cannot actively control. All the software can do is monitor it's output on a serial port & import the data readings. The user must configure it with NanoCorr front panel & then make the HMSGHAT2000DDI software settings match the NanoCorr settings.

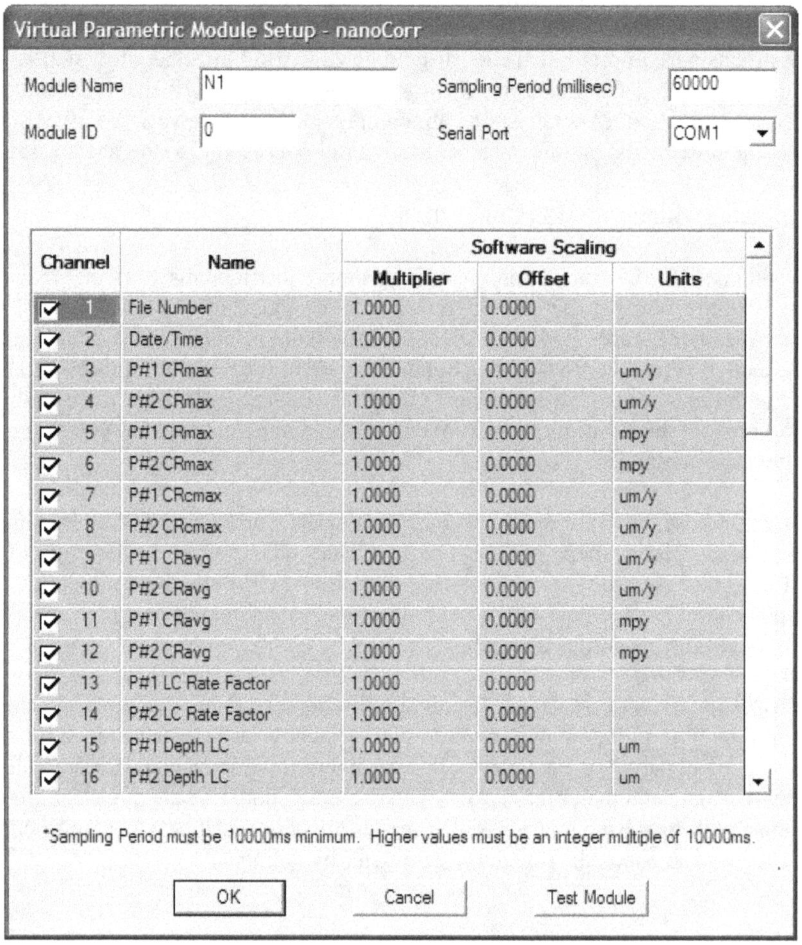

Figure 164...NanoCorr Setup Dialog Box.

Common to All Modules

Module Name, Module ID, Channel On/Off Toggle, Channel Name, Multiplier & Offset – These are described in the section on the Measurement Computing boards. Please refer back to that section for detailed descriptions.

Sampling Period – For NanoCorr as for the Weather Station, the Sampling Period must be at least 10000 ms. You can set a higher value, but not a lower one. Higher values should be integer multiples of 10s (10000ms), so valid entries are 10000, 20000, 30000, but not 15000, 25000.

Specific to NanoCorr

Serial Port – The NanoCorr is designed to communicate over an RS-232 or RS-485 serial connection. It can also be configured to communicate over an Ethernet connection masquerading as a serial connection. Select the port number of the serial port that the NanoCorr is hooked up to. If using an Ethernet connection, refer to Chapter 5 of the NanoCorr manual for details on how to set up the connection & determine the serial port number to use here. HMSGHAT2000DDI expects the instrument to communicate at the defaults of 115200 baud, 8 data bits, no parity, 1 stop bit.

Test Module – You may use the Test Module button to verify that HMSGHAT2000DDI can properly communicate with the NanoCorr. As for the Weather Station, you may receive the cryptic Windows error message "The system cannot find the file specified.". Again Windows treats serial ports like a file for reading & writing information which is why it says it cannot find the file. In this case the file is the serial port & what Windows is saying is that it cannot open that port. Most likely you've used a serial port that does not exist on your computer. Assuming the port exists, you will receive the following message:

> COM Port opened. Must wait for nanoCorr to transmit a new reading.
> Click on OK when ready to check for a data string.

The NanoCorr test capabilty is very primitive due to the limitation that HMSGHAT2000DDI can only monitor the output of the NanoCorr over its serial port connection. It cannot send a command to the NanoCorr & have it respond. The Test Module button just checks for incoming data the chosen serial port. At this point, you will have to wait until you feel the NanoCorr has transmitted data to the PC. If you used the front panel of the NanoCorr to tell it to collect new readings every 2 minutes, you will have to wait 2 minutes before pressing the OK button. If you press the button, before data has been transmitted to the PC, you will receive a see a "No data received from nanoCorr." message. If data has been received, you will see a message that shows you the text received. It should be a series of numbers that looks something similar to this NanoCorr data record:

```
NanoCorr data string read =

"186 2008-07-23 15:22:54    591.0344848    23.269074      591.0344848    122.956886
     4.840822       4.806843       0.0962117      0.0202236      4.757399        -0.554344
     -109.181907    6.709134       0.0209924      -       0      722047  -27     -89     -17531
     602829  -220275 25      -73     49      -293030 -315945 -399937 -40374  -43860  -82     47
```

If you don't end up seeing any data, you might want to try using HyperTerminal to monitor the serial port that the NanoCorr is connected to. Set up HyperTerminal as described in the "**Troubleshooting Problems with an RS-485 Connection**" section for the Weather Station, but use 115200 as the baud rate instead of 19200. The NanoCorr by default communicates at 115200 baud rate, 8 data bits, no parity, 1 stop bit. You still want to turn on "Append line feeds to incoming line ends." Just leave HyperTerminal open till you see data strings come in.

If you leave the window open long enough to see 2 data records received, you can tell how long the time is between data readings. Note in the example data string above, the time of the data reading is the 3rd number in the string. Just compare the 2 readings to figure out the time interval. That's how long you need to leave the message box open when it is waiting on data from the NanoCorr. If HyperTerminal nevers receives any data, neither will HMSGHAT2000DDI. Either you are using the wrong serial port or the NanoCorr is not actively collecting data.

Mistas Group 9380-2214 4-Channel Strain Gauge Module

As noted above, the Strain Gauge Module is only supported by the Sensor Highway Smart Monitor system. It is not available for the SAMOS, PCI2 or DiSP systems.

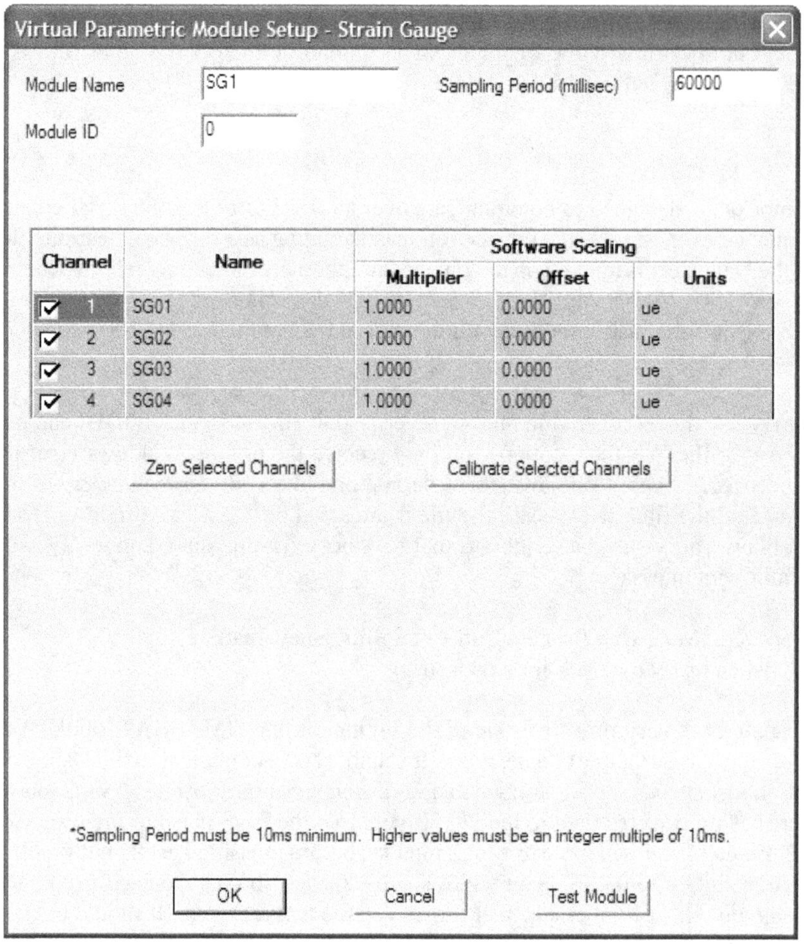

*Figure 165...Strain Gauge **Module** Setup Dialog Box.*

Common to All Modules

Module Name, Channel On/Off Toggle, Channel Name, Multiplier & Offset – These are described in the section on the Measurement Computing boards. Please refer back to that section for detailed descriptions.

Sampling Period – For the Strain Gauge module, the Sampling Period must be at least 10 ms. You set a higher value, but not a lower one. Higher values should be integer multiples of 10s (10000ms), so valid entries are 10000, 20000, 30000, but not 15000, 25000.

Specific to Strain Gauge Module

Test Module – You may use the Test Module button to verify that HMSGHAT2000DDI can properly find the Strain Gauge hardware. If there is a problem, you will get a "Strain Gauge 0 could not be detected." If all goes well, you will get a "Strain Gauge 0 successfully detected." message & it will give you information on the firmware & PCB versions of the hardware.

Zero Selected Channels and *Calibrate Selected Channels* – These buttons work with the channels that are checked on in the grid. They just run thru the routines responsible for zeroing & recalibrating the channel readings.

Virtual Parametrics Page in Hardware Setup

The Configure Virtual Parametric Modules dialog box can also be accessed thru the Virtual Parametrics tab in Hardware Setup. This tab in Hardware Setup (shown below) will let you adjust the setup of all the virtual parametric modules without having to jump in & out of the individual setup dialog boxes for each module. You cannot adjust all the settings for all modules, but you can control the main ones. You can turn channels on & off, revise the channel Name, Multiplier, Offset & Units string. The Module Name is displayed, but it is a read only column. To change the Module Name, you will have to display the Configure Virtual Parametric Modules dialog box & use it to bring up the setup dialog box for the module you want to rename. The Sampling Period also requires accessing the individual setup dialog boxes as some modules have 1 setting for the whole module & others allow individual settings for each channel.

Virtual Channel	Module	Channel	Name	Software Scaling		
				Multiplier	Offset	Units
66	N1	☐ 2	Date/Time	1.000000	0.000000	
67	N1	☑ 3	P#1 CRmax	1.000000	0.000000	um/y
68	N1	☑ 4	P#2 CRmax	1.000000	0.000000	um/y
69	N1	☑ 5	P#1 CRmax	1.000000	0.000000	mpy
70	N1	☑ 6	P#2 CRmax	1.000000	0.000000	mpy
71	N1	☑ 7	P#1 CRcmax	1.000000	0.000000	um/y
72	N1	☑ 8	P#2 CRcmax	1.000000	0.000000	um/y
73	N1	☑ 9	P#1 CRavg	1.000000	0.000000	um/y
74	N1	☑ 10	P#2 CRavg	1.000000	0.000000	um/y
75	N1	☑ 11	P#1 CRavg	1.000000	0.000000	mpy
76	N1	☑ 12	P#2 CRavg	1.000000	0.000000	mpy
77	N1	☑ 13	P#1 LC Rate Factor	1.000000	0.000000	
78	N1	☑ 14	P#2 LC Rate Factor	1.000000	0.000000	
79	N1	☑ 15	P#1 Depth LC	1.000000	0.000000	um
80	N1	☑ 16	P#2 Depth LC	1.000000	0.000000	um
81	N1	☑ 17	P#1 Depth Avg	1.000000	0.000000	um

*Figure 166...Virtual Parametrics Page of **Hardware** Setup.*

There is one setting on this tab that you cannot adjust anywhere else. At the bottom of the tab is a checkbox labelled:

> "When scaling parametric values during data replay, use the multiplier and offset from the data file, not the ones shown in the grid."

During data acquisition, the virtual parametric settings that are in use are stored in the DTA file. This checkbox controls whether you use the settings in your layout or from the DTA file when replaying the data. This allows you flexibility in the units that are displayed during a replay. If you use the settings from the DTA file, you will see the data in the same format as it was collected. If, however you want to changes some units, you would want to use the settings from your layout file, so that you can rescale the data using the Multiplier & Offset values & display the correct units string. If you don't need to change any units, you should normally stick with using the settings in the DTA file, so that your data displays properly.

If you use the settings from your layout, you should start with the layout that you used to collect the data, because it will have the correct modules already defined. When you create a layout & tell HMSGHAT2000DDI which modules you are using, it keeps track of the data using a virtual channel number. These are shown in the lefmost column in the grid. They range from 1 to the total number of virtual parametric channels available for the modules in use. If you are only using 1 module, the virtual channel numbers are identical to the module channel numbers. If you are using 2 or more modules, though, the virtual channel numbers for the 2nd module start where the 1st modules channels end.

For example, the Measurement Computing DAS6033 & USB2533 have 64 channels of data available, while the Weather Station has 22. When they are both being used, the virtual channel numbers for the Weather Station start at the end of the 64 from the Measurement Computing module, so channel 1 for the Weather Station corresponds to virtual channel number 65. These are important, because you need them to graph virtual parametric data. In this case, if you want to graph channel 1 for the Weather Station, you need to look for Virtual Parametric 65 in Graph Setup. These numbers can also show up in the Line Display. More on this in the next section.

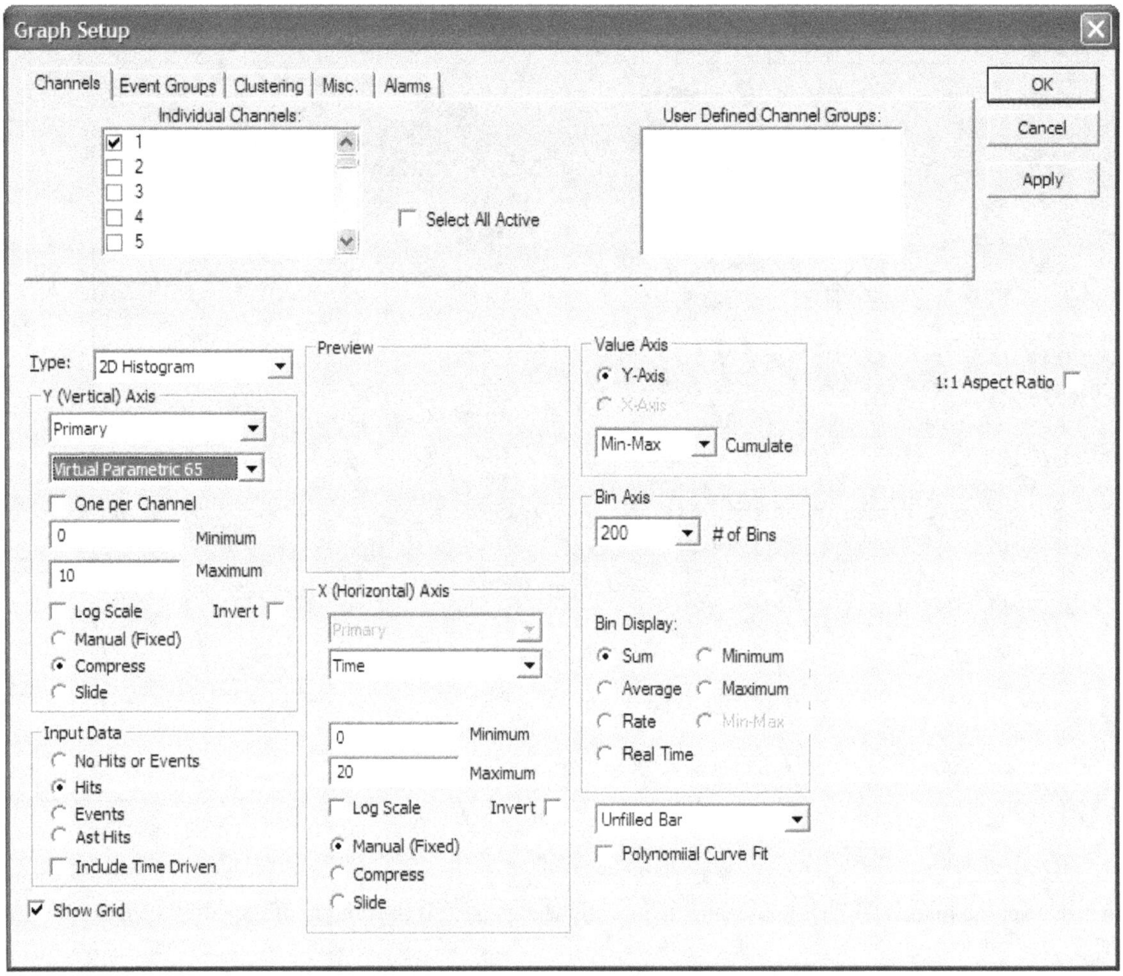

Figure 167...Accessing Virtual Parametric in Graph Setup.

Line Display & Virtual Parametric Data

To get virtual parametric data to show up in the Line Display, you will need to turn on both the Line Display & make sure virtual parametric is being output. Use the Line Display Setup dialog box shown below. Check "Enable Line Display" in the upper lefthand corner to turn on the line display.

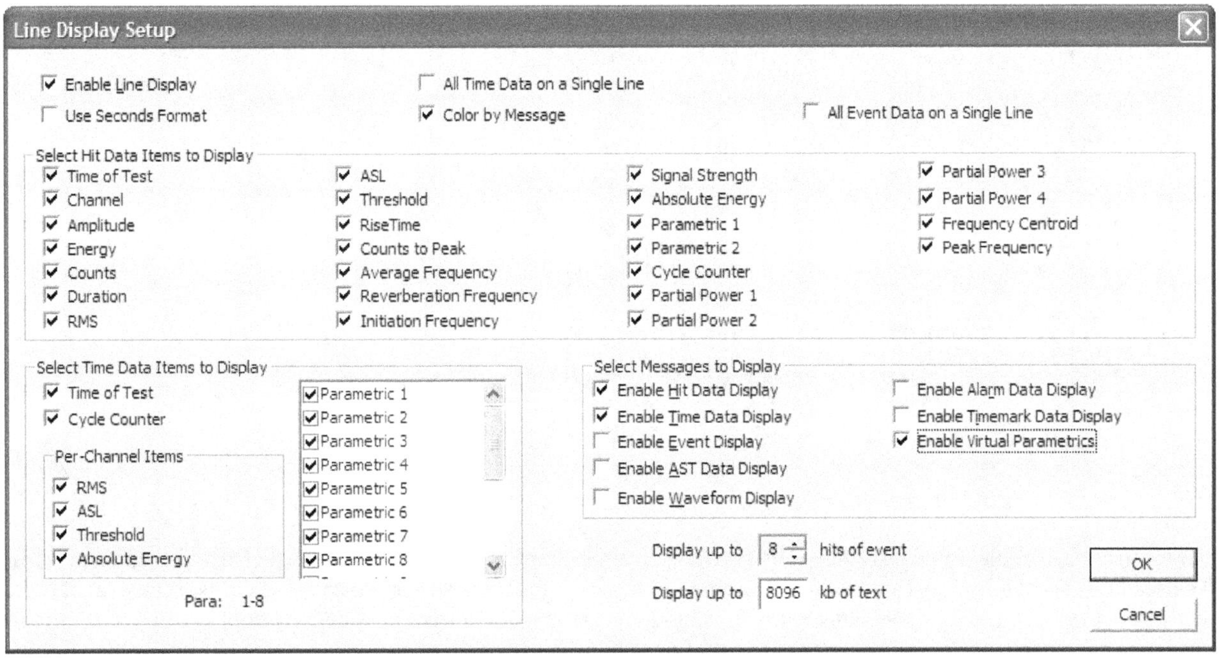

Figure 168...Enabling Virtual Parametrics in Line Display Setup.

Check "Enable Virtual Parametrics" in the "Select Messages to Display" section in the lower right hand side of the dialog box. Note there is a difference between the line display being on & being visible. You can have it on, but not have it visible. You can use the F7 key to show or hide the Line Display window itself. When you replay a DTA file with virtual parametric data in it, it will show up in the Line Display as message ID 9. Here is a sample for data collected from a NanoCorr instrument:

```
9   0 00:00:18.2061840
    N1:P#1 CRmax           =   0.3219 um/y     N1:P#2 CRmax           =   0.1372 um/y
    N1:P#1 CRmax           =   0.0127 mpy      N1:P#2 CRmax           =   0.0054 mpy
    N1:P#1 CRcmax          =   0.3219 um/y     N1:P#2 CRcmax          =   0.1372 um/y
    N1:P#1 CRavg           =   0.0522 um/y     N1:P#2 CRavg           =   0.0207 um/y
    N1:P#1 CRavg           =   0.0021 mpy      N1:P#2 CRavg           =   0.0008 mpy
    N1:P#1 LC Rate Factor  =   6.1660          N1:P#2 LC Rate Factor  =   6.6439
    N1:P#1 Depth LC        =   0.0000 um       N1:P#2 Depth LC        =   0.0000 um
    N1:P#1 Depth Avg       =   0.0000 um       N1:P#2 Depth Avg       =   0.0000 um
    N1:P#1 LC Depth Factor =   1.8404          N1:P#2 LC Depth Factor =   1.2792
    N1:P#1E                =   0.0013 V        N1:P#1E                =   0.0013 V
    N1:Temperature         =  24.8901 oC       N1:pH                  =   7.0001
    N1:ORP                 =   0.0072 V        N1:Remarks             =   0.0000
    N1:P#1I Offset         =  18.0000 pA       N1:P#2I Offset         =-170.0000 pA
    N1:P#1E#1              = -75.0000 pA       N1:P#1E#2              =-217.0000 pA
    N1:P#1E#3              = -10.0000 pA       N1:P#1E#4              =  61.0000 pA
    N1:P#1E#5              =  16.0000 pA       N1:P#1E#6              = -13.0000 pA
    N1:P#1E#7              = 130.0000 pA       N1:P#1E#8              =  83.0000 pA
    N1:P#1E#9              =  26.0000 pA       N1:P#2E#1              =  30.0000 pA
    N1:P#2E#2              = -99.0000 pA       N1:P#2E#3              =  30.0000 pA
    N1:P#2E#4              =  -7.0000 pA       N1:P#2E#5              = -27.0000 pA
    N1:P#2E#6              =  10.0000 pA       N1:P#2E#7              =   2.0000 pA
    N1:P#2E#8              =  50.0000 pA       N1:P#2E#9              =  10.0000 pA
```

Figure 169...Sample Line ***Display*** *Output for the NanoCorr.*

If you replay a DTA file with virtual parametric data in it & choose to scale the data using the settings in the layout file, but aren't using the same virtual parametric modules as were used to collect the data, the software may not be able to correctly display the virtual parametric data. Here is the same NanoCorr data as above replayed with a layout that did not have any virtual parametric modules defined to illustrate how that looks:

```
9   0 00:00:18.2061840
    ??:VCh1   = 202.0000     ??:VCh2   =   0.0000
    ??:VCh3   =   0.3219     ??:VCh4   =   0.1372
```

```
??:VCh5    =    0.0127        ??:VCh6    =    0.0054
??:VCh7    =    0.3219        ??:VCh8    =    0.1372
??:VCh9    =    0.0522        ??:VCh10   =    0.0207
??:VCh11   =    0.0021        ??:VCh12   =    0.0008
??:VCh13   =    6.1660        ??:VCh14   =    6.6439
??:VCh15   =    0.0000        ??:VCh16   =    0.0000
??:VCh17   =    0.0000        ??:VCh18   =    0.0000
??:VCh19   =    1.8404        ??:VCh20   =    1.2792
??:VCh21   =    0.0013        ??:VCh22   =    0.0013
??:VCh23   =   24.8901        ??:VCh24   =    7.0001
??:VCh25   =    0.0072        ??:VCh26   =    0.0000
??:VCh27   =   18.0000        ??:VCh28   =-170.0000
??:VCh29   =  -75.0000        ??:VCh30   =-217.0000
??:VCh31   =  -10.0000        ??:VCh32   =   61.0000
??:VCh33   =   16.0000        ??:VCh34   =  -13.0000
??:VCh35   =  130.0000        ??:VCh36   =   83.0000
??:VCh37   =   26.0000        ??:VCh38   =   30.0000
??:VCh39   =  -99.0000        ??:VCh40   =   30.0000
??:VCh41   =   -7.0000        ??:VCh42   =  -27.0000
??:VCh43   =   10.0000        ??:VCh44   =    2.0000
??:VCh45   =   50.0000        ??:VCh46   =   10.0000
```

***Figure 170...**Line Display Output Without Any Virtual Parametrics Modules Defined.*

Notice that the Module Name is displayed as ?? & the Channel Name is VCh#n where #n is the virtual channel number. The readings are the raw unscaled values from the DTA file. You can also see a display like this, if you rely on the settings in the DTA file & it's an older DTA file that does not have the virtual parametric settings stored in it. In this case you just need to define your virtual parametric modules & change to scaling the data using the settings in the layout.

The number of channels displayed will allow you to guess at the virtual parametric modules in use when the data was collected, because the total number of channels differs for each type of virtual parametric data. The Measurement Computing boards have 64 channels of data, the NanoCorr has 46, the Weather Station 22 & the Strain Gauge Module only 4 channels. If you were to mistakenly use a Weather Station module & replay NanoCorr data, you would get a display like the following one where some channels are defined & some aren't:

```
9   0 00:00:18.2061840
    W1:Wind Speed Min          = 202.0000 m/s          W1:Wind Speed Avg          =    0.0000 m/s
    W1:Wind Speed Max          =   0.3219 m/s          W1:Wind Direction Min      =    0.1372 deg
    W1:Wind Direction Avg      =   0.0127 deg          W1:Wind Direction Max      =    0.0054 deg
    W1:Air Pressure            =   0.3219 bar          W1:Air Temperature         =    0.1372 C
    W1:Internal Temperature    =   0.0522 C            W1:Humidity                =    0.0207 %RH
    W1:Rain Accumulation       =   0.0021 mm           W1:Rain Duration           =    0.0008 sec
    W1:Rain Intensity          =   6.1660 mm/hr        W1:Rain Peak Intensity     =    6.6439 mm/hr
    W1:Hail Accumulation       =   0.0000 hits/cm2     W1:Hail Duration           =    0.0000 sec
    W1:Hail Intensity          =   0.0000 hits/cm2h    W1:Hail Peak Intensity     =    0.0000 hits/cm2h
    W1:Heating Temperature     =   1.8404 C            W1:Heating Voltage         =    1.2792 V
    W1:Supply Voltage          =   0.0013 V            W1:Reference Voltage       =    0.0013 V
    ??:VCh23                   =  24.8901              ??:VCh24                   =    7.0001
    ??:VCh25                   =   0.0072              ??:VCh26                   =    0.0000
    ??:VCh27                   =  18.0000              ??:VCh28                   =-170.0000
    ??:VCh29                   = -75.0000              ??:VCh30                   =-217.0000
    ??:VCh31                   = -10.0000              ??:VCh32                   =   61.0000
    ??:VCh33                   =  16.0000              ??:VCh34                   =  -13.0000
    ??:VCh35                   = 130.0000              ??:VCh36                   =   83.0000
    ??:VCh37                   =  26.0000              ??:VCh38                   =   30.0000
    ??:VCh39                   = -99.0000              ??:VCh40                   =   30.0000
    ??:VCh41                   =  -7.0000              ??:VCh42                   =  -27.0000
    ??:VCh43                   =  10.0000              ??:VCh44                   =    2.0000
    ??:VCh45                   =  50.0000              ??:VCh46                   =   10.0000
```

***Figure 171...**Line Display Output Without All Virtual Parametrics Modules Defined.*

Anytime you see virtual parametric data in the Line Display showing ?? for the Module Name & virtual channel numbers, you know that your virtual parametric setup is incorrect. Just modify the modules in use in your layout & replay the data again till you get the number of channels defined in the layout to match the data in the DTA file. Most often the simplest solutions to these problems is to display the data using the settings stored in the DTA file. As stated earlier, though, this does not work for early version files that do not contain the virtual parametric settings used to collect the data.

OPTIONS FOR HMSGHAT2000DDI

Supervisor Option

The 'Supervisor Option' allows the user to restrict access to the software by the use of a password. With this option installed the software will start up in ***Operator Mode*** by default.

In ***Operator Mode*** the user has limited access to the system. He can:

- Load a layout file.
- Print/Export graphs or copy them to the clipboard.
- Change the Test Title.
- Edit the Logistics Information.
- Setup the Audio Monitor (if present).
- Perform a test or AST.
- Replay an existing data file.
- Toggle the Line Listing Display on/off.

The user can switch to supervisor mode by clicking 'Switch to Supervisor Mode' from the File menu and entering the password.

In ***Supervisor Mode*** the user has full access to the system. He can perform all the functions possible in operator mode plus he can:

- Change the password.
- Create and save layout files.
- Access the various Test Setup menus.
- Create, modify and delete graphs.
- Modify the arrangement of the graphs, create and delete pages, etc.

How to set the password the first time. The first time the user tries to switch to supervisor mode the software will ask for an initial password. Enter a password of your choice which must be 4-8 alphanumeric characters in length. It is not case sensitive. From this point on the user must enter this password to switch to supervisor mode.

Note: If your system came with HMSGHAT2000DDI pre-installed and the program doesn't request an initial password, this means that the initial password was set at the factory. The password in this case is 'HMSGHAT2000DDI'.

How to change the password: The user can change the password by switching to supervisor mode and clicking 'Change Password' from the file menu.

What to do if you forget the password: If you forget the password the only way to recover is to uninstall the software and reinstall it. This will allow you to enter a new password.

LeakTEC Alarms

LeakTEC Alarms (part of the 'LeakTEC/On-Line-Crack Alarm' Option) are used to monitor ASL or RMS levels (using Time Data) during acquisition and alert the user if they stray beyond certain pre-set lower and upper limits. If ASL or RMS levels stray beyond these limits for a certain period of time an Alarm is generated.

Setup

If you have the 'LeakTEC/On-Line-Crack Alarm' Option installed then you can enable and customize these alarms by using the LeakTEC Alarm Setup dialog box located on the Acquisition Setup menu. Its controls are described below.

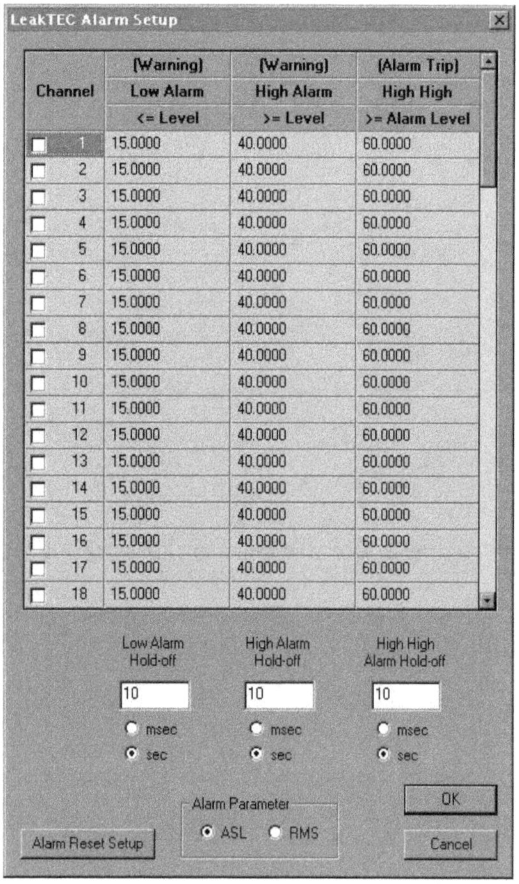

Figure 172. LeakTEC Alarm Setup

Channel – Each channel can be enabled/disabled and customized independently. Click a channel's checkbox to enable that channel.
Low Alarm – Set the Low Alarm Level for a channel in this cell. An alarm 'warning' is triggered if ASL/RMS levels drop to this level or below for a duration of time equal to the 'Low Alarm Hold-off' period.
High Alarm – Set the High Alarm Level for a channel in this cell. An alarm 'trip' is triggered if ASL/RMS levels equal or exceed this level for a duration of time equal to the 'High Alarm Hold-off' period.
High High Alarm – Set the High High Alarm Level for a channel in this cell. An alarm 'trip' is triggered if ASL/RMS levels equal or exceed this level for a duration of time equal to the 'High High Alarm Hold-off' period.
Low Alarm Hold-off – This is the period of time that a channel's ASL/RMS must remain at or below the Low Alarm Level before a Low alarm 'warning' is triggered.
High Alarm Hold-off – This is the period of time that a channel's ASL/RMS must remain at or above the High Alarm Level before a High alarm 'trip' is triggered.
High High Alarm Hold-off – This is the period of time that a channel's ASL/RMS must remain at or above the High High Alarm Level before a High High alarm 'trip' is triggered.
ASL/RMS – Select which Time Dataset feature to monitor using this control.
Alarm Reset Setup – Click this button to enter the Auto Alarm-Reset dialog box. This can be used to set the system to automatically clear the alarms periodically.

Operation

If the ASL/RMS for a particular channel violates an alarm level (low, high or high high) and remains continuously in violation of that level for a duration of time equal to the appropriate 'Hold-Off' then an alarm is triggered.

What happens when an alarm is triggered

5) An alarm message (16) is generated. If autodump is on then this message is saved to the datafile. If the line display is active then it will be displayed there with the time was triggered and the conditions that triggered it.
6) The Alarm Display dialog box is opened (or updated if already open). The details of the alarm and the conditions that triggered it are shown. The alarm can be cleared by clicking the 'Clear Alarms' button.
7) If the alarm is an alarm 'warning' then the Alarm Warning output line is set high. If the alarm is an alarm 'trip' then both the Alarm Warning and Alarm Trip lines are set high.
8) An audible alarm is triggered if you have an audio board installed and enabled.

On-Line Crack Alarms

On-Line Crack Alarms (part of the 'LeakTEC/On-Line-Crack Alarm' Option) are used to monitor hits or events during acquisition and alert the user if the hit/event or energy rates exceed certain pre-set levels. If these rates exceed these limits within a certain period of time an Alarm is generated.

Setup

If you have the 'LeakTEC/On-Line Crack Alarm' Option installed then you can enable and customize these alarms by using the On-Line Crack Alarm Setup dialog box located on the Acquisition Setup menu. Its controls are described below.

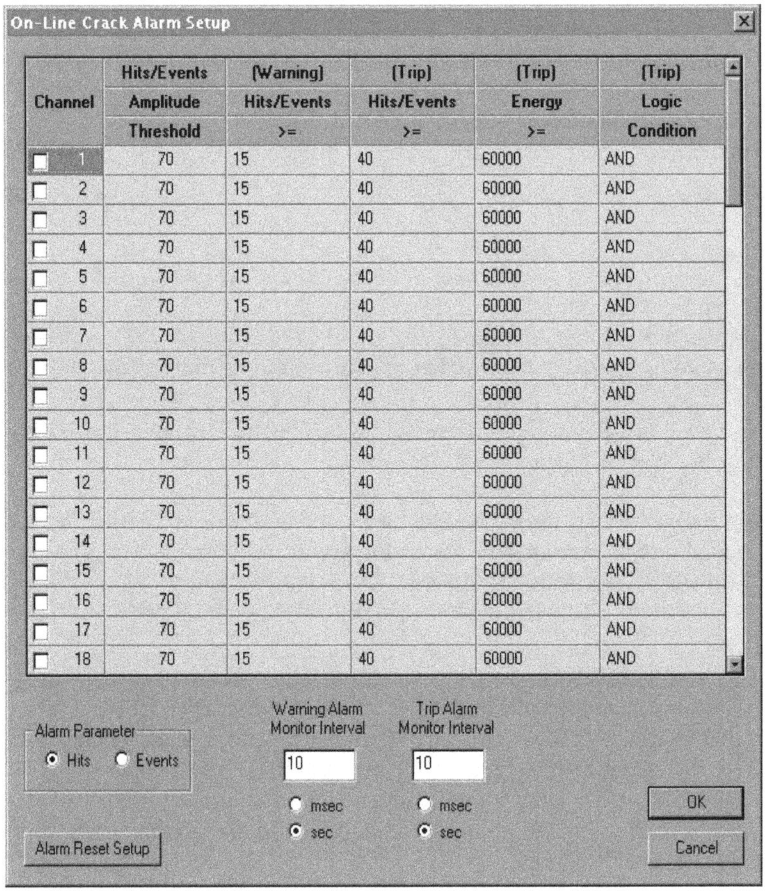

Figure 173. On-Line Crack Alarm Setup

Channel – Each channel can be enabled/disabled and customized independently. Click a channel's checkbox to enable that channel.
Hits/Events Amplitude Threshold – This is the amplitude that a hit (or first hit of an event) must equal or exceed in order to be counted towards a 'Warning Hit/Events' or 'Trip Hit/Events' alarm. This does not affect energy alarms.
[Warning] Hits/Events – A 'Hits/Events Warning' alarm will be triggered if at least this many hits/events that meet the Amplitude Threshold are acquired within a period of time not exceeding the Warning Alarm Monitor Interval.
[Trip] Hits/Events – A 'Trip Hits/Events' alarm will be triggered if at least this many hits/events that meet the amplitude threshold are acquired within a period of time not exceeding the Trip Alarm Monitor Interval. If the logic condition is 'ENERGY' then this alarm type is disabled and will not trigger. If the logic condition is 'AND' then the Trip Energy condition must be met as well before an alarm will be triggered.

[Trip] Energy – A 'Trip Energy' alarm will be triggered if the total energy of any number of hits/events (with any amplitude) are acquired within a period of time not exceeding the Trip Alarm Monitor Interval meets or exceeds this value. If the logic condition is 'Hits/Events' then this alarm type is disabled and will not trigger. If the logic condition is 'AND' then the 'Trip Hits/Events' condition must be met as well before an alarm will be triggered.

[Trip] Logic Condition – The logic condition controls how the Trip alarms are calculated. If it is set to 'Hits/Events' then the 'Trip Energy' condition is ignored and no alarms will be triggered using it. If it is set to 'Energy' then the 'Trip Hits/Events' condition is ignored and no alarms will be triggered using it. If it is set to AND then both the 'Trip Hits/Events' and 'Trip Energy' conditions must be met before an alarm is triggered. If it is set to OR then meeting either condition will trigger a Trip alarm.
Warning Alarm Monitor Interval – This is the period of time in which hits/events (passing the threshold) are accumulated towards a Warning alarm.
Trip Alarm Monitor Interval – This is the period of time in which hits/events (passing the threshold) and energy are accumulated towards a Trip alarm.
Alarm Parameter – Select whether to use hits or the first hit of events for alarm determination.
Alarm Reset Setup – Click this button to enter the Auto Alarm-Reset dialog box. This can be used to set the system to automatically clear the alarms periodically.

Operation

Four kinds of alarms can be generated by this option.

- Warning Hits/Event – A number of hits/events that met the Amplitude Threshold were acquired within a period of time not exceeding the Warning Alarm Monitor Interval. The number of hits/events met or exceeded the [Warning] Hits/Events value.
- Trip Hits/Event – A number of hits/events that met the Amplitude Threshold were acquired within a period of time not exceeding the Trip Alarm Monitor Interval. The number of hits/events met or exceeded the [Trip] Hits/Events value.
- Trip Energy – The total energy of all hits/events (with any amplitude) that were acquired within a period of time not exceeding the Trip Alarm Monitor Interval met or exceeded the [Trip] Energy value.
- Trip Hits/Event and Energy – Both the 'Trip Hits/Event' and 'Trip Energy' conditions were met.

What happens when an alarm is triggered

5) An alarm message (16) is generated. If autodump is on then this message is saved to the datafile. If the line display is active then it will be displayed there with the time was triggered and the conditions that triggered it.
6) The Alarm Display dialog box is opened (or updated if already open). The details of the alarm and the conditions that triggered it are shown. The alarm can be cleared by clicking the 'Clear Alarms' button.
7) If the alarm is an alarm 'warning' then the Alarm Warning output line is set high. If the alarm is an alarm 'trip' then both the Alarm Warning and Alarm Trip lines are set high.
8) An audible alarm is triggered if you have an audio board installed and enabled.

APPENDIX

UTILITIES MENU HMSGHAT2000DDI™ SOFTWARE

Index

A

B

C

D

E

F

G

H

I

O

P

R

S

T

W

www.ingramcontent.com/pod-product-compliance
Lightning Source LLC
LaVergne TN
LVHW080013210926
848481LV00011B/3014
* 9 7 9 8 3 4 9 2 1 1 6 0 7 *